Manufacturing Technology for Aerospace Structural Materials

Manufacturing Technology for Aerospace Structural Materials

F.C. Campbell

ELSEVIER

AMSTERDAM • BOSTON • HEIDELBERG • LONDON • NEW YORK • OXFORD
PARIS • SAN DIEGO • SAN FRANCISCO • SINGAPORE • SYDNEY • TOKYO

Elsevier B.V.
Radarweg 29
P.O. Box 211, 1000 AE
Amsterdam
The Netherlands

Elsevier Inc.,
525 B Street, Suite 1900
San Diego,
CA 92101-4495
USA

Elsevier Ltd.
The Boulevard, Langford Lane
Kidlington, Oxford
OX5 1GB
UK

Elsevier Ltd.
84 Theobalds Road
London
WC1Z 8RR
UK

© 2006 Elsevier Ltd. All rights reserved.

This work is protected under copyright by Elsevier Ltd., and the following terms and conditions apply to its use:

Photocopying
Single photocopies of single chapters may be made for personal use as allowed by national copyright laws. Permission of the Publisher and payment of a fee is required for all other photocopying, including multiple or systematic copying, copying for advertising or promotional purposes, resale, and all forms of document delivery. Special rates are available for educational institutions that wish to make photocopies for non-profit educational classroom use.

Permissions may be sought directly from Elsevier's Rights Department in Oxford, UK: phone (+44) 1865 843830, fax (+44) 1865 853333, e-mail: permissions@elsevier.com. Requests may also be completed on-line via the Elsevier homepage (http://www.elsevier.com/locate/permissions).

In the USA, users may clear permissions and make payments through the Copyright Clearance Center, Inc., 222 Rosewood Drive, Danvers, MA 01923, USA; phone: (+1) (978) 7508400, fax: (+1) (978) 7504744, and in the UK through the Copyright Licensing Agency Rapid Clearance Service (CLARCS), 90 Tottenham Court Road, London W1P 0LP, UK; phone: (+44) 20 7631 5555; fax: (+44) 20 7631 5500. Other countries may have a local reprographic rights agency for payments.

Derivative Works
Tables of contents may be reproduced for internal circulation, but permission of the Publisher is required for external resale or distribution of such material. Permission of the Publisher is required for all other derivative works, including compilations and translations.

Electronic Storage or Usage
Permission of the Publisher is required to store or use electronically any material contained in this work, including any chapter or part of a chapter.

Except as outlined above, no part of this work may be reproduced, stored in a retrieval system or transmitted in any form or by any means, electronic, mechanical, photocopying, recording or otherwise, without prior written permission of the Publisher.

Address permissions requests to: Elsevier's Rights Department, at the fax and e-mail addresses noted above.

Notice
No responsibility is assumed by the Publisher for any injury and/or damage to persons or property as a matter of products liability, negligence or otherwise, or from any use or operation of any methods, products, instructions or ideas contained in the material herein. Because of rapid advances in the medical sciences, in particular, independent verification of diagnoses and drug dosages should be made.

First edition 2006
Library of Congress Control Number: 2006927672

ISBN-13: 978-1-85-617495-4
ISBN-10: 1-85-617495-6

∞ The paper used in this publication meets the requirements of ANSI/NISO Z39.48-1992 (Permanence of Paper).

Printed in Great Britain.

06 07 08 09 10   10 9 8 7 6 5 4 3 2 1

**Working together to grow
libraries in developing countries**

www.elsevier.com | www.bookaid.org | www.sabre.org

ELSEVIER   BOOK AID International   Sabre Foundation

Contents

	Preface	xiii
Chapter 1	**Introduction**	**1**
	1.1 Aluminum	4
	1.2 Magnesium and Beryllium	6
	1.3 Titanium	7
	1.4 High Strength Steels	8
	1.5 Superalloys	8
	1.6 Composites	9
	1.7 Adhesive Bonding and Integrally Cocured Structure	10
	1.8 Metal and Ceramic Matrix Composites	11
	1.9 Assembly	12
	Summary	12
	References	13
Chapter 2	**Aluminum**	**15**
	2.1 Metallurgical Considerations	17
	2.2 Aluminum Alloy Designation	23
	2.3 Aluminum Alloys	25
	2.4 Melting and Primary Fabrication	31
	2.4.1 Rolling Plate and Sheet	33
	2.4.2 Extrusion	37
	2.5 Heat Treating	37
	2.5.1 Solution Heat Treating and Aging	37
	2.5.2 Annealing	42
	2.6 Forging	43
	2.7 Forming	46
	2.7.1 Blanking and Piercing	47
	2.7.2 Brake Forming	48
	2.7.3 Deep Drawing	49
	2.7.4 Stretch Forming	50

		2.7.5 Rubber Pad Forming	51
		2.7.6 Superplastic Forming	51
	2.8	Casting	57
		2.8.1 Sand Casting	60
		2.8.2 Plaster and Shell Molding	62
		2.8.3 Permanent Mold Casting	63
		2.8.4 Die Casting	64
		2.8.5 Investment Casting	64
		2.8.6 Evaporative Pattern Casting	64
		2.8.7 Casting Heat Treatment	65
		2.8.8 Casting Properties	65
	2.9	Machining	66
		2.9.1 High Speed Machining	68
		2.9.2 Chemical Milling	76
	2.10	Joining	76
	2.11	Welding	77
		2.11.1 Gas Metal and Gas Tungsten Arc Welding	78
		2.11.2 Plasma Arc Welding	80
		2.11.3 Laser Welding	81
		2.11.4 Resistance Welding	82
		2.11.5 Friction Stir Welding	83
	2.12	Chemical Finishing	88
		Summary	89
		Recommended Reading	90
		References	90
Chapter 3	**Magnesium and Beryllium**		**93**
		MAGNESIUM	95
	3.1	Magnesium Metallurgical Considerations	95
	3.2	Magnesium Alloys	97
		3.2.1 Wrought Magnesium Alloys	97
		3.2.2 Magnesium Casting Alloys	99
	3.3	Magnesium Fabrication	103
		3.3.1 Magnesium Forming	103
		3.3.2 Magnesium Sand Casting	104
		3.3.3 Magnesium Heat Treating	106
		3.3.4 Magnesium Machining	107
		3.3.5 Magnesium Joining	107
	3.4	Magnesium Corrosion Protection	108
		BERYLLIUM	109
	3.5	Beryllium Metallurgical Considerations	109
	3.6	Beryllium Alloys	110

	3.7	Beryllium Powder Metallurgy	111
	3.8	Beryllium Fabrication	114
		3.8.1 Beryllium Forming	114
		3.8.2 Beryllium Machining	115
		3.8.3 Beryllium Joining	116
	3.9	Aluminum–Beryllium Alloys	116
		Summary	116
		References	118
Chapter 4	**Titanium**		**119**
	4.1	Metallurgical Considerations	120
	4.2	Titanium Alloys	126
		4.2.1 Commercially Pure Titanium	126
		4.2.2 Alpha and Near-Alpha Alloys	127
		4.2.3 Alpha–Beta Alloys	128
		4.2.4 Beta Alloys	131
	4.3	Melting and Primary Fabrication	132
	4.4	Forging	137
	4.5	Directed Metal Deposition	140
	4.6	Forming	143
	4.7	Superplastic Forming	145
	4.8	Heat Treating	150
		4.8.1 Stress Relief	151
		4.8.2 Annealing	152
		4.8.3 Solution Treating and Aging	152
	4.9	Investment Casting	154
	4.10	Machining	158
	4.11	Joining	165
	4.12	Welding	165
	4.13	Brazing	170
		Summary	171
		Recommended Reading	172
		References	173
Chapter 5	**High Strength Steels**		**175**
	5.1	Metallurgical Considerations	176
	5.2	Medium Carbon Low Alloy Steels	182
	5.3	Fabrication of Medium Carbon Low Alloy Steels	186
	5.4	Heat Treatment of Medium Carbon Low Alloy Steels	191
	5.5	High Fracture Toughness Steels	198
	5.6	Maraging Steels	200
	5.7	Precipitation Hardening Stainless Steels	202

		Summary	207
		Recommended Reading	207
		References	208

Chapter 6 Superalloys 211

6.1	Metallurgical Considerations		213
6.2	Commercial Superalloys		219
	6.2.1	Nickel Based Superalloys	221
	6.2.2	Iron–Nickel Based Superalloys	224
	6.2.3	Cobalt Based Superalloys	225
6.3	Melting and Primary Fabrication		225
6.4	Powder Metallurgy		228
	6.4.1	Powder Metallurgy Forged Alloys	228
	6.4.2	Mechanical Alloying	230
6.5	Forging		232
6.6	Forming		236
6.7	Investment casting		238
	6.7.1	Polycrystalline Casting	239
	6.7.2	Directional Solidification (DS) Casting	240
	6.7.3	Single Crystal (SC) Casting	242
6.8	Heat Treatment		243
	6.8.1	Solution Strengthened Superalloys	243
	6.8.2	Precipitation Strengthened Nickel Base Superalloys	244
	6.8.3	Precipitation Strengthened Iron–Nickel Base Superalloys	246
	6.8.4	Cast Superalloy Heat Treatment	247
6.9	Machining		248
	6.9.1	Turning	251
	6.9.2	Milling	252
	6.9.3	Grinding	254
6.10	Joining		256
	6.10.1	Welding	256
	6.10.2	Brazing	260
	6.10.3	Transient Liquid Phase (TLP) Bonding	263
6.11	Coating Technology		264
	6.11.1	Diffusion Coatings	264
	6.11.2	Overlay Coatings	265
	6.11.3	Thermal Barrier Coatings	266
	Summary		266
	Recommended Reading		270
	References		270

Chapter 7 Polymer Matrix Composites — 273

- 7.1 Materials — 276
    - 7.1.1 Fibers — 277
    - 7.1.2 Matrices — 280
    - 7.1.3 Product Forms — 282
- 7.2 Fabrication Processes — 286
- 7.3 Cure Tooling — 286
    - 7.3.1 Tooling Considerations — 286
- 7.4 Ply Collation — 291
    - 7.4.1 Manual Lay-up — 291
    - 7.4.2 Flat Ply Collation and Vacuum Forming — 294
- 7.5 Automated Tape Laying — 295
- 7.6 Filament Winding — 298
- 7.7 Fiber Placement — 304
- 7.8 Vacuum Bagging — 307
- 7.9 Curing — 311
    - 7.9.1 Curing of Epoxy Composites — 313
    - 7.9.2 Theory of Void Formation — 314
    - 7.9.3 Hydrostatic Resin Pressure — 318
    - 7.9.4 Resin and Prepreg Variables — 322
    - 7.9.5 Condensation Curing Systems — 323
    - 7.9.6 Residual Curing Stresses — 324
- 7.10 Liquid Molding — 327
- 7.11 Preform Technology — 328
    - 7.11.1 Fibers — 329
    - 7.11.2 Woven Fabrics — 330
    - 7.11.3 Multiaxial Warp Knits — 331
    - 7.11.4 Stitching — 331
    - 7.11.5 Braiding — 333
    - 7.11.6 Preform Handling — 334
- 7.12 Resin Injection — 336
    - 7.12.1 RTM Curing — 338
    - 7.12.2 RTM Tooling — 338
- 7.13 Vacuum Assisted Resin Transfer Molding — 339
- 7.14 Pultrusion — 341
- 7.15 Thermoplastic Composites — 343
    - 7.15.1 Thermoplastic Consolidation — 345
    - 7.15.2 Thermoforming — 351
    - 7.15.3 Thermoplastic Joining — 355
- 7.16 Trimming and Machining Operations — 361
- Summary — 364
- Recommended Reading — 366
- References — 366

Chapter 8 Adhesive Bonding and Integrally Cocured Structure 369
8.1	Advantages of Adhesive Bonding	370
8.2	Disadvantages of Adhesive Bonding	371
8.3	Theory of Adhesion	372
8.4	Joint Design	372
8.5	Adhesive Testing	377
8.6	Surface Preparation	378
8.7	Epoxy Adhesives	383
	8.7.1 Two-part Room Temperature Curing Epoxy Liquid and Paste Adhesives	384
	8.7.2 Epoxy Film Adhesives	385
8.8	Bonding Procedures	385
	8.8.1 Prekitting of Adherends	385
	8.8.2 Prefit Evaluation	386
	8.8.3 Adhesive Application	387
	8.8.4 Bond Line Thickness Control	388
	8.8.5 Bonding	388
8.9	Sandwich Structures	390
	8.9.1 Honeycomb Core	393
	8.9.2 Honeycomb Processing	399
	8.9.3 Balsa Wood	403
	8.9.4 Foam Cores	404
	8.9.5 Syntactic Core	406
	8.9.6 Inspection	407
8.10	Integrally Cocured Structure	408
	Summary	415
	Recommended Reading	416
	References	417

Chapter 9 Metal Matrix Composites 419
9.1	Discontinuously Reinforced Metal Matrix Composites	424
9.2	Stir Casting	424
9.3	Slurry Casting – Compocasting	427
9.4	Liquid Metal Infiltration (Squeeze Casting)	427
9.5	Pressure Infiltration Casting	430
9.6	Spray Deposition	431
9.7	Powder Metallurgy Methods	432
9.8	Secondary Processing of Discontinuous MMCs	434
9.9	Continuous Fiber Aluminum Metal Matrix Composites	435
9.10	Continuous Fiber Reinforced Titanium Matrix Composites	440

	9.11	Secondary Fabrication of Titanium Matrix Composites	447
	9.12	Fiber Metal Laminates	452
		Summary	455
		Recommended Reading	456
		References	456

Chapter 10 Ceramic Matrix Composites   459

10.1	Reinforcements	464
10.2	Matrix Materials	467
10.3	Interfacial Coatings	470
10.4	Fiber Architectures	471
10.5	Fabrication Methods	472
10.6	Powder Processing	472
10.7	Slurry Infiltration and Consolidation	474
10.8	Polymer Infiltration and Pyrolysis (PIP)	476
10.9	Chemical Vapor Infiltration (CVI)	482
10.10	Directed Metal Oxidation (DMO)	487
10.11	Liquid Silicon Infiltration (LSI)	488
	Summary	490
	Recommended Reading	492
	References	492

Chapter 11 Structural Assembly   495

11.1	Framing		496
11.2	Shimming		498
11.3	Hole Drilling		499
	11.3.1	Manual Drilling	500
	11.3.2	Power Feed Drilling	504
	11.3.3	Automated Drilling	505
	11.3.4	Automated Riveting Equipment	508
	11.3.5	Drill Bit Geometries	509
	11.3.6	Reaming	514
	11.3.7	Countersinking	514
11.4	Fastener Selection and Installation		515
	11.4.1	Special Considerations for Composite Joints	518
	11.4.2	Solid Rivets	520
	11.4.3	Pin and Collar Fasteners	523

	11.4.4	Bolts and Nuts	525
	11.4.5	Blind Fasteners	527
	11.4.6	Fatigue Improvement and Interference Fit Fasteners	528
11.5	Sealing		533
11.6	Painting		534
	Summary		535
	Recommended Reading		537
	References		537

Appendix A Metric Conversions — 539

Appendix B A Brief Review of Materials Fundamentals — 541

B.1	Materials	542
B.2	Metallic Structure	543
B.3	Ceramics	555
B.4	Polymers	556
B.5	Composites	562
	Recommended Reading	565
	References	566

Appendix C Mechanical and Environmental Properties — 567

C.1	Static Strength Properties	568
C.2	Failure Modes	570
C.3	Fracture Toughness	572
C.4	Fatigue	576
C.5	Creep and Stress Rupture	581
C.6	Corrosion	582
C.7	Hydrogen Embrittlement	584
C.8	Stress Corrosion Cracking	586
C.9	High Temperature Oxidation and Corrosion	587
C.10	Polymeric Matrix Composite Degradation	587
	Recommended Reading	591
	References	591

Index — 593

Preface

This book is intended for anyone wishing to learn more about the materials and manufacturing processes used to fabricate and assemble advanced aerospace structures. The remarkable performance characteristics of modern aerospace vehicles are, to a large degree, a result of the high performance materials and manufacturing technology used in both the airframes and propulsion systems. To obtain continual performance increases, designers are constantly searching for lighter, stronger and more durable materials.

Chapter 1 gives a brief overview of the structural materials that are used in aerospace structures. The next five chapters are then devoted to the important metals, namely aluminum, magnesium, beryllium, titanium, high strength steels and superalloys.

Aluminum alloys (Chapter 2) have been the main airframe material since they started replacing wood in the early 1920s. Even though the role of aluminum in future commercial aircraft will probably be somewhat eroded by the increasing use of composite materials, high strength aluminum alloys are and will remain important airframe materials.

Although both magnesium and beryllium (Chapter 3) are extremely lightweight materials, they both have serious drawbacks that limit their applications. The biggest obstacle to the use of magnesium alloys is its extremely poor corrosion resistance. Beryllium is also a very lightweight metal with an attractive combination of properties. However, beryllium must be processed using powder metallurgy technology that is costly, and beryllium powder and dust are toxic, which further increases its cost through the requirement for controlled environments.

Titanium (Chapter 4) is often used to save weight by replacing heavier steel alloys in the airframe and superalloys in the low temperature portions of gas turbines. Titanium is becoming even more important as an airframe material due to its outstanding resistance to fatigue, its high temperature capability and its resistance to corrosion.

While high strength steels (Chapter 5) normally account for only about 5–15% of the airframe structural weight, they are often used for highly critical parts such as landing gear components. The main advantages of high strength steels are their extremely high strengths and stiffness. This can be extremely important in landing gear applications where it is critical to minimize the volume of the gear components.

Superalloys (Chapter 6) are another enabling material for modern flight where they are used extensively in the jet turbine engines. Some superalloys are capable of being used in load bearing applications in excess of 80% of their incipient melting temperatures while

Preface

exhibiting high strength, good fatigue and creep resistance, good corrosion resistance and the ability to operate at elevated temperatures for extended periods of time.

Chapters 7 through 10 deal with the important field of composite materials. Chapter 7 covers polymer matrix composites. This is followed by Chapter 8 on Structural Adhesives and Cocured Structure. Metal Matrix Composites and Ceramic Matrix Composites are covered in Chapters 9 and 10 respectively.

The advantages of high performance polymer matrix composites (Chapter 7) are many, including lighter weight; the ability to tailor lay-ups for optimum strength and stiffness; improved fatigue life; corrosion resistance; and with good design practice, reduced assembly costs due to fewer detail parts and fasteners. The specific strength (strength/density) and specific modulus (modulus/density) of high strength fiber composites, especially carbon, are higher than comparable aerospace metallic alloys. This translates into greater weight savings resulting in improved performance, greater payloads, longer range and fuel savings.

Adhesive Bonding and Cocured Structure (Chapter 8) covers how parts can be combined into a single cured assembly during either initial cure or by secondarily adhesive bonding. Large one piece composite structures have demonstrated the potential for impressive reductions in part counts and assembly costs.

Metal matrix composites (Chapter 9) offer a number of advantages compared to their base metals, such as higher specific strengths and moduli, higher elevated temperature resistance, lower coefficients of thermal expansion and, in some cases, better wear resistance. On the down side, they are more expensive than their base metals and have lower toughness. Due to their high cost, commercial applications for metal matrix composites are limited.

Similar to metal matrix composites, there are very few commercial applications for ceramic matrix composites (Chapter 10) due to their high costs and concerns for reliability. Carbon–carbon has found applications in aerospace for thermal protection systems. However, metal and ceramic matrix composites remain an important material class, because they are considered enablers for future hypersonic flight vehicles.

Assembly (Chapter 11) represents a significant portion of the total airframe manufacturing cost, as much as 50% of the total delivered airframe cost. In this chapter, the emphasis is on mechanical joining including the hole preparation procedures and fasteners used for structural assembly. Sealing and painting are also briefly discussed.

This book is intended for the engineer or student who wants to learn more about the materials and processing used in aerospace structure. It would be useful to designers, structural engineers, material and process engineers and manufacturing engineers involved with advanced materials. A first course in Materials Science would be helpful in understanding the material in this book; however, a brief review of some of the fundamentals of materials science is included as Appendix B. There is also an Appendix C, which gives a brief explanation of some of the mechanical property terms and environmental degradation mechanisms that are encountered throughout this book.

The reader is hereby cautioned that the data presented in this book are not design allowables. The reader should consult approved design manuals for statistically derived design allowables.

Preface

I would like to thank a number of my colleagues for reviewing the chapters in this book, in particular Dr J.A. Baumann, D.R. Bolser, D.M. Furdek, T.L. Hackett, N. Melillo, and Dr K.T. Slattery. In addition, I would like to acknowledge the help and guidance of Dr Geoff Smaldon, Elsevier Advanced Technology, and Priyaa H. Menon, Integra Software Services Pvt. Ltd, and their staffs for their valuable contributions.

F.C. Campbell
St. Louis, Missouri
January 2006

Chapter 1

Introduction

The remarkable performance characteristics of a modern fighter aircraft (Fig. 1.1) are, to a large degree, a result of the high performance materials used in both the airframe and propulsion systems. A commercial aircraft will fly over 60 000 h. during its 30-year life, with over 20 000 flights, and will taxi over 100 000 miles.[1] To obtain continual performance increases, designers are constantly searching for lighter and stronger materials. Reducing material density is recognized as the most efficient way of reducing airframe weight and improving performance. It has been estimated that reductions in material density are about 3 to 5 times more effective than increasing tensile strength, modulus or damage tolerance.[2] For gas turbine jet engines, advances in materials have allowed significantly higher operating temperatures, which result in increases in thrust levels, again increasing performance.

Over the next 20 years, it has been forecast that there will be a significant increase in the demand for air travel, especially in the large population centers in Asia. It is possible that the number of air travelers will double with the demand for new aircraft increasing between 13 500 and 17 000, with yearly deliveries between 675 and 850 aircraft, at a total estimated value of approximately 1.25 trillion dollars.[3]

Airframe durability is becoming a greater concern since the life of many aircraft, both commercial and military, are being extended far beyond their intended design lives. Will the B-52 bomber, which first flew in 1954, become the first 100-year airframe? Even for an aircraft with "only" a 30-year lifetime, it has been estimated that the cost of service and maintenance over the 30-year life of the aircraft exceeds the original purchase price by a factor of two.[1]

Since the early-1920s, airframes have been built largely out of metal, aluminum in particular has been the material of choice. When high performance composites (i.e., first boron and then carbon fibers) started being developed in the mid-1960s and early-1970s, the situation started changing. The earliest developers, and users, of composites were the military. The earliest production usage of high performance composites were on the empennages of the F-14 and F-15 fighter aircraft. Boron/epoxy was used for the horizontal stabilators on both of these aircraft, and for the rudders and vertical fins on the F-15. In the mid-1970s, with the maturity of carbon fibers, a carbon/epoxy speedbrake was implemented on the F-15. While these early applications resulted in significant weight savings (~20%), they accounted for only small amounts of the airframe structural weight.

However, as shown in Fig. 1.2, composite usage quickly expanded from only 2% of the airframe on the F-15 to as much as 27% on the AV-8B Harrier by the early-1980s. Significant applications included the wing (skins and substructure), the forward fuselage, and the horizontal stabilator, all fabricated from carbon/epoxy. While the amount of composites used on the AV-8B was somewhat on the high side, most modern fighter aircraft contain over 20% composite

Introduction

Fig. 1.1. F/A-18 Fighter Aircraft
Source: U.S. Navy & The Boeing Company

Fig. 1.2. Composites Evolution in Fighter Aircraft
Source: The Boeing Company

structure. Typical weight savings usually range from 15 to 25%, depending on the particular piece of structure, with about 20% being a good rule of thumb.

Similar trends have been followed for commercial aircraft, although at a slower and more cautious pace. Until recently, Airbus has been somewhat more aggressive in using composites than Boeing, primarily for horizontal stabilizers and vertical fins on their A300 series of aircraft. However, Boeing recently made a major commitment to composites, when it decided to use upwards of 50% on its new 787, which includes both a composite wing and fuselage, as depicted in Fig. 1.3.

1.1 Aluminum

Aluminum alloys have been the main airframe material since they started replacing wood in the early 1920s. The dominance of aluminum alloys in airframe applications is shown in Fig. 1.4. Even though the role of aluminum in future

Introduction

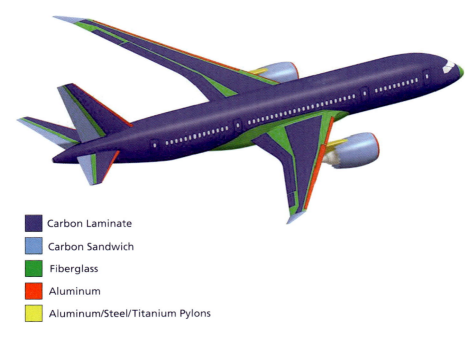

Fig. 1.3. *Boeing 787 Material Distribution*
Source: The Boeing Company

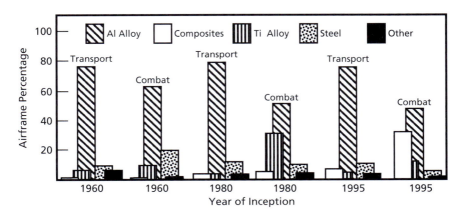

Fig. 1.4. *Dominance of Aluminum in Airframes*[1]

commercial aircraft will probably be somewhat eroded by the increasing use of composite materials, high strength aluminum alloys are, and will remain, an important airframe material. Even in fighter aircraft, which already has composite material percentages in the range of 20–30%, aluminum still plays a significant role. The attractiveness of aluminum is that it is a relatively low cost, light weight metal that can be heat treated to fairly high strength levels, and it is one of the most easily fabricated of the high performance materials, which usually correlates directly with lower costs.

The aluminum–copper (2XXX series) and aluminum–zinc (7XXX series) alloys are the primary alloys used in airframe structural applications. The 2XXX alloys are used in damage tolerance applications, such as the lower wing skins and fuselage structure of commercial aircraft, while the 7XXX alloys are used where higher strength is required, such as the upper wing skins. The 2XXX alloys also have slightly higher temperature capability (300 vs. 250° F). Improvements in compositional control and processing have continually produced improved alloys. Reducing impurities, in particular iron and silicon, has resulted in higher fracture toughness and better resistance to fatigue crack initiation and crack growth. Examples of these newer alloys are 2524-T3, 7150-T77 and 7055-T77, which are used on the Boeing 777. The venerable alloy 2024-T3 has been one of the most widely used alloys in fuselage construction. While it only has a moderate yield strength, it has very good resistance to fatigue crack growth and good fracture toughness. However, the newer alloy 2524-T3 has a 15–20% improvement in fracture toughness and twice the fatigue crack growth resistance of 2024-T3.

The 7XXX alloys have higher strengths than the 2XXX alloys and are used in sheet, plate, forgings and extrusions. Like 2024-T3, 7075-T6 has been used for a great many years in airframe construction; however, stress corrosion cracking has been a recurring problem. Newer alloys, such as 7055-T77, have higher strength and damage tolerance than 7050-T7451, while 7085-T7651 has higher thick section toughness. Along with tightening compositional controls and eliminating unwanted impurities, the development of improved aging heat treatments for the 7XXX alloys has resulted in greatly reduced stress corrosion cracking susceptibility and improved fracture toughness, with only a minimal impact on strength.

Improvements in aluminum manufacturing technology include high speed machining and friction stir welding. Although higher metal removal rates are an immediate benefit of high speed machining, an additional cost saving is the ability to machine extremely thin walls and webs. This allows the design of weight competitive high speed machined assemblies, in which sheet metal parts that were formally assembled with mechanical fasteners can now be machined from a single, or several blocks, of aluminum plate. Another recent development, called friction stir welding, is a new solid state joining process, which has the ability to weld the difficult, or impossible, to fusion weld 2XXX and 7XXX alloys with less distortion, fewer defects, and better durability than achievable using conventional welding techniques.

1.2 Magnesium and Beryllium

Although both magnesium and beryllium are extremely lightweight materials, they both have serious drawbacks that limit their applications. Magnesium alloys usually compete with aluminum alloys for structural applications. Compared to high strength aluminum alloys, magnesium alloys are not normally as strong and have a lower modulus of elasticity. However, magnesium alloys are significantly lighter and are therefore competitive on a specific strength (strength/density) and specific modulus (modulus/density) basis. The biggest obstacle to the use of magnesium alloys is their extremely poor corrosion resistance. Magnesium occupies the highest anodic position on the galvanic series, and, as such, there is always the strong potential for corrosion.

Beryllium is also a very lightweight metal with an attractive combination of properties. However, beryllium must be processed using powder metallurgy technology that is costly, and beryllium powder and dust are toxic, which further increases its cost through the requirement for controlled manufacturing environments and the concern for safety during the repair/service of deployed structures.

1.3 Titanium

Titanium can often be used to save weight by replacing heavier steel alloys in the airframe and superalloys in the low temperature portions of gas turbines. Titanium is also used instead of aluminum when the temperature requirements exceed aluminum's capabilities ($\sim 300°$ F), or in areas where fatigue or corrosion has been a recurring problem.

Titanium is becoming more important as an airframe material. Due to their outstanding resistance to fatigue, high temperature capability and resistance to corrosion, titanium alloys comprise approximately 42% of the structural weight of the new F-22 fighter aircraft, over 9000 lb. in all.[4] In commercial aircraft, the Boeing 747-100 contained only 2.6% titanium, while the newer Boeing 777 contains 8.3%.[5] New applications for titanium include landing gears, traditionally made from high strength steels. For example, to save weight and eliminate the risk of hydrogen embrittlement, the beta alloy Ti-10V-2Fe-3Al is used for landing gear components on the Boeing 777. Titanium alloys are also used extensively in the lower temperature regions of jet turbine engines. In commercial aircraft engines, titanium alloys are used in the fan, the low pressure compressor, and about 2/3 of the high pressure compressor.[5]

Although many new alloys have been developed since the mid-1980s, the alpha–beta alloy Ti-6Al-4V is still the most widely used titanium alloy. However, stronger alpha–beta alloys and beta alloys are starting to replace Ti-6Al-4V in some applications. In addition, as for the compositional and processing control developed for aluminum alloys, similar controls have been developed for titanium alloys, with improved melting practices such as multiple vacuum arc melting. Cold hearth melting is another new melting practice that can now be used to produce even cleaner and more homogeneous ingots. Alloys produced by vacuum melting, in combination with the cold hearth process, have proven to be essentially free of melt-related inclusions.

Near net shape processes can lead to savings in materials, machining costs and cycle times over conventional forged or machined parts. Investment casting, in combination with hot isostatic pressing (HIP), can produce aerospace quality titanium near net shaped parts that can offer significant cost savings over forgings and built-up structure. Titanium is also very amenable to superplastic forming, and, when combined with diffusion bonding (SPF/DB), is capable of producing complex unitized structures. In another process, called directed metal deposition, a focused laser or electron beam is used to melt titanium powder and deposit the melt in a predetermined path on a titanium substrate plate. The deposited preform is then machined to the final part shape.

1.4 High Strength Steels

While high strength steels normally account for only about 5–15% of the airframe structural weight, they are often used for highly critical parts such as landing gear components. The main advantages of high strength steels are their extremely high strengths and stiffness. This can be extremely important in landing gear applications, where it is important to minimize the volume of the gear components. The disadvantages are their high densities and susceptibility to brittle fracture. As a result of their high strength levels, they are often susceptible to both hydrogen embrittlement and stress corrosion cracking, which can cause sudden brittle failures. Work is underway to develop ultrahigh strength stainless steels to further alleviate some of the corrosion and embrittlement problems with these materials.

1.5 Superalloys

Superalloys are another enabling material for modern aircraft, where they are used extensively in the jet turbine engines. Some superalloys are capable of being used in load-bearing applications in excess of 80% of their incipient melting temperatures, while exhibiting high strength, good fatigue and creep resistance, good corrosion resistance, and the ability to operate at elevated temperatures for extended periods of time. As a general class of materials, superalloys include nickel, iron–nickel and cobalt based alloys that operate at temperatures exceeding 1000° F. In a jet engine, specific thrust is defined as thrust divided by weight (thrust/weight), and the most effective way of increasing thrust is through increasing the operating temperature of the engine. The remarkable role superalloy technology has played in allowing higher engine operating temperatures is illustrated in Fig. 1.5.

Due to the extremely high operating temperatures, about one third of the high pressure compressor, the combustor, and both the high and low pressure turbines require superalloys. Superalloys were originally developed using conventional ingot melting and wrought technology, but the highest temperature parts (i.e., turbine blades) are now castings. Originally, turbine blades

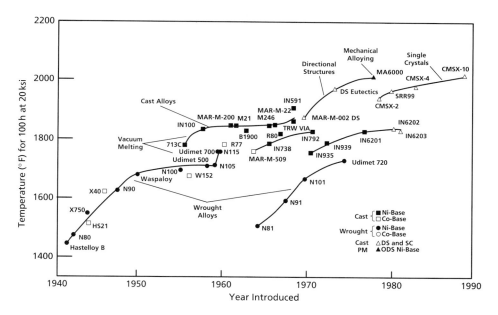

Fig. 1.5. Temperature Capability of Superalloys[6]

were made from conventional wrought materials such as forgings. However, the high temperature creep mechanism of grain boundary sliding limited the operating capability of these fine grained equiaxed structures. Higher operating temperatures were achieved with higher alloy contents and coarse equiaxed grain castings. Even further improvements were made with directionally solidified (DS) castings, which eliminated the transverse grain boundaries. Finally, single crystal (SC) casting was developed that eliminated all of the grain boundaries. The development of single crystal castings also allowed alloy designers to remove alloying elements that had been added to prevent grain boundary cracking but were detrimental to creep strength. To obtain even higher operating temperatures, the hollow blades contain strategically located rows of small holes to allow film cooling from bleed air supplied by the compressor. They are also coated with thermal barrier coatings (TBC), which consist of ceramic coatings that reduce the heat flux through the airfoil, allowing greater gas–metal temperature differences.

1.6 Composites

The advantages of high performance composites are many, including lighter weight, the ability to tailor lay-ups for optimum strength and stiffness, improved fatigue life, corrosion resistance, and, with good design practice, reduced assembly costs due to fewer detail parts and fasteners. The specific strength and

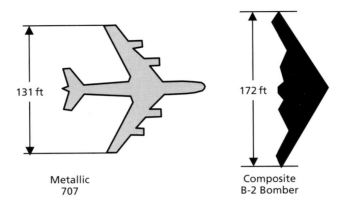

Fig. 1.6. Comparative Planform Areas[7]

specific modulus of high strength fiber composites, especially carbon fibers, are higher than other comparable aerospace metallic alloys. This translates into greater weight savings resulting in improved performance, greater payloads, longer range and fuel savings.

The U.S. military was an early developer and adapter of composite materials. As mentioned earlier, the AV-8B aircraft had an airframe with 27% composite structure by the early 1980s. The first large scale usage of composites in commercial aircraft occurred in 1985, when the Airbus A320 first flew with composite horizontal and vertical stabilizers. Airbus has applied composites in up to 15% of the overall airframe weight for their A320, A330 and A340 family of aircraft.[5] While the percentages are lower for commercial aircraft, the part sizes are much larger. For example, the Boeing 777 horizontal stabilizer has approximately the same surface area as a Boeing 737 wing. The B-2 bomber (Fig. 1.6) has the largest composite parts manufactured to date and is proof that at least some of the issues with fabricating large commercial structures have been addressed. Channel laminators and large tape laying machines, capable of making one piece skins in excess of 65 ft in length, demonstrated automated composite part fabrication in a large factory environment.[6] However, to realize this type of structure in the commercial aircraft world, the cost of composite structure still needs to be reduced through innovative design and manufacturing technologies. The cost of composites is the number one deterrent to their broader application.[7]

1.7 Adhesive Bonding and Integrally Cocured Structure

Assembly costs usually account for about 50% of the cost of an airframe.[8] Composites offer the opportunity to significantly reduce the amount of assembly labor and fasteners. Detail parts can be combined into a single cured assembly during either initial cure or by secondarily adhesive bonding. Large one piece composite structures have demonstrated the potential for impressive reductions in part count. Lower part count on a wide-body aircraft could allow the elimination of several hundred thousand fasteners.[9]

Cocuring is a process in which uncured composite plies are cured and bonded simultaneously during the same cure cycle to either core materials or to other composite parts, while cobonding is a process in which precured parts (e.g., stiffeners) are bonded to another part (e.g., a skin) at the same time the skin is cured. The ability to make large cocured or cobonded unitized structure can eliminate a significant portion of the assembly costs.

Adhesive bonding is a more traditional method of joining structure that eliminates some, or all, of the cost and weight of mechanical fasteners. In adhesive bonding, cured composites or metals are adhesively bonded to other cured composites, honeycomb core, foam core or metallic pieces. The advantages of adhesive bonding include a more uniform stress distribution by eliminating the individual stress concentrations caused by mechanical fasteners. Bonded joints are usually lighter than mechanically fastened joints, and bonded joints enable the design of smooth external surfaces with minimum sensitivity to fatigue crack propagation.

1.8 Metal and Ceramic Matrix Composites

Metal matrix composites offer a number of advantages compared to their base metals, such as higher specific strengths and moduli, higher elevated temperature resistance, lower coefficients of thermal expansion and, in some cases, better wear resistance. On the down side, they are more expensive than their base metals and have lower toughness. Metal matrix composites also have some advantages compared to polymer matrix composites, including higher matrix dependent strengths and moduli, higher elevated temperature resistance, no moisture absorption, higher electrical and thermal conductivities, and non-flammability. However, metal matrix composites are normally more expensive than even polymer matrix composites, and the fabrication processes are much more limited, especially for complex structural shapes. Due to their high cost, commercial applications for metal matrix composites are sparse. However, because they are considered enablers for future hypersonic flight vehicles, both metal and ceramic matrix composites remain important materials. Fiber metal laminates, in particular glass fiber reinforced aluminum laminates (GLARE), are another form of composite material that offers fatigue performance advantages over monolithic aluminum structure.

Ceramics contain many desirable properties, such as high moduli, high compression strengths, high temperature capability, high hardness and wear resistance, low thermal conductivity, and chemical inertness. However, due to their very low fracture toughness, ceramics are limited in structural applications. They have a very low tolerance to crack-like defects, which can occur either during fabrication or in-service. Even a very small crack can quickly grow to critical size, leading to sudden failure. While reinforcements such as fibers, whiskers or particles are used to strengthen polymer and metal matrix composites, reinforcements in ceramic matrix composites are primarily used to increase toughness.

Due to their high costs and concerns for reliability, there are also very few commercial applications for ceramic matrix composites. However, carbon–carbon (CC) composites[10] have found applications in aerospace for thermal protection systems. Carbon–carbon composites are the oldest and most mature of the ceramic matrix composites. They were developed in the 1950s for use as rocket motor casings, heat shields, leading edges and thermal protection. The most recognized application is the Space Shuttle leading edges. For high temperature applications, carbon–carbon composites offer exceptional thermal stability, provided they are protected with oxidation resistant coatings.

1.9 Assembly

Assembly represents a significant portion of the total manufacturing cost, as much as 50% of the total delivered part cost.[1] Labor costs account for more than 60% of the recurring costs of each airplane for production runs up to 300 units, and remain above 50% even for runs up to 600 units.[11] While assembly operations will be with us for the foreseeable future, it is possible to reduce this cost element by fabricating more large one piece unitized structures. Additional cost reduction opportunities include automated hole drilling and fastening.

Summary

A relative comparison of the different material classes is shown in Table 1.1. It should be noted that no single material fulfills all of the needs of current or future aerospace vehicles. All of them have some shortcomings, either in properties or in cost. With the increasing emphasis on cost, today's and tomorrow's

Table 1.1 Relative Comparison of Material Groups

Material Class	Tension Strength	Compression Strength	Stiffness	Ductility	Temperature Capability	Density	Cost
Metals	High	High	Medium	High	High	High	$$
Ceramics	Low	High	Very High	Nil	Very High	Medium	$$$
Polymers	Very Low	Very Low	Very Low	High	Low	Low	$
PMC	Very High	High	Very High	Low	Medium	Low	$$$
MMC	High	High	Very High	Low	High	Medium	$$$$
CMC	Medium	High	Very High	Low	Vey High	Medium	$$$$$

PMC – Polymer Matrix Composite
MMC – Metal Matrix Composite
CMC – Ceramic Matrix Composite

materials will need to be processed and manufactured in a very cost competitive environment. While much progress has been made in the twentieth century, even more improvements in both the materials and their associated manufacturing technologies will be required in the twenty-first century.

References

[1] Peel, C.J., Gregson, P.J., "Design Requirements for Aerospace Structural Materials", in *High Performance Materials in Aerospace*, Chapman & Hall, 1995, pp. 1–48.
[2] Ekvall, J.C., Rhodes, J.E., Wald, G.G., "Methodology of Evaluating Weight Savings From Basic Material Properties", in *Design of Fatigue and Fracture Resistant Structures*, ASTM STP 761, American Society for Testing and Materials, 1982, pp. 328–341.
[3] Barington, N., Black, M., "Aerospace Materials and Manufacturing Processes at the Millenium", in *Aerospace Materials*, Institute of Physics Publishing, 2002, pp. 3–14.
[4] Cotton, J.D., Clark, L.P., Phelps, H.R., "Titanium Alloys on the F-22 Fighter Aircraft", *Advanced Materials & Processes*, May 2002, pp. 25–28.
[5] Williams, J.C., Starke, E.A., "Progress in Structural Materials for Aerospace Systems", *Acta Materialia*, Vol. 51, 2003, pp. 5775–5799.
[6] Molloy, W.J., "Investment-Cast Superalloys – A Good Investment", *Advanced Materials & Processes*, October 1990, pp. 23–30.
[7] Freeman, W.T., "The Use of Composites in Aircraft Primary Structure", *Composites Engineering*, Vol. 3, Nos 7–8, 1993, pp. 767–775.
[8] Taylor, A., "RTM Material Developments for Improved Processability and Performance", *SAMPE Journal*, Vol. 36, No. 4, July/August 2000, pp. 1–24.
[9] Piellisch, R., "Composites Roll Sevens", *Aerospace America*, Vol. 30, No. 10, pp. 26–43.
[10] Buckley, J.D., "Carbon-Carbon Composites", in *Handbook of Composites*, Chapman & Hall, 1998, pp. 333–351.
[11] "Transport Construction Costs Studied", *Aviation Week and Space Technology*, Vol. 107, No. 24, pp. 68–70.

Chapter 2

Aluminum

A typical material distribution for a modern commercial airliner, shown in Fig. 2.1, illustrates the heavy dominance of aluminum alloys. The attractiveness of aluminum is that it is a relatively low cost, light weight metal that can be heat treated to fairly high strength levels, and it is one of the more easily fabricated of the high performance materials, which usually results in lower costs. The advantages of aluminum as a high performance material can be summarized:

- High strength-to-weight ratio. The high strength 2XXX and 7XXX alloys are competitive on a strength-to-weight ratio with the higher strength but heavier titanium and steel alloys, and thus have traditionally been the predominate structural material in both commercial and military aircraft.
- Cryogenic properties. Aluminum alloys are not embrittled at low temperatures and become even stronger as the temperature is decreased without significant ductility losses, making them ideal for cryogenic fuel tanks for rockets and launch vehicles.
- Fabricability. Aluminum alloys are among the easiest of all metals to form and machine. The high strength 2XXX and 7XXX alloys can be formed in a relatively soft state and then heat treated to much higher strength levels after forming operations are completed.

Aluminum is also a consumer metal of great importance. In addition to the advantages cited above, other properties of commercial importance include corrosion resistance to natural atmospheres, suitability for food and beverage storage, high electrical and thermal conductivity, high reflectivity, and ease of recycling. As a result of a naturally occurring tenacious surface oxide

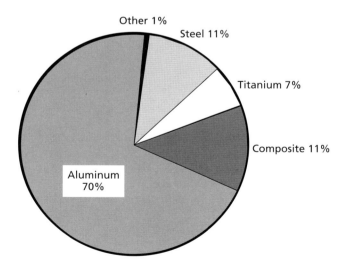

Fig. 2.1. Material Distribution for Boeing 777 Aircraft[1]

film (Al_2O_3), a great number of aluminum alloys provide exceptional resistance to corrosion in many atmospheric and chemical environments. Pure aluminum and some of its alloys have exceptionally high electrical conductivities, second only to copper for alloys commonly used as conductors. In addition, aluminum and its alloys are among the easiest to recycle of any of the structural materials.

Disadvantages of high strength aluminum alloys include a low modulus of elasticity, rather low elevated temperature capability, and susceptibility to corrosion. The modulus of elasticity of aluminum alloys is generally between 10 and 11 msi, which is lower than competing metals, such as titanium (16 msi) and steel (29 msi). Although aluminum alloys can be used for short times at temperatures as high as 400–500° F, their long-term usage temperatures are usually restricted to 250–300° F. Finally, although commercially pure aluminum and many aluminum alloys are very corrosion resistant, corrosion can be a problem for the highly alloyed high strength aluminum alloys used in aerospace.

This chapter is organized as follows. First, some of the general metallurgical considerations for aluminum alloys are discussed along with an introduction to precipitation hardening, the main method of strengthening the aerospace structural alloys. The next section covers the designation system and tempers for the various series of aluminum alloys. Then, some of the specific alloys used for aerospace applications are discussed, primarily the 2XXX, 6XXX, 7XXX, and 8XXX aluminum–lithium alloys. Melting and primary fabrication, which are mill processes, are then briefly discussed. The remainder of the chapter then covers the main fabrication processes, i.e. heat treatment, forging, forming, casting, machining, and joining.

2.1 Metallurgical Considerations

Aluminum is a light weight material with a density of 0.1 lb/in.^{3}. Pure aluminum and its alloys have the face centered cubic (FCC) structure, which is stable up to its melting point at 1215° F. Since the FCC structure contains multiple slip planes, this crystalline structure greatly contributes to the excellent formability of aluminum alloys. Only a few elements have sufficient solid solubility in aluminum to be major alloying elements. These include copper, magnesium, silicon, zinc, and more recently lithium. Important elements with lower solid solubility are the transition metals chromium, manganese, and zirconium, which normally form compounds that help to control the grain structure. Some of the important properties of each of the alloy series are given in Table 2.1. Aluminum alloys are normally classified into one of three groups: wrought non-heat treatable alloys, wrought heat treatable alloys, and casting alloys.

The wrought non–heat treatable alloys cannot be strengthened by precipitation hardening; they are hardened primarily by cold working. The wrought non–heat

Table 2.1 Major Attributes of Wrought Aluminum Alloys[2]

1XXX: Pure Al. The major characteristics of the 1XXX series are:

- Strain hardenable
- Exceptionally high formability, corrosion resistance, and electrical conductivity
- Typical ultimate tensile strength range: 10–27 ksi
- Readily joined by welding, brazing, soldering

2XXX: Al–Cu Alloys. The major characteristics of the 2XXX series are:

- Heat treatable
- High strength, at room and elevated temperatures
- Typical ultimate tensile strength range: 27–62 ksi
- Usually joined mechanically but some alloys are weldable
- Not as corrosion resistant as other alloys

3XXX: Al–Mn Alloys. The major characteristics of the 3XXX series are:

- High formability and corrosion resistance with medium strength
- Typical ultimate tensile strength range: 16–41 ksi
- Readily joined by all commercial procedures
- Hardened by strain hardening

4XXX: Al–Si Alloys. The major characteristics of the 4XXX series are:

- Some heat treatable
- Good flow characteristics, medium strength
- Typical ultimate tensile strength range: 25–55 ksi
- Easily joined, especially by brazing and soldering

5XXX: Al–Mg Alloys. The major characteristics of the 5XXX series are:

- Strain hardenable
- Excellent corrosion resistance, toughness, weldability, moderate strength
- Building and construction, automotive, cryogenic, marine applications
- Typical ultimate tensile strength range: 18–51 ksi

6XXX: Al–Mg–Si Alloys. The major characteristics of the 6XXX series are:

- Heat treatable
- High corrosion resistance, excellent extrudability; moderate strength
- Typical ultimate tensile strength range: 18–58 ksi
- Readily welded by GMAW and GTAW methods
- Outstanding extrudability

7XXX: Al–Zn Alloys. The major characteristics of the 7XXX series are:

- Heat treatable
- Very high strength; special high toughness versions
- Typical ultimate tensile strength range: 32–88 ksi
- Mechanically joined

8XXX: Alloys with Al-Other Elements (Not Covered by Other Series).

The major characteristics of the 8XXX series are:
- Heat treatable
- High conductivity, strength, hardness
- Typical ultimate tensile strength range: 17–60 ksi
- Common alloying elements include Fe, Ni and Li

treatable alloys include the commercially pure aluminum series (1XXX), the aluminum–manganese series (3XXX), the aluminum–silicon series (4XXX), and the aluminum–magnesium series (5XXX). While some of the 4XXX alloys can be hardened by heat treatment, others can only be hardened by cold working. The 4XXX alloys are mainly used as welding and brazing filler metals. Since the wrought non–heat treatable alloys are hardened primarily by cold working, they are not adequate for load-bearing structural applications even at moderately elevated temperature, since the cold worked structure could start softening (i.e., annealing) in service. Although the wrought non–heat treatable alloys are of great commercial importance, they are not generally candidates for structural airframe applications.

The wrought heat treatable alloys can be precipitation hardened to develop quite high strength levels. These alloys include the 2XXX series (Al–Cu and Al–Cu–Mg), the 6XXX series (Al–Mg–Si), the 7XXX series (Al–Zn–Mg and Al–Zn–Mg–Cu), and the aluminum–lithium alloys of the 8XXX alloy series. The 2XXX and 7XXX alloys, which develop the highest strength levels, are the main alloys used for metallic airframe components; however, there are some minor applications for some of the 6XXX and 8XXX alloys. The Al–Cu alloys of the 2XXX series, the Al–Mg–Si (6XXX), and the Al–Zn–Mg alloys of the 7XXX series are medium strength alloys and some are classified as being fusion weldable, while the Al–Cu–Mg alloys of the 2XXX series and the Al–Zn–Mg–Cu alloys of the 7XXX series are generally higher strength but are not fusion weldable. In reality, except for a limited number of the 2XXX alloys that are used for welded fuel tanks for launch vehicles, aluminum welded structure is not widely used for aerospace structures. However, as friction stir welding technology matures, this situation could change in the future.

The process of strengthening by precipitation hardening plays a critical role in high strength aluminum alloys. Precipitation hardening consists of three steps: (1) solution heat treating, (2) rapidly quenching to a lower temperature, and (3) aging. In solution heat treating, the alloy is heated to a temperature that is high enough to put the soluble alloying elements in solution. After holding at the solution treating temperature for some period of time, it is quenched to a lower temperature (e.g., room temperature) to keep the alloying elements trapped in solution. During aging, the alloying elements trapped in solution precipitate to form a uniform distribution of very fine particles. This fine distribution of precipitates strengthens and hardens the alloy by creating obstacles to dislocation movement. Some aluminum alloys will harden after a few days at room temperature – a process called natural aging, while others are artificially aged by heating to an intermediate temperature.

Consider the aluminum–copper system shown in Fig. 2.2. If an alloy of aluminum containing 4% copper is heated to 940° F and held for 1 h, the copper will go into solution in the aluminum. After solution heat treating, the alloy is quenched in cold water to room temperature to keep the copper in solution.

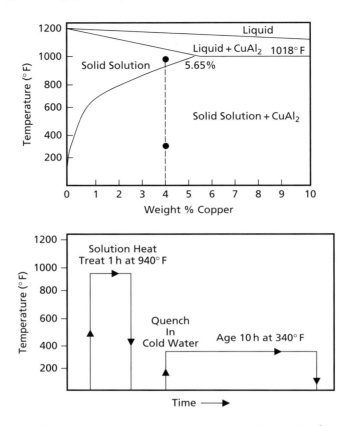

Fig. 2.2. Precipitation Hardening of an Aluminum–Copper Alloy[3]

The alloy is then artificially aged at 340° F for 10 h. During the aging process, very fine particles of aluminum–copper are precipitated and the strength and hardness increases dramatically.

Precipitation hardening strengthens the alloy due to the formation of sub-microscopic precipitates that severely strain the matrix lattice. The strengthening effect is maximized when the precipitate is coherent with the matrix. A coherent precipitate is one in which the atomic arrangement of both the precipitate and the matrix is the same with no discontinuity in the lattice; however, the atomic spacings are different enough to distort the crystal lattice in the manner shown in Fig. 2.3. This causes an increase in strength by obstructing and retarding dislocation movement. In the aluminum–copper system, these solute clusters of precipitate are called Guinier-Preston (GP) zones which are solute rich domains that are fully coherent with the matrix. The GP zones are extremely fine with sizes in the range of tens of angstroms. The exact shape, size, and distribution of the GP zones depend on the specific alloy and on the thermal and

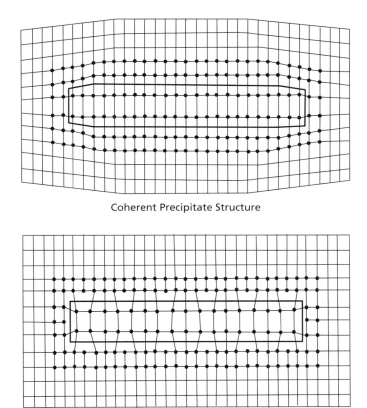

Fig. 2.3. Coherent and Incoherent Precipitates[3]

mechanical history of the product. The progression of precipitation hardening in the aluminum–copper system is:

Super Saturated Solid Solution → Clustering → GP Zones → $\theta'' \rightarrow \theta' \rightarrow \theta$

The GP zones will normally develop on aging at room temperature. During heating, the GP zones develop an intermediate precipitate (θ''), which has a tetragonal structure that forms as plates that maintains coherency with the matrix and further increases the strain in the matrix, providing peak strength levels. On still further heating, θ'' is replaced by a second intermediate precipitate θ' which is not coherent with the matrix and the strength starts to decrease, and the alloy is now termed overaged. However, in the highest strength condition, both θ'' and θ' are generally present. Both the precipitate particles themselves and the strains they produce in the lattice structure inhibit dislocation motion, and thus both contribute to strengthening. Further heating of the alloy causes θ' to transform to the equilibrium precipitate θ, which is stoichiometric $CuAl_2$. The other heat treatable aluminum alloys behave similarly but the precipitates are, of course, different. The progression of the aging process is shown schematically in Fig. 2.4. Both the underaged and overaged conditions have lower strengths and hardness levels than the peak aged condition.

The wrought heat treatable 2XXX alloys generally contain magnesium in addition to copper as an alloying element; the significance being that these alloys can be aged at either room temperature or at elevated temperature. Other significant alloying additions include titanium to refine the grain structure during ingot casting and transition element additions (manganese, chromium, and/or zirconium) that form dispersoid particles ($Al_{20}Cu_2Mn_3$, $Al_{18}Mg_3Cr_2$, and Al_3Zr) which help control the wrought grain structure. Iron and silicon are considered impurities and are held to an absolute minimum, because they form intermetallic

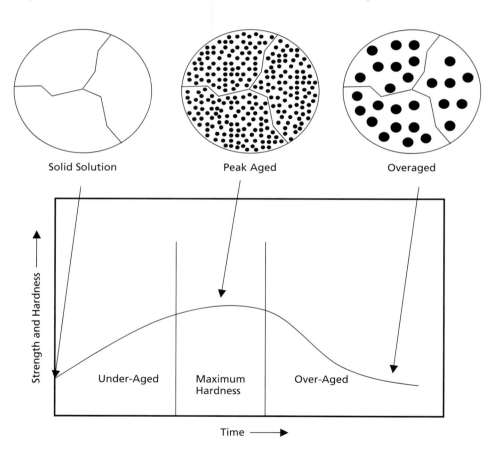

Fig. 2.4. Typical Aging Curve for Aluminum Alloys

compounds (Al_7Cu_2Fe and Mg_2Si) that are detrimental to both fatigue and fracture toughness.

The wrought heat treatable 7XXX alloys are even more responsive to precipitation hardening than the 2XXX alloys and can obtain higher strength levels. These alloys are based on the Al–Zn–Mg(–Cu) system. The 7XXX alloys can be naturally aged but are not because they are not stable if aged at room temperature, i.e. their strength will gradually increase with increasing time and can continue to do so for years. Therefore, all 7XXX alloys are artificially aged to produce a stable alloy.

2.2 Aluminum Alloy Designation

Wrought aluminum alloys are designated by a four digit numerical system developed by the Aluminum Association (Table 2.2). The first digit defines the major alloying class of the series. The second digit defines variations in the original basic alloy; that digit is always a zero (0) for the original composition, a one (1) for the first variation, a two (2) for the second variation, and so forth. Variations are typically defined by differences in one or more alloying elements of 0.15–0.50% or more, depending upon the level of the added element. The third and fourth digits designate the specific alloy within the series; there is no special significance to the values of those digits except in the 1XXX series, nor are they necessarily used in sequence.[2]

For cast alloys (Table 2.3), the first digit again refers to the major alloying element, while the second and third digits identify the specific alloy. The zero after a period identifies the alloy as a cast product. If the period is followed by the number 1, it indicates an ingot composition that would be supplied to a casting house. A letter prefix is used to denote either an impurity level or the presence of a secondary alloying element. These letters are assigned in alphabetical sequence starting with A but omitting I, O, Q, and X (X is reserved for experimental alloys). For example, the designation AXXX.0 would indicate a higher purity level than the original alloy XXX.0.

Table 2.2 Designations for Aluminum Wrought Alloys

Series	Al Content or Main Alloying Element
1XXX	99.00% Minimum
2XXX	Copper
3XXX	Manganese
4XXX	Silicon
5XXX	Magnesium
6XXX	Magnesium and Silicon
7XXX	Zinc
8XXX	Others
9XXX	Unused

Table 2.3 Designations for Aluminum Casting Alloys

Series	Al Content or Main Alloying Element
1XX.0	99.00% Minimum
2XX.0	Copper
3XX.0	Silicon with Copper and/or Magnesium
4XX.0	Silicon
5XX.0	Magnesium
6XX.0	Unused
7XX.0	Zinc
8XX.0	Tin
9XX.0	Other

The temper designations for aluminum alloys are shown in Table 2.4. Alloys in the as-fabricated condition are designated by an F; those in the annealed condition are designated with an O; those in the solution treated condition that have not attained a stable condition are designated with a W; and those that have been hardened by cold work are designated with an H. If the alloy has

Table 2.4 Temper Designation for Aluminum Alloys[4]

Suffix Letter F, O, H, T or W Indicates Basic Treatment Condition	First Suffix Digit Indicates Secondary Treatment Used to Influence Properties	Second Suffix Digit for Condition H Only Indicates Residual Hardening
F – As Fabricated		
O – Annealed-Wrought Products Only		
H – Cold Worked Strain Hardened	1 – Cold Worked Only	2 – $1/4$ Hard
		4 – $1/2$ Hard
	2 – Cold Worked and Partially Annealed	6 – $3/4$ Hard
		8 – Hard
	3 – Cold Worked and Stabilized	9 – Extra Hard
W – Solution Heat Treated		
T – Heat Treated Stable		
T1 – Cooled from an Elevated Temperature Shaping Operation + Natural Age		
T2 – Cooled from an Elevated Temperature Shaping Operation + Cold Worked + Natural Age		
T3 – Solution Treated + Cold Worked + Natural Age		
T4 – Solution Treated + Natural Age		
T5 – Cooled from an Elevated Temperature Shaping Operation + Artificial Age		
T6 – Solution Treated + Artificially Aged		
T7 – Solution Treated + Overaged		
T8 – Solution Treated + Cold Worked + Artificial Aged		
T9 – Solution Treated + Artificial Aged + Cold Worked		
T10 – Cooled from an Elevated Temperature Shaping Operation + Cold Worked + Artificial Age		

been solution treated and then aged, by either natural or artificial aging, it is designated by a T with the specific aging treatment designated by the number 1 through 10.

To redistribute residual stresses after quenching, stress relieving by deformation is often applied to wrought products. This makes the product less susceptible to warping during machining and improves both the fatigue strength and stress corrosion resistance. A summary of these stress relieving tempers is given in Table 2.5. When stress relieving is used, it is indicated by the number 5 following the last digit for precipitation hardening tempers (e.g., Tx5x). The number designation 51 indicates stress relief by stretching, while the designation 52 indicates compression stress relief. If the product is an extrusion, a third digit may be used. The number 1 for extruded products indicates the product was straightened by stretching, while the number 0 indicates that it was not mechanically straightened.

2.3 Aluminum Alloys[5-7]

Due to their superior damage tolerance and good resistance to fatigue crack growth, the 2XXX alloys are used for fuselage skins. As shown in Fig. 2.5, the 2XXX alloys are also used for lower wing skins on commercial aircraft, while the 7XXX alloys are used for upper wing skins, where strength is the primary design driver. The superior fatigue performance of 2024-T3 compared

Table 2.5 Stress Relieving Tempers for Aluminum Alloys[4]

Stress relieved by stretching
T-51: Applies to plate and rolled or cold-finished rod or bar, die or ring forgings, and rolled rings when stretched after solution heat treatment or after cooling from an elevated temperature shaping process. The products receive no further straightening after stretching.

T-510: Applies to extruded rod, bar, profiles, and tubes and to drawn tube when stretched after solution heat treatment or after cooling from an elevated temperature shaping process.

T-51: Applies to extruded rod, bar, profiles, and tubes and to drawn tube when stretched after solution heat treatment or after cooling from an elevated temperature shaping process. These products may receive minor straightening after stretching to comply with standard tolerances.

These stress-relieved temper products usually have larger tolerances on dimensions than products of other tempers.

Stress relieved by compressing
T-52: Applies to products that are stress relieved by compressing after solution heat treatment or cooling from an elevated temperature shaping process to produce a permanent set of 1–5%.
Stress relieved by combined stretching and compressing
T-54: Applies to die forgings that are stress relieved by restriking cold in the finish die.

For wrought products heat treated from annealed or F temper (or other temper when such heat treatments result in the mechanical properties assigned to these tempers)
T-42: Solution heat treated from annealed or F temper and naturally aged to a substantially stable condition (Example: 2024-T42)

T-62: Solution heat treated from annealed or F temper and artificially aged.

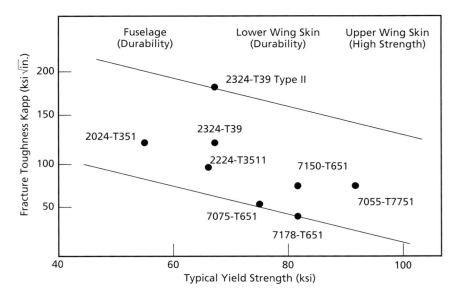

Fig. 2.5. Fracture Toughness vs. Yield Strength[1]

to 7075-T6 in the 10^5 cycle range has led to the widespread use of the 2XXX alloys in tension–tension applications. Note that as the yield strength increases, fracture toughness decreases – a phenomenon observed with not only aluminum alloys but all structural metallic materials.

Dramatic improvements in aluminum alloys have occurred since they were first introduced in the 1920s. These improvements, shown in Fig. 2.6, are a result of a much better understanding of chemical composition, impurity control, and the effects of processing and heat treatment. The chemical compositions of the prevalent aerospace grade aluminum alloys are given in Table 2.6, and the mechanical properties of a number of alloys are shown in Table 2.7.

Alloy 2024 has been the most widely used alloy in the 2XXX series. It is normally supplied in the T3 temper (i.e., solution heat treated, cold worked and then naturally aged). Typical cold working operations include roller or stretcher leveling to achieve flatness that introduces modest strains in the range of 1–4%. Although 2024-T3 only has a moderate yield strength, it has very good resistance to fatigue crack growth and good fracture toughness. Alloy 2024-T3 sheet is commonly used for fuselage skins where it is Alclad for corrosion protection. The T8 heat treatment is also frequently used with 2XXX alloys in which the alloy is solution heat treated, cold worked, and then artificially aged. Cold working before aging helps to nucleate precipitates, decrease the number and size of grain boundary precipitates, and reduce the aging time required to

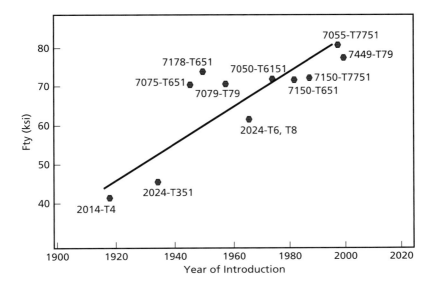

Fig. 2.6. Yield Strength vs. Year of Introduction[5]

obtain peak strength. The T8 temper also reduces the susceptibility to stress corrosion cracking.

Alloys developed for lower wing skin structure include 2324-T39 plate and 2224-T351 extrusions. Compared to 2024, both compositional and processing changes for these two alloys resulted in improved properties. A lower volume fraction of intermetallic compounds improved fracture toughness. For example, the maximum iron content is 0.12% and silicon is 0.10% in 2224 as compared to 0.50% for both impurities in 2024. The tensile yield strength of the newer plate materials was also increased by increasing the amount of cold work (stretching) after quenching.

Although copper is the main alloying element that contributes to the high strength of the 2XXX alloys, the corrosion resistance of aluminum alloys is usually an inverse function of the amount of copper in the alloy; therefore, the 2XXX alloys, which usually contain around 4% copper, are the least corrosion resistant alloys.[4] Therefore, 2XXX alloys are often Alclad with a thin coating of pure aluminum or an aluminum alloy (e.g., Al–1% Zn) to provide corrosion protection. The Alclad is chosen so that it is anodic to the core alloy and corrodes preferentially instead of the underlying core alloy. The Alclad, which is applied during rolling operations, is usually in the range of 1.5–10% of the alloy thickness. Since the Alclad material is usually not as strong as the core material, there is a slight sacrifice in mechanical properties.

The 7XXX alloys are also heat treatable, and the Al–Zn–Mg–Cu versions, in particular, provide the highest strengths of all aluminum alloys. Some of the

Table 2.6 Chemical Composition of Aerospace Aluminum Alloys[7]

Alloy	Zn	Mg	Cu	Mn	Cr	Zr	Fe	Si	Li	O	Other
1420@	–	5.2	–	–	–	0.1	–	–	2.0	–	–
2004	–	–	6.0	–	–	0.4	–	–	–	–	–
2014	–	0.5	4.4	0.8	–	–	0.7*	0.8	–	–	–
2017	–	0.6	4.0	0.7	–	–	0.7*	0.5	–	–	–
2020	–	–	4.5	0.55	–	–	0.4*	0.4*	1.3	–	0.25 Cd
2024	–	1.5	4.4	0.6	–	–	0.5*	0.5*	–	–	–
X2080	1.85	3.7	–	–	0.2	0.20*	0.10*	–	0.2	–	–
2090	–	–	2.7	–	–	0.1	0.12*	0.10*	2.2	–	–
2091	–	1.5	2.1	–	–	0.1	0.30*	0.20*	2.0	–	–
X2095	0.25*	0.4	4.2	–	–	0.1	0.15*	0.12*	1.3	–	0.4 Ag
2219	–	–	6.3	0.3	–	0.2	0.3*	0.2*	–	–	0.1 V
2224	–	1.5	4.1	0.6	–	–	0.15*	0.12*	–	–	–
2324	–	1.5	4.1	0.6	–	–	0.12*	0.10*	–	–	–
2519	–	0.2	5.8	–	–	0.2	0.3*	0.2*	–	–	0.1 V
6013	–	1.0	0.8	0.35	–	–	0.30*	0.8	–	–	–
6113	–	1.0	0.8	0.35	–	–	0.30*	0.8	–	0.2	–
7010	6.2	2.35	1.7	–	–	0.1	0.15*	0.12*	–	–	–
7049	7.7	2.45	1.6	–	0.15	–	0.35*	0.25*	–	–	–
7050	6.2	2.25	2.3	–	–	0.1	0.15*	0.12*	–	–	–
7055	8.0	2.05	2.3	–	–	0.1	0.15*	0.1*	–	–	–
7075	5.6	2.5	1.6	–	0.23	–	0.4*	0.4*	–	–	–
7079	4.3	3.2	0.6	0.2	0.15	–	0.4*	0.3*	–	–	–
X7093	9.0	2.5	1.5	–	–	0.1	0.15*	0.12*	–	0.2	–
7150	6.4	2.35	2.2	–	–	0.1	0.15*	0.12*	–	–	–
7178	6.8	2.8	2.0	–	0.23	–	0.5*	0.4*	–	–	–
7475	5.7	2.25	1.6	–	0.21	–	0.12*	0.10*	–	–	–
8009	–	–	–	–	–	–	8.65	1.8	–	0.30*	1.3 V
X8019	–	–	–	–	–	–	8.3	0.2*	–	0.2	4.0 Ce
8090	–	0.9	1.3	–	–	0.1	0.30*	0.20*	2.4	–	–

*Maximum.
Most ingot metallurgy aluminum alloys contain about 0.05–0.1% Ti to refine the ingot grain size.
@ Not an Aluminum Association Designation.

7XXX alloys contain about 2% copper in combination with magnesium and zinc to develop their strength. These alloys, such as 7049, 7050, 7075, 7175, 7178, and 7475, are the strongest but least corrosion resistant of the series. However, the addition of copper does help in stress corrosion resistance, because it allows precipitation hardening at higher temperatures. The copper-free 7XXX alloys (e.g., 7005 and 7029) have lower strengths but are tougher and exhibit better weldability. However, the majority of the 7XXX alloys are not considered weldable and are therefore joined with mechanical fasteners.

There are several alloys in the 7XXX series that are produced especially for their high fracture toughness, such as 7050, 7150, 7175, 7475, and 7085. Similar to some of the newer 2XXX alloys, controlled impurity levels, particularly iron and silicon, maximize the combination of strength and fracture toughness. For

Table 2.7 Mechanical Properties of Several Aluminum Alloys

Alloy and Temper	YS (ksi)	UTS (ksi)	Elongation (%)
2024-O	8	24	12
2024-T3	42	64	15
2024-T81	58	67	5
2024-T42	38	62	15
2024-T62	50	64	5
2024-T72	46	60	5
2024-T3 Alclad	39	59	15
2024-T81 Alclad	56	65	5
2124-T351	40	60	7
2124-T851	57	66	5
6061-O	8	18	16
6061-T6	35	42	10
7050-T7451	64	74	10
7075-O	21	40	10
7075-T6	68	78	8
7075-T76	62	73	8
7475-T61	64	75	9
7475-T7351	55	67	10

example, while the total iron and silicon content is 0.90% maximum in 7075, the combined total is limited to 0.22% in 7475. Therefore, the size and number of intermetallic compounds that assist crack propagation are much reduced in 7475. The importance of controlling impurity levels is illustrated in Table 2.8. Note the improvement in fracture toughness of 7149 and 7249 compared to the original 7049 composition.

In the peak aged T6 condition, thick plate, forgings, and extrusions of many of the 7XXX alloys are susceptible to stress corrosion cracking (SCC), particularly when stressed through-the-thickness. For example, SCC of 7075-T6 has frequently occurred in service. To combat the SCC problem, a number of overaged T7 tempers have been developed for the 7XXX alloys. Although there is usually some sacrifice in strength, these tempers have dramatically reduced

Table 2.8 Effect of Impurity Content on High Strength Aluminum Extrusions[3]

Alloy and Temper	Si Max	Fe Max	Mn Max	Fty (ksi)	Ftu (ksi)	Elong. (%)	K_{IC} (ksi $\sqrt{in.}$)
7049-T73511	0.25	0.35	0.35	73	80	11.6	24
7149-T73511	0.15	0.20	0.20	75	82	13.3	30
7249-T73511	0.20	0.12	0.12	77	84	13.3	34

the susceptibility to SCC. An additional benefit is that they improve the fracture toughness, especially the through-the-thickness (i.e., short transverse) fracture toughness of thick plate. For example, the overaged T73 temper, which was originally developed to reduce the susceptibility to SCC, reduces the yield strength of 7075 by 15% but increases the threshold stress for SCC for 7075 by a factor of six. Other overaged tempers, such as the T74, T75, and the T77 tempers, were then developed to provide trade-offs in strength and SCC resistance between the T6 and T73 tempers.[6]

To further enhance corrosion resistance, finished parts are frequently anodized before being placed in service. Anodizing is conducted in either chromic acid or sulfuric acid bath to increase the thickness of the Al_2O_3 oxide layer on the surface. The part is the anode in an electrolytic cell while the acid bath serves as the cathode. Anodizing consists of degreasing, chemical cleaning, anodizing, and then sealing the anodized coating in a slightly acidified hot water bath. Depending on the temperature and time of the anodizing operation, the oxide layer can range from 0.2 to 0.5 mil to provide enhanced corrosion protection to the underlying metal.

The 6XXX series alloys have much better corrosion resistance than the 2XXX alloys and excellent fabricability but, in general, they traditionally have not attained the overall balance of properties to compete with the 2XXX alloys.[6] However, a relatively new alloy, 6013-T6, has 12% higher strength than Alclad 2024-T3 with comparable fracture toughness and resistance to fatigue crack growth rate, and it does not have to be clad. There is renewed interest in the 6XXX alloys because they can be laser welded, while most of the 2XXX and 7XXX alloys have very limited weldability. Some of the 6XXX alloys, such as 6013-T6 and 6056-T6, which contain appreciable copper, tend to form precipitate free zones at the grain boundaries during precipitation hardening, making them susceptible to intergranular corrosion. A new temper (T78) reportedly desensitizes 6056 to intergranular corrosion while maintaining a tensile strength close to that of the T6 temper.

The 8XXX series is reserved for those alloys with lesser used alloying elements, such as iron, nickel, and lithium. Iron and nickel provides strength with only a little loss in electrical conductivity and are used in a series of alloys for conductors. Aluminum–iron alloys have also been developed for potential elevated temperature applications. The 8XXX series also contains some of the high strength aluminum–lithium alloys which are potential airframe materials.

Aluminum–lithium alloys are attractive for aerospace applications because the addition of lithium increases the modulus of aluminum and reduces the density. Each 1 wt% of lithium increases the modulus by about 6% while decreasing the density about 3%. However, the early promise of property improvements with Al–Li alloys has not been realized. Even the second generation alloys, which contained around 2% lithium, that came out in the 1980s, experienced a number of serious technical problems: excessive anisotropy in the mechanical properties; lower than desired fracture toughness and ductility; hole cracking and delamination during drilling; and low stress corrosion cracking thresholds.[5] The anisotropy experienced by these alloys is a result of the strong crystallographic textures that develop during processing, with the fracture toughness problem being one of primarily low strength in the short transverse direction. The second generation alloys included 2090, 2091, 8090, and 8091 in the western world and 1420 in Russia. The Al–Li alloys 2090, 2091, 8090, and 8091 contain 1.9–2.7% lithium, which results in about a 10% lower density and 25% higher specific stiffness than the 2XXX and 7XXX alloys.[6] To circumvent some of these problems, a third generation of alloys have been developed with lower lithium contents. Alloy 2195 also has a lower copper content and has replaced 2219 for the cryogenic fuel tank on the Space Shuttle, where it provides a higher strength, higher modulus, and lower density than 2219. Other alloys, including 2096, 2097, and 2197, also have lower copper contents but also have slightly higher lithium contents than 2195.

2.4 Melting and Primary Fabrication

Aluminum production starts with the mineral bauxite, which contains approximately 50% alumina (Al_2O_3). In the Bayer process, pure alumina is extracted from bauxite using a sodium hydroxide solution to precipitate aluminum hydroxide which is then subjected to calcination to form alumina. The Hall–Heroult process is then used to reduce the alumina to pure aluminum. This is an electrolytic process in which alumina, dissolved in a bath of cryolite, is reduced to pure aluminum by high electrical currents.

During casting of aluminum ingots, it is important to remove as many oxide inclusions and as much hydrogen gas as possible. Oxides originate primarily from moisture on the furnace charge being melted; therefore, every effort is made to insure that the materials are dry and free of moisture. Hydrogen gas can cause surface blistering in sheets and is a primary cause of porosity in castings. Fluxing with chlorine, inert gases, and salts is used to remove the oxides and hydrogen from the melt before casting.

Semi-continuous direct chill casting is the primary method for producing ingots for high performance aluminum alloys. In this process (Fig. 2.7), the molten metal is extracted through the bottom of a water-cooled mold producing fine grained ingots with a minimum amount of segregation. Low melting point alloying elements, such as magnesium, copper, and zinc, are added to the molten charge as pure elements, while high melting elements (e.g., titanium, chromium, zirconium, and manganese) are added in the form of master alloys. To prevent hot cracking and refine the grain size, inoculants such as Ti and Ti-B are added to the melt.

The molten aluminum is poured into a shallow water-cooled cross-sectional shape of the ingot desired. When the metal begins to freeze in the mold, the false

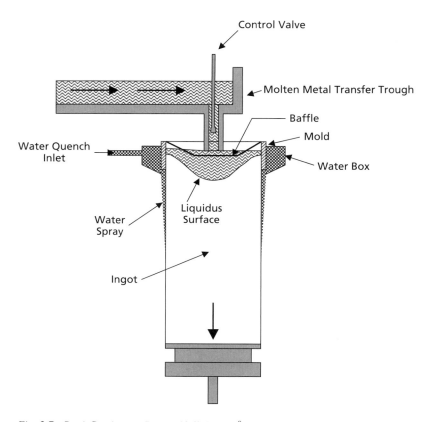

Fig. 2.7. Semi-Continuous Direct Chill Casting[8]

bottom of the mold is slowly lowered and water is sprayed on the surface of the freshly solidified metal as it comes out of the mold. The temperature and flow rate of the water are controlled so that it will wet the surfaces and then cascade down the surfaces. Typical casting speeds are in the range of 1–5 in./min.

Because the liquid metal freezing front is almost horizontal and the metal freezes from the bottom to the top of the ingot, the direct chill casting process produces fine grained ingots with a minimum of segregation. It can also produce fairly large ingots at slow speeds, a necessary requirement for the high strength alloys to prevent cracking. Also, metal can be transferred to the mold slowly, uniformly, and at relatively low temperatures. Low temperatures, only about 50° F above the liquidus temperature, are used to minimize hydrogen pickup and oxide formation. Also, if the temperature of the aluminum is too high, coarse grained structures will result.

Typical problems encountered in the direct-chill process can include ingot cracking or splitting, segregation, liquation, bleeding, and cold shutting. The higher strength aluminum alloys, and the 7XXX series in particular, are susceptible to splitting during or after casting. The first metal to freeze during casting forms a solid outer shell filled with liquid metal. After the outer shell has contracted from freezing, the inner metal tries to contract as it freezes, setting up large internal tensile stresses in the ingot that can cause immediate cracking or delayed cracking up to several weeks later. Slow casting speeds help minimize this problem since they minimize thermal gradients during freezing. Due to their relatively high solidification rates, direct-chill ingots are prone to inverse segregation. In inverse segregation, a high concentration of lower melting constituents is found at the surfaces of the casting rather than at the center due to the large shrinkage of aluminum during freezing. Lower melting constituents which are squeezed to the surface of the cast ingots due to liquation must be removed by scalping before hot rolling. Bleeding occurs when the hot liquid metal in the center of the ingot melts the already frozen outer shell and runs along the outer surface. Slower casting speeds are used to eliminate surface bleeding. However, slow casting speeds can cause cold shutting in which deep wrinkles form on the surface of the ingot as the frozen surface layer contracts toward the center of the ingot. Therefore, to successfully cast an ingot using the direct-chill process, the ingot must be cast slow enough to eliminate splitting, bleeding, and liquation, while fast enough to eliminate cold shutting. Nevertheless, the use of the direct-chill process results in higher mechanical properties than the previous tilt mold process.[9]

2.4.1 Rolling Plate and Sheet

Hot rolling is conducted at temperatures above the recrystallization temperature to create a finer grain size and less grain directionality. The upper temperature is determined by the lowest melting point eutectic in the alloy, while the lower temperature is determined by the lowest temperature that can safely be passed through the rolling mill without cracking. Sheet is defined as material that is 0.006–0.249 in. thick, while plate is material that is over 0.250 in. thick. For aircraft applications, typical sheet and plate thicknesses are 0.04–0.40 in. for fuselage skins and stringers, 1 to 2 in. for wing skins and up to 8 in. for bulkheads and wing spars. Extrusions of various angles and shapes (L, J, T, and H) are commonly used for wing stringers and fuselage frames. Hot rolling of as-cast ingots consists of:

1. Scalping of ingots
2. Homogenizing the ingots
3. Reheating the ingots to the hot rolling temperature, if necessary
4. Hot rolling to form a slab
5. Intermediate annealing and
6. Cold rolling and annealing for sheet.

Scalping of the ingot, in which approximately 0.25–0.38 in. of material is removed from each surface, is conducted so that surface defects will not be rolled into the finished sheet and plate. Homogenizing of the ingots is conducted to remove any residual stresses in the ingots and improve the homogeneity of the as-cast structure by reducing the coring experienced during casting. Good temperature control is required during homogenization because the ingots are heated to within 20–40° F of the lowest melting eutectic in the alloy. The 2XXX alloys are usually soaked for 4–12 h at 900–950° F, while the 7XXX alloys are soaked for 8–24 h at 850–875° F. Since homogenization is a diffusion controlled process, the long times are necessary to allow time for the alloying elements to diffuse from the grain boundaries and other solute-rich regions to the grain centers. An important function of homogenization is to remove non-equilibrium low melting point eutectics that could cause ingot cracking during subsequent hot working operations. Electric heaters and fans are used to circulate the air to insure temperature uniformity and produce maximum heat transfer. If the alloy is going to be Alclad for corrosion protection, the scalping operation is usually done after homogenization to remove the heavy oxide layer that builds up during the rather long homogenization soaks, making it easier to obtain a good bond between the Alclad and the core during hot rolling.

The ingots can be hot rolled right after removing them from the soaking pits, or if cooled to room temperature after homogenization, they must be reheated for hot rolling. Ingots as large as 20 ft long by 6 ft wide by 2 ft thick weighing over 20 tons are initially hot rolled back and forth through the rolling mill into plate between 0.250 and 8.0 in. thick. Modern rolling mill facilities can heat the plate, roll it to the desired thickness, spray quench it to harden it, and then stretch it to relieve stresses. The 2XXX alloys are hot rolled in the temperature range of 750–850° F, while the 7XXX alloys are rolled at 750–825° F. Hot rolling helps to break up the as-cast structure and to provide a more uniform grain size and a better distribution and size of constituent particles. During hot rolling, the grain structure becomes elongated in the rolling direction, as shown in Fig. 2.8. This grain directionally can have a substantial effect on some of the mechanical properties, especially fracture toughness and corrosion resistance, in which the properties are lowest in the through-the-thickness or short transverse direction.

Initial rolling is done in a four-high reversing mill to breakdown the ingot. As shown in Fig. 2.9, a four-high mill uses four rolls in which the two smaller center rolls contact the workpiece and the two larger outer rolls provide support for the inner rolls. As the ingot is run back and forth through the mill, it rapidly becomes longer as the thickness is reduced. If wide plate or sheet is required, the ingot is removed after the first few passes and rotated 90° and then cross-rolled. After the slab has been reduced in thickness, it is removed from the mill, given an intermediate anneal, and then placed in a five stand four-high mill to roll to thinner plate or sheet with successive reductions at each station.

Aluminum

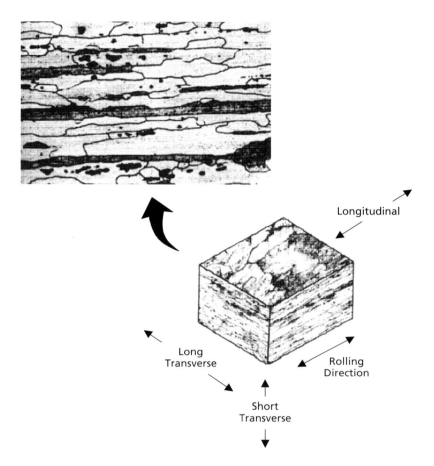

Fig. 2.8. Grain Directionality Due to Rolling

The cold work put into the aluminum during hot rolling must be sufficient to cause recrystallization during annealing. Intermediate anneals are required after cold reductions in the range of 45–85%. Intermediate anneals are required to keep the sheet from cracking during cold rolling; however, the amount of cold work must be sufficient to cause a fine grain size during annealing. If the final product form is sheet, it is annealed and then sent to a four-high cold rolling mill for further reduction. The number of intermediate anneals required during cold rolling depends on the alloy and the final gage required.

During spray quenching from the solution heat treating temperature, the surface cools much quicker than the center resulting in residual stresses. The faster cooling surface develops compressive stresses, while the slower cooling center develops tensile stresses. This residual stress pattern with compressive

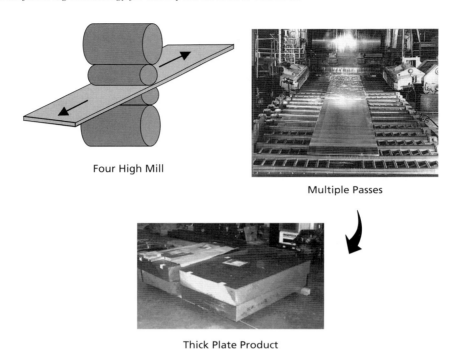

Four High Mill

Multiple Passes

Thick Plate Product

Fig. 2.9. Hot Rolling Aluminum Plate

stresses on the surface helps in preventing fatigue and stress corrosion cracking. However, during machining operations, if the surface material containing the compressive stresses is removed, the interior material with tensile residual stresses is exposed and the part is even more susceptible to warping, fatigue, and stress corrosion cracking. To minimize these problems, aluminum plate and extrusions are often stress relieved by stretching 1/2 to 5%; a temper designated as Tx5x or Tx5xx.

Cold working during rolling results in highly directional grain structures which can affect stress corrosion resistance. The longitudinal direction is the most resistant, followed by the long transverse direction, with the short transverse being the most susceptible. Thick 7XXX plate is therefore supplied in stress corrosion resistant tempers, such as the T73, T74, and T77 tempers, and the 2XXX alloys are given the T6 and T8 tempers. For example, 7075-T6 resists stress corrosion cracking at tensile stresses up to only 7 ksi, while 7075-T73 resists stress corrosion cracking up to 44 ksi when tested under similar conditions.[4] For thinner sheet, which is not as affected by through-the-thickness effects, the 7XXX alloys can be used in the higher strength T6 temper and the 2XXX alloys are given the T3 or T4 tempers.

2.4.2 Extrusion

Extruded structural sections are produced by hot extrusion in which a heated cylindrical billet is pushed under high pressure through a steel die to produce the desired structural shape. The extrusion is then fed onto a run-out table where it is straightened by stretching and cut to length. During extrusion, metal flow occurs most rapidly at the center of the ingot resulting in oxides and surface defects being left in the last 10–15% of the extrusion which is discarded.

In general, the stronger the alloy, the more difficult it is to extrude. One of the advantages of the 6XXX alloys is that they exhibit good extrudability. On the other hand, the 2XXX and 7XXX alloys are referred to as "hard" alloys because they are more difficult to extrude. A profile's shape factor (the ratio of the perimeter of the profile to its area) is an approximate indicator of its extrudability, i.e. the higher the ratio, the more difficult it is to extrude. Unsymmetric shapes, shapes with sharp corners, profiles with large thickness variations across their cross section, and those that contain fine details are all more difficult to extrude. Generous fillets and rounded corners help to reduce extrusion difficulties.

2.5 Heat Treating[10]

The term "heat treating" refers to the heating and cooling operations that are performed in order to change the mechanical properties, metallurgical structure or residual stress state of a metal product. For aluminum alloys, the term "heat treating" usually refers to precipitation hardening of the heat treatable aluminum alloys. Annealing, a process that reduces strength and hardness while increasing ductility, can also be used for both the non–heat treatable and heat treatable grades of wrought and cast alloys.

2.5.1 Solution Heat Treating and Aging

The importance of precipitation hardening of aluminum alloys can be appreciated by examining the data presented in Fig. 2.10 for naturally aged 2024 and the artificially aged 7075. Note the dramatic increase in strength of both due to precipitation hardening with only a moderate reduction in elongation.

For an aluminum alloy to be precipitation hardened, certain conditions must be satisfied. First, the alloy must contain at least one element or compound in a sufficient amount that has a decreasing solid solubility in aluminum with decreasing temperature. In other words, the elements or compounds must have an appreciable solubility at high temperatures and only minimal solubility at lower temperatures. Elements that have this characteristic are copper, zinc, silicon, and magnesium, with compounds such as $CuAl_2$, Mg_2Si, and $MgZn_2$. While this is a requirement, it is not sufficient; some aluminum systems that display this behavior cannot be strengthened by heat treatment. The second

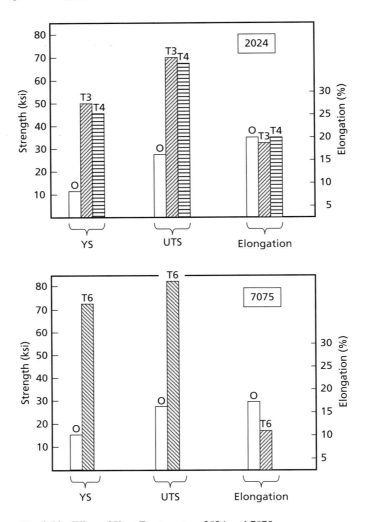

Fig. 2.10. Effect of Heat Treatment on 2024 and 7075

requirement is that the element or compound that is put into solution during the solution heat treating operation must be capable of forming a fine precipitate that will produce lattice strains in the aluminum matrix. The precipitation of these elements or compounds progressively hardens the alloy until a maximum hardness is obtained. Alloys that are not aged sufficiently to obtain maximum hardness are said to be underaged, while those that are aged past peak hardness are said to be overaged. Underaging can be a result of not artificially aging at a high enough temperature or an aging time that is too short, while overaging is usually a result of aging at too high a temperature. An example of the aging

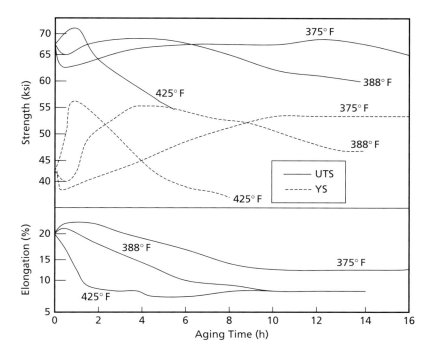

Fig. 2.11. Artificial Aging Curves for 2024-T4 Sheet[3]

behavior of artificially aged 2024-T4 is shown in Fig. 2.11. Note the large reduction in strength when it is overaged at 425° F.

Aluminum alloys that satisfy both of these conditions are classified as heat treatable, namely the 2XXX, 6XXX, 7XXX, and some of the 8XXX wrought alloys. The precipitation hardening process is conducted in three steps:

1. Heating to the solution heat treating temperature and soaking for long enough to put the elements or compounds into solution.
2. Quenching to room or some intermediate temperature (e.g., boiling water) to keep the alloying elements or compounds in solution; essentially this creates a supersaturated solid solution.
3. Aging at either room temperature (natural aging) or a moderately elevated temperature (artificial aging) to cause the supersaturated solution to form a very fine precipitate in the aluminum matrix.

The solution heat treating temperature is as high above the solid solubility curve as possible without melting the lowest melting point eutectic constituents. Therefore, close temperature control, normally ±10° F, is required for the furnaces used to heat treat aluminum alloys. If the alloy is heated too high and incipient grain boundary melting occurs, the part is ruined and must be scrapped. For example, 2024 is solution treated in the range of 910–930° F, while a low melting eutectic forms at 935° F, only 5° F higher than the upper range of the solution heat treating temperature. On the other hand, if the temperature is too low, solution will be incomplete and the aged alloy will not develop as high a strength as expected. The solution heat treating time should be long enough to allow diffusion to establish an equilibrium solid solution. The product form can determine the time required for solution treating, i.e. castings require more time than wrought products to dissolve their relatively large constituents into solution. The time required can vary anywhere from less than a minute for thin sheet to up to 20 h for large sand castings. While longer than required soak times are not usually detrimental, caution needs to be applied if the part is Alclad, since excessive times can result in alloying elements diffusing through the Alclad layer and reducing the corrosion resistance.

The oven used to conduct solution heat treating should be clean and free of moisture. The presence of moisture can cause hydrogen to be absorbed into the aluminum parts, while sulfur compounds can decompose the protective surface oxide, making the part even more susceptible to hydrogen absorption. Absorbed moisture can result in internal voids or surface blisters, and the 7XXX alloys are the most susceptible to this type of attack followed by the 2XXX alloys. Moisture can be minimized by thoroughly cleaning and drying parts and racks before placing them into the oven.

After the elements are dissolved into solution, the alloy is quenched to a relatively low temperature to keep the elements in solution. Quenching is perhaps the most critical step in the heat treating operation. The problem is to quench the part fast enough to keep the hardening elements in solution, while at the same time minimizing residual quenching stresses that cause warpage and distortion. In general, the highest strength levels, and the best combinations of strength and toughness, are obtained by using the fastest quench rate possible. Resistance to corrosion and SCC are usually improved by faster quenching rates; however, the resistance to SCC of certain copper-free 7XXX alloys is actually improved by slow quenching. While fast quenching rates can be achieved by cold water, slower quenching rates (e.g., hot or boiling water) are often used to sacrifice some strength and corrosion resistance for reduced warpage and distortion.

If premature precipitation during quenching is to be avoided, two requirements must be met. First, the time required to transfer the part from the furnace to the quench tank must be short enough to prevent slow cooling through the critical temperature range where very rapid precipitation takes place. The high strength 2XXX and 7XXX alloys should be cooled at rates exceeding 800° F/sec. through the temperature range of 750–550° F. The second requirement is that the volume of the quenching tank must be large enough so that the quench tank temperature does not rise appreciably during quenching and allow premature precipitation.

Both cold and hot water are common quenching media for aluminum alloys. Cold water, with the water maintained below 85° F, is used with the requirement that the water temperature does not rise by more than 20° F during the quenching operation. The quench rate can be further increased by agitation that breaks up the insulating steam blanket that forms around the part during the early stages of quenching. Parts that distort during quenching require straightening before aging, so hot water quenching, with water maintained between 150 and 180° F or at 212° F, is a less drastic quench resulting in much less distortion and is often used for products where it is impracticable to straighten after quenching. Polyalkylene glycol solutions in water are also used to quench aluminum alloys because they produce a stable film on the surface during quenching, resulting in more uniform cooling rates and less distortion. The time from removing from the solution treating furnace until placing in the quench media is also critical; maximum quench delays are on the order of 5 s for 0.016 in. thick material, 7 s for material between 0.017 and 0.031 in., 10 s for material between 0.031 and 0.090 in., and 15 s for thicker material.

Aging is conducted at either room temperature (natural aging) or elevated temperature (artificial aging). The 2XXX alloys can be aged by either natural or artificial aging. The 2XXX alloys can be naturally aged to obtain full strength after 4–5 days at room temperature with about 90% of their strength being obtained within the first 24 h. Natural aging of the 2XXX alloys consists of solution heat treating, quenching, and then aging at room temperature to give the T4 temper. Naturally aged alloys are often solution treated and quenched (W temper), refrigerated until they are ready to be formed, and then allowed to age at room temperature to peak strength (T3 temper). To prevent premature aging, cold storage temperature needs to be in the range of -50 to $-100°$ F. It should be noted that artificial aging of the 2XXX alloys to the T6 temper produces higher strengths and higher tensile-to-yield strength ratios but lower elongations than natural aging.

The T8 temper (i.e., solution treating, quenching, cold working, and then artificial aging) produces high strengths in many of the 2XXX alloys. Alloys such as 2011, 2024, 2124, 2219, and 2419 are very responsive to cold working by stretching and cold rolling; the cold work creates additional precipitation sites for hardening. The T9 temper is similar except that the cold work is introduced after artificial aging (i.e., solution treating, quenching, artificial aging, and cold working). Since the 7XXX alloys do not respond favorably to cold working during the precipitation hardening process, they are not supplied in the T8 or T9 tempers.

Artificial aging treatments are generally low temperature, long time processes; temperatures range from 240 to 375° F for times of 5–48 h. The 7XXX alloys, although they will harden at room temperature, are all given artificial aging treatments. The 7XXX alloys are usually aged at 250° F for times up to 24 h, or longer, to produce the T6 temper. Many of the thick product forms for the 7XXX alloys that contain more than 1.25% copper are provided in the T7 oveaged condition. While overaging does reduce the strength properties, it improves the corrosion resistance, dimensional stability, and fracture toughness, especially in the through-the-thickness short transverse direction. There are a number of T7 tempers that have been developed that trade-off various amounts of strength for improved corrosion resistance. Most involve aging at a lower temperature to develop strength properties followed by aging at a higher temperature to improve corrosion resistance. For example, the T73 aging treatment consists of an aging temperature of 225° F followed by a second aging treatment at 315–350° F. The T76 temper is similar with an initial age at 250° F followed by a 325° F aging treatment. The T76 temper has a little higher strength than the T73 temper but is also a little less corrosion resistant and produces a lower fracture toughness. As shown in Table 2.9, the T77 aging treatment developed by Alcoa, which is a variation of the retrogression and re-aging treatment, produces the best combination of mechanical properties and corrosion resistance. Although it depends on the specific alloy, in the T77 treatment, the part is solution treated, quenched, and aged. It is then re-aged for 1 h at 390° F and water quenched, and finally aged again for 24 h at 250° F.

Verification of heat treatment is usually conducted by a combination of hardness and electrical conductivity.

2.5.2 Annealing

Cold working results in an increase in internal energy due to an increase in dislocations, point defects, and vacancies. The tensile and yield strengths increase with cold working, while the ductility and elongation decrease. If cold worked aluminum alloys are heated to a sufficiently high temperature for a sufficiently long time, annealing will occur in three stages: recovery, recrystallization, and grain growth. During recovery, the internal stresses due to cold work are reduced with some loss of strength and a recovery of some ductility. During recrystallization, new unstrained nuclei form and grow until they impinge on each other to form a new recrystallized grain structure. Although heating for longer times or at higher temperatures will generally result in grain growth, aluminum

Table 2.9 Effect of T7 Heat Treatments on 7XXX Aluminum Alloys[3]

Alloy and Temper	YS (ksi)	UTS (ksi)	Elong. (%)	K_{IC} (ksi $\sqrt{in.}$)
7075-T6	69	80	11	23
7075-T73	65	74	13	27
7055-T74	68	76	17	36
7055-T76	78	83	13	29
7055-T77	86	87	11	20

alloys contain dispersoids of manganese, chromium, and/or zirconium that help to suppress grain growth.

Annealing treatments are used during complex cold-forming operations to allow further forming without the danger of sheet cracking. The softest, most ductile and most formable condition for aluminum alloys is produced by full annealing to the O condition. Strain hardened products normally recrystallize during annealing, while hot worked products may or may not recrystallize, depending upon the amount of cold work present. Full annealing of both the 2XXX and the majority of the 7XXX alloys can be accomplished by heating to 775°F for 2–3 h followed by cooling at 50°F/h or slower to 500°F. This treatment will also remove the effects of precipitation hardening. Heating to 650°F will remove the effects of cold work but only partially remove the effects of precipitation hardening.

2.6 Forging

Forgings are often preferred for aircraft bulkheads and other highly loaded parts because the forging process allows for thinner cross-section product forms prior to heat treat and quenching, enabling superior properties. It can also create a favorable grain flow pattern which increases both fatigue life and fracture toughness when not removed by machining. Also, forgings generally have less porosity than thick plate and less machining is required.

Aluminum alloys can be forged using hammers, mechanical presses, or hydraulic presses. Hammer forging operations can be conducted with either gravity or power drop hammers and are used for both open and closed die forgings. Hammers deform the metal with high deformation speed; therefore, it is necessary to control the length of the stroke, the speed of the blows, and the force being exerted. Hammer operations are frequently used to conduct preliminary shaping prior to closed die forging. Both mechanical and screw presses are used for forging moderate size parts of modest shapes and are often used for high volume production runs. Mechanical and screw presses combine impact with a squeezing action that is more compatible with the flow characteristics of aluminum than hammers. Hydraulic presses are the best method for producing large and thick forgings, because the deformation rate is slower and more controlled than with hammers or mechanical/screw presses. The deformation or strain rate can be very fast ($>10\,\text{s}^{-1}$) for processes such as hammer forging or very slow ($<0.1\,\text{s}^{-1}$) for hydraulic presses. Since higher strain rates increase the flow stress (decrease forgeability) and the 2XXX and 7XXX alloys are even more sensitive than other aluminum alloys, hydraulic presses are usually preferred for forging these alloys. Hydraulic presses are available in the range of 500–75 000 tons (biggest press is in Russia) and can produce forgings up to around 3000 lb.

For the 7XXX alloys, the forging pressure is actually higher than that for low carbon steels. As shown in Fig. 2.12, the flow stress of 7075 is considerably

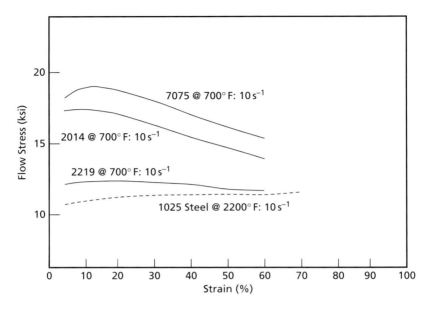

Fig. 2.12. Flow Curves for Aluminum Forging Alloys[11]

higher than that of 1025 plain carbon steel at the same strain rate, while the flow stress of 2219 is about the same as 1025. Although flow stress represents the lower bound required for forging, it does illustrate that the 7XXX series, such as 7010, 7049, 7050, and 7075, require substantial pressure. As the temperature increases toward the maximum in the range, the flow stress decreases and the forgeability increases.

Dies for forging aluminum alloys are usually made from die steels such as the hot work steels H11, H12, and H13 heat treated to 44–50 Rc. The surfaces may be hardened for additional durability by carburizing, nitriding, or carbonitriding. The die cavities are normally polished to produce a good surface finish on the forging. The dies are usually heated for forging of aluminum alloys, normally from as low as 200° F to as high as 800° F. Since the higher temperatures are used for hydraulic press forging, this operation is essentially conducted in the isothermal range in which the dies and part are at or near the same temperature.

Aluminum alloys are heated to the forging temperature with a wide variety of heating equipment, including electric furnaces, muffle furnaces, oil furnaces, induction heating units, fluidized beds, and resistance heating units. Regardless of the heating method, it is very important to minimize the absorption of hydrogen, which can result in surface blistering or internal porosity, referred to as bright flakes, in the finished forging. The 2XXX alloys are normally forged at 800–860° F, while the 7XXX alloys are forged at 750–825° F. Since the temperature range for forging is generally less than 100° F, and always less than 150° F, temperature control during forging is very important. Soak times of 1–2 h are usually sufficient with temperatures controlled to ±10° F. Before forging, graphite based lubricants are sprayed on the dies.

Open die forgings, also know as hand forgings, are produced in dies that do not provide lateral restraint during the forging operation. In this process, the metal is forged between either flat or simply shaped dies. This process is used to produce small quantities where the small quantities do not justify the expense of matched dies. Although open die forgings somewhat improve the grain flow of the material, they offer minimal economic benefit in reduced machining costs. Open die forging is often used to produce preforms for closed die forging.

Most aluminum forgings are produced in closed dies. Closed die forgings are produced by forging ingots, plates, or extrusions between a matched set of dies. Closed die forging uses progressive sets of dies to gradually shape the part to near net dimensions. Die forgings can be subdivided into four categories from the lowest cost, least intricate to the highest cost, most intricate. A comparison of the relative amount of part definition for these different forging processes[12] is shown in Fig. 2.13.

- Blocker forgings may be chosen if the total quantities are small (e.g., 200). Since they have large fillet and corner radii, they require extensive machining to produce a finished part. The fillets are about 2 times the radius and the corner radii about 1.5 times the radii of conventional forgings. Therefore, a blocker forging costs less than a conventional forging but requires more machining. Finish only forgings are similar to blocker forgings in that only one set of dies are used; however, since they have one more squeeze applied, they have somewhat better part definition. Fillets

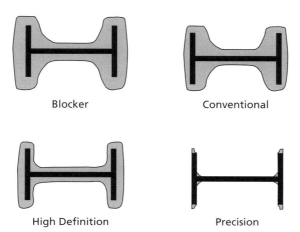

Fig. 2.13. Part Definition Produced by Different Forging Methods

are about 1.5 times the radius of conventional forgings with corner radii about the same as conventional forgings. A quantity of about 500 might justify the use of finish only forgings.
- Conventional forgings require two to four sets of dies with the first set producing a blocker type forging that is subsequently finished in the other sets. This is the most common type of aluminum forging and is usually specified for quantities of 500 or more. Conventional forgings have more definition and require less machining than blocker forgings but the die cost is higher.
- High definition forgings contain even better definition and tolerance control than conventional forgings with less machining costs. These forgings are near net shape forgings produced on multiple die sets. In some applications, some of the forged surfaces may not require machining.
- Precision forgings produce the best part definition and highest quality but are, of course, the most expensive. These forgings have tighter tolerances than those produced by even high definition forgings with better grain flow. Minimal or no machining is required to finish these forgings.

Other common forging methods for aluminum alloys include upset forging, roll forging, orbital or rotary forging, spin forging, ring rolling, and mandrel forging. The choice of a particular forging method depends on the shape required and the economics of the number of pieces required traded off against higher quality and lower machining costs.

All 2XXX and 7XXX forged alloys are heat treated after forging. To minimize distortion during quenching, the racking procedures are important in obtaining uniform cooling rates. Aluminum forgings are often straightened between solution treatment and aging. Compressive stress relieving with a permanent set of 1–5% is used to reduce internal stresses and minimize distortion during subsequent machining operations. When this is conducted in the finishing dies, it is designated as TXX54 temper. When it is conducted in a separate set of cold dies, it is designated as the TXX52 temper.

Residual stresses generated during forging and subsequent heat treatment can cause significant problems when the part is machined. The 7050-T7452 forging shown in Fig. 2.14 bowed almost a foot after machining. A study[13] of the residual stresses in this type of part revealed that the sequence used to conduct compression stress relief can be important. If a large number of small "bites" are taken along the length during the incremental compression stress relief, the residual stress pattern is not as uniform than if a smaller number of larger bites is taken.

2.7 Forming

Due to their FCC structure and their relatively slow rate of work hardening, aluminum alloys are highly formable at room temperature. The 2XXX and 7XXX high strength alloys can be readily formed at room temperature, provided

Aluminum

Fig. 2.14. Distortion of Machined Aluminum Forging[13]

the alloy is in either the O or W temper. The choice of the temper for forming depends on the severity of the forming operation and the alloy being formed. Although the annealed or O condition is the most formable condition, it is not always the best choice because of the potential for warping during subsequent heat treatment. The solution treated and quenched condition (W temper) is nearly as formable as the O condition and requires only aging after forming to obtain peak strengths, without the potential of warping after forming during the quenching operation. Both the 2XXX and 7XXX alloys must be formed immediately after quenching or be refrigerated after heat treating to the W temper prior to forming. Since aluminum has a relatively low rate of work hardening, a fair number of forming operations are possible before intermediate anneals are required. Tools for forming aluminum alloys require good surface finishes to minimize surface marking. The oxide film on aluminum is highly abrasive and many forming tools are therefore made of hardened tool steels.

2.7.1 Blanking and Piercing

As shown in Fig. 2.15, blanking is a process in which a shape is sheared from a larger piece of sheet, while piercing produces a hole in the sheet by punching out a slug of metal. Both blanking and piercing operations are usually preformed in a punch press. The clearance between the punch and die must be controlled to obtain a uniform shearing action. Clearance is the distance between the mating surfaces of the punch and die, usually expressed as a percentage of sheet thickness. The recommended clearance for the 2XXX and

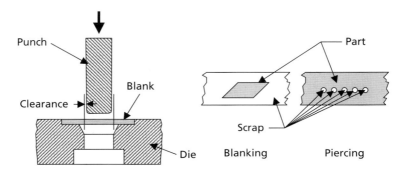

Fig. 2.15. Blanking and Piercing

7XXX alloys in the O condition is 6.5%. The walls of the die opening are tapered to minimize sticking, and the use of lubricants such as mineral oil mixed with small quantities of fatty oils also reduces sticking tendencies. A tolerance of 0.005 in. is normal in blanking and piercing of aluminum; however, wider tolerances, when permissible, will help in reducing costs.

Dull cutting edges on punches and dies have effects similar to excessive clearance with burrs becoming excessive. With sharp tools and proper clearance, the fractures are clean without evidence of secondary shearing or excessive burring. When the clearance is too small, secondary shearing can occur, and if the clearance is too large, the sheared edge will have a large radius and a stringy burr. Cast zinc tools, which are much less expensive than steel tools, are often used for runs of up to about 2000 parts.

2.7.2 Brake Forming

In brake forming (Fig. 2.16), the sheet is placed over a die and pressed down by a punch that is actuated by the hydraulic ram of a press brake. Springback is the partial return of the part to its original shape after forming. The amount of springback is a function of the yield strength of the material being formed, the bend radius, and the sheet thickness. Springback is compensated by over bending the material beyond the final angle so that it springbacks to the desired angle. The springback allowance (i.e., the amount of over bend) increases with increasing yield strength and bend radius, but varies inversely with sheet thickness. The spring back allowance ranges from 1 to 12° for 2024-O and 7075-O and from 7.25 to 33.5° for 2024-T3. Since aluminum sheet tends to develop anisotropy during rolling operations, there is less tendency for cracking during forming if the bend is made perpendicular to the rolling or extrusion direction. The smallest angle that can be safely bent, called the minimum bend radius, depends on the yield strength and on the design, dimensions, and conditions of the tooling. The most severe bends can be made across the rolling direction. If similar bends are

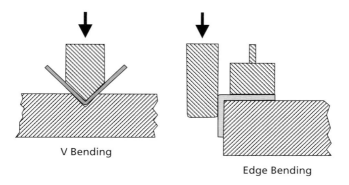

Fig. 2.16. Brake Forming

to be made in two or more directions, it is best to make all bends at an angle to the direction of rolling. Relatively simple and long parts can usually be press brake formed to a tolerance of 0.030 in., while for larger more complex parts, the tolerance may be as much as 0.063 in.

2.7.3 Deep Drawing

Punch presses are used for most deep drawing operations. In a typical deep drawing operation, shown in Fig. 2.17, a punch or male die pushes the sheet into the die cavity while it is supported around the periphery by a blankholder. Single action presses can be operated at 90–140 ft/min., while double action presses operate at 40–100 ft/min. for mild draws and at less than 50 ft/min. for deep draws. For the higher strength aluminum alloys, speeds of 20–40 ft/min. are more typical. Clearances between the punch and die are usually equal to the sheet thickness plus an additional 10% per side for the intermediate strength alloys, while an additional 5–10% clearance may be needed for the high strength

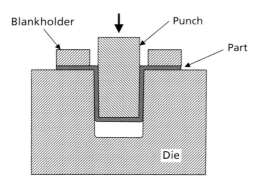

Fig. 2.17. Deep Drawing

alloys. Excessive clearance can result in wrinkling of the sidewalls of the drawn shell, while insufficient clearance increases the force required for drawing and tends to burnish the part surfaces. The draw radius on tools is normally equal to four to eight times the stock thickness. If the punch radius is too large, wrinkling can result, and if the radius is too small, sheet fracture is a possibility. Draw punches and dies should have a surface finish of 16 μin. or less for most applications. Tools are often chrome plated to minimize friction and dirt that can damage the part finish. Lubricants for deep drawing must allow the blank to slip readily and uniformly between the blankholder and die. Stretching and galling during drawing must be avoided. During blank preparation, excessive stock at the corners must be avoided because it obstructs the uniform flow of metal under the blankholder leading to wrinkles or cracks.

The rate of strain hardening during drawing is greater for the high strength aluminum alloys than for the low to intermediate strength alloys. For high strength aluminum alloys, the approximate reductions in diameter during drawing are about 40% for the first draw, 15% for the second draw, and 10% for the third draw. Local or complete annealing for 2014 and 2024 is normally required after the third draw. Severe forming operations of relatively thick or large blanks of high strength aluminum alloys generally must be conducted at temperatures around 600°F, where the lower strength and partial recrystallization aids in forming. This is possible with alloys such as 2024, 2219, 7075, and 7178, but the time at temperature should be minimized to limit grain growth.

2.7.4 Stretch Forming

In stretch forming (Fig. 2.18), the material is stretched over a tool beyond its yield strength to produce the desired shape. Large compound shapes can be formed by stretching the sheet both longitudinally and transversely. In addition, extrusions are frequently stretch formed to moldline curvature. Variants of

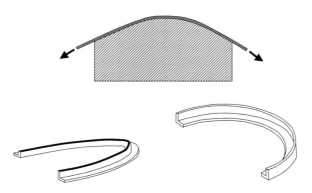

Fig. 2.18. Stretch Forming[14]

stretch forming include stretch draw forming, stretch wrapping, and radial draw forming. Forming lubricants are recommended except when self-lubricating smooth faced plastic dies are used; however, the use of too much lubricant can result in workpiece bucking. The 2XXX and 7XXX alloys can be stretch formed in either the O or W condition. Material properties that help in stretch forming are a high elongation, a large spread between the yield and ultimate strengths (called the forming range), toughness, and a fine grain structure. Alloys with a narrow spread between the yield and ultimate strengths are more susceptible to local necking and failure. For example, 7075-W has a yield strength of 20 ksi, an ultimate strength of 48 ksi, a forming range of 28 ksi (48 ksi – 20 ksi) and a stretchability rating of 100, while 7075-T6 has a yield strength of 67 ksi, an ultimate strength of 76 ksi, a forming range of only 9 ksi, and a stretchability rating of only 10.

2.7.5 Rubber Pad Forming

In rubber pad forming, a rubber pad is used to exert nearly equal pressure over the part as it is formed down over a form block. Rubber pad forming and a closely related process, fluid cell forming, are shown in Fig. 2.19. The rubber pad acts somewhat like a hydraulic fluid, spreading the force over the surface of the part. The pad can either consist of a solid piece or may be several pieces laminated together. The pad is usually in the range of 6–12 in. thick and must be held in a sturdy retainer as the pressures generated can be as high as 20 ksi. The 2XXX and 7XXX alloys are formed in either the O or W temper. Rubber pad forming can often be used to form tighter radii and more severe contours than other forming methods because of the multidirectional nature of the force exerted on the workpiece. The rubber acts somewhat like a blankholder helping to eliminate the tendency for wrinkling. This process is very good for making sheet parts with integral stiffening beads. Most rubber pad forming is conducted on sheet 0.063 in. or less in thickness; however, material as thick as 0.625 in. thick has been successfully formed. Although steel tools are normally used for long production runs, aluminum or zinc tools will suffice for short or intermediate runs. Fluid cell forming, which uses a fluid cell to apply pressure through an elastomeric membrane, can form even more severe contours than rubber pad forming. Due to the high pressures employed in this process, as high as 15–20 ksi, many parts can be formed in one shot with minimal or no springback. However, fluid cell forming presses are usually expensive.

2.7.6 Superplastic Forming

Superplasticity is a property that allows sheet to elongate to quite large strains without localized necking and rupture. In uniaxial tensile testing, elongations to failure in excess of 200% are usually indicative of superplasticity. Micrograin

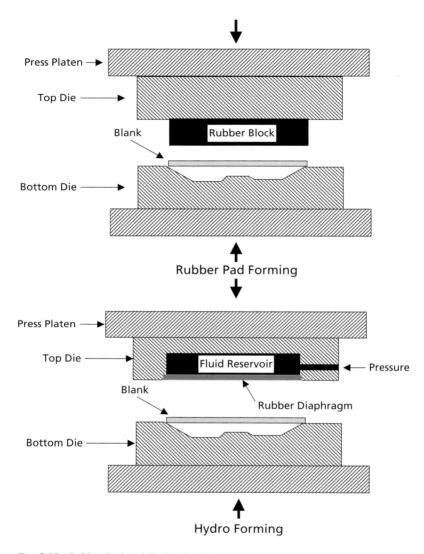

Fig. 2.19. Rubber Pad and Hydro Forming

superplasticity occurs in some materials with a fine grain size, usually less than 10 μm, when they are deformed in the strain range of 0.00005–0.01/s at temperatures greater than $0.5T_m$, where T_m is the melting point in degrees Kelvin. Although superplastic behavior can produce strains in excess of 1000%, superplastic forming (SPF) processes are generally limited to about 100–300%. The advantages of SPF include the ability to make part shapes not possible with conventional forming, reduced forming stresses, improved formability with essentially no springback and reduced machining costs. The disadvantages are that the process is rather slow and the equipment and tooling can be relatively expensive.

The main requirement for superplasticity is a high strain rate sensitivity. In other words, the strain rate sensitivity "m" should be high where m is defined as:[15]

$$m = d(\ln \sigma)/d(\ln \dot{\varepsilon})$$

where

$m =$ strain rate sensitivity
$\sigma =$ flow stress
$\dot{\varepsilon} =$ strain rate

The strain rate sensitivity describes the ability of a material to resist plastic instability or necking. For superplasticity, m is usually greater than 0.5 with the majority of superplastic materials having an m value in the range of 0.4–0.8, where a value of 1.0 would indicate a perfectly superplastic material. The presence of a neck in a material undergoing a tensile strain results in a locally high strain rate and, for a high value of m, to a sharp increase in the flow stress within the necked region, i.e. the neck undergoes strain hardening which restricts its further development. Therefore, a high strain rate sensitivity resists neck formation and leads to the high tensile elongations observed in superplastic materials. The flow stress decreases and the strain rate sensitivity increases with increasing temperature and decreasing grain size. The elongation to failure tends to increase with increasing m.

Superplasticity depends on microstructure and exists only over certain temperature and strain rate ranges. A fine grain structure is a prerequisite since superplasticity results from grain rotation and grain boundary sliding, and increasing grain size results in increases in flow stress. Equiaxed grains are desirable because they contribute to grain boundary sliding and grain rotation. A duplex structure also contributes to superplasticity by inhibiting grain growth at elevated temperature. Grain growth inhibits superplasticity by increasing the flow stress and decreasing m.

The Ashby and Verrall model for superplasticity, based on grain boundary sliding with diffusional accommodation, is shown in Fig. 2.20 in which grains switch places with their neighbors to facilitate elongation. However, in real metals, since the grains are not all the same size, some rotation must also take place. Slow strain rates are necessary to allow the diffusion mechanisms time to allow this rearrangement. Since a fine grain size is a prerequisite for superplasticity, a fine dispersion of the metastable cubic phase Al_3Zr can be used in aluminum alloys to help prevent grain growth during SPF at temperatures up to 930° F. The other option is to use thermomechanical processing to achieve

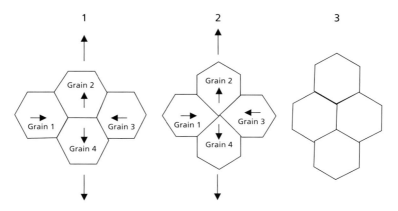

Fig. 2.20. Grain Boundary Rotation[16]

a very fine grain size which, as a result of residual "cold" work, will cause dynamic recrystallization during SPF.

The original superplastic aluminum alloy, Supral 100 (Alloy 2004), was developed especially for its superplastic characteristics. Supral 100 is a medium strength alloy with mechanical properties similar to 6061 and 2219 and is normally used in lightly loaded or non-structural applications. Alloy 7475 is a higher strength aluminum alloy capable of superplasticity that derives its superplasticity from thermomechanical processing. In addition, a number of the aluminum–lithium alloys (e.g., 8090) exhibit superplasticity.

In the single sheet SPF process, illustrated in Fig. 2.21, a single sheet of metal is sealed around its periphery between an upper and lower die. The lower die is either machined to the desired shape of the final part or a die inset is placed in the lower die box. The dies and sheet are maintained at the SPF temperature, and gas pressure is used to form the sheet down over the tool. The lower cavity is maintained under vacuum or can be vented to the atmosphere. After the sheet is heated to its superplastic temperature range, gas pressure is injected through inlets in the upper die. This pressurizes the cavity above the metal sheet forcing it to superplastically form to the shape of the lower die. Gas pressurization is applied slowly so that the strains in the sheet are maintained in the superplastic range, and the pressure is varied during the forming process to maintain the required slow strain rate. As shown in Table 2.10, typical forming cycles for aluminum alloys are 700–900 psi at 840–975° F.

During the forming operation, the metal sheet is being reduced uniformly in thickness; however, wherever the sheet makes contact with the die it sticks and no longer thins-out. This results in a part with non-uniform thickness. To reduce these thickness variations, overlay forming can be used. In overlay forming, the sheet that will become the final part is cut slightly smaller than the tool

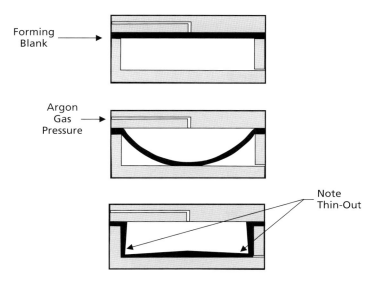

Fig. 2.21. Single Sheet Superplastic Forming

periphery. A sacrificial overlay sheet is then placed on top of it and clamped to the tool periphery. As gas is injected into the upper die cavity, the overlay sheet forms down over the lower die, forming the part blank simultaneously with it. While overlay forming does help to minimize thickness variations, it requires a sacrificial sheet for each run that must be discarded. Two other forming methods, shown in Fig. 2.22, were developed to reduce thickness non-uniformity during forming. However, both of these methods require moving rams within the pressure chamber which increases capital equipment costs.

The hard particles at the grain boundaries that help control grain growth may contribute to the formation of voids in aluminum alloys, a process called cavitation. Cavitation on the order of 3% can occur after about 200% of superplastic deformation. Cavitation can be minimized, or eliminated, by applying a hydrostatic back pressure to the sheet during forming, as shown schematically

Table 2.10 Superplastic Forming Parameters for Aluminum[1]

Alloy	Forming Temperature (°F)	Strain Rate (s^{-1})	Forming Pressure (psi)	Back Pressure (psi)	Strain (%)
2004	840	5×10^4	870	–	1000
7475	960	2×10^4	580	435	500–1000
5083	950	1.5×10^4	870	580	500
8090	970	3×10^4	725	435	200–400

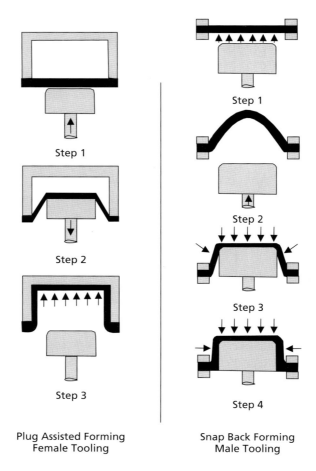

Fig. 2.22. *SPF Methods for Reducing Non-Uniform Thin-Out*[17]

in Fig. 2.23. Back pressures of 100–500 psi are normally sufficient to suppress cavitation.

Gas pressure is an effective pressure medium for SPF for several reasons: (1) it permits the application of a controlled uniform pressure; (2) it avoids the local stress concentrations that are inevitable in conventional forming where a tool contacts the sheet; and (3) it requires relatively low pressures (<1000 psi). Forming parameters (time, temperature, and pressure) have traditionally been determined empirically by trial and error; however, there are now a number of finite element programs that greatly aid in reducing the development time. The disadvantages of SPF are that the process is rather slow and the equipment and tooling can be expensive. For example, a part undergoing 100% strain at

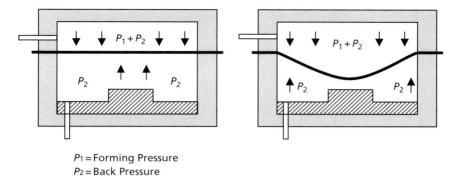

Fig. 2.23. Back Pressure Forming to Suppress Cavitation

0.000 1/s would require almost 3 h at temperature, including the time required for heat-up and cool-down.

For titanium alloys, SPF can be combined with diffusion bonding (SPF/DB) to form one piece unitized structure. Titanium is very amendable to DB because the thin protective oxide layer (TiO_2) dissolves into the titanium above 1150° F leaving a clean surface. However, the aluminum oxide (Al_2O_3) on aluminum does not dissolve and must either be removed, or ruptured, to promote DB. Although DB of aluminum alloys has successfully been demonstrated in the laboratory, SPF/DB of aluminum alloys is not yet a commercial process.

2.8 Casting

Due to their lower properties and higher variability than wrought product forms, aluminum castings are not used for primary structural applications. But for lightly loaded secondary structures, castings can offer significant cost savings by reducing part count and the associated assembly cost. Aluminum casting alloys have different compositions than the wrought alloys, i.e. they are tailored to increase the fluidity of the molten metal, be resistant to hot tearing during solidification, and reproduce the details of the mold shape.

The chemical compositions of two casting alloys used in aerospace are given in Table 2.11. Note that since A357.0 contains an appreciable amount of silicon, it is easier to cast than A201.0. On the other hand, since A201.0 contains copper and silver, it is capable of higher strength levels through precipitation hardening. The alloy A357.0 can also be hardened by the precipitation of Mg_2Si, but the strength is not as high as for the copper and silver containing alloy A201.0.

Silicon is by far the most important alloying addition for aluminum castings. It greatly improves the fluidity of molten aluminum, especially when the amount approaches the eutectic. It is used in amounts up to about 25% in some commercial alloys. Alloys that contain 12% are referred to as eutectic alloys, those

Table 2.11 Chemical Composition of Aerospace Cast Aluminum Alloys[7]

Alloy	Si	Fe	Cu	Mn	Mg	Ti	Be	Ag
A201.0	0.1	0.15	4.6	0.35	0.35	0.25	–	0.7
A357.0	7.0	0.20	0.2	0.10	0.55	0.15	0.05	–

containing less than 12% are hypoeutectic alloys, and those with greater than 12% are hypereutectic alloys. When combined with small amounts of phosphorus, small insoluble particles of AlP are formed that serve as nuclei to help refine the grain size. The silicon containing 3XX castings alloys account for about 80% of all sand and permanent mold castings. Copper is another alloying element that is frequently used in casting alloys because it makes them respond to precipitation hardening, but copper reduces fluidity, promotes hot shortness, and increases the susceptibility to stress corrosion cracking. For example, the copper and silver containing alloy A201.0 is often overaged to the T7 temper to improve its resistance to stress corrosion cracking. Magnesium is frequently added to improve strength, machinability, and corrosion resistance. Grain refiners, such as titanium and boron, are used that form small crystals ($TiAl_2$ or TiB_2) which serve as nucleation sites. The 2XX casting alloys, although harder to cast than the 3XX alloys, can be heat treated to the highest strength levels of the aluminum casting alloys. In order of decreasing castability, the alloy groups can be classified in the order 3XX, 4XX, 5XX, 2XX, and 7XX.

Premium quality castings provide higher quality and reliability than conventional cast products. Important attributes include high mechanical properties determined by test coupons machined from representative parts, low porosity levels as determined by radiography, dimensional accuracy, and good surface finishes. Premium casting alloys include A201.0, A206.0, 224.0, 249.0, 354.0, A356.0, A357.0, and A358.0. The requirements for premium quality castings are usually negotiated through special specifications such as SAE-AMS-A-21180.

Grain size control for castings is important because fine grain sizes result in higher strengths and greater ductility. However, the grain size of aluminum castings can be as small as 0.005 in. and as large as 0.5 in. Normally, grain sizes no larger than 0.04 in. are desired for premium quality castings. Since the size of porosity in an aluminum casting scales somewhat with grain size, finer porosity goes with finer grain sizes. In addition, shrinkage and hot cracking are more prevalent in castings with a coarse grain size. Grain size is a function of pouring temperature, solidification rate, and the presence or absence of a grain refiner. Low pouring temperatures, faster solidification rates, and grain refiners, such as titanium and boron, all produce finer grain sizes.

Aluminum ingots for casting are usually reheated and melted using one of three types of furnaces: direct fuel fired furnaces, indirect fuel fired furnace, or electric furnaces. Direct fuel fired furnaces use hydrocarbon fuels to heat the metal which places the hot combustion gases in direct contact with the charge being melted. Indirect fuel fired furnaces also use hydrocarbon fuels, but the charge is separated by a crucible from direct contact with the hot combustion gases. The advantage of the indirect method is that it helps prevent combustion products, hydrogen in particular, from being absorbed into the melt. Electric furnaces consist of low frequency induction furnaces, high frequency induction furnaces, and electric resistance furnaces. Low frequency induction is by far the most common; a typical furnace operates at 60 cycles, 20–200 kW with capacities of 700–3000 lb. The induced electromagnetic field stirs and mixes the melt, thus aiding in maintaining uniform melt temperatures.

Proper temperature control during melting and pouring is critical; many casting problems have eventually been traced to poor temperature control. The equipment must be capable of holding a temperature tolerance of $\pm 10°$ F to insure satisfactory results. If the pouring or casting temperature is too low, misruns and cold shuts can occur, while if the pouring temperature is too high, coarse grains, excessive porosity, excessive shrinkage, and hot tearing are all possible.

Molten aluminum is an extremely reactive metal that readily combines with other metals, gases, and even some refractories. It experiences both a large solidification shrinkage (6%) and contraction shrinkage (10%). It also has a high surface tension that, when combined with an oxide film, makes it difficult to obtain sound castings in thin sections. Rather than attempt to alloy the melt themselves, most foundries purchase prealloyed ingot; however, alloys containing magnesium are prone to loss by oxidation and evaporation in the melt and must be replenished prior to casting. To prevent segregation in the melt, the melt must be stirred; however, excessive stirring promotes oxidation of the melt.

Molten aluminum is also subject to contamination by iron, oxides, and hydrogen; therefore, proper steps must be taken to control all three. Iron reduces the ductility and toughness of the casting and promotes the formation of sludge that accumulates in the bottom of the crucible. To prevent sludging, the iron equivalent, $\%Fe + 2(\%Mn) + 3(\%Cr)$, must generally be held below 1.9%. Sludge that accidentally gets poured into a casting is often discovered as hard spots during machining, or worse yet, as stress cracks in service.

Oxides of aluminum and magnesium form as a thin film on the bath surface that actually prevents further oxidation as long as the film is not disturbed. Oxidation can result from moisture introduced by the furnace charge, excessive stirring, pouring from too high a temperature, or pouring from too high a height. Surface oxides are removed by fluxing and then skimming the surface.

If hydrogen is not effectively removed from the melt, the likely result will be a casting containing excessive porosity. Hydrogen originates from moisture on the furnace charge and from hydrocarbon combustion products. At temperatures less than 1250° F, hydrogen absorption is minimal but increases rapidly at higher temperatures. Hydrogen is removed from the melt by using degassing fluxes after the oxides are removed. Degassing fluxes include chlorine gas, nitrogen–chlorine mixtures, and hexachloroethane. The removal of hydrogen is a mechanical, not chemical, process in which the hydrogen attaches itself to the fluxing gas.

Sludge formation and settling is a problem with alloys containing 5% or more of silicon. Three practices that minimize sludge formation are: keeping the iron content as low as possible; keeping the melting furnace at temperatures lower than 1350° F; and keeping the holding furnace at temperatures of 1200° F or less. If sludge does build up in the bottom of the crucible, it is necessary to scoop it out.

2.8.1 Sand Casting

Sand casting is perhaps the oldest casting process known. The molten metal is poured into a cavity shaped inside a body of sand and allowed to solidify. Advantages of sand casting are low equipment costs, design flexibility, and the ability to use a large number of aluminum casting alloys. It is often used for the economical production of small lot sizes and is capable of producing fairly intricate designs. The biggest disadvantages are that the process does not permit close tolerances, and the mechanical properties are somewhat lower due to larger grain sizes as a result of slow cooling rates. However, the mechanical properties are improving as a result of improvements in casting materials and procedures. The steps involved in sand casting are shown in Fig. 2.24 and consist of the following:

1. Fabricate a pattern, usually wood, of the desired part and split it down the centerline.
2. Place the bottom half of the pattern, called the drag, in a box called a flask.
3. Apply a release coating to the pattern, fill the flask with sand and then compact the sand by ramming.
4. Turn the drag half of the mold over and place the top half of the flask on top of it. The top half of the pattern, called the cope, is then placed over the drag half of the pattern and release coated.
5. Risers and a sprue are then installed in the cope half of the flask. The sprue is where liquid metal enters the mold. In a complex casting, the sprue is usually gated to different positions around the casting. The risers are essentially reservoirs for liquid metal that keep the casting supplied with liquid metal as the metal shrinks and contracts on freezing.
6. The cope half is then packed with sand and rammed.
7. The two halves are separated and the patterns are removed. If hollow sections are required, a sand core is placed in the drag half of the mold. A gating system is then cut into the sand on the cope half of the mold.
8. The two halves are reassembled and clamped or bolted shut for casting.

Aluminum

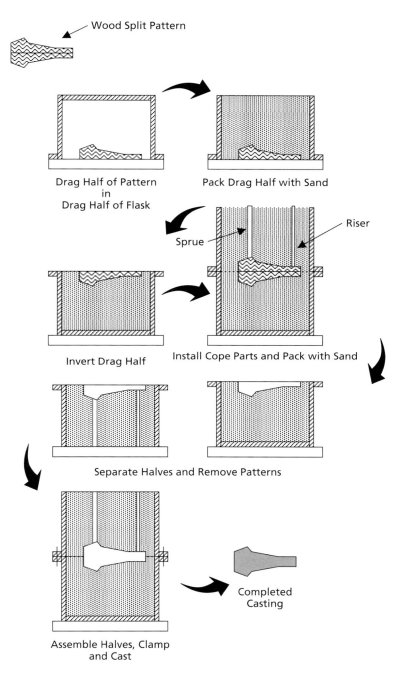

Fig. 2.24. Sand Casting Process

Molding sands usually consist of sand grains, a binder, and water. The properties that are important are good flowability or the ability to be easily worked around the pattern, sufficient green strength, and sufficient permeability to allow gas and steam to escape during casting. Sand cores for molded-in inserts can be made using either heat cured binder systems or no-bake binder systems. No-bake binder systems are usually preferred since they provide greater dimensional accuracy, have higher strengths, are more adaptable to automation, and can be used immediately after fabrication. The no-bake systems typically consist of room temperature curing sodium silicate sands, phenolic-urethanes, or furan acids combined with sand.

Gates are used to evenly distribute the metal to the different locations in the casting. The objective is to have progressive solidification from the point most distant from the gate toward the gate, i.e. the metal in the casting should solidify before the metal in the gates solidifies. Normally, the area of the gates and runner system connecting the gates should be about four times larger than the sprue. When feeding needs to be improved, it is better to increase the number of gates rather than increase the pouring temperature.

2.8.2 Plaster and Shell Molding

Plaster mold casting is basically the same as sand casting except gypsum plasters replace the sand in this process. The advantages are very smooth surfaces, good dimensional tolerances, and uniformity due to slow uniform cooling. However, as a result of the slow solidification rates, the mechanical properties are not as good as with sand castings. In addition, since plaster can absorb significant moisture from the atmosphere, it may require slow drying prior to casting.

Shell molding can also be used in place of sand casting when a better surface finish or tighter dimensional control is required. Surfaces finishes in the range of 250–450 μin. are typical with shell molding. Since it requires precision metal patterns and more specialized equipment, shell molding should be considered a higher volume process than sand casting. Shell molding, shown in Fig. 2.25, consists of the following:

1. A fine silica sand coated with a phenolic resin is placed in a dump box that can be rotated.
2. A metal pattern is heated to 400–500° F, mold released and placed in the dump box.
3. The pattern and sand are inverted allowing the sand to coat the heated pattern. A crust of sand fuses around the part as a result of the heat.
4. The dump box is turned right side up, the pattern with the shell crust is removed and cured in an oven at 650–750° F.
5. The same process is repeated for the other half of the mold.
6. The two mold halves are clamped together and placed in a flask supported with either sand or metal shot.

Aluminum

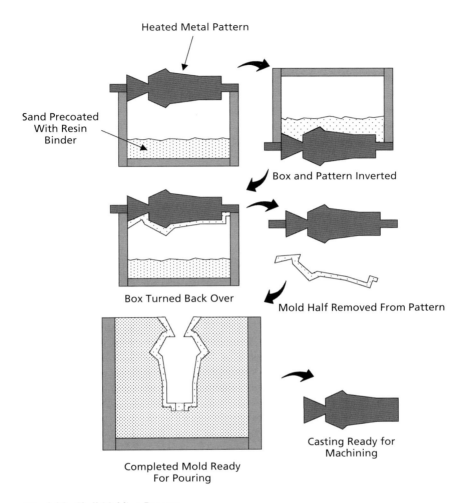

Fig. 2.25. Shell Molding Process

2.8.3 Permanent Mold Casting

In permanent mold casting, liquid metal is poured into a metal mold and allowed to solidify. This method is second only to die casting in the number of aluminum castings produced annually. However, due to the tooling costs, it is usually reserved for high volume applications. The castings produced are normally small compared to sand casting and rather simple in shape. The process produces fairly uniform wall thicknesses but, unless segmented dies are used, is not capable of undercuts. Compared to sand castings, permanent mold castings are more uniform and have better dimensional tolerances, superior surfaces finishes (275–500 μin. are typical), and better mechanical properties due to the faster solidification rates. Mold materials include gray cast iron and hot work die steels such as H11 and H13. When a disposable sand or plaster core is used with this process, it is referred to as semipermanent mold casting. Another variant of the permanent mold process is low-pressure permanent mold casting. Here the casting is done inside a pressure vessel and an inert gas is used to apply 5–10 psi pressure on the liquid metal. This results in shorter cycle times and excellent mechanical properties.

2.8.4 Die Casting

Die casting is a permanent mold casting process in which the liquid metal is injected into a metal die under high pressure. It is a very high rate production process using expensive equipment and precision matched metal dies. Since the solidification rate is high, this process is amendable to high volume production. It is used to produce very intricate shapes in the small to intermediate part size range. Characteristics of the process include extremely good surface finishes and the ability to hold tight dimensions; however, die castings should not be specified where high mechanical properties are important because of the inherently high porosity level. The high pressure injection creates a lot of turbulence that traps air resulting in high porosity levels. In fact, die cast parts are not usually heat treated because the high porosity levels can cause surface blistering. To reduce the porosity level, the process can be done in vacuum (vacuum die casting) or the die can be purged with oxygen just prior to metal injection. In the latter process, the oxygen reacts with the aluminum to form an oxide dispersion in the casting.

2.8.5 Investment Casting

Investment casting is used where tighter tolerances, better surface finishes, and thinner walls are required than can be obtained with sand casting. Although the investment casting process is covered in greater detail in Chapter 4 on Titanium, a brief description of the process is that investment castings are made by surrounding, or investing, an expendable pattern, usually wax, with a refractory slurry that sets at room temperature. The wax pattern is then melted out and the refractory mold is fired at high temperatures. The molten metal is cast into the mold and the mold is broken away after solidification and cooling.

2.8.6 Evaporative Pattern Casting

A process that is used quite extensively in the automotive industry is evaporative pattern casting. Part patterns of expandable polystyrene are produced in metal dies. The patterns may consist of the entire part or several patterns may be assembled together. Gating patterns are attached and the completed pattern is coated with a thin layer of refractory slurry which is allowed to dry. The slurry must still be permeable enough to allow mold gases to escape during casting. The slurry coated pattern is then placed in a flask supported by sand. When the molten metal is poured, it evaporates the polystyrene pattern. This process is capable of producing very intricate castings with close tolerances, but the mechanical properties are low due to the large amounts of entrapped porosity.

2.8.7 Casting Heat Treatment

The major differences between heat treating cast aluminum alloys,[18] as compared with wrought alloys, are the longer soak times during solution heat treating and the use of hot water quenches. Longer soak times are needed because of the relatively coarse microconstituents present in castings that do not have the benefit of the homogenization treatments given to wrought products before hot working. Boiling water is a common quenchant to reduce distortion for castings that normally contain more complex configurations than wrought products. Aluminum–copper castings are usually solution treated at 950–960° F and then quenched in hot water maintained at 150–212° F to minimize quenching stresses and distortion. Cast aluminum alloys are usually supplied in either the T6, T7, or T5 tempers. The T6 temper is used where maximum strength is required. If low internal stresses, dimensional stability, and resistance to stress corrosion cracking are important, then the casting can be overaged to a T7 temper. The T5 temper is produced by aging the as-cast part without solution heat treating and quenching. This treatment is possible because most of the hardening elements are retained in solid solution during casting; however, the strengths obtained with the T5 temper will be lower than those with the T6 heat treatment.

2.8.8 Casting Properties

Since the mechanical properties of castings are not as consistent as wrought products, it is normal practice to use a casting factor (CF) for aluminum castings. The CF usually ranges from 1.0 to 2.0 depending on the end usage of the casting. For example, if the casting factor is 1.25 and the material has a yield strength of 30 ksi, the maximum design strength would be $30\,\text{ksi}/1.25 = 24\,\text{ksi}$. In addition, sampling is used during production in which a casting is periodically selected from the production lot and cut-up for tensile testing. The sampling plan depends on the criticality of the casting. All premium castings are subjected to both radiographic and penetrant inspection. Premium castings can also be hot isostatic pressed (HIP) to help reduce internal porosity. HIP is usually conducted using argon pressure at 15 ksi and temperatures in the range of 900–980° F. HIP usually results in improved mechanical properties, especially fatigue strength but, of course, it adds to the cost and cycle time. The improvement in fatigue life for A201.0-T7 as a result of HIP is shown in Fig. 2.26. Surface defects in

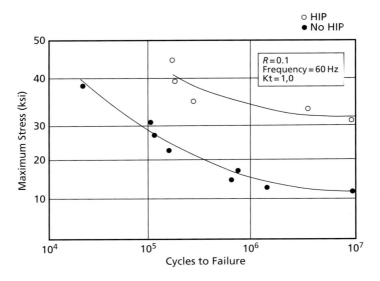

Fig. 2.26. Effect of HIP on Fatigue Life of A201.0-T7 Casting[19]

casting are normally repaired by gas tungsten arc welding (GTAW) using filler wire cast from the same alloy as the casting.

2.9 Machining

Aluminum exhibits extremely good machinability. Its high thermal conductivity readily conducts heat away from the cutting zone allowing high cutting speeds, usually expressed in surface feet per minute (SFM):

$$\text{SFM} = \pi DN/12$$

where

D = tool diameter in inches
N = tool rotation in revolutions per minute (RPM)

The objective of any machining operation is to maximize the metal removal rate (MRR) in in.^{3}/min which for milling is equal to:

$$\text{MRR} = \text{RDOC} \times \text{ADOC} \times F$$

where

RDOC = radial depth of cut in inches
ADOC = axial depth of cut in inches
F = feed rate in in./min

Another important concept in machining is feed per tooth, also called chip load, which is:

$$f_t = F/nN$$

where

f_t = feed per tooth in in./tooth
F = feed rate in in./min
n = number of teeth on cutter
N = tool rotation in RPM

Cutting speeds for aluminum alloys are high, with speeds approaching, or exceeding, 1000 sfm common. In fact, as the cutting speed is increased from 100 to 200 sfm, the probability of forming a built-up edge on the cutter is reduced; the chips break more readily and the surface finish on the part is improved. The depth of cut should be as large as possible to minimize the number of cuts required. During roughing, depths of cuts range from 0.250 in. for small parts to as high as 1.5 in. for medium and large parts, while finishing cuts are much lighter with depths of cuts less than 0.025 in. commonplace. Feed rates for roughing cuts are in the range of 0.006–0.080 in./rev., while lighter cuts are used for finishing, usually in the range of 0.002–0.006 in./rev.

Because aluminum alloys have a relatively low modulus of elasticity, they have a tendency to distort during machining. Also, due to aluminum's high coefficient of thermal expansion, dimensional accuracy requires that the part be kept cool during machining; however, the high thermal conductivity of aluminum allows most of the heat to be removed with the chips. The flushing action of a cutting fluid is generally effective in removing the remainder of the heat. The use of stress relieved tempers, such as the TX51 tempers, stress relieved by stretching, also helps to minimize distortion during machining.

Excessive heat during machining can cause a number of problems when machining aluminum alloys. Friction between the cutter and the workpiece can result from dwelling, dull cutting tools, lack of cutting fluid, and heavy end mill plunge cuts rather than ramping the cutter into the workpiece. Inadequate backup fixtures, poor clamping, and part vibration can also create excessive heat. Localized overheating of the high strength grades can even cause soft spots which are essentially small areas that have been overaged due to excessive heat experienced during machining. These often occur at locations where the cutter is allowed to dwell in the work, for example during milling in the corners of pockets.

Standard high speed tool steels, such as M2 and M7 grades, work well when machining aluminum. For higher speed machining operations, conventional C-2 carbides will increase tool life, resulting in less tool changes and allowing higher cutting speeds. For example, typical peripheral end milling parameters for the 2XXX and 7XXX alloys are 400–800 sfm with high speed tool steel and 800–1300 sfm for carbide tools. For higher speed machining operations, large cuttings, such as inserted end mills, should be dynamically balanced. Diamond tools are often used for the extremely abrasive hypereutectic silicon casting alloys.

2.9.1 High Speed Machining

High speed machining is somewhat an arbitrary term. It can be defined for aluminum as: (1) machining conducted at spindle speeds greater than 10000 rpm; (2) machining at 2500 sfm or higher where the cutting force falls to a minimum as shown in Fig. 2.27; and (3) machining at speeds in which the impact frequency of the cutter approaches the natural frequency of the system.

In an end milling operation, assume the following roughing parameters:

- A 1.5 in. diameter end mill with 3 teeth
- A speed of 3600 rpm maximum
- A feed rate of 40 ipm
- An axial depth of cut of 1 in. and
- A radial depth of cut of 1 in.

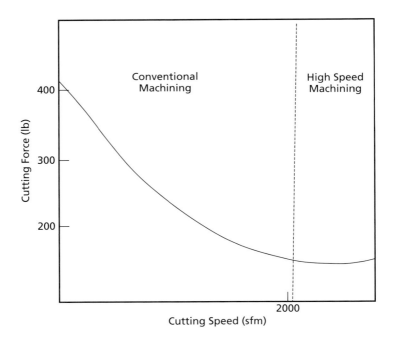

Fig. 2.27. *Cutting Force vs. Cutting Speed for Aluminum*

Then

$$\text{MRR} = (1.0\,\text{in.})(1.0\,\text{in.})(40\,\text{in./min}) = 40\,\text{in.}^3/\text{min}$$

This is a fairly respectable and easily achievable metal removal rate using conventional machining parameters for aluminum. However, with the advent of high speed machining of aluminum, even higher metal removal rates are obtainable. It should be emphasized that while higher metal removal rates are good, another driver for developing high speed machining of aluminum is the ability to machine extremely thin walls and webs. For example, the minimum machining gage for conventional machining might be 0.060–0.080 in. or higher without excessive warpage, while with high speed machining, wall thicknesses as thin as 0.020–0.030 in. without distortion are readily achievable. This allows the design of weight competitive high speed machined "assemblies" in which sheet metal parts that were formally assembled with mechanical fasteners can now be machined from a single or several blocks of aluminum plate. For example, as shown in Fig. 2.28, an avionics rack that originally consisted of 44 formed sheet metal parts was replaced by a single high speed machined part and five other pieces, resulting in a 73% cost reduction. The final shelf weighs 8.5 lb, needs only five tools, and takes only 38.6 h of manufacturing time, as opposed to the original built-up design that weighed 9.5 lb, required 53 individual tools, and took 1028 h to manufacture.[20]

Successful high speed machining requires an integrated approach between the cutter, workpiece, machine tool, and cutting strategy. For example, step cutting, as shown in Fig. 2.29, uses the workpiece to help provide rigidity during the machining process. For thin ribs (i.e., side cutting), this means using a large

Sheet Metal Assembly	High Speed Machined

Was		Now	
Number of Pieces	44	Number of Pieces	6
Number of Tools	53	Number of Tools	5
Design and Fabrication h (Tools)	965	Design and Fabrication h (Tools)	30
Fabrication h	13.0	Fabrication h	8.6
Assembly Manhours	50	Assembly Manhours	5.3
Weight (lb)	9.58	Weight (lb)	8.56

Fig. 2.28. Comparison between Assembled and High Speed Machined Avionics Rack
Source: The Boeing Company

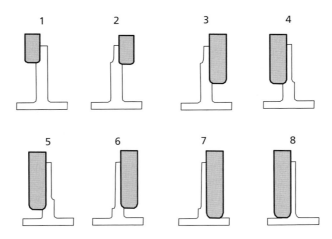

Fig. 2.29. Step Cutting Methodology

radial depth of cut and a small axial depth of cut, while for thin floor webs (i.e., end cutting), a large axial depth of cut should be combined with a small radial depth of cut. Although the depths of cut are small compared to conventional machining, the feed rates used in high speed machining are so much higher that they allow higher metal removal rates. Another fundamental associated with step cutting is that all cutters have a maximum depth of cut before they chatter; therefore, taking small depths of cut with the step cutting technique helps to prevent chatter, i.e. the cutter remains in the stability zone where chatter does not occur.

The comparison between conventional and high speed machining in Table 2.12 illustrates the small depths of cuts used in high speed machining. Again, since the feed rates are so much higher in high speed machining, the metal removal rate is still much higher.

Attempts to cut extremely thin wall structures using conventional machining are generally unsuccessful. In conventional machining, cutter and part deflection result in cutting through the thin walls as illustrated in Fig. 2.30. As shown,

Table 2.12 Comparison of Conventional and High Speed Machining

	Conventional Machining	High Speed Machining
Cutter	1.5 Diameter × 2 in. long	1.0 Diameter × 2 in. long
Speed (RPM)	3600	20000
Feed (ipm)	40	280
Axial Depth of Cut (in.)	1	0.4
Radial Depth of Cut (in.)	1	1
Metal Removal Rate (in.^{3}/min)	40	112

Aluminum

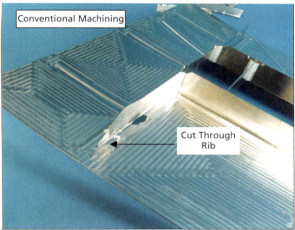

Fig. 2.30. Thin Wall Ribs: High Speed vs. Conventional Machining
Source: The Boeing Company

the exact same part configuration was successfully cut from a two-inch thick plate of 7050-T7451 using high speed machining techniques. The two-inch high stiffeners ranged from 0.029 to 0.035 in. in thickness. Another example of thin wall machining capability is shown in the space structure in Fig. 2.31. This rather remarkable large part was machined to a wall thickness of only 0.015 in. and floor thickness of 0.020 in. The wall heights are almost 5 in.

High speed machining of aluminum was originally implemented on the F/A-18E/F fighter to save weight; 150 high speed machined "assemblies" saved approximately 80 lb. of weight. It soon became apparent that the higher metal removal rates could also save costs by eliminating multiple parts and assembly

Fig. 2.31. High Speed Machined Thin Wall Space Structure
Source: AUT Inc.

costs. An example of a high speed machining replacing an existing composite part is shown in Fig. 2.32. Although the machined part is slightly heavier, the cost savings were so significant that the change was justified. Another example is the large bulkhead shown in Fig. 2.33. Note that the holding fixture is actually the periphery of the plate, eliminating the need for holding fixtures. Finally, in Fig. 2.34 is a one piece unitized substructure that replaced multiple machined spars and ribs assembled with mechanical fasteners.

In conventional machining, roughing is usually followed by finishing using lighter depths of cut at much slower speeds. In high speed machining, it is often possible to combine the roughing and finishing cuts in the same operation. An enabler for high speed machining has been the development of porosity-free thick plate in the 7XXX series of alloys. For example, 7050-T7351 is available in plate stock up to 8 in. thick.

Solid two flute carbide cutters are normally used for high speed machining. Carbide tools reduce the number of tool changes required, due to reduced wear. However, more importantly, they also provide much greater cutter stiffness than high speed steel. Since a stiff cutter can take a much larger depth of cut than a flexible cutter, cutters with small length-to-diameter ratios are also used to maintain maximum cutter stiffness. Smaller diameter cutters may be required for finishing if tight corner radii are required. To provide further stiffness, the shortest possible tool holder is recommended. Since high speed machining is frequently used to produce thin walls, runout should be held to 0.001 in. Since high speeds are used, both the tool and holder need to be balanced. Due to the high rpms involved, inserted tools should not be used for safety reasons.

Vibration is a natural concern when machining at high speeds. Two types of vibration can occur: cutter vibration (chatter) and workpiece, or part, vibration.

Aluminum

Attribute	Conventional	HSM
No. Parts	8 Machined Parts 7 Honeycomb Core Parts	3
No. Fasteners	372	20
Part Weight	92 lb	110 lb
No. Tools	438	6

Fig. 2.32. High Speed Machined Unitized Speedbrake
Source: The Boeing Company

Chatter appears as a series of uniform continuous series of marks on the workpiece surface, while part vibration causes deeper marks that are more randomly distributed. Chatter occurs when the impact frequency of the cutter begins to vibrate near its natural frequency. Part vibration occurs in two slightly different ways depending on whether a web or floor is being cut or a rib or flange is being machined. Webs or floors excited by the cutting process can start to vibrate at their natural frequency giving the appearance of a bouncing motion. Ribs and flanges vibrate by forced resonant vibration in which the natural frequencies of the ribs match the natural frequency of the cutter.

Manufacturing Technology for Aerospace Structural Materials

Fig. 2.33. *High Speed Machined Unitized Bulkheads*
Source: *The Boeing Company*

Fig. 2.34. *High Speed Machined Unitized Substructure*
Source: *The Boeing Company*

Matching the cutter impact frequency to the workpiece vibration frequency produces a stable non-chattering cut and a smooth surface. As shown in Fig. 2.35, if the teeth are hitting the workpiece at one frequency and the cutter is vibrating at another frequency, the cutter produces chatter because the teeth will hit the workpiece at different points in the vibration. On the other hand, if the teeth hit at the same rate the cutter is vibrating, chatter is eliminated and a smooth surface finish results. Chatter can be detected by using a microphone along with a data acquisition system during machining. The system records the data and chatter detection software uses a Fast Fourier Transform to produce a plot like the one shown in Fig. 2.36. The extra spike indicates the chatter frequency.

Aluminum

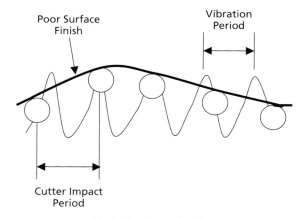

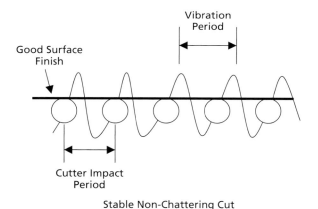

Fig. 2.35. Stable and Unstable Cutting

If the tool system has a chatter frequency of 2000 Hz for a two flute cutter, then the optimum spindle speed would be 2000 Hz × 60 s/min/2 teeth = 60 000 rpm. If the spindle only has a maximum speed of 40 000 rpm, then another region of stability and optimum speed can be determined by dividing the optimum spindle speed by an integer. For example, in our case, 60 000 rpm/2 = 30 000 rpm. To find the optimum spindle speed for highest metal removal rate (MRR) possible, gradually increase the depth of cut until a new chatter limit is encountered and repeat the process of measuring the chatter frequency and then calculating the best spindle speed. Machining can be conducted safely at or below this speed. Reducing spindle speed is another effective method for controlling chatter and improving part quality by damping out vibrations. It should be noted that this assumes that vibration due to the workpiece has been eliminated by proper

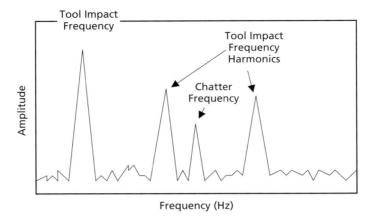

Fig. 2.36. Frequency Response to Detect Chatter

workpiece fixturing and proper tool path programming.[20] Machine tool and spindle builders that supply high speed machining equipment have the capability to help in establishing high speed machining parameters.

2.9.2 Chemical Milling

Shallow pockets are sometimes chemically milled into aluminum skins for weight reduction. The process is used mainly for parts having large surface areas requiring small amounts of metal removal. Rubber maskant is applied to the areas where no metal removal is desired. In practice, the maskant is placed over the entire skin and allowed to dry. The maskant is then scribed according to a pattern and the maskant removed from the areas to be milled. The part is then placed in a tank containing sodium hydroxide heated to $195 \pm 5°F$ with small amounts of triethanolamine to improve the surface finish. The etchant rate is in the range of 0.0008–0.0012 in./min. Depths greater than 0.125 in. are generally not cost competitive with conventional machining, and the surface finish starts to degrade. After etching, the part is washed in fresh water and the maskant is stripped.

2.10 Joining

Aluminum alloys can be joined by a variety of commercial methods including welding, brazing, adhesive bonding, and mechanical fastening. Adhesive bonding is covered in Chapter 8 on Adhesive Bonding and Integrally Cocured Structure and mechanical fastening is covered in Chapter 11 on Structural Assembly. Brazing is not used as a joining method for the 2XXX and 7XXX alloys, because their melting points are too low to be successfully brazed.

2.11 Welding

Weldability can be defined as the ability to produce a weld free of discontinuities and defects that results in a joint with acceptable mechanical properties, either in the as-welded condition or after a post-weld heat treatment. Although aluminum has a low melting point, it can be rather difficult to weld for several reasons:[21] (1) the stable surface oxide must be removed by either chemical methods or more typically by thoroughly wire brushing the joint area; (2) the high coefficient of thermal expansion of aluminum can result in residual stresses leading to weld cracking or distortion; (3) the high thermal conductivity of aluminum requires high heat input during welding further leading to the possibility of distortion or cracking; (4) aluminum's high solidification shrinkage with a wide solidification range also contributes to cracking; (5) aluminum's high solubility for hydrogen when in the molten state leads to weld porosity; and (6) the highly alloyed, high strength 2XXX and 7XXX alloys are especially susceptible to weld cracking.

For aircraft structural joints, mechanical fastening is the preferred joining method; however, for some launch vehicles, the 2XXX alloys are fusion welded to fabricate large cryogenic fuel tanks. The 2XXX alloys that can be fusion welded include 2014, 2195, 2219, and 2519. These alloys have lower magnesium contents, reducing their propensity for weld cracking. Alloy 2219 is readily weldable, and 2014 can also be welded to somewhat less extent. Alloy 2319 is also commonly used as a filler metal when welding 2219. Although the weldability of the aluminum–lithium alloy 2195 approaches that of 2219, it tends to crack more than 2195 during repair welding. The 6XXX alloys can also be prone to hot cracking but are successfully welded in many applications. Post-weld heat treatments can be used to restore the strength of 6XXX weldments. For the 7XXX alloys, those with a low copper content, such as 7004, 7005, and 7039, are somewhat weldable, while the remainder of the 7XXX series are not fusion weldable due to weld cracking and excessive strength loss.

The ability to fusion weld aluminum is often defined as the ability to make sound welds without weld cracking. Two types of weld cracking can be experienced – solidification cracking and liquation cracking. Solidification cracking, also called hot tearing, occurs due to the combined influence of high levels of thermal stresses and solidification shrinkage during weld solidification. Solidification cracking occurs in the fusion zone, normally along the centerline of the weld or at termination craters. Solidification cracking can be reduced by minimal heat input and by proper filler metal selection. The 4XXX alloys, with their narrow solidification range, are often used as filler metals when welding the 2XXX and 7XXX alloys. Liquation cracking occurs in the grain boundaries next to the heat affected zone (HAZ). Highly alloyed aluminum alloys typically contain low melting eutectics that can melt in the adjacent metal during the welding operation. Similar to solidification cracking, liquation cracking can be minimized by lower heat input and proper filler wire selection.

Due to its high solubility in molten aluminum and low solubility in solid aluminum, hydrogen can enter the molten pool and, with its decreasing solubility during freezing, form porosity in the solidified weld. Hydrogen is approximately 20 times more soluble in the liquid than the solid. Hydrogen normally originates from three sources: (1) hydrogen from the base metal, (2) hydrogen from the filler metal and (3) hydrogen from the shielding gas. Hydrogen from the base metal and filler wire can be minimized by ensuring that there is no moisture present and that all hydrocarbon residues and the surface oxide are thoroughly removed prior to welding.

Welding processes that produce a more concentrated heat source result in smaller HAZs and lower post-weld distortions; however, the capital cost of the equipment is roughly proportional to the intensity of the heat source. The nature of welding in the aerospace industry is characterized by low unit production, high unit cost, extreme reliability, and severe service conditions; therefore, the more expensive and more concentrated heat sources such as plasma arc, laser beam, and electron beam welding are often selected for welding of critical components.[22]

2.11.1 Gas Metal and Gas Tungsten Arc Welding

Gas metal arc welding (GMAW), as shown in Fig. 2.37, is an arc welding process that creates the heat for welding by an electric arc that is established between a consumable electrode wire and the workpiece. The consumable electrode wire is fed through a welding gun that forms an arc between the electrode and the workpiece. The gun controls the wire feed, the current, and the shielding gas. In GMAW, the power supply is direct current with a positive electrode. The positive electrode is hotter than the negative weld joint ensuring complete fusion of the wire in the weld joint. In addition, the direct current electrode positive (DCEP) arrangement provides cathodic cleaning of the oxide layer during welding. When the electrode is positive and a direct current is used, there is a flow of electrons from the workpiece to the electrode with ions traveling in the opposite direction and bombarding the workpiece surface. The ion bombardment breaks up and disperses the oxide film to create a clean surface for welding. The DCEP arrangement also provides good arc stability, low spatter, a good weld bead profile, and the greatest depth of penetration. A shielding gas, such as argon, is used to protect the liquid metal fusion zone; however, the addition of helium to argon provides deeper penetration. GMAW has the advantage of good weld metal deposit per unit time.

Gas tungsten arc welding (GTAW) uses a non-consumable tungsten electrode to develop an arc between the electrode and the workpiece. A schematic of the GTAW process is shown in Fig. 2.38. Although it has lower metal deposition rates than GMAW, it is capable of higher quality welds. However, when the joint thickness exceeds 0.375 in., GMAW is probably a more cost-effective

Aluminum

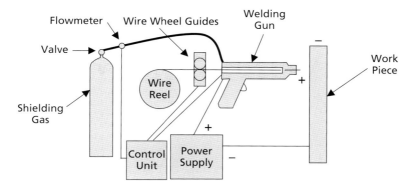

Fig. 2.37. *Gas Metal Arc Welding*

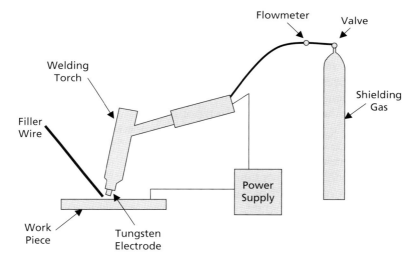

Fig. 2.38. *Gas Tungsten Arc Welding*

method. For welding aluminum with GTAW, an alternating current arrangement is used, which like the DCEP arrangement for GMAW, provides cleaning of the oxide layer during the welding process. The alternating current causes rapid reversing of the polarity between the workpiece and the electrode at 60 Hz. For this welding arrangement, tungsten electrodes and argon shielding gas are used. In general, material less than 0.125 in. thick can be welded without filler wire addition if solidification cracking is not a concern.

The reduction of strength and hardness in a fusion welded HAZ is illustrated in Fig. 2.39. The degradation of properties within the HAZ usually dictates joint strength. High heat inputs and preheating prior to welding increase the

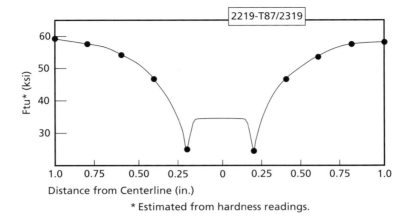

Fig. 2.39. Strength Across Fusion Weld Joint[21]

width of the HAZ and the property loss. The property loss within the HAZ can be minimized by the elimination of preheating, minimizing heat input, using high welding speeds where possible, and by using multi-pass welding. Post-weld heat treatments can also be employed to improve properties, either by complete solution heat treating and aging or by post-weld aging only. Solution heat treating and aging will restore the highest properties but water quenching may result in excessive distortion; therefore, polymer quenchants that produce slower cooling rates are often used for post-weld heat treatments. While post-weld aging at moderate temperatures does not achieve as high properties as full heat treating, it is often used because it does not result in excessive distortion or warpage. It should be noted that both full heat treatment and post-weld aging result in decreased joint ductility.

2.11.2 Plasma Arc Welding

Automated variable polarity plasma arc (VPPA) welding is often used to weld large fuel tank structures. Plasma arc welding, shown in Fig. 2.40, is a shielded arc welding process in which heat is created between a tungsten electrode and the workpiece. The arc is constricted by an orifice in the nozzle to form a highly collimated arc column with the plasma formed through the ionization of a portion of the argon shielding gas. The electrode positive component of the VPPA process promotes cathodic etching of the surface oxide allowing good flow characteristics and consistent bead shape. Pulsing times are in the range of 20 ms for the electrode negative component and 3 ms for the electrode positive polarity. A keyhole welding mode is used in which the arc fully penetrates the workpiece, forming a concentric hole through the thickness. The molten metal then flows around the arc and resolidifies behind the keyhole as the torch

Aluminum

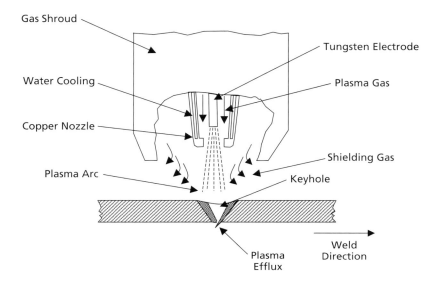

Fig. 2.40. Plasma Arc Welding Process

traverses through the workpiece. The keyhole process allows deep penetration and high welding speeds while minimizing the number of weld passes required. VPPA welding can be used for thicknesses up to 0.50 in. with square grooved butt joints and even thicker material with edge beveling. While VPPA welding produces high integrity joints, the automated equipment used for this process is expensive and maintenance intense.

2.11.3 Laser Welding

There is considerable interest in laser beam welding (LBW) of high strength aluminum alloys, particularly in Europe,[23–25] where limited aerospace production has been announced. The process is attractive because it can be conducted at high speeds with excellent weld properties. No electrode or filler metal is required and narrow welds with small HAZs are produced. Although the intensity of the energy source is not quite as high as that in electron beam (EB) welding, EB welding must be conducted in a vacuum chamber. The coherent nature of the laser beam allows it to be focused on a small area leading to high energy densities. Since the typical focal point of the laser beam ranges from 0.004 to 0.040 in., part fit-up and alignment are more critical than conventional welding methods. Both high power continuous wave carbon dioxide (CO_2) and neodymium-doped yttrium aluminum garnet (Nd:YAG) lasers are being evaluated. The wavelength of light from a CO_2 laser is 10.6 μm, while that of Nd:YAG laser is 1.06 μm. Since the absorption of the beam energy by the material being welded increases with decreasing wavelength, Nd:YAG lasers are better suited for welding aluminum.[25] In addition, the solid state Nd:YAG lasers use fiber optics for beam delivery, making it more amenable to automated robotic welding. Laser welding produces a concentrated high energy density heat source that results in very narrow heat affected zones, minimizing both distortion and loss of strength in the HAZ. In 0.080 in. sheet, speeds up to 6.5 ft/min are achievable with a 2 kW Nd:YAG laser and 16–20 ft/min with a 5 kW CO_2 laser.

2.11.4 Resistance Welding

Resistance welding can produce excellent joint strengths in the high strength heat treatable aluminum alloys. Resistance welding is normally used for fairly thin sheets where joints are produced with no loss of strength in the base metal and without the need for filler metals. In resistance welding, the faying surfaces are joined by heat generated by the resistance to the flow of current through workpieces held together by the force of water-cooled copper electrodes. A fused nugget of weld metal is produced by a short pulse of low voltage, high amperage current. The electrode force is maintained while the liquid metal rapidly cools and solidifies. In spot welding, as shown in Fig. 2.41, the two parts to be joined are pressed together between two electrodes during welding. In seam welding, the two electrodes are replaced with wheels. While the 2XXX and 7XXX alloys are easy to resistance weld, they are more susceptible to shrinkage cracks and

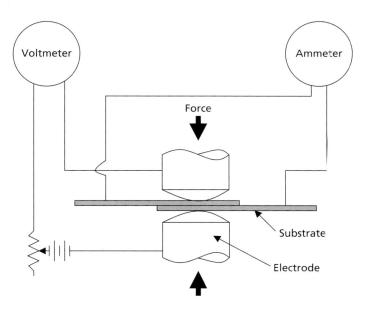

Fig. 2.41. Resistance Spot Welding

porosity than lower strength aluminum alloys.[26] Alclad materials are also more difficult to weld due to the lower electrical resistance and higher melting point of the clad layers. Removal of the surface oxide is important to produce good weld quality. Both mechanical and chemical methods are used, with surface preparation being checked by measuring the surface resistivity.

2.11.5 Friction Stir Welding

A new welding process which has the potential to revolutionize aluminum joining has been developed by The Welding Institute in Cambridge, UK.[27] Friction stir welding is a solid state process that operates by generating frictional heat between a rotating tool and the workpiece, as shown schematically in Fig. 2.42. The welds are created by the combined action of frictional heating and plastic deformation due to the rotating tool.

A tool with a knurled probe of hardened steel or carbide is plunged into the workpiece creating frictional heating that heats a cylindrical column of metal around the probe, as well as a small region of metal underneath the probe. As shown in Fig. 2.43, a number of different tool geometries have been developed, which can significantly affect the quality of the weld joint. The threads on the probe cause a downward component to the material flow, inducing a counterflow extrusion toward the top of the weld, or an essentially circumferential flow around the probe.[28] The rotation of the probe tool stirs the material into a plastic

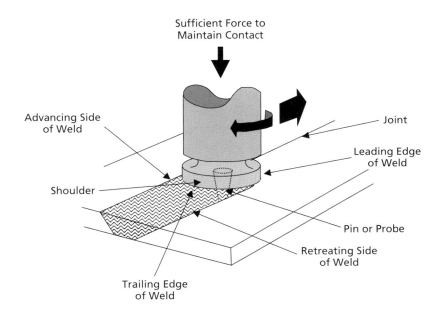

Fig. 2.42. Friction Stir Welding Process[27]

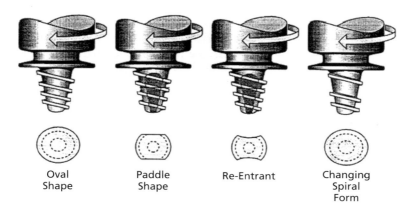

Fig. 2.43. Sample Friction Stir Welding Tool Geometries
Source: TWI

state creating a very fine grain microstructural bond. The tool contains a larger diameter shoulder above the knurled probe which controls the depth of the probe and creates additional frictional heating on the top surface of the workpiece. It also prevents the highly plasticized metal from being expelled from the joint. Prior to welding, the workpieces have to be rigidly fixed with the edges butted to each other and must have a backing plate to withstand the downward forces exerted by the tool. A typical welding operation is shown in Fig. 2.44.

The larger the diameter of the shoulder, the greater is the frictional heat it can contribute to the process. Once the shoulder makes contact, the thermally softened metal forms a frustum shape corresponding to the tool geometry with the top portion next to the shoulder being wider and then tapering down to the probe diameter. The maximum temperature reached is of the order of 0.8 of the melting temperature.[29] Material flows around the tool and fuses behind it. As the tool rotates, there is some inherent eccentricity in the rotation that allows the hydromechanically incompressible plasticized material to flow more easily around the probe.[27] Heat transfer studies[30] have shown that only about 5% of the heat generated in friction stir welding flows into the tool with the rest flowing into the workpieces; therefore, the heat efficiency in FSW is very high relative to traditional fusion welding processes where the heat efficiency is only about 60–80%.

Once the tool has penetrated the workpieces, the frictional heat caused by the rotating tool and rubbing shoulder results in frictional heating and plasticization of the surrounding material. Initially, the material is extruded at the surface but as the tool shoulder contacts the workpieces, the plasticized metal is compressed between the shoulder, workpieces, and backing plate. As the tool moves down the joint, the material is heated and plasticized at the leading edge of the tool

Aluminum

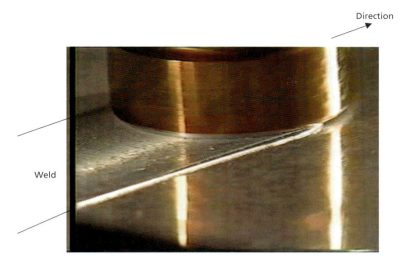

Fig. 2.44. Friction Stir Welding
Source: ESAB Welding Equipment AB

and transported to the trailing edge of the probe, where it solidifies to form the joint.[31]

The advantages of friction stir welding include (1) the ability to weld butt, lap and T joints, (2) minimal or no joint preparation, (3) the ability to weld the difficult to fusion weld 2XXX and 7XXX alloys, (4) the ability to join dissimilar alloys, (5) the elimination of cracking in the fusion and HAZs, (6) lack of weld porosity, (7) lack of required filler metals, and (8) in the case of aluminum, no requirement for shielding gases.[30] In general, the mechanical properties are better than for many other welding processes. For example, the static properties of 2024-T351 are between 80 and 90% of the parent metal, and the fatigue properties approach those of the parent metal.[32] In a study of lap shear joints, friction stir welded joints were 60% stronger than comparable riveted or resistance spot welded joints.[33]

The weld joint does not demonstrate many of the defects encountered in normal fusion welding and the distortion is significantly less. A typical weld joint, as shown in Fig. 2.45, contains a well-defined nugget with flow contours that are almost spherical in shape but are somewhat dependent on tool geometry. TWI has recommended the microstructural classification shown in Fig. 2.45 be used for friction stir welds. The fine-grained recrystallized weld nugget and the adjacent unrecrystallized but plasticized material is referred to as the thermomechanically affect zone (TMAZ); therefore, the TMAZ results from both thermal exposure and plastic deformation and extends from the width of the shoulder at the top surface to a narrower region on the backside. A series

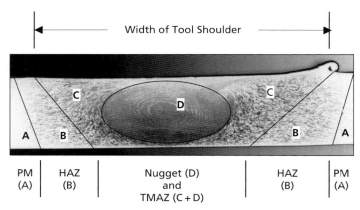

Fig. 2.45. Friction Stir Fusion Weld[29]

of concentric rings, called onion rings, are frequently observed within the weld nugget, possibly as a result of the swirling motion of the plasticized material behind the advancing tool probe. The unrecrystallized portion of the TMAZ, which has also undergone thermal exposure and plastic deformation, has a grain size similar to that of the parent metal. The HAZ is typically trapezoidal in shape for a single pass weld with a greater width at the tool shoulder due to the heat generated between the shoulder and the top of the workpieces. The HAZ results primarily from thermal exposure with little or no plastic deformation present.

Grain sizes in the weld nugget are extremely fine, much finer than in the base metal. Some preliminary investigations have indicated that the extremely fine grain size may promote superplasticity in friction stir "processed" material, possibly leading to some interesting applications such as friction stir welded/SPF aluminum unitized structure.[34–36] On the other hand, under certain conditions, it has been reported that abnormal grain growth can result in the center nugget area and on the surface under the shoulder during post-weld solution heat treatment. However, this may be a function of welding parameters. In another study,[37] it was reported that both tool rotation and feed rate influenced grain growth during heat treatment, with tool rotational speed being the predominate variable. Higher rotational speeds (i.e., 1000 rpm vs. 500 rpm) and slower feeds (i.e., 0.08 in./s vs. 0.20 in./s) decreased abnormal grain growth in 2095 that was subsequently solution heat treated and naturally aged to the T4 condition. It should be noted that speeds and feeds will influence the heat input during welding; at a constant

Aluminum

feed rate, the amount of heat input will increase with increasing tool rotational speed, while at a constant tool rotational speed, the heat input will decrease with slower feed rates.

The friction stir welding process has already been adapted to a number of industrial applications. In 1999, the fuel tanks on the Boeing Delta II rocket were launched with friction stir longitudinal welds. Based on this early success, Boeing purchased the large friction stir welding unit shown in Fig. 2.46 to weld

Fig. 2.46. Friction Stir Welding Applications for Delta
Source: ESAB Welding Equipment AB and The Boeing Company

Fig. 2.47. Large Friction Stir Welder Being Used in Marine Industry
Source: ESAB Welding Equipment AB

Delta IV fuel tanks. The machine is equipped with one milling head and one welding head traveling on a vertical main beam. The parts to be welded are loaded on an indexing fixture, the edges are milled, welding is carried out, and the final length is then milled. Tanks up to 40 ft long by 16 ft in diameter can be welded. Another early adapter of friction stir welding has been the marine industry as shown in the welded ship skin in Fig. 2.47. Construction of high speed trains is another important potential application for friction stir welding.

2.12 Chemical Finishing

Aluminum forms a natural alumina (Al_2O_3) oxide on its surface that helps prevent corrosion. However, this oxide (0.0002–0.0006 in. thick) is thin and is also a poor base for paint. Two types of coatings, chemical conversion coatings and anodizing, are used to form a more uniform and thicker oxide for enhanced corrosion protection. Chemical conversion coatings produce a porous and absorptive oxide (0.002–0.003 in. thick) that is very uniform and morphologically tailored to bond well with paint primers. The oxides are chromate or phosphate based which further aids in corrosion protection.

Anodizing is an electrolytic process that produces thicker (0.002–0.005 in.) and more durable oxides than those produced by conversion coatings; therefore, it provides better corrosion resistance. Both sulfuric and chromic acid baths are used along with an electrical current to deposit a porous oxide layer on the surfaces. After anodizing, the pores in the oxide film are sealed by placing the part in slightly acidified 180–200° F water. This sealing treatment converts the aluminum oxide to aluminum monohydrate. When the aluminum oxide is hydrated, it expands to fill the pores.

Summary

Aluminum alloys have played a critical role in modern aviation. Aluminum is a relatively low cost, lightweight metal that can be heat treated to fairly high strength levels, and its ease of fabrication leads to low costs. It has a high strength-to-weight ratio and performs well from cryogenic temperatures to moderate temperatures. The disadvantages of high strength aluminum alloys include a low modulus of elasticity, rather low elevated temperature capability, and susceptibility to corrosion.

The wrought heat treatable alloys can be strengthened by precipitation hardening to develop quite high strength levels. These alloys include the aluminum–copper series (2XXX), the aluminum–silicon–magnesium series (6XXX), the aluminum–zinc series (7XXX), and some of the 8XXX Al-Li alloys. The 2XXX and 7XXX alloys, which develop the highest strength levels, are the dominate alloys for metallic airframe components. Due to their superior damage tolerance and good resistance to fatigue crack growth, the 2XXX alloys are used for fuselage skins and lower wing skins on commercial aircraft, while the 7XXX alloys are used for upper wing skins, where strength is the primary design driver.

Both the 2XXX and 7XXX alloys are very responsive to precipitation hardening. Precipitation hardening consists of three steps: (1) solution heat treating, (2) rapidly quenching to a lower temperature, and (3) aging. The 2XXX alloys are often used in the T3 or T4 naturally aged condition, while the 7XXX are artificially aged to the peak strength T6 or the overaged T7 conditions. Overaging to the T7 condition improves corrosion resistance and produces higher fracture toughness at some sacrifice in strength properties.

Most aluminum forgings are produced in closed dies. Closed die forgings are produced by forging ingots, plates, or extrusions between a matched set of dies. Progressive sets of dies gradually shape the part to near net dimensions. High strength alloys are normally forged in hydraulic presses because more uniform structures can be generated with a slow controlled squeezing pressure.

Aluminum alloys are highly formable at room temperature due to their FCC structure and their relatively slow rate of work hardening. The choice of the temper for forming depends on the severity of the forming operation and the alloy being formed. The 2XXX and 7XXX high strength alloys can be readily formed at room temperature, provided the alloy is in either the O or W temper.

Due to their lower properties and higher variability than wrought product forms, aluminum castings are not used for primary structural applications, but for lightly loaded secondary structures, they can offer significant cost savings by reducing part count and the associated assembly cost. Aluminum castings can be produced using a variety of casting processes with sand casting the most prevalent method in the aerospace industry.

Aluminum exhibits extremely good machinability with cutting speeds approaching as high as 1000 sfm common. With the implementation of high speed machining, even higher metal removal rates are obtainable; however, another driver for developing high speed machining is the ability to machine extremely thin walls and webs. This allows the design of weight competitive high speed machined assemblies in which parts that were formally assembled with mechanical fasteners can now be machined from a single or several blocks of aluminum plate.

The 2XXX and 7XXX alloys can be rather difficult to weld with weld cracking and distortion creating problems. However, a number of these alloys are weldable by conventional methods such as GMAW, GTAW, and resistance welding. For critical load-bearing applications, VPPA is often specified. A relatively new process, friction stir welding, offers some distinct advantages over other fusion welding processes including the ability to weld the difficult to fusion weld 2XXX and 7XXX alloys.

Recommended Reading

[1] Metallic Materials Properties Development and Standardization, U.S. Department of Transportation, Federal Aviation Agency, DOT/FAA/AR-MMPDS-01, January 2004.
[2] Davis, J.R., *ASM Specialty Handbook: Aluminum and Aluminum Alloys*, ASM International, 1993.
[3] *ASM Handbook Volume 2: Properties and Selection: Nonferrous Alloys and Special Purpose Materials*, ASM International, 1990.
[4] *ASM Handbook Volume 4: Heat Treating*, ASM International, 1991.
[5] *ASM Handbook Volume 6: Welding, Brazing, and Soldering*, ASM International, 1993.
[6] *ASM Handbook Volume 14: Forming and Forging*, ASM International, 1988.
[7] *ASM Handbook Volume 15: Casting*, ASM International, 1988.
[8] *ASM Handbook Volume 16: Machining*, ASM International, 1990.

References

[1] Smith, B., "The Boeing 777", *Advanced Material & Process*, September 2003, pp. 41–44.
[2] Kaufman, J.G., "Aluminum Alloys", in *Handbook of Materials Selection*, John Wiley & Sons, Inc., 2002, pp. 89–134.
[3] Smith, W.F., "Precipitation Hardening of Aluminum Alloys", ASM Course *Aluminum and Its Alloys*, Lesson 9, ASM International, 1979.
[4] Kissell, J.R., "Aluminum Alloys", in *Handbook of Materials for Product Design*, McGraw-Hill, 2001, pp. 2.1–2.178.
[5] Williams, J.C, Starke, E.A., "Progress in Structural Materials for Aerospace Systems", *Acta Materialia*, Vol. 51, 2003, pp. 5775–5799.
[6] Immarigeon, J.-P, Holt, R.T., Koul, A.K., Zhao, L., Wallace, W., Beddoes, J.C. "Lightweight Materials for Aircraft Applications", *Materials Characterization*, Vol. 35, 1995, pp. 41–67.
[7] Starke, E.A., Staley, J.T., "Application of Modern Aluminum Alloys to Aircraft", *Progress Aerospace Science*, Vol. 32, 1996, pp. 131–172.
[8] "Aluminum Casting Principles", ASM Course *Aluminum and Its Alloys*, Lesson 5, ASM International, 1979.

[9] Sanders, R.E., "Technology Innovation in Aluminum Products", *Journal of Metals*, February 2001, pp. 21–25.
[10] "Heat Treating of Aluminum Alloys", in *Heat Treating*, ASM Handbook, Vol. 4, ASM International, 1991.
[11] Kuhlman, G.W., "Forging of Aluminum Alloys", in *Forming and Forging*, ASM Handbook, Vol. 14, ASM International, 1988.
[12] "Forging of Aluminum Alloys", in *Forging and Forming*, ASM Handbook, Vol. 14, ASM International, 1988.
[13] Bowden, D.M., Sova, B.J., Biesiegel, A.L., Halley, J.E., "Machined Component Quality Improvements Through Manufacturing Process Simulation", SAE 2001-01-2607, 2001 Aerospace Conference, 10–14 September 2001, Seattle, Washington.
[14] "Forming of Aluminum Alloys", in *Forging and Forming*, ASM Handbook, Vol. 14, ASM International, 1988.
[15] Yue, S., Durham, S., "Titanium", in *Handbook of Materials for Product Design*, McGraw-Hill, 2001, pp. 3.1–3.61.
[16] Ashby, M.F., Verrall, R.A., "Diffusion Accommodated Flow and Superplasticity", *Acta Metallurgia*, Vol. 21, 1973, p. 149.
[17] Hamilton, C.H., Ghosh, A.K., "Superplastic Sheet Forming", in *Forming and Forging*, ASM Handbook, Vol. 14, ASM International, 1988.
[18] Rooy, E.L., "Aluminum and Aluminum Alloys", in *Casting*, ASM Handbook, Vol. 15, ASM International, 1988.
[19] Davis, J.R., *ASM Specialty Handbook: Aluminum and Aluminum Alloys*, ASM International, 1993, p. 108.
[20] Koelsch, J.R., "High-Speed Machining: A Strategic Weapon", *Machine Shop Guide*, November 2001.
[21] Martukanitz, R.P., "Selection and Weldability of Heat-Treatable Aluminum Alloys", in *Welding, Brazing and Soldering*, ASM Handbook, Vol. 6, ASM International, 1993.
[22] Mendez, P.F., Eager, T.W., "Welding Processes for Aeronautics", *Advanced Materials & Processes*, May 2001, pp. 39–43.
[23] Schubert, E., Klassen, M., Zerner, I., Waltz, C., Sepold, G., "Light-Weight Structures Produced by Laser Beam Joining for Applications in Automobile and Aerospace Industry", *Journal of Materials Processing Technology*, Vol. 115, 2001, pp. 2–8.
[24] Li, Z., Gobbi, S.L., "Laser Welding for Lightweight Structures", *Journal of Materials Processing Technology*, Vol. 70, 1997, pp. 137–144.
[25] Behler, K., Berkmanns, J., Ehrhardt, A., Frohn, W. "Laser Beam Welding of Low Weight Materials and Structures", *Materials & Design*, Vol. 18, Nos 4/6, 1998, pp. 261–267.
[26] Mathers, G., Chapter 8 "Other Welding Processes", in *The Welding of Aluminum and Its Alloys*, Woodhead Publishing Limited and CRC Press, 2002.
[27] Thomas, W.M., Nicholas, E.D., "Friction Stir Welding for the Transportation Industries", *Materials & Design*, Vol. 18, Nos 4/6, 1997, pp. 269–273.
[28] Fonda, R.W., Bingert, J.F., Colligan, K.J., "Development of Grain Structure During Friction Stir Welding", *Scripta Materialia*, Vol. 51, 2004, pp. 243–248.
[29] Bhadeshia, H.K.D.H., "Joining of Commercial Aluminum Alloys", *Proceedings of the International Conference on Aluminum*, 2003, pp. 195–204.
[30] Chao, Y.J., Qi, X., Tang, W., "Heat Transfer in Friction Stir Welding-Experimental and Numerical Studies", *Transactions of the ASME*, Vol. 125, 2003, pp. 138–145.
[31] Bradley, G.R., James, M.N., "Geometry and Microstructure of Metal Inert Gas and Friction Stir Welded Aluminum Alloy 5383-H321", 2000, Published on the Internet.
[32] Williams, S.W., "Welding of Airframes using Friction Stir", *Air & Space Europe*, Vol. 3, No. 3/4, 2001, pp. 64–66.

[33] Cederqvist, L., Reynolds, A.P., "Factors Affecting the Properties of Friction Stir Welded Aluminum Lap Joints", Reprint from *The Welding Journal*.

[34] Charit, I., Mishra, R.S., "High Strain Superplasticity in Commercial 2024 Al Alloy via Friction Stir Processing", *Materials and Engineering*, A359, 2003, pp. 290–296.

[35] Ma, Z.Y., Mishra, R.S., Mahoney, M.W., Grimes, R. "High Strain Rate Superplasticity in Friction Stir Processed Al-Mg-Zr Alloy", *Materials Science and Engineering*, A351, 2003, pp. 148–153.

[36] Charit, I., Mishra, R.S., Mahoney, M.W., "Multi-Sheet Structures in 7475 Aluminum by Friction Stir Welding in Concert with Post-Weld Superplastic Forming", *Scripta Materialia*, Vol. 47, 2002, pp. 631–636.

[37] Moataz, M.A., Salem, H.G., "Friction Stir Welding Parameters: A Tool for Controlling Abnormal Grain Growth During Subsequent Heat Treatment", *Materials Science and Engineering*, A391, 2005, pp. 51–59.

Chapter 3

Magnesium and Beryllium

Considering their location on the periodic table of elements, one would think that both magnesium and beryllium would play major roles as aerospace structural materials. However, both of these extremely lightweight metals have serious limitations that make them only minor players. Magnesium alloys usually compete with aluminum alloys for structural applications. Compared to high strength aluminum alloys, magnesium alloys are not as strong (tensile strength of 20–50 ksi vs. 40–80 ksi) and have a lower modulus of elasticity (6.5 msi vs. 10–11 msi). However, magnesium is significantly lighter (0.063 vs. 0.100 lb/in.^{3}) and therefore its alloys are competitive on a specific strength and modulus basis. Magnesium alloys, with their hexagonal close-packed structure, must usually be formed at elevated temperatures, while aluminum can be readily formed at room temperature. In addition, magnesium alloys are normally more expensive than comparable aluminum alloys. However, the biggest obstacle to the use of magnesium alloys is their extremely poor corrosion resistance. Magnesium occupies the highest anodic position on the galvanic series, and, as such, there is always the strong potential for corrosion as shown in the example of Fig. 3.1. However, some of the newer alloys have much better corrosion resistance than the older alloys. As shown in Fig. 3.2, some of the newer cast alloys approach the corrosion resistance of competing aluminum casting alloys. Magnesium alloys do have very good damping capacity and castings have found application in high vibration environments, such as helicopter gear boxes.[1]

Beryllium is also a very lightweight metal (0.068 lb/in.^{3}) that has good strength properties and a very high tensile modulus (44 msi). These properties, combined with its attractive electrical and thermal properties, have led to its use in high value aerospace electronic and guidance system applications. However, beryllium must be processed using powder metallurgy that makes it costly. Like magnesium, its hexagonal close-packed (HCP) crystalline structure greatly impairs its formability. Finally, beryllium powder and dust are toxic, which

Fig. 3.1. Severely Corroded Magnesium Part

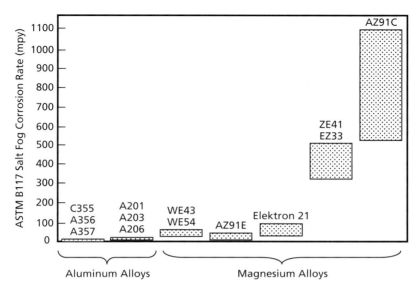

Fig. 3.2. Corrosion Comparison – Aluminum and Magnesium Casting Alloys[1]

further increases its cost through the requirement for a controlled manufacturing environment.

MAGNESIUM

The use of magnesium alloys in aerospace is fairly limited. In fact, the use of magnesium alloys in aerospace has decreased steadily since World War II. Most current applications are sand cast alloys for secondary structural applications, such as helicopter gear boxes. Low weight and good damping properties are the motivation for their selection.

3.1 Magnesium Metallurgical Considerations

Pure magnesium has a melting point of 1202° F and a HCP crystalline structure. Since the HCP crystalline structure restricts slip to the basal planes, magnesium is difficult to plastically deform at room temperature, i.e. the work hardening rate at room temperature is high and the ductility is low. At elevated temperatures, other slip planes become operative so magnesium alloys are normally formed at temperatures greater than 440° F, usually in the range of 650–950° F. Another consequence of the HCP structure is the mechanical property anisotropy (or directionality) in cold rolled sheet due to its crystallographic texture. For example, the yield strength in compression of wrought products is only about 40–70% of that in tension.[2] Because of the difficulty of cold forming magnesium alloys, castings are the more prevalent product form than wrought products. One consequence of its rather low melting point is its susceptibility to creep at moderately elevated temperatures. However, alloys have been developed with improved creep performance.

Aluminum and zinc provide solid solution strengthening. Aluminum, in addition to providing strength and hardness, widens the freezing range and makes the alloy easier to cast. Aluminum in amounts greater than 6% promotes precipitation hardening. Zinc is the second most important alloying element. Zinc helps to refine the precipitate in aluminum-containing alloys. Zinc can also be used in combination with zirconium, rare earths (RE), or thorium to produce precipitation hardening alloys. Silver also improves the response to precipitation hardening. Zirconium is used in casting alloys for grain refinement. The powerful grain refining effect of zirconium is shown in Fig. 3.3. However, zirconium is not used in alloys containing aluminum because brittle compounds are formed. Manganese additions improve the corrosion resistance in sea water by removing iron from solution. Silicon increases fluidity for casting alloys but decreases the corrosion resistance if iron is present. Thorium and yttrium improve creep resistance; however, the use of thorium, which is mildly radioactive, has decreased due to increasing regulations on its use. Although much less soluble than aluminum and zinc, the RE elements are potent solid solution strengtheners. The rare earths are usually added as natural mixtures of either mischmetal or as didymium. Mischmetal contains about 50% cerium with the remainder as mainly lanthanum and neodymium, while didymium contains

Fig. 3.3. Grain Refinement with Zirconium[1]

approximately 85% neodymium and 15% praseodymium. The impurity elements nickel, iron, and copper must be held to low levels to minimize corrosion.

Magnesium alloys are produced in both the wrought and cast conditions. Some alloys are strengthened by cold working while others can be precipitation hardened by heat treatment. The alloys themselves can be divided into two broad classes: those that contain 2–10% aluminum with additions of zinc and manganese, and those containing zinc with additions of RE metals, thorium, silver, and zirconium for strength, creep resistance, and grain refinement. The tensile properties of magnesium alloys generally range from 10 to 50 ksi yield strength and 20–55 ksi tensile strength with elongations of 1–15%.[3]

3.2 Magnesium Alloys

Magnesium alloys are designated by a combination of letters and numbers. The first two letters indicate the two major alloying elements in the alloy, while the following two numbers give the approximate amounts for the first and second alloying elements respectively. For example, the alloy AZ91 contains approximately 9% aluminum and 1% zinc. There is also a letter that follows the basic alloy designation; A is the original composition, B is the second modification, C is the third modification, D indicates a high purity version, and E is a corrosion resistant composition. In our example, AZ91C would indicate the third modification to AZ91. The magnesium alloy designation system is shown in Table 3.1 and the composition of a number of magnesium alloys are given in Table 3.2.

The heat treatment temper designations for magnesium alloys are the same as for aluminum alloys. This designation system is again shown in Table 3.3; however, since most aerospace applications use cast magnesium alloys, the most prevalent tempers are the T4, T5, and T6 tempers. The typical tensile properties of a number of wrought and cast alloys are given in Table 3.4.

3.2.1 Wrought Magnesium Alloys

Wrought magnesium alloys are available as bars, billets, sheet, plate, and forgings. The principal sheet and plate alloy is AZ31. Since AZ31 is strengthened by a combination of solid solution strengthening, grain size control, and cold working, it is not really a candidate for aerospace structural applications due to the possibility of softening (i.e., annealing) at elevated temperature. Alloy AZ31 is available in several tempers but all are limited to approximately 200° F. For higher temperature applications, the thorium-containing alloys HK31 and HM21 are available. For maximum creep resistance, alloy HK31 requires a T6 heat treatment, while HM21 is cold worked prior to aging (T8 temper). In addition, due to the HCP structure, all but the most mild forming operations must be done at elevated temperature.

Table 3.1 ASTM Designation for Magnesium Alloys[4]

First Part	Second Part	Third Part
Indicates the two principal alloying elements	Indicates the amounts of the two principal elements	Distinguishes between different alloys with the same percentages of the two principal alloying elements
Consists of two code letters representing the two main alloying elements arranged in order of decreasing percentage (or alphabetically if percentages are equal)	Consists of two numbers corresponding to rounded-off percentages of the two main alloying elements and arranged in same order as alloy designations in first part	Consists of a letter of the alphabet assigned in order as compositions become standard
A – **A**luminum C – **C**opper E – Rare **E**arth H – T**H**orium K – Zir**K**onium M – **M**anganese Q – Silver (**Q**uick Silver) S – **S**ilicon T – **T**in W – yttrium Z – **Z**inc	Whole numbers	A – First composition registered with ASTM B – Second composition registered with ASTM C – Third composition registered with ASTM D – High purity registered with ASTM E – High corrosion registered with ASTM X1 – Not registered with ASTM

For extrusions, one of the Al–Zn alloys is often selected such as AZ31, AZ61, or AZ80. A number of alloys, such as AZ80 and ZK60, respond to precipitation hardening. Since the extrusion process is carried out at approximately the solution heat treating temperature, and the extrusion cools fairly rapidly in air, it is only necessary to age these alloys to produce the T5 temper. For example, ZK60, in the T5 condition, is often specified where higher strength and toughness is required. Other high strength extrusion alloys include ZK61 and ZCM711. For high temperature applications, the alloys HK31 and HM21 can be specified. An important factor in extrusion is symmetry, preferably around both axes. The optimum width-to-thickness ratio for magnesium extrusions is normally less than 20.

Forging alloys include AZ31, AZ61, AZ80, and ZK60. AZ31 can be hammer forged whereas the others are usually press forged. Magnesium alloys are heated to 650–950° F for forging. ZK60 has slightly better forgeability than the other alloys. Although forgings have the highest strengths of the various magnesium product forms, they are sometimes specified because of their pressure tightness, machinability, and lack of warpage rather than for their strength.

Due to anisotropy, or texture, produced by mechanical working, the compression yield strength of wrought magnesium alloys can be appreciably less than the tensile yield strength. The compression yield strength varies from about

Table 3.2 Nominal Compositions for Select Magnesium Alloys

Alloy	Nominal Composition (% by Weight)					
	Al	Zn	Mn	Zr	RE	Other
Wrought Alloys						
AZ31	3.0	1.0	0.30	–	–	–
AZ61	6.5	1.0	0.30	–	–	–
AZ80	8.5	0.5	0.20	–	–	–
HK31	–	–	–	0.7	–	3.25 Th
HM21	–	–	0.80	–	–	2.0 Th
ZCM711	–	6.5	0.75	–	–	1.25 Cu
ZK60	–	5.5	–	0.45	–	–
ZK61	–	6.0	–	0.8	–	–
Casting Alloys						
AM100	10.0	–	0.10	–	–	–
AZ63	6.0	3.0	0.30	–	–	–
AZ81	7.6	0.7	0.30	–	–	–
AZ91	8.7	0.7	0.30	–	–	–
AZ92	9.0	2.0	0.10	–	–	–
EZ33	–	2.7	–	0.7	3.3	–
QE22	–	–	–	0.7	2.5	2.5 Ag
QH21	–	–	–	0.7	1.0	2.5 Ag, 1.0 Th
WE43	–	–	–	0.7	3.4	4.0 Y
WE54	–	–	–	0.7	3.0	5.1 Y
ZE41	–	4.2	0.15	0.7	1.2	–
ZH62	–	5.7	–	0.7	–	1.8 Th
ZK51	–	4.6	–	0.7	–	–
ZK61	–	6.0	–	0.7	–	–
Electron 21	–	0.3	–	0.5	4.0	–

0.4 to 0.7 of the tension yield strength. Since castings do not develop texture, the compression yield strength of castings is about equal to the tensile yield strength.

3.2.2 Magnesium Casting Alloys[3]

Magnesium castings are used in aerospace secondary structural applications because of their low weight and good damping characteristics. However, since magnesium alloys are subject to galvanic corrosion, proper surface treatments and coatings must be used to prevent corrosion. Chemical composition developments have also significantly helped in reducing their corrosion potential.

Mg–Al and Mg–Al–Zn casting alloys. Aluminum is alloyed with magnesium because it increases strength, castability, and corrosion resistance. Since aluminum has a maximum solid solubility in magnesium of 12.7% at 808° F that decreases to about 2% at room temperature, it would at first appear that this system could

Table 3.3 Temper Designations for Magnesium Alloys

Designation	Explanation
F	As-Fabricated
O	Annealed, Recrystallized (Wrought Products Only)
H	Strain Hardened (Wrought Products Only)
H1	Strain Hardened Only
H2	Strain Hardened and Partially Annealed
H3	Strain Hardened and Stabilized
W	Solution Heat Treated
T	Solution Heat Treated to Produce Stable Tempers
T2	Annealed (Cast Products Only)
T3	Solution Heat Treated, Cold Worked, Naturally Aged
T4	Solution Heat Treated, Naturally Aged
T5	Artificially Aged
T6	Solution Heat Treated, Artificially Aged
T7	Solution Heat Treated, Overaged
T8	Solution Heat Treated, Cold Worked, Artificially Aged
T9	Solution Heat Treated, Artificially Aged, Cold Worked
T10	Artificially Aged, Cold Worked

Table 3.4 Typical Mechanical Properties of Representative Magnesium Alloys[4]

Alloy	Temper	UTS (ksi)	YS (ksi)	Elongation (%)
Sand and Permanent Mold Castings				
AM100A	T61	40	22	1
AZ91C and E	T6	40	21	6
AZ92A	T6	40	22	3
EZ33A	T5	23	16	2
QE22A	T6	38	28	3
WE43A	T6	36	24	2
ZE41A	T5	30	20	3.5
ZK-61A	T6	45	28	10
Extrusions				
AZ31B-C	F	38	29	15
AZ61A	F	45	33	16
ZK60A	T5	53	44	11
Sheet and Plate				
AZ31B	H24	42	32	15

be strengthened by precipitation hardening. However, the resulting precipitate is rather coarse and causes only moderate hardening. When zinc is added to the composition, it refines the precipitate and increases the strength by a combination of solid solution strengthening and precipitation hardening, as shown in Fig. 3.4. Even then, the degree of strengthening is minimal compared to that achievable in the heat treatable aluminum alloys. With tensile strengths in the range of

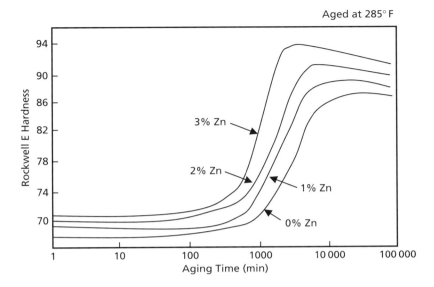

Fig. 3.4. Aging Curves for Mg–9%Al Alloy with Various Zn Additions

31–35 ksi and elongations of 1–8%, the Mg–Al–Zn alloys are not particularly strong or ductile but have low densities and are easy to cast.

More corrosion resistant Mg–Al–Zn alloys were developed in the mid-1980s by using higher purity starting materials and by limiting the amounts of iron (<0.005%), nickel (<0.001%), and copper (<0.015%). The low levels of nickel and copper are controlled by the purity of the starting materials, while the low iron levels are controlled with $MnCl_2$ additions. For example, the high purity alloy AZ91E, due to its lower iron content, has improved corrosion resistance compared to the earlier alloy AZ91C. Magnesium alloys are limited to a total aluminum and zinc level of less than 10%; at higher levels, the ductility is drastically reduced due to the formation of intermetallic compounds. Thus, if the zinc content of an Mg–Al–Zn alloy is raised to 3%, the aluminum content must be reduced to about 6%, as in AZ63. However, as the zinc contents increase in the Mg–Al–Zn alloys, there is an increase in microporosity and shrinkage. The Mg–Al and Mg–Al–Zn alloys are limited to usage temperatures of about 200°F.

Mg–Zn–Zr and Mg–Zn–Rare Earth–Zr casting alloys. Alloys, such as ZK51 and ZK61, were developed as sand casting alloys by combining 5–6% zinc for increased strength with about 0.7% zirconium for grain refinement. Although these are relatively high strength alloys, they are not widely used because of their susceptibility to microporosity during casting, and they cannot be weld repaired due to their high zinc contents. RE additions to the Mg–Zn–Zr casting alloys improved their castability due to the formation of low melting point eutectics that form at the grain boundaries during solidification, which tends to suppress microporosity and hot cracking, while improving strength and creep resistance. However, the room temperature tensile strengths of EZ33-T5 of 20 ksi and ZE41-T5 of 29 ksi are low due to the removal of zinc from solid solution to form the Mg–Zn–RE phases in the grain boundaries. At low stress levels, these alloys do have respectable creep resistance up to 320° F.

Mg–Ag–Rare Earth casting alloys. The addition of 2.5% silver and 2.5% REs produces better precipitation hardening with good tensile properties up to 400° F in the alloy QE22, which has tensile strength of 35 ksi in the T6 condition. Casting alloys with about 4–5% yttrium have also been developed which have better elevated temperature properties. For example, alloy WE43 has a room temperature tensile strength of 36 ksi when heat treated to the T6 condition. This alloy maintains a tensile strength of 36 ksi after long-term aging (5000 h) at 400° F. The effect of 400° F exposure on the room temperature strength of WE43 is shown in Fig. 3.5. A relatively new alloy, Elektron 21 as specified in

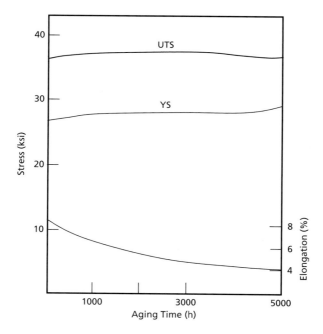

Fig. 3.5. Effect of 400° F Aging on Tensile Properties of WE43A-T6
Source: Magnesium Electron, Ltd

AMS 4429, with design data available in Metallic Materials Properties Development and Standardization (MMPDS), offers many of the advantages of the WE43; however, the cost is lower and the castability is better. Instead of using yttrium, neodymium and gadolinium are used along with zinc and zirconium.

3.3 Magnesium Fabrication

Metallic magnesium can be produced by several extractive metallurgy processes; however, the most widely used process involves precipitating magnesium in dolomite $[CaMg(CO_3)_2]$ and seawater as insoluble magnesium hydroxide $[Mg(OH)_2]$, which is then treated with hydrochloric acid to produce magnesium chloride. The $MgCl_2$ is fed into electrolytic cells where electricity is used to convert it to magnesium metal and chlorine gas.[3]

3.3.1 Magnesium Forming[5]

Wrought magnesium alloys, like other alloys with the HCP structure, are much more formable at elevated temperatures than at room temperature. Wrought alloys are usually formed at elevated temperatures; room temperature forming is used only for mild deformations around generous radii. Minimum bend radii for annealed sheet formed at room temperature are $5-10T$, and $10-20T$ for work hardened sheet, where T is the sheet thickness.

Forming magnesium alloys at elevated temperatures has several advantages: (1) forming operations can usually be conducted in one step without the need for intermediate anneals; (2) parts can be made to closer tolerances with less springback; and (3) hardened steel dies are not necessary for most forming operations. The approximate formability of magnesium alloy sheet is indicated by its ability to withstand bending over a 90° mandrel without cracking. The formability depends on composition and temper, material thickness, and forming temperature. With correct temperatures and forming parameters, all magnesium sheet alloys can be deep drawn to about equal reductions.

Since magnesium is a rather soft metal, both the part and forming tools should be clean and free of scratches, and a forming lubricant, such as colloidal graphite, must be used. In most hot forming operations, both the sheet and the tools are heated. Acceptable heating methods include electric cartridge heaters embedded in dies, radiant heating, infrared, gas, and heat transfer fluids. Lubrication is more important in hot forming than cold forming, because the tendency of galling increases with increasing temperature.

For severe forming operations, the annealed O temper is preferred. Sheet in the partially hardened temper, such as H24, can be formed to a considerable extent but the time at temperature will cause softening and a reduction in properties. It should also be noted that the time at temperature is cumulative if multiple forming operations are involved. AZ31-H24 sheet is normally formed at temperatures less than 325°F to avoid excessive annealing and lower than desired properties.

For deep drawing operations, magnesium alloys can be cold drawn to a maximum reduction of 15–25% in the annealed condition. The cold drawability of AZ31-O is about 20%. Cold drawn parts are stress relieved at 300°F for 1 h after the final draw to eliminate the possibility of cracking from residual stresses. Although both hydraulic and mechanical presses can be used, hydraulic presses are preferred because they are slower and easier to control. Hot drawing has the advantage that the operation can usually be conducted in one step. For example, AZ31-O can be hot drawn up to 68% in a single operation.

Both magnesium sheet and extrusions can be stretch formed. Sheet is usually heated to 325–550°F and slowly stretched to the desired contour. AZ31-O sheet is usually stretch formed at 550°F without a change in mechanical properties, while AZ31-H24 is usually formed at 325°F for times less than 1 h to prevent an appreciable strength loss.

3.3.2 Magnesium Sand Casting[4]

Sand casting is the most economical method of producing low volume castings, which explains why it is often the process of choice for the aerospace industry. The reactivity of magnesium causes reactions between the liquid metal and water in the sand of "green" sand molds or oxygen in dry sand molds. These reactions cause a blackening of the casting skin to an appreciable depth with local porosity and gray oxide powder effects, called burning. To avoid these defects, which adversely affects strength, the sand is mixed with inhibitors such as 0.4–0.8% potassium fluoroborate or sodium silica fluoride. Magnesium casting alloys are normally melted in a low carbon steel crucible. The metal can be either poured from the steel crucible or transferred to a ladle for pouring. Molten magnesium alloys tend to oxidize and burn in air; therefore, molten surfaces must be protected from air. Although there are both flux and flux-less processes, the flux-less process is the most widely used. In the flux-less process, either a protective atmosphere of air/sulfur hexafluoride gas or air/carbon dioxide/sulfur hexafluoride gas mixture is used to eliminate the contamination problems inherent in solid chloride fluxes.

Grain refinement is an important aspect of magnesium alloy sand castings. The Mg–Al and Mg–Al–Zn alloys are usually grain refined by carbon inoculation with hexachloroethane or hexachlorobenzene compressed tablets. Grain refinement is attained due to the formation of aluminum carbide (Al_4C_3), which provides heterogeneous nucleation sites. The release of chlorine from the tablets also helps to remove hydrogen gas from the melt. Zirconium is added to the non–aluminum containing magnesium castings alloys to refine grain size. Zirconium cannot be used for grain refinement in the aluminum-containing magnesium alloys because it forms a brittle intermetallic alloy with aluminum. Along with grain refiners, manganese chloride ($MnCl_2$) is added to the melt to precipitate out iron impurities. After the alloy melt is stirred, the molten metal is allowed to stand for about 15 min to allow the Al–Mn–Fe intermetallic compounds to form and settle to the bottom of the melting pot. After pouring, the precipitated sludge of Al–Mn–Fe compounds is removed from the bottom of the melting pot. Grain size during casting is often checked by pouring a small bar along with the casting and then fracturing it and comparing the fracture appearance to a known set of samples that have different grain sizes. For aerospace castings, tensile test bars are also poured during casting. In addition, a complete destructive analysis of castings may be required on a sampling basis.

Gravity pouring is normally used for magnesium sand castings. The metal flows down a sprue and into the runner system. The sprues are tapered to help keep air from entering the casting. Gating practices are important because turbulence during pouring can result in surface oxides and dross being folded into the flowing metal causing inclusions or surface pitting. Screens or filters are frequently used to remove oxide films and dross. Advancements in sand composition and core manufacturing now allow quite complex magnesium alloy castings to be cast. They are frequently used for complex gear box housings that contain integral small diameter oil cooling holes surrounded by thin wall compartments. A typical cast gearbox housing is shown in Fig. 3.6.

Magnesium alloys are cast in permanent molds when the number of parts justifies the higher cost of the tooling. The mechanical properties of sand and permanent mold castings are comparable, but tighter dimensional control and better surface finishes result from permanent mold casting. As a result of the rather slow solidification rates in both sand and permanent mold castings, a heat treatment is usually required to obtain acceptable properties.

Fig. 3.6. Cast Magnesium Gearbox Housing
Source: Hitchcock Industries, Inc.

3.3.3 Magnesium Heat Treating[6]

Wrought magnesium alloys can be annealed by heating to 550–850° F for 1 to 4 h to produce the maximum anneal practical. Because most forming operations are done at elevated temperatures, the need for full annealing is less than with many other metals.

Stress relieving is used to remove or reduce residual stresses in wrought magnesium alloys produced by cold and hot working, shaping and forming, straightening, and welding. Stress relieving is generally conducted at 300–800° F for times ranging from 15 to 180 min. Castings are also stress relieved for a variety of reasons: (1) to prevent stress corrosion cracking for magnesium castings containing more than 1.5% aluminum, especially if the casting has been weld repaired; (2) to allow precision machining of castings to close dimensional tolerances; and (3) to avoid warpage and distortion in service. Although magnesium castings do not normally contain high residual stresses, even moderate residual stresses can cause large elastic strains due to magnesium's low modulus of elasticity. Residual stresses can result from non-uniform contraction during solidification, non-uniform cooling during heat treatment, machining operations, and weld repair. Welded Mg–Al–Zn castings that do not require solution heat treatment after welding should be stress relieved 1 h at 500° F to eliminate the possibility of stress corrosion cracking. Likewise, Mg–Al–Zn wrought alloys require stress relieving after cold forming to prevent stress corrosion cracking.

Although magnesium alloys do not attain the high strengths that aluminum alloys experience during precipitation hardening, there is some strength benefit to heat treatment for a number of the casting alloys. The solution heat treatment helps to reduce or eliminate the brittle interdendritic networks in the as-cast structure. Thus, solution-treated castings have better ductility than as-cast alloys with some increase in strength. The most prevalent precipitation hardening treatments for cast magnesium alloys are solution treating and naturally aging (T4), naturally aging only after casting (T5), and solution treating and artificially aging (T6).

For solution heat treatment, the parts are usually placed in a preheated furnace (500° F) and slowly heated to 735–980° F. Solution heat treating furnaces are usually electrically heated or gas-fired controlled to ±10° F and are equipped with fans to maximize circulation. To prevent excessive surface oxidation during solution heat treating, protective atmospheres of sulfur hexafluoride, sulfur dioxide, or carbon dioxide are used. The furnaces are also equipped to handle a fire in case the furnace malfunctions and overheating occurs. In the event of a fire, boron trifluoride gas can be pumped into the furnace. Although there are exceptions, slow heating to the solution treating temperature is recommended to avoid melting of eutectic compounds with the subsequent formation of grain boundary voids. The parts are held at the solution heat treating temperature for times in the range of 16–24 h. These hold times are long because the solution treatment also serves the purpose of homogenizing the cast structure. Castings often require support fixtures during solution heat treating to prevent them from sagging under their own weight. Some magnesium alloys are subject to excessive grain growth during solution heat treating; however, there are special heat treatments available to minimize grain growth.

Magnesium is normally quenched in air following the solution treatment. Still air is usually sufficient but forced air cooling is recommended for dense loads or parts that have thick sections. Hot water quenching is used for the alloys QE22 and QH21 to develop the best mechanical properties. Glycol quenchants can also be used to help prevent distortion. Artificial aging consists of reheating to 335–450° F and holding for 5–25 h. Hardness cannot be used for verification of heat treatment. For cast products, tensile test specimens can be either cut from a portion of the casting or cast as separate tensile test bars.

3.3.4 Magnesium Machining[7]

Magnesium is extremely easy to machine at high speeds using greater depths of cuts and higher feed rates than other structural metals. Dimensional tolerances of a few thousandths of an inch are possible with surfaces finishes as fine as 3–5 μin. Machining is usually conducted dry; however, cutting fluids can be used to reduce the chances of distortion and minimize the danger of fire when chips are fine. Fine finishing cuts are a greater fire hazard than heavier roughing cuts. When magnesium chips ignite, they burn with a brilliant white light. To reduce the fire hazard when machining magnesium: (1) use only sharp tools; (2) use heavy feeds to produce thick chips; (3) use mineral oil coolants, especially during fine finishing cuts; (4) remove chips frequently from the work area and store in clean covered metal containers; and (5) keep an adequate supply of recommended magnesium fire extinguishers at all work areas.

3.3.5 Magnesium Joining

Magnesium alloys can be welded by gas shielded arc welding and by resistance spot welding. The GTAW process uses a tungsten electrode, magnesium alloy filler wire, and an inert gas, such as argon or helium, for shielding. In the GMAW process, a continuously fed magnesium alloy wire acts as the electrode for maintaining the arc while an argon gas shield prevents oxidation of the weld puddle. No flux is required and welding operations are similar to those for aluminum alloys. Welds in magnesium alloys are characterized by a fine grain size, averaging less than 0.01 in. Welding problems due to residual stresses and the tendency for certain alloys to crack can be minimized by preheating, post-weld heating, and stress relieving. In the Mg–Al–Zn alloys (AZ31, AZ61, AZ63, AZ80, AZ81, AZ91 and AZ92), aluminum contents of up to 10% aids weldability by refining the grain structure, while zinc contents of more than 1% increases hot shortness which can cause weld cracking. Weld joints in the Mg–Al–Zn alloys and alloys containing more than 1% aluminum require stress relieving, because they are subject to stress corrosion cracking if not stress relieved. Stress relief is usually conducted by heating to 300–800° F for times ranging from 15 to 120 min. Alloys with high zinc contents, such as ZH62, ZK51, ZK60, and ZK61, are very susceptible to weld cracking and have poor weldability. Weld repaired castings are normally heat treated after welding to either the T4, T5, or T6 tempers. If the casting is not heat treated after weld repair, it is usually stress relieved.

Although magnesium sheet can be spot welded, it is usually not because the fatigue strength of spot welded joints is lower than that for either riveted or adhesive bonded joints. Thus, spot welding should not be used for joints in fatigue or vibration environments. In riveted joints, only galvanically compatible rivets, such as 5056 aluminum, should be used. Quarter-hard 5056-H32 aluminum rivets are satisfactory for normal riveting.

3.4 Magnesium Corrosion Protection[8]

With magnesium alloys, corrosion is an ever present reality. For optimum corrosion protection, a chemical conversion coating or anodizing treatment is required prior to the application of an organic paint system. These treatments roughen and chemically modify the surface for maximum paint adhesion. A typical anodize treatment (Dow 17) is shown in Fig. 3.7. In this treatment, the part is first alkaline cleaned and then anodized in a solution of NH_4HF_2, $Na_2Cr_2O_7 \cdot 2H_2O$, and H_3PO_4 heated to 160–180° F using either an AC or DC current (5–50 A/ft^2). This method produces a two-layer coating; the first layer is a thin light-green coating (0.2 mil) done at lower voltages that is followed by a thicker dark-green coating (1.2 mil) done at higher voltages. The thicker coating enhances

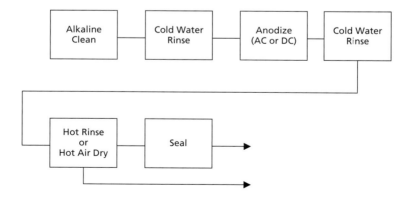

Fig. 3.7. Anodize Treatment for Magnesium Alloys[8]

corrosion protection and forms an excellent base for paint but can be susceptible to spalling under impact, deformation, or flexing.

Porous casting surfaces are normally filled with a penetrating resin prior to paint application. Primers usually contain zinc chromate or titanium dioxide pigments for improved corrosion resistance. Both air drying and baked-on paints are used, with the baked-on paints being harder and more resistant to solvents. Depending on the application, vinyl alkyds provide resistance to alkali, acrylics for resistance to salt spray, alkyd enamels for exterior durability, and epoxies for abrasion resistance. Vinyls can withstand temperatures up to about 300° F. Higher temperature finishes include modified vinyls, epoxies, modified epoxies, epoxy-silicones, and silicones. It is important to maintain the integrity of the paint system when the part is placed in-service, as the chemical conversion and anodized surfaces will readily corrode if exposed to the atmosphere.

BERYLLIUM

Due to its combination of low density, high stiffness, and attractive electrical and thermal properties, beryllium is used in high value space craft electronic and guidance system applications. It is limited to wider use due to its high cost, very limited formability, and safety concerns over its toxicity.

3.5 Beryllium Metallurgical Considerations

Beryllium has a low density ($0.068\,lb/in.^3$), a moderately high melting point ($2341°\,F$), a high tensile modulus ($44\,msi$), good electrical conductivity (40% IACS), and thermal conductivity ($109\,Btu/ft\,h°\,F$). Unalloyed beryllium is used in aerospace applications requiring its excellent stiffness and low thermal expansion, which provides physical stability and low distortion during inertial or thermal stressing. High value optical components and precision instruments are examples. Structural designs utilizing beryllium sheet should allow for anisotropy, particularly the very low short transverse properties.

Although beryllium can be melted and cast, the resulting grain structure is coarse ($>100\,\mu m$) making the product difficult to process. Therefore, beryllium is made using powder metallurgy (PM) methods, which yields grain sizes in the range of $5-15\,\mu m$. Powder consolidation is normally done by either vacuum hot pressing (VHP) or hot isostatic pressing (HIP). In spite of this fine grain size, the elongation at failure is still only about 3% at room temperature, primarily because of a large covalent component in the atomic bonding in the c-axis direction of the HCP structure.[9] This low ductility is not improved by controlling impurities to as low as 10 ppm.

Beryllium forms a thin protective oxide (BeO) that provides excellent corrosion resistance in moist environments at temperatures up to 570° F. However, the corrosion resistance will be impaired by the presence of impurities. The presence of carbides or chlorides can cause a reaction with moist air at room temperature.

Beryllium is a toxic metal when inhaled. Inhalation of respirable beryllium and its compounds must be avoided. Users should comply with the occupational safety and health standards applicable to beryllium in Title 29, Part 1910, Code of Federal Regulations. The main concern with beryllium handling is the effect on the lungs when beryllium fumes, powder, or dust is inhaled. Two forms of lung disease associated with beryllium are acute berylliosis and chronic berylliosis. The acute form, which can appear suddenly, resembles pneumonia or bronchitis. On the other hand, chronic berylliosis occurs slowly over a period of time. It seems to result from an allergic reaction and depends on the individual.[10]

3.6 Beryllium Alloys

Commercial grades of beryllium are characterized by their impurity levels and BeO content. The metal normally contains 0.7–4.25% BeO as a result of the powder metallurgy process that increases with increasing fineness of the powder. Standard grade beryllium bars, rods, tubing, and machined shapes are produced from VHP powder with a 1.5% maximum beryllium oxide content. Sheet and plate are fabricated from VHP powder with a 2% maximum beryllium oxide content. Beryllium has a limited solubility for oxygen and readily forms particles at grain boundaries that help in controlling grain growth without harming ductility at contents below 1.5% BeO. Of the structural grades, indicated by the prefix S, S-65B with a 99.0% minimum Be and 0.7% BeO has the best ductility (3% minimum elongation) at room temperature with a tensile strength of 42 ksi and a yield strength of 30 ksi. Grade S-200F, which contains 98.5% minimum Be and 1.5% BeO, is stronger with a tensile strength of 47 ksi and a yield strength of 35 ksi but has low elongation (2%) due to the higher BeO content. As shown in Fig. 3.8, Grade S-200F has usable strength up to 1200°F.

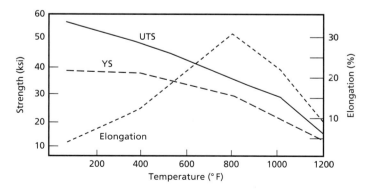

Fig. 3.8. Tensile Properties of S-200F Beryllium[11]

3.7 Beryllium Powder Metallurgy

Almost all beryllium in use is a PM product.[12] PM is required for a number of reasons, the primary one being that beryllium castings contain too much porosity and other casting defects to allow their use in critical applications. In addition to the casting defects, the grain size of cast beryllium is too coarse ($>100\,\mu m$). Since the strength and ductility of beryllium depends on a fine grain size, grain sizes of less than $15\,\mu m$ are required to obtain satisfactory mechanical properties. The effects of grain size on beryllium properties are shown in Fig. 3.9. PM techniques can produce grain sizes as fine as $1-10\,\mu m$ when required. Beryllium powders are consolidated into near net shapes by either VHP or HIP to obtain parts within 99.5% of theoretical density.

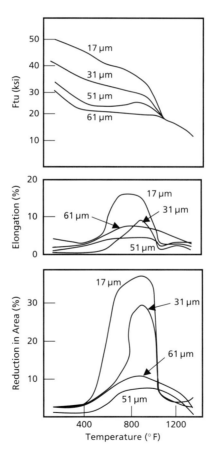

Fig. 3.9. *Effect of Grain Size on Beryllium Mechanical Properties*[9]

To produce beryllium powder, castings are first poured and then turned into chips using a lathe with multi-head cutting tools. The chips are ground into powder using impact milling to produce powders of differing sizes and shapes. Impact milling is the most prevalent method because it produces the best powder product. In impact milling, beryllium particles are driven against a beryllium target using a high velocity gas stream. A blocky powder particle is produced that has less tendency to align preferentially during powder compaction than those produced by ball or attrition milling. When the powder is compacted under heat and pressure, it is more uniform, resulting in greater ductility. Since the low ductility of beryllium has always been a concern, improvements in room temperature ductility have been achieved by control of preferred orientation, improved purity, reduction of inclusions, control of inclusion distribution, and by reducing grain size. Elongation values of 4–5% are currently achievable. The strength of beryllium is also a function of grain size, which is determined by particle size, oxide content, and consolidation temperature.

The impact milling system shown in Fig. 3.10 is used to produce beryllium powder. Coarse powder is fed from the feed hopper into a gas stream. As the gas borne powder is carried through a nozzle, it accelerates and impacts a beryllium target. The debris is then carried to the primary classifier where the particles drop out and fines go to a secondary classifier and are discarded. This cycle continues until the desired particle size is achieved. Impact milling enables consistent control of powder composition by reducing impurity contamination and oxidation of powder particles. The process also yields improved powder

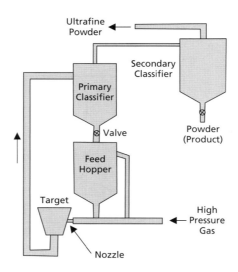

Fig. 3.10. Impact Milling System[9]

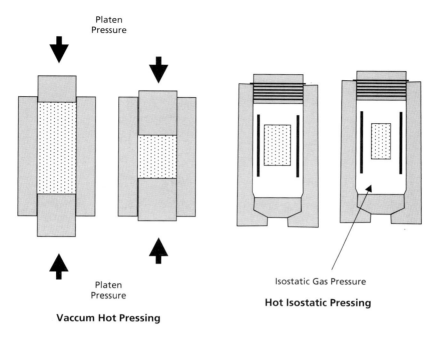

Fig. 3.11. Vacuum Hot Pressing and Hot Isostatic Pressing[13]

configuration and morphology, resulting in improved isotropy and a cleaner microstructure of the final consolidated product.[9]

Powder compaction is conducted by either VHP or HIP, as shown schematically in Fig. 3.11. In VHP, the powder is loaded into a graphite die and vibrated, the die is placed into a vacuum hot press, a vacuum is established, and the powder is consolidated under 500–2000 psi at temperatures between 1830 and 2020° F. Densities in excess of 99% of theoretical are obtained in diameters of 8–72 in. The HIP process is similar except that the powder must be enclosed in a mild steel can prior to consolidation. After canning, the can is degassed under vacuum at 1100–1300° F to remove all air and gases absorbed on the particle surfaces. The can is then sealed and put through the HIP consolidation cycle at 15 ksi and 1400–2010° F. Although HIP is generally more expensive than VHP, it allows the best control of grain size because it allows greater latitude in temperature selection. HIP is capable of attaining 100% of theoretical density. A comparison of the two methods on the properties of grade S-200 is shown in Fig. 3.12. The improvement in anisotropy for the HIP-processed material is evident and is primarily a result of being able to use lower temperatures to keep the grain size fine and apply uniform and high pressures in all directions during consolidation.

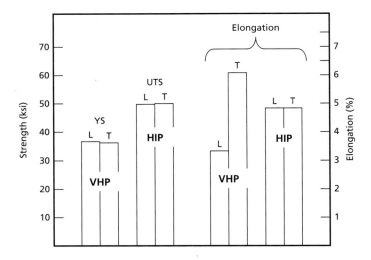

Fig. 3.12. Impact of Compaction Method on Properties of S-200[9]

After powder consolidation, beryllium can be processed by conventional metal working technologies such as rolling and extrusion. The mechanical properties are improved by metal working in the predominant direction of metal flow. Biaxial and/or triaxial deformation can be used to obtain more balanced properties. The directionality developed during hot working is a direct result of beryllium's HCP crystalline structure. Beryllium is rolled and extruded into shapes at elevated temperatures in the range of 1470–2110° F. Recrystallization annealing can be conducted at 1340–1650° F. As typical with metals containing a HCP structure, beryllium displays anistropic mechanical properties because of the crystallographic texture it develops during hot working. Hot pressed blocks are normally encased in mild steel jackets for protection during hot rolling of sheet and plate, which is cross-rolled to produce uniform biaxial properties. Sheet and plate are then surface ground and pickled. Due to the nature of the extrusion process, the properties of extrusions are higher in the direction of the extrusion than in the transverse direction.

3.8 Beryllium Fabrication

3.8.1 Beryllium Forming[14]

Compared to most other metals, the formability of beryllium is low due to the limited number of slip planes available on its HCP crystalline structure. In fact, the elevated temperature formability of beryllium is lower than other HCP structural metals, such as magnesium and titanium. Even elevated temperature forming operations must be done with great care to prevent sheet cracking.

Beryllium must be formed at relatively slow speeds at temperatures in the range of 1000–1450° F. Beryllium sheet should be formed at 1300–1350° F, holding at temperature no more than 1.5 h, for minimum springback. Hydraulic or air-operated presses must be used instead of brake presses to insure that the forming is done slow enough to prevent sheet cracking. Both the workpiece and the die must be heated during forming operations. Dies are usually constructed from hot work die steels due to the high forming temperatures required. Lubricants, such as colloidal graphite in oil, are also required. When forming thick sheet into complex shapes, an intermediate stress relief for 30 min at 1300–1400° F can be employed.

3.8.2 Beryllium Machining[15]

Because of its high cost and toxicity potential, machining operations should be minimized where possible by the use of near net shapes. For near net shapes produced by powder metallurgy, as little as 0.10 in. of material is all that must be removed by machining. Since the chips from machining operations are normally recycled, the majority of machining operations are conducted dry. In addition, due to the toxic nature of beryllium dust, the chips are collected by vacuum as close to the cutting edge as possible. Generally, the exhaust system is located within 0.50–0.75 in. of the cutting tool. Beryllium, being extremely brittle, must be securely clamped during machining operations to minimize any vibration or chatter. Carbide tools are most often used in machining beryllium. Finishing cuts are usually 0.002–0.005 in. in depth to minimize surface damage. Although most machining operations are performed without coolant to avoid chip contamination, the use of coolant can reduce the depth of damage and give longer tool life. The drilling of sheet can lead to delamination and breakout unless the drill head is of the controlled torque type and the drills are carbide burr type.

Machining beryllium results in surface damage that extends from 0.001 to 0.020 in. deep, depending on the machining process. This surface damage consists of a dense network of intersecting twins and microcracks. Since surface damage adversely affects the mechanical properties, the machining process must be closely controlled; however, even with controlled machining, some surface damage will occur. To restore the mechanical properties, the surface damage must be either removed by chemical etching or "healed" by heat treatment. Chemical milling of beryllium usually removes about 0.002–0.004 in. from the surface, which is about the typical thickness of the damaged surface. Chemical milling can be conducted using sulfuric acid, nitric/hydrofluoric acid, and ammonium bifluoride. One disadvantage of chemical milling is that the removal is sometimes uneven and can impact precision tolerances. Vacuum heat treating at 1470° F for 2 h will remove most of the machining damage and restore all most all of the mechanical properties. Heat treating removes the thermally induced twins and does not impact part tolerances. However, heat treating will not remove microcracks.

3.8.3 Beryllium Joining

Parts may be joined mechanically by riveting, but only by careful squeeze riveting to avoid damage to the beryllium, by bolting, by threading, or by using press fittings specifically designed to avoid damage. Parts can also be joined by brazing, soldering, braze welding, adhesive bonding, and diffusion bonding. Fusion welding is not recommended. Brazing can be accomplished with zinc, aluminum–silicon, or silver base filler metals. Many elements, including copper, can cause embrittlement when used as brazing filler metals. However, specific manufacturing techniques have been developed by various beryllium fabricators to use many of the common braze materials.

3.9 Aluminum–Beryllium Alloys[16]

In addition to the beryllium alloys, there are a limited number of aluminum–beryllium alloys that are attractive in stiffness driven designs. An example is the alloy AlBeMet AM162 which contains 62% beryllium and 38% aluminum by weight. The alloy is made by powder metallurgy followed by cold isostatic pressing (CIP) to approximately 80% of its final density. The material is then rolled or extruded to its final shape and density. A flow chart of the processing is given in Fig. 3.13.

The mechanical properties are compared with high strength aluminum and titanium in Table 3.5. With its low density and high elastic modulus, AlBeMet AM 162 would be 3.9 times stiffer than 7075-T73 Al and Ti-6Al-4V in a tension tie application. The disadvantages of AlBeMet AM 162 are the limitations on extrusion and rolled sheet sizes available and high cost (approximately 400 times the cost of aluminum). While this alloy is somewhat more difficult to machine than aluminum (20% lower speeds and feeds), it is easier to machine than titanium. Like all beryllium-containing alloys, the machining process needs to be controlled due to the hazardous nature of beryllium.

Summary

At a density of only $0.063 \, lb/in.^3$, magnesium is the lightest weight of the structural alloys. However, for most structural applications, magnesium alloys cannot compete with aluminum when all of its characteristics are taken into account. It is more expensive than aluminum and the fabrication costs are higher because most forming operations must be conducted at elevated temperature. Although a number of the magnesium alloys will respond to heat treatment, the extent of hardening and strengthening is not as high as that obtained with

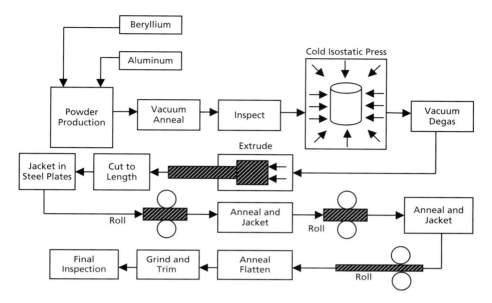

Fig. 3.13. AlBeMet AM 162 Manufacturing Flow[9]

Table 3.5 Property Comparison for AlBeMet AM 162[16]

Property	AlBeMet AM 162	7075-T73 Al	Ti-6Al-4V
Density (lb/in.^{3})	0.076	0.098	0.160
UTS (ksi)	49	56	130
YS (ksi)	41	44	120
Elongation (%)	10	7	10
E (msi)	29	10	16
Endurance Limit (ksi)	24	23	80
Fracture Toughness (ksi $\sqrt{in.}$)	22	42	97
CTE (μin./in.°F)	7.7	13	4.9
Thermal Conductivity (BTU/h-ft-°F)	139.0	93.8	4.2

- Extrusions
- Transverse Properties

the aluminum alloys. In addition, corrosion is an ever present concern with magnesium alloys.

Beryllium is also a very lightweight metal (0.068 lb/in.^{3}) that has fairly good strength properties and a very high tensile modulus (44 msi). These properties, combined with its attractive electrical and thermal properties, have led to its use in high value aerospace electronic and guidance system applications. However, beryllium must be processed using PM technology that leads to high costs. Like magnesium, its HCP crystalline structure greatly impairs its room temperature formability. Finally, beryllium powder and dust are toxic, which further increases its cost through the requirement for a controlled manufacturing environment.

References

[1] Lyon, P., "Electron 21 for Aerospace and Specialty Applications", *AeroMet Conference & Exposition*, Seattle, WA, 7–10 June 2004.

[2] Caron, R.N., Staley, J.T., "Magnesium and Magnesium Alloys: Effects of Composition, Processing, and Structure on Properties of Nonferrous Alloys", in *ASM Handbook Vol. 20, Materials Selection and Design*, ASM International, 1997.

[3] Smith, W.F., Chapter 12: "Magnesium and Zinc Alloys", in *Structure and Properties of Engineering Alloys*, 2nd edition, McGraw-Hill Inc., 1993, pp. 537–560.

[4] Proffitt, H., "Magnesium Alloys", in *ASM Handbook Vol. 15, Casting*, ASM International, 1988.

[5] Housh, S., Mikucki, B., Stevenson, A., "Formability: Selection and Application of Magnesium and Magnesium Alloys", in *ASM Handbook Vol. 2, Properties and Selection: Nonferrous Alloys and Special-Purpose Materials*, ASM International, 1990.

[6] Stevenson, A., "Heat Treating of Magnesium Alloys", in *ASM Handbook Vol. 4, Heat Treating*, ASM International, 1991.

[7] Housh, S., Mikucki, B., Stevenson, A., "Machinability: Selection and Application of Magnesium and Magnesium Alloys", in *ASM Handbook Vol. 2, Properties and Selection: Nonferrous Alloys and Special-Purpose Materials*, ASM International, 1990.

[8] Hillis, J.E., "Surface Engineering of Magnesium Alloys", in *ASM Handbook Vol. 5, Surface Engineering*, ASM International, 1994.

[9] Marder, J.M., "Powder Metallurgy Beryllium", in *ASM Handbook Vol. 7, Powder Metal Technologies and Applications*, ASM International, 1998.

[10] Stonehouse, A.J., Marder, J.M., "Beryllium: Health and Safety Considerations", in *ASM Handbook Vol. 2, Properties and Selection: Nonferrous Alloys and Special-Purpose Materials*, ASM International, 1990.

[11] Stonehouse, A.J., Marder, J.M., "Beryllium Grades and Their Designations", in *ASM Handbook Vol. 2, Properties and Selection: Nonferrous Alloys and Special-Purpose Materials*, ASM International, 1990.

[12] Caron, R.N., Staley, J.T., "Beryllium: Effects of Composition, Processing, and Structure on Properties of Nonferrous Alloys", in *ASM Handbook Vol. 20, Materials Selection and Design*, ASM International, 1997.

[13] Stonehouse, A.J., Marder, J.M., "Beryllium: Powder Consolidation Methods", in *ASM Handbook Vol. 2, Properties and Selection: Nonferrous Alloys and Special-Purpose Materials*, ASM International, 1990.

[14] Grant, L., "Forming of Beryllium", in *ASM Handbook Vol. 14, Forming and Forging*, ASM International, 1988.

[15] Gallagher, D.V., Hardesty, R.E., "Machining of Beryllium", in *ASM Handbook Vol. 16: Machining*, ASM International, 1998.

[16] Speer, W., Es-Said, O.S., "Applications of an Aluminum-Beryllium Composite for Structural Aerospace Components", *Engineering Failure Analysis*, Vol. 11, 2004, pp. 895–902.

Chapter 4

Titanium

Titanium is an attractive structural material due to its high strength, low density, and excellent corrosion resistance. However, even though titanium is the fourth most abundant element in the earth's crust, due to its high melting point and extreme reactivity, the cost of titanium is high. The high cost includes both the mill operations (sponge production, ingot melting, and primary working) as well as many of the secondary operations conducted by the user. The primary reasons for which titanium alloys are used in aerospace applications include:[1]

- Weight savings. The high strength-to-weight ratio of titanium alloys allows them to replace steel in many applications requiring high strength and fracture toughness. With a density of 0.16 lb/in.^3, titanium alloys are only about ½ as heavy as steel and nickel based superalloys, yielding excellent strength-to-weight ratios.
- Fatigue strength. Titanium alloys have much better fatigue strength than aluminum alloys and are frequently used for highly loaded bulkheads and frames in fighter aircraft.
- Operating temperature capability. When the operating temperature exceeds about 270° F, aluminum alloys lose too much strength and titanium alloys are often required.
- Corrosion resistance. The corrosion resistance of titanium alloys is superior to both aluminum and steel alloys.
- Space savings. Titanium alloys are used for landing gear components on commercial aircraft where the size of aluminum components would not fit within the landing gear space envelope.

Due to their outstanding resistance to fatigue, high temperature capability, and resistance to corrosion, titanium alloys comprise approximately 42% of the structural weight of the new F-22 fighter aircraft, over 9000 lb in all.[2] In commercial aircraft, the Boeing 747-100 contained only about 2.6% titanium, while the newer Boeing 777 contains about 8.3%.[3] In commercial passenger aircraft engines, the fan, the low pressure compressor, and about 2/3 of the high pressure compressor are made from titanium alloys.[3]

4.1 Metallurgical Considerations

Titanium exists in two crystalline forms as shown in Fig. 4.1: the one stable at room temperature is called alpha (α) and has a HCP structure, while the one that is stable at elevated temperature is BCC and is called beta (β). In pure titanium, the alpha phase is stable up to 1620° F where it transforms to the beta phase; the transition temperature is known as the beta transus. The beta phase is then stable from 1620° F up the melting point (3130° F).

At room temperature, commercially pure titanium is composed primarily of the alpha phase. As alloying elements are added to titanium, they tend to change the amount of each phase present and the beta transus temperature in the manner

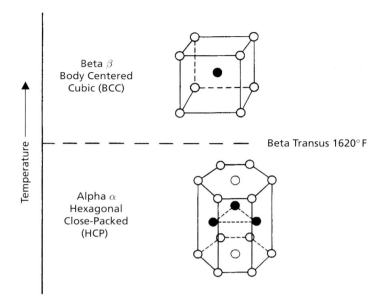

Fig. 4.1. Crystal Structures of Pure Titanium

shown in Fig. 4.2. Those elements that increase the beta transus temperature through stabilizing the alpha phase are called alpha stabilizers and include aluminum, oxygen, nitrogen, and carbon. Those elements that decrease the beta transus temperature are called beta stabilizers. The beta stabilizers are further subdivided into beta isomorphous elements, which have a high solubility in titanium, and beta eutectoid elements, which have only limited solubility and tend to form intermetallic compounds. The beta isomorphous elements are molybdenum, vanadium, niobium, and tantalum, while beta eutectoid elements include manganese, chromium, silicon, iron, cobalt, nickel, and copper. Finally, tin and zirconium are considered neutral because they neither raise nor lower the beta transus temperature, but since both contribute to solid solution strengthening, they are frequently used as alloy additions.

Titanium alloys are classified according to the amount of alpha and beta retained in their structures at room temperature. Classifications include alpha, near-alpha, alpha–beta, and metastable beta. As their name implies, alpha alloys do not contain beta phase at room temperature, while near-alpha alloys contain mostly alpha with only a small amount of beta. Alpha–beta alloys are two-phase alloys containing both alpha and beta phases. Metastable beta alloys contain mostly beta with small amounts of alpha present. The beta alloys are called metastable because thermodynamically, if given the time, the beta would transform to equilibrium phases of alpha and intermetallic compounds; however,

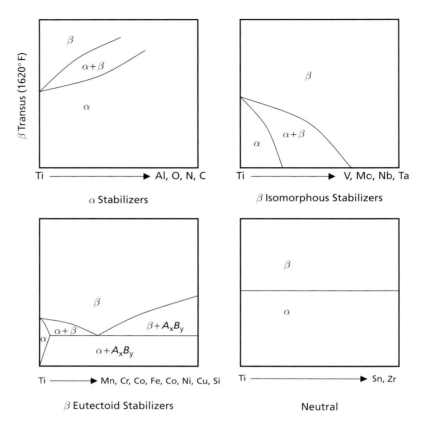

Fig. 4.2. Phase Diagrams for Binary Titanium Alloys

for all practical purposes, this will not happen so they are often referred to as just beta alloys.

Alpha and near-alpha alloys usually contain 5–6% aluminum as the main alloying element with additions of the neutral elements tin and zirconium and some beta stabilizers. Since these alloys retain their strength at elevated temperatures and have the best creep resistance of the titanium alloys, they are often specified for high temperature applications. The addition of silicon, in particular, improves the creep resistance by precipitating fine silicides, which hinders dislocation climb. Aluminum also contributes to oxidation resistance. These alloys also perform well in cryogenic applications. Due to the limitation of slip systems in the HCP structure, the alpha phase is less ductile and more difficult to deform than the BCC beta phase. Because these are single phase alloys containing only alpha, they cannot be strengthened by heat treatment.

The alpha–beta alloys have the best balance of mechanical properties and are the most widely used. In fact, one alpha–beta alloy, Ti-6Al-4V, is by far and away the most widely utilized alloy. In this alloy, aluminum stabilizes the alpha phase while vanadium stabilizes the beta phase; therefore, alpha–beta alloys contain both alpha and beta phases at room temperature. In contrast to the alpha and near-alpha alloys, the alpha–beta alloys can be heat treated to higher strength levels, although their heat treat response is not as great as that for the beta alloys. In general, the alpha–beta alloys have good strength at room temperature and for short times at elevated temperatures, although they are not noted for their creep resistance. The weldability of many of these alloys is poor due to their two-phase microstructures.

The metastable beta alloys contain a sufficient amount of beta stabilizing elements that the beta phase is retained to room temperature. Since the BCC beta phase exhibits much more deformation capability than the HCP alpha phase, these alloys exhibit much better formability than the alpha or alpha–beta alloys. Where the alpha and alpha–beta alloys would require hot forming operations, some of the beta alloys can be formed at room temperature. The beta alloys can be solution treated and aged (STA) to higher strength levels than the alpha–beta alloys while still retaining sufficient toughness. The biggest drawbacks of the beta alloys are increased densities due to alloying elements such as molybdenum, vanadium, and niobium; reduced ductility when heat treated to peak strength levels; and some have limited weldability.

Titanium has a great affinity for interstitial elements, such as oxygen and nitrogen, and readily absorbs them at elevated temperature. Oxygen tends to increase the strength and decrease the ductility. As the amount of oxygen and nitrogen increases, the yield and ultimate strengths increase and the ductility and fracture toughness decreases. Titanium absorbs oxygen at temperatures above 1300° F, which complicates the processing and increases the cost, since many hot working operations are conducted at temperatures exceeding 1300° F. However, oxygen, in controlled amounts, is actually used to strengthen the commercially pure (CP) grades. Some alloys are available in an extra low interstitial (ELI) grade that is used for applications requiring maximum ductility and fracture toughness. Hydrogen is always minimized in titanium alloys because it causes hydrogen embrittlement by the precipitation of hydrides, so the maximum limit allowed is about 0.015%.

Titanium alloys derive their strength from the fine microstructures produced by the transformation from beta to alpha. If alpha–beta or beta alloys are solution heat treated and aged, titanium martensite can form during the quenching operation; however, the martensite formed in titanium alloys is not like the extremely hard and strong martensite formed during the heat treatment of steels. For example, the tensile strength of Ti-6-4 only increases from 130 to 170 ksi on STA, while the tensile strength of 4340 steel can be increased from 110 to 280 ksi by heat treatment. While the grain size does not normally affect the

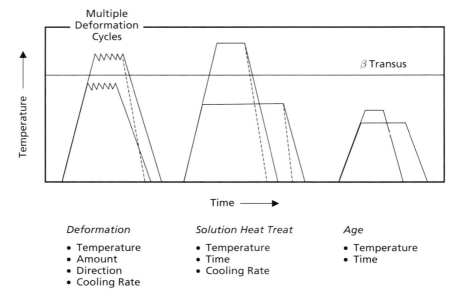

Fig. 4.3. Thermomechanical Treatment of Titanium Alloys

ultimate tensile strength, the finer the grain size, the higher the yield strength, ductility, and fatigue strength. However, to resist grain boundary sliding and rotation, larger grain sizes are actually preferred for some applications requiring creep resistance.

The microstructure of titanium alloys is determined by both alloy composition and the thermomechanical treatment (TMT) history that the alloy undergoes during processing. TMT, as shown in Fig. 4.3, is a combination of mechanical deformation sequences and heat treatments. TMT for wrought alloys consists of three processing steps: (1) a series of elevated temperature deformations, (2) solution heat treatment, and (3) aging. This is a highly simplified explanation of TMT processing in that the elevated temperature deformation may occur in several steps and many titanium alloys, Ti-6Al-4V being a good example, are used in the mill annealed condition more than they are in the solution treated and aged condition.

A key element in microstructure and mechanical property development is whether or not the TMT treatments are conducted above or below the beta transus temperature. Lamellar structures (Fig. 4.4) are a result of cooling from the beta phase field in which the alpha phase nucleates at prior beta grain boundaries and then grows into the beta grains. If the cooling rate is fairly fast, then the microstructure will be fine (Widmanstatten or basketweave), whereas if it is slow, it will be coarse (colony structure). Lamellar structures offer

Titanium

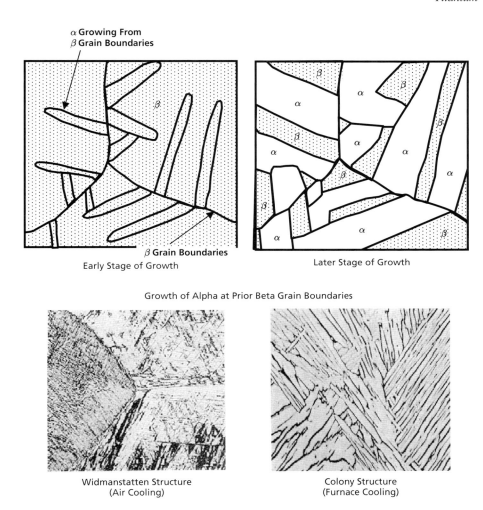

Fig. 4.4. Lamellar Structures Produced by Cooling from above the Beta Transus[4]

the highest fracture toughness and the lowest fatigue crack growth rates. On the other hand, equiaxed structures, as shown in Fig. 4.5, are a result of a recrystallization process and therefore require previous cold work to initiate recrystallization. This can be accomplished by deformation in the two-phase alpha + beta field followed by solution treating in the two-phase field. Equiaxed structures produce the highest strength and ductility and the best fatigue strength. The difference between the terms "fatigue strength" and "crack growth rate" should be explained. Fatigue strength is the total fatigue life before failure. It consists of crack initiation and then crack growth until final failure. The solution

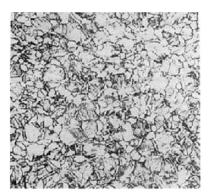

Fig. 4.5. Equiaxed Grain Structure in Titanium[4]

treating temperature determines the volume fraction of the primary alpha phase with temperatures just below the beta transus leading to duplex structures in which equiaxed alpha is dispersed in an alpha + beta lamellar matrix. Duplex structures produce a balance of properties between the equiaxed and lamellar structures.

Although the melting point of titanium is higher than 3000° F, it is not possible to create titanium alloys that operate at temperatures approaching their melting point. Titanium alloys are usually restricted to maximum temperatures of 600–1100° F depending upon alloy composition.

4.2 Titanium Alloys

As discussed in the previous section, titanium alloys are classified according to the amount of alpha and beta retained in their structures at room temperature. Classifications include commercially pure, alpha and near-alpha, alpha–beta, and metastable beta. The properties of a number of commercially important alloys are given in Table 4.1.

4.2.1 Commercially Pure Titanium

Commercially pure grades are available with yield strengths ranging from 25 to 70 ksi, with the higher strength grades containing more oxygen. The CP grades have good formability, are readily weldable, and have excellent corrosion resistance. The CP grades are supplied in the mill annealed condition which permits extensive forming at room temperature, while severe forming operations can be conducted at 300–900° F. However, property degradation can be experienced after severe forming if the material is not stress relieved. One of the largest usages of CP alloys is for corrosion resistant tubing, tanks, and

Table 4.1 Properties of Selected Titanium Alloys

Alloy	Nominal Composition	Condition	UTS (ksi)	YS (ksi)	E (Msi)	Elong. (%)	Reduction in Area (%)
		Commercially Pure					
Grade 1	0.03N, 0.20Fe, 0.18O	Annealed	35	25	14.9	30	55
Grade 4	0.05N, 0.50Fe, 0.40O	Annealed	80	70	14.9	20	40
		Alpha and Near-Alpha					
Ti-5-2.5	Ti-5Al-2.5Sn	Annealed	115	110	16.0	16	40
Half 6-4	Ti-3Al-2.5V	Annealed	90	75	15.5	20	–
Ti-6242S	Ti-6Al-2Sn-4Zr-2Mo-0.25Si	Annealed	130	120	16.5	15	35
Ti-8-1-1	Ti-8Al-1Mo-1V	Annealed	130	120	18.0	15	28
		Alpha–Beta					
Ti-6-4	Ti-6Al-4V	Annealed	130	120	16.5	14	30
		STA	170	160	16.5	10	25
Ti-6-4 ELI	Ti-6Al-4V	Annealed	120	110	16.5	15	35
Ti-6-6-2	Ti-6Al-6Sn-2V	Annealed	150	145	16.0	14	30
		STA	185	170	16.0	10	20
Ti-6246	Ti-6Al-2Sn-4Zr-6Mo	STA	189	170	16.5	10	23
Ti-6-22-22S	Ti-6Al-2Sn-2Zr-2Mo-2Cr-0.25Si	Annealed	150	140	17.7	–	–
		STA	185	165	17.7	11	33
		Beta					
Ti-10-2-3	Ti-10V-2Fe-3Al	STA	170	160	16.2	10	19
Ti-15-3	Ti-15V-3Al-3Cr-3Sn	Annealed	114	112	–	22	–
		STA	159	143	–	12	–

fittings in the chemical processing industry. Palladium can be added to further enhance corrosion resistance. A large titanium heat exchanger is shown in Fig. 4.6.

4.2.2 Alpha and Near-Alpha Alloys

The alpha and near-alpha alloys are not heat treatable, have medium formability, are weldable, and have medium strength, good notch toughness, and good creep resistance in the range of 600–1100° F. Aluminum is the most important alloying element in titanium, because it is a potent strengthener and also reduces density. However, the aluminum content is usually restricted to about 6% because higher contents run the risk of forming the brittle intermetallic compound Ti_3Al. The only true alpha alloy that is commercially available is the alloy Ti-5Al-2.5Sn. Ti-5-2.5 is used in cryogenic applications, because it retains its ductility and fracture toughness down to cryogenic temperatures. The remainder of the commercially available alloys in this class are classified as near-alpha alloys.

Fig. 4.6. Large Titanium Heat Exchanger

Near-alpha alloys are those which contain some beta phase dispersed in an otherwise all alpha matrix. The near-alpha alloys generally contain 5–8% aluminum, some zirconium, and tin, along with some beta stabilizer elements. Because these alloys retain their properties at elevated temperature and posses good creep strength, they are often specified for elevated temperature applications. Silicon in the range of 0.10–0.25% enhances the creep strength. High temperature near-alpha alloys include Ti-6242S (Ti-6Al-2Sn-4Zr-2Mo-0.25Si) and IMI 829 (Ti-5.5Al-3.5Sn-3Zr-1Nb-0.3Si), which can be used to 1000°F, and IMI 834 (Ti-5.8Al-4Sn-3.5Zr-0.7Nb-0.5Mo-0.35Si) and Ti-1100 (Ti-6Al-2.8Sn-4Zr-0.4Mo-0.4Si), a modification of Ti-6242S, which can be used to 1100°F. Ti-8Al-1Mo-1V is an older alloy that has rather poor stress corrosion resistance (due to its extremely high aluminum content) but has a low density and a higher modulus than other titanium alloys.

4.2.3 Alpha–Beta Alloys

The alpha–beta alloys are heat treatable to moderate strength levels but do not have as good elevated temperature properties as the near-alpha alloys. Their weldability is not as good as the near-alpha alloys but their formability is better. The alpha–beta alloys, which include Ti-6Al-4V, Ti-6Al-6V-2Sn, and Ti-6Al-2Sn-4Zr-6Mo, are all capable of higher strengths than the near-alpha alloys. They have a good combination of mechanical properties, rather wide processing windows, and can be used in the range of about 600–750°F. They can be strengthened by STA but the strength obtainable decreases with section thickness. The lean alloys, such as Ti-6-4, are weldable. Alpha–beta alloys contain elements, in particular aluminum, to strengthen the alpha phase and beta stabilizers to provide solid solution strengthening and response to heat treatment. As the percentage of beta stabilizing elements increases, the hardenability increases and the weldability decreases. Ti-6-4 accounts for approximately 60% of the titanium used in aerospace and up to 80–90% for airframes. There are four different heat treatments that are often used for wrought Ti-6-4:[1]

(1) *Mill Annealed.(MA)*. Mill annealed is the most common heat treatment. It produces a tensile strength of approximately 130 ksi, good fatigue properties, moderate fracture toughness ($K_{IC} = 60$ ksi $\sqrt{\text{in.}}$), and reasonable fatigue crack growth rate.
(2) *Recrystallization Anneal (RA)*. A recrystallization anneal can be used for parts requiring increased damage tolerance, as the K_{IC} value for Ti-6-4 goes from around 60 to 70 ksi $\sqrt{\text{in}}$. This heat treatment produces slightly lower strength and fatigue properties and improved fracture toughness and slower fatigue crack growth rates.
(3) *Beta Anneal (BA)*. Beta annealing is used where it is important to maximize fracture toughness (minimum $K_{IC} \sim 80$ ksi $\sqrt{\text{in.}}$) and minimize fatigue crack grow rate. However, the fatigue strength is significantly degraded. Alpha–beta alloys for fracture critical applications are often beta annealed to develop a transformed beta structure. A transformed beta structure produces tortuous crack paths with secondary cracking which enhances fracture toughness and slows fatigue crack growth; however, the equiaxed alpha–beta microstructure has twice the ductility and better fatigue life than the transformed beta microstructure.
(4) *Solution Treated and Aged (STA)*. STA provides maximum strength but full hardenability is limited to sections 1 in. thick or less. This heat treatment is used for mechanical fasteners with a minimum tensile strength of 160 ksi. The STA treatment is not normally used for structural components due to its limited hardenability and warping problems during heat treating.

Ti-6-4 ELI, with a maximum oxygen content of 0.13%, is used for fracture critical structures and for cryogenic applications. Oxygen is a potent strengthening element in Ti-6-4 and must be held to low limits in order to develop high fracture toughness. The commercial grade of Ti-6-4 has an oxygen content of 0.16–0.18% while the ELI grade is limited to 0.10–0.13%. With higher oxygen content, the commercial grade has higher strength and slightly lower ductility, while the ELI grade has about a 25% higher fracture toughness. A comparison of the properties of commercial and ELI Ti-6-4 is given in Fig. 4.7. The beta transus temperature is also influenced by the oxygen content with the

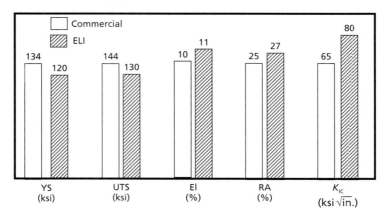

Fig. 4.7. Room Temperature Properties of Commercial vs. ELI Ti-6Al-4V[5]

beta transus being in the range of 1850–1870° F for the commercial grade and 1760–1800° F for the ELI grade.

Ti-6-22-22S (Ti-6Al-2Sn-2Zr-2Mo-2Cr-0.25Si) is similar to Ti-6-4 but is a higher strength alloy with a tensile strength of 150 ksi in the MA condition and 170 ksi in the STA condition. Ti-6-22-22S[2] was originally developed as a deep-hardening high strength alloy for moderate service temperatures. Due to its higher alloying content, it has better room and elevated temperature static and fatigue strength than Ti-6-4. It can be given a triplex heat treatment to maximize damage tolerant properties. Typical properties are a tensile strength of 150 ksi with a minimum fracture toughness of $K_{IC} = 70$ ksi $\sqrt{in}$. The heat treatment consists of solution treating above the beta transus followed by a solution treatment below the beta transus and then a lower temperature aging treatment. This alloy also exhibits excellent SPF characteristics and can be formed at temperatures lower than that for Ti-6-4 and yet have higher strengths.

SP-700 (Ti-4.5Al-3V-2Mo-2Fe) was developed as a lower temperature SPF alloy which can be formed at temperatures as low as 1300° F compared to around 1650° F for Ti-6-4. This reduces energy costs, lengthens die life, and reduces the amount of alpha case that must be removed after part fabrication. Since alpha case is an oxygen-enriched surface layer that is extremely brittle and causes a significant reduction in fatigue life, it must be removed by either chemical milling or machining. Ti-3Al-2.5V is a lean version of Ti-6-4 that is often used for aircraft hydraulic tubing in sizes ranging from 0.25 to 1.5 in. diameter and for honeycomb core. Ti-6246 (Ti-6Al-2Sn-4Zr-6Mo) and Ti-17 (Ti-5Al-2Sn-2Zr-4Mo-4Cr) are engine alloys that are used up to 600 and 750° F respectively, where higher strengths than Ti-6-4 are needed. Where higher temperature creep resistance is required, the near-alpha alloy Ti-6242S is normally specified.

4.2.4 Beta Alloys

Beta alloys contain high percentages of the BCC beta phase that greatly increases their response to heat treatment, provides higher ductility in the annealed condition, and provides much better formability than the alpha or alpha–beta alloys. In general, they exhibit good weldability, high fracture toughness, and a good fatigue crack growth rate; however, they are limited to about 700° F due to creep.

The beta alloys, including Ti-10-2-3 (Ti-10V-2Fe-3Al), Ti-15-3 (Ti-15V-3Al-3Cr-3Sn), and Beta 21S (Ti-15Mo-3Al-2.7Nb-0.25Si), are high strength alloys that can be heat treated to tensile strength levels approaching 200 ksi. In general, they are highly resistant to stress corrosion cracking. The fatigue strength of these alloys depends on the specific alloy; for example, the fatigue strength is of Ti-10-2-3 is very good but that of Ti-15-3 is not so good. The beta alloys possess fair cold formability which can eliminate some of the hot forming operations normally required for the alpha–beta alloys.

Ti-10-2-3 is the mostly widely used of the beta alloys in airframes. Ti-10-2-3 is a popular forging alloy because it can be forged at relatively low temperatures, offering flexibility in die materials and forging advantages for some shapes. It is used extensively in the main landing gear of the Boeing 777.[6] It is used at three different tensile strength levels; 140, 160, and 170 ksi. At the 170 ksi level, it has a fairly narrow processing window to meet the strength, ductility, and toughness requirements. The lower strength levels have wider processing windows and improved toughness. It exhibits excellent fatigue strength but only moderate fatigue crack growth rates. Although Ti-10-2-3 is not as deep hardening or as fatigue resistant as Ti-6-22-22S, it is capable of being heat treated to higher strengths than Ti-6-22-22S. A Russian near-beta alloy, Ti-5-5-5-3 (Ti-5Al-5V-5Mo-3Cr), is receiving quite a bit of consideration for new applications. Forged bar that has been STA heat treated has a tensile strength of 175 ksi, a yield strength of 162 ksi, and an elongation of 10%. Although the fracture toughness is somewhat lower than comparatively processed Ti-10-2-3, it is not a concern for the applications being considered. Although primarily a wrought alloy, it has also been evaluated for investment castings with promising results.

Heat treated Beta C (Ti-3Al-8V-6Cr-4Mo-4Zr) is often used for springs. Ti-15-3 can also be used for springs when it is heat treated to a tensile strength of 150 ksi. An advantage of Ti-15-3 is the ability to cold form the material in thin gages and then STA to high strengths. Ti-15-3 can also be used for castings where higher strength (UTS = 165 ksi) than Ti-6-4(UTS = 120 ksi) is required. Beta 21S (Ti-15Mo-3Al-2.7Nb-0.25Si) was originally developed as an oxidation resistant alloy for high temperature metal matrix composites for the National Aerospace Plane. Even though it is a beta alloy, it has reasonable creep properties, better than Ti-6-4. It can be heat treated to a tensile strength of 125 ksi for temperature uses between 900 and 1050° F or to higher strength levels (UTS = 150 ksi) for lower temperature usage. One of the distinguishing features of Beta 21S is its resistance to hydraulic fluids, presumably due to the synergistic effects of molybdenum and niobium.

4.3 Melting and Primary Fabrication

Titanium can be extracted from rutile (TiO_2) ore by chemical reduction, using either the Kroll process, which uses magnesium, or by the Hunter process, which uses sodium. Tickle ($TiCl_4$) is first produced by reacting TiO_2 with coke and chlorine. Generation of $TiCl_4$ allows purification of the titanium by distillation. Then, in the Kroll process, $TiCl_4$ is reacted with magnesium to form a porous mass of titanium called sponge.

$$TiO_2 + 2Cl_2 + C \rightarrow TiCl_4 + CO_2$$
$$TiCl_4 + 2Mg \rightarrow Ti + 2MgCl_2$$

Both of these processes, while capable of producing high grade titanium sponge, are expensive and significantly contribute to the cost of titanium. However, recent research[7] shows great promise in reducing the cost of titanium extraction. The product flow for various titanium product forms is shown in Fig. 4.8.

Due to their high reactivity with refractory lined vacuum induction furnaces, titanium alloys are made by consumable vacuum arc melting.[8] Vacuum arc melting, actually called vacuum arc remelting (VAR) even though it is the initial melting operation, is used to remove hydrogen and other volatiles and to reduce alloy segregation. An electrode is made by blending small diameter (less than 0.4 in.) sponge particles with master alloy granules of the same size. Master alloy is a pre-prepared mixture that contains the alloying elements at a much higher concentration than the final alloy composition, which helps to reduce segregation and achieve the correct alloy chemistry. The blended sponge and master alloy are pressed together into briquettes and then welded together to form the electrode. Recycled titanium, or revert, can also be added to reduce costs if permitted by the specification. A typical electrode size is 14 ft long by 2 ft in diameter. The minimization of residual elements that could form refractory or high density inclusions is important in both the sponge and the electrode fabrication processes. As shown in Fig. 4.9, during melting, the electrode is the anode and the water cooled copper crucible is the cathode, i.e. the titanium electrode is consumably arc remelted in a water cooled copper crucible in a vacuum arc melt furnace. The ingot produced by this process becomes the starting electrode for additional VAR. The ingot is then VAR once or twice more or remelted using either a plasma torch or electron beam in a cold hearth furnace.

Titanium

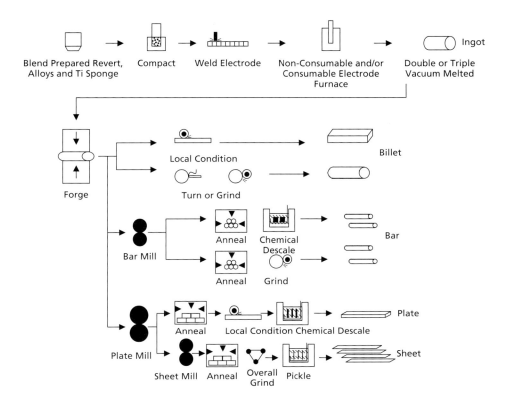

Fig. 4.8. Titanium Product Flow

Double and triple melting practices are used to help eliminate segregation. Ingots range from 26 to 36 in. in diameter and weigh between 8000 and 15 000 lb.

A relatively new method, called cold hearth melting, is conducted in a water cooled copper hearth. By balancing the heat input from a plasma or electron beam with the heat extracted by water cooling, the melt can be held in the hearth for any period of time with only a solid layer of titanium, called a skull, separating it from the copper hearth. Since this method allows the titanium to remain molten for longer periods of time, it has the advantage of being able to reduce alloy segregation in the final product. Cold hearth melting has proven very successful in reducing the incidence of heavy metal high density inclusions because they sink to the bottom and become trapped in the mushy skull. A final VAR step is still required to improve the chemical homogeneity and the surface condition. Cold hearth melting, in combination with VAR, is now being used for fatigue critical jet engine components. Alloys produced by the VAR in combination with the cold hearth process have proven to be essentially free of melt-related inclusions.[9]

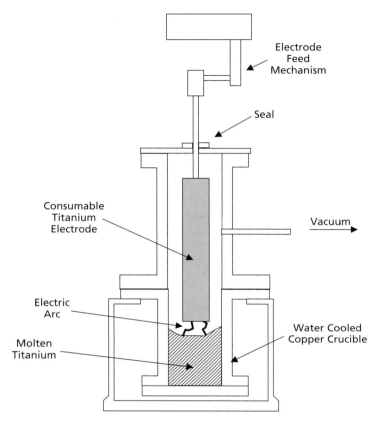

Fig. 4.9. Vacuum Arc Melting of Titanium Ingots

The majority of ingot defects that cause structural concerns are a result of segregation of alloying elements. Low density inclusions (LDI), hard-α, or α-I inclusions are interstitial defects resulting from high local concentrations of nitrogen in the original sponge that stabilizes the alpha phase and creates local hard spots with low ductility, usually associated with cracks that form during thermomechanical processing. These defects, Fig. 4.10, are hard brittle inclusions containing as much as 10 wt% nitrogen, typically in the form of TiN[3]. Hard-α is essentially an incipient crack that will propagate under stress, which adversely affects the fatigue strength of titanium alloys and has resulted in in-service failures.[10]

Another type of defect is high density inclusions (HDI), which can result from contamination introduced during the electrode preparation process. HDIs can be introduced by heavy metal contamination from sources such as tungsten carbide cutting tool edges or tungsten welding electrode tips inadvertently mixed with recycled machining chips. Since HDIs are essentially non-deforming at the stress

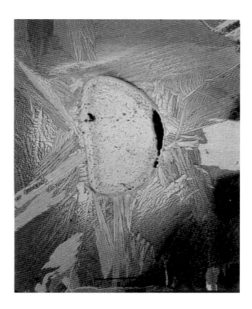

Fig. 4.10. Hard Alpha Inclusion

levels encountered in titanium alloys, they are sources of strain incompatibility that leads to fatigue cracking. Strict management of chips and turnings, including 100% radiographic inspection, has reduced the incidence of HDIs to relatively low levels.

Beta flecks, caused by solute segregation during solidification of the ingot, are regions of microsegregation that have a lower concentration of alpha than the surrounding matrix and are therefore high in beta stabilized material. They are most prevalent in large diameter ingots and with alloys that contain strong beta stabilizing elements, particularly the beta eutectoid elements chromium, iron, and copper. Beta flecks have different mechanical properties than the surrounding matrix and can be either stronger or weaker than the matrix. During cyclic loading, these regions cause steep local strain gradients and can cause early crack initiation.[11] It should be noted that titanium producers go to great lengths to control the entire process to insure high quality ingots, including clean starting materials and a high level of process control throughout the entire process. All as-cast ingots are carefully ultrasonically inspected for the presence of defects.

Titanium alloys are available in most mill product forms: billet, bar, plate, sheet, strip, foil, extrusions, wire, and tubing; however, not all alloys are available in all product forms. Primary fabrication includes the operations performed at the mill to convert ingot into products. Besides producing these shapes, primary fabrication hot working is used to refine the grain size, produce a uniform micro structure, and reduce segregation. It has long been recognized that these initial hot working operations will significantly affect the properties of the final product.

Prior to thermomechanical processing, the as-cast ingot is conditioned by grinding to remove surface defects. The first step in deforming as-cast ingots is a series of slow speed steps including upset forging, side pressing, and press cogging to help homogenize the structure and break-up the transformed beta structure. Cogging is a simple open die forging process between flat dies conducted in slow speed machines such as a hydraulic press. The ingot is fed through the press in a series of short bites that reduces the cross-sectional area.[12] Electrically heated air furnaces are used to preheat the ingot to 1300–1400° F and then it is forged at 1700–2150° F in large presses so that the deformation can be applied slowly to avoid cracking. These operations are initially done above the beta transus but significant amounts of deformation are also done below the beta transus, but high in the alpha + beta field, to further refine the microstructure to a fine equiaxed alpha–beta structure while avoiding surface rupturing. Working as the temperature falls through the beta transus is also an effective way of eliminating grain boundary alpha which has an adverse effect on fatigue strength. Final hot working must be carried out in the alpha + beta field to develop a microstructure that has better ductility and fatigue properties than if all of the hot working were conducted above the beta transus. In alpha–beta alloys, slow cooling from above the beta transus must be avoided or alpha will precipitate at the prior beta grain boundaries leading to a decrease in strength and ductility.

To obtain an equiaxed structure in near-alpha and alpha–beta alloys, the structure is sufficiently worked to break-up the lamellar structure and then annealed to cause recrystallization of the deformed structure into an equiaxed structure. The equiaxed structure obtained is a function of the prior microstructure, the temperature of deformation, the type of deformation, the extent of deformation, the rate of deformation, and the annealing temperature and time. The most important variable is to obtain sufficient deformation to cause recrystallization. In general, the finer the initial microstructure and the lower the deformation temperature (i.e., greater percent of cold work), the more efficient is the deformation in causing recrystallization. After forging, billets and bars are straightened, annealed, finished by turning or surface grinding to remove surface defects and alpha case, and ultrasonically inspected.

Slabs from the forging operations are hot rolled into plate and sheet products using two-and three-high mills. For thin sheet, pack rolling is often used to maintain the temperatures required for rolling. Four or five sheets are coated with parting agent and sandwiched together during rolling. Cross-rolling can be used to reduce the texture affect in plate and sheet. Specific hot rolling procedures for bar, plate, and sheet are proprietary to the individual producers; however, typical hot rolling temperatures for Ti-6-4 are 1750–1850° F for bar, 1700–1800° F for plate, and 1650–1700° F for sheet. Typical finishing operations for hot rolled material are annealing, descaling in a hot caustic bath, straightening, grinding, pickling, and ultrasonic inspection.

4.4 Forging[4]

Titanium alloys are difficult to forge but not as difficult as refractory metals and superalloys. Titanium alloy forgings can be produced by a wide variety of forging processes including both open and closed die methods. In addition to providing a structural shape, the forging process improves the mechanical properties. Tensile strength, fatigue strength, fracture toughness, and creep resistance can all be improved by forging. The forging process selected depends on the shape to be produced, the cost, and the desired mechanical properties. Often, two or more forging processes are used to produce the final part. For example, open die forging may be conducted prior to closed die forging to conserve material and produce more grain flow. Since titanium alloys are quite a bit more difficult to forge than either aluminum or steel, the final forging will normally have more stock that will have to be removed by machining.

The forging conditions for titanium alloys have a greater effect on the final microstructure and mechanical properties than for other metals such as steel or aluminum. While forging of many metals is primarily a shaping process, forging of titanium alloys is used to produce specific shapes but is also a method used to control the microstructure. To help control microstructure and minimize contamination, titanium is forged at only about 60–70% of its melting point, while steels are normally forged at 80–90% of their melting points and nickel alloys are forged at 85–95% of their melting points. Whether the part is forged above or below the beta transus temperature will have a pronounced affect on both the forging process and the resultant microstructure and mechanical properties. The properties of the alpha–beta alloys can be varied significantly depending on whether they are forged above or below the beta transus as shown in Table 4.2. It should be noted that while beta forging can be used to increase the fracture toughness and fatigue crack growth rate, there is a significant penalty

Table 4.2 Effect of Beta Forging on Ti-6Al-4V[9]

Property	Beta Forging
Strength	Lower
Ductility	Lower
Fracture Toughness	Higher
Fatigue Life	Lower
Fatigue Crack Growth Rate	Lower
Creep Strength	Higher
Aqueous Stress Corrosion Cracking Resistance	Higher
Hot Salt Stress Corrosion Cracking Resistance	Lower

on fatigue strength; as much as a 50% reduction compared to alpha + beta processed material. Alpha–beta forging is used to develop optimal combinations of strength and ductility and optimal fatigue strength. Forging in the two-phase alpha + beta field is a process in which all or most of the deformation is conducted below the beta transus. The resultant microstructure will contain either deformed or equiaxed alpha and transformed beta. For alpha–beta forging, the starting forging temperature is usually 50° F lower than the beta transus. This is not as simple as it may seem because small compositional changes will affect the actual beta transus temperature. For example, if Ti-6-4 ELI material is forged at the same temperature as the standard grade Ti-6-4, microstructural defects can occur because the beta transus can be as much as 50–110° F lower for the ELI grade.[13] Since the deformation history and forging parameters have a significant impact on both the microstructure and the mechanical properties of the finished part, close control over the forging process and any subsequent heat treatment is required to obtain a satisfactory product.

Hot worked billet and bar are the primary product forms used as starting stock for forging operations. Titanium alloys, and in particular the beta and near-beta alloys, are highly strain rate sensitive in deformation processes such as forging. Therefore, relatively slow strain rates are used to reduce the resistance to deformation. Open die forging is often used when the number of parts does not warrant the investment in closed dies; however, the majority of titanium parts are produced as closed die forgings. Closed die forging includes blocker forging with a single die set, conventional forging with two or more die sets, high definition forging with two or more die sets, and precision forging with two or more die sets under isothermal conditions. Conventional closed die forgings are more expensive than blocker forgings but yield better properties and reduce machining costs.

The dies are heated to reduce the pressure required for forging and to reduce surface chilling that can lead to inadequate die filling and surface cracking. Conventional alpha–beta forging is conducted with die temperatures in the range 400–500° F for hammer forging and up to 900° F for press forging. Typical die materials are H11 and H12 die steels. The billets are normally coated with a glass to provide surface contamination protection and to act as lubricant. Lubricants are also applied to the dies to prevent sticking and galling. In conventional forging, the temperature is allowed to fall during the forging process, while in isothermal forging, the dies and part are kept at a constant forging temperature. The pressure requirements for forging are dependent on the specific alloy composition, the forging process being used, the forging temperature range, and the strain rate of the deformation. In general, the pressures are somewhat higher than those required for steel.

During alpha–beta forging, defects such as wedge cracks and cavities, known as strain-induced porosity (SIP), can develop as a result of the coarse beta structure previously developed during ingot breakdown. Wedge cracks are microcracks that develop at the beta grain boundary triple points, while cavitation is microvoiding that forms along the grain boundaries. Slow strain rates and high temperatures help to reduce these defects. The presence of grain boundary alpha appears to promote these defects. Non-uniform deformation, which can result from die chilling and/or excessive friction, can lead to bulging during upsetting/pancake forging of cylindrical preforms or ingots.[14] Shear banding and cracking have been observed in both conventional and isothermal forgings.

In beta forging, at least part of the forging process will be conducted above the beta transus. As shown in Fig. 4.11, the percentage of beta phase increases as the alloy is heated toward the beta transus. Normally, the first part of the forging is done above the beta transus followed by some final deformation below the transus. For beta forging, the starting temperatures are about 75°F higher than the beta transus. Due to the higher temperatures employed, lower pressures can be used and there is less tendency for surface cracking. However, the higher starting temperatures used in beta forging leads to more scale and oxygen enriched alpha case. It is important to continue to forge the alloy as it cools through the beta transus to avoid the formation of grain boundary alpha.

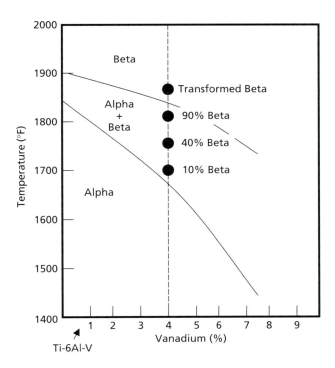

Fig. 4.11. Phase Transformations on Heating Through Alpha + Beta Region[4]

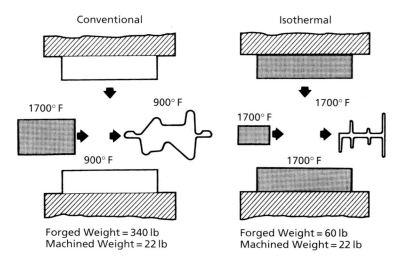

Fig. 4.12. Comparison of Conventional and Isothermal Forging

Forging through the beta transus results in continuous recrystallization of the beta phase with little or no grain boundary alpha formation.

Hot die and isothermal forging are near net shape processes in which the dies are maintained at significantly higher temperatures than in conventional forging. This reduces die chill and increases metal flow. Isothermal forging, as shown in Fig. 4.12, is done at a constant temperature where the flow stress is constant resulting in more uniform microstructures and less property variation. Isothermal forging requires less die pressure and helps insure that the dies are filled during forging. However, expensive high temperature die materials, such as molybdenum alloys, and vacuum or inert atmospheres are required to prevent excessive die oxidation. While conventional die forging can require two or three separate operations, isothermal forging can often be accomplished in a single operation. Like so many other advanced manufacturing processes, forging has greatly benefited from automated process control and computer modeling.[15]

4.5 Directed Metal Deposition[16]

Directed metal deposition, also known as laser powder deposition, laser direct manufacturing, and electron beam free form fabrication, is a rather recent development that can help to reduce the cost of titanium parts. A focused laser beam, shown in Fig. 4.13, is used to melt titanium powder and deposit the melt in a predetermined path on a titanium substrate plate. The metal deposited preform is then machined to the final part shape. This near net process leads to

Titanium

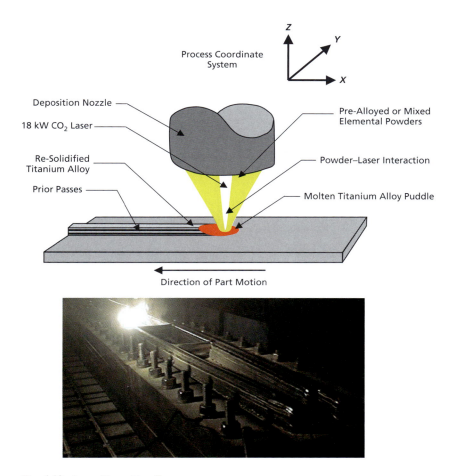

Fig. 4.13. Laser Deposition Process
Source: Aeromet

savings in materials, machining costs, and cycle times over conventional forged or machined parts.

Laser forming is conducted in a chamber that is constantly being purged with high purity argon to prevent atmospheric contamination, such as the one shown in Fig. 4.14. CAD files are used to generate the desired trajectory paths. The trajectory paths are then transmitted as machine instructions to the system which contains a high power CO_2 laser heating source. The laser beam traces out the desired part by moving the titanium substrate plate beneath the beam in the appropriate x–y trajectories. Titanium pre-alloyed powder is introduced into the molten metal puddle, and provides for build-up of the desired shape, as the laser is traversed over the target plate. A 3-D preform is fabricated by repeating

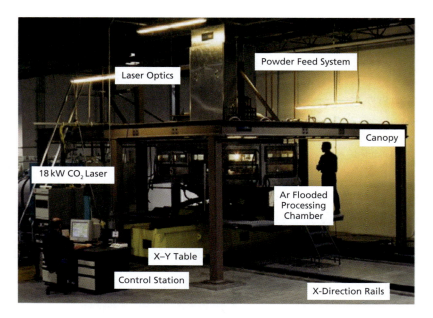

Fig. 4.14. Laser Deposition System
Source: Aeromet

the pattern, layer-by-layer, over the desired geometry and indexing the focal point up one layer for each repeat pattern. Preforms generally are deposited with 0.030–0.20 in. of excess material which is removed by final machining. The process is capable of depositing between two and ten pounds of material per hour, depending on part complexity.[17] Preforms have been laser formed from a number of titanium alloys, including Ti-6Al-4V, Ti-5-2.5, Ti-6242S, Ti-6-22-22S, and CP Ti. Depending on the exact process used, the static properties will be 0–5 ksi lower than equivalent wrought properties. The fatigue properties are essentially equivalent to wrought material; however, some processes require HIP after forming to obtain equivalence by closing internal pores.

Potential savings with this technology include much better material utilization. While the buy-to-fly ratio for forgings is in the range of 5 to 1 for simple shapes and 20 to 1 for complex shapes,[18] the ratio for laser forming is about 3 to 1 which includes the substrate that forms 2/3 of the as-deposited part. Machining times are reduced by as much as 30%. In addition, there are no expensive dies or tools required. Finally, the part lead times are much shorter; 12–18 months for forgings versus only several months for laser formed preforms. This process can also be used to conduct local repairs on damaged or worn parts. Some of the part features that can be achieved are shown in Fig. 4.15, and a typical rib section is shown in Fig. 4.16.

Titanium

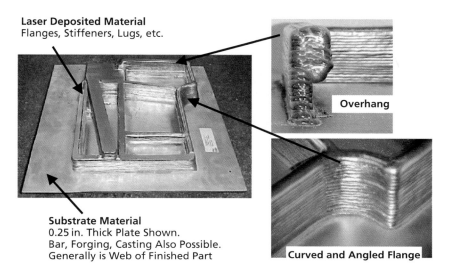

Fig. 4.15. Laser Deposited Part Features

4.6 Forming[4]

The yield strength-to-modulus ratio in titanium results in a significant amount of springback, as much as 15–25° after cold forming. To compensate for springback, titanium is normally overformed or hot sized after cold forming. Besides springback, cold forming takes more force, requires stress relieving between forming operations, and must be done at slow forming speeds to prevent cracking. Strain hardening also increases the yield and tensile strengths while causing a slight drop in ductility. Cold formed titanium alloys also experience a decrease in the compressive yield strength in one direction when a tensile strain is applied in the other direction, a phenomena known as the Bauschinger effect. For example, a 2% tensile strain applied to solution treated Ti-6-4 causes the compressive yield strength to drop to less than half; however, a full stress relief will restore the properties as shown in Fig. 4.17.

Except for thin gages, near-alpha and alpha–beta alloys are usually hot formed. Beta alloys, which contain the BCC structure, are much amendable to cold forming and one of them (Ti-15-3), along with the CP grades, can be successfully formed at room temperature in sheet form. In general, the bends must be of a larger radii than in hot forming. Titanium can also be stretched formed at room temperature at slow speeds with dies heated to around 300° F. Cold forming is usually followed by hot sizing and stress relieving to reduce residual stresses, restore the compressive yield strength, improve dimensional accuracy, and make the part more resistant to delayed cracking. During hot sizing, the part is held in fixtures or dies to prevent distortion. All sheet products that are going to

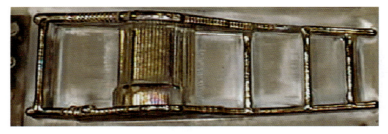

Deposited Preform

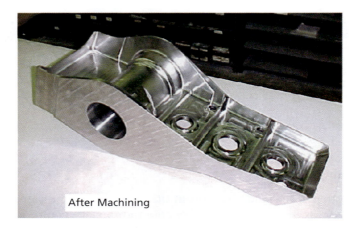

After Machining

Fig. 4.16. Titanium Laser Deposited Rib Section
Source: The Boeing Company

be formed should be free of scratches, and gouges and burrs and sharp edges should be filed smooth to prevent edge cracking.

Hot forming, conducted at 1100–1500° F, greatly improves formability, reduces springback, and eliminates the need for stress relieving; however, the requirement to hot form increases the cost of titanium structures by increasing the cost by having to heat the material; the need for more expensive tools capable of withstanding the temperature; and the requirement to remove surface contamination (i.e., alpha case) after forming. In addition, the material needs to be free of all grease and any residue that could cause stress corrosion cracking. Severe forming operations are done in hot dies with preheated stock. Due to the tendency of titanium to gall, forming lubricants containing graphite or molybdenum sulfide will reduce both part and tool damage, especially for severe forming operations such as drawing. Since forming is usually done above 1200° F, alpha case will form on the surface of unprotected titanium and must be removed by either machining or chemical milling. Although forming can be done in a

Titanium

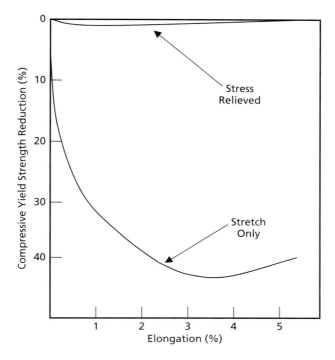

Fig. 4.17. Bauschinger Effect on Ti-6Al-4V

vacuum, under a protective atmosphere, or with oxidation resistant coatings, the normal practice is to conduct the forming in air and then remove the alpha case by chemical milling.

Vacuum, or creep, forming can also be used to form sheet and plate to mild contours. In this process, the plate is placed on a ceramic tool that contains integral heaters. Insulation is then placed over the sheet followed by a silicone rubber vacuum bag. The part is slowly heated to temperature while a vacuum is applied to the bag. The part slowly creeps to shape under the combined influence of heat and pressure.

4.7 Superplastic Forming

The advantages of superplastic forming include the ability to make part shapes not possible with conventional forming, reduced forming stresses, improved formability with essentially no springback and reduced machining costs. The mechanism of superplasticity was covered in Chapter 2 on Aluminum. In general, titanium alloys exhibit much higher superplastic elongations than aluminum alloys, and there are a much wider variety of titanium alloys that exhibit superplasticity. In addition, for titanium, SPF can be combined with diffusion bonding

(DB) to make large one piece unitized structures. Since superplasticity depends on microstructure, the fine grain equiaxed two-phase alpha–beta alloys exhibit inherent grain stability and are therefore resistant to grain growth at elevated temperature. Although the optimum structure depends on the specific alloy, the optimum volume fraction of beta is about 20% in Ti-6-4. While internal void formation by cavitation is a concern with aluminum and some other alloys, it has not been a problem in titanium alloys; therefore, the additional complication of back pressure to suppress cavitation is not required. Superplastic titanium alloys include Ti-6-4, Ti-6242S, Ti-6-22-22S, Ti-3-2.5, Ti-8-1-1, Ti-1100, Timetal 550, and SP700.

In the single-sheet SPF process, illustrated in Fig. 4.18, a single sheet of metal is sealed around its periphery between an upper and lower die. The lower die is either machined to the desired part shape or a die inset is placed in the lower die box. The dies and sheet are heated to the SPF temperature, and argon gas pressure is used to slowly form the sheet down over the tool. The lower cavity is maintained under vacuum to prevent atmospheric contamination. After the sheet is heated to its superplastic temperature range, argon gas is injected through inlets in the upper die. This pressurizes the cavity above the metal sheet forcing it to superplastically form to the shape of the lower die. Gas pressurization is slowly applied so that the strains in the sheet are maintained in the superplastic

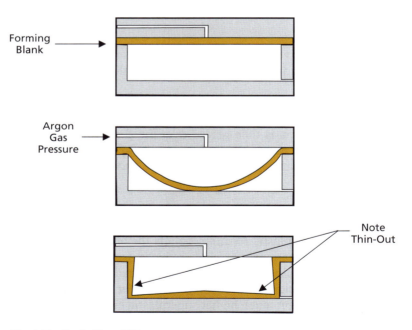

Fig. 4.18. Single Sheet SPF

range. Typical forming cycles for Ti-6-4 are 100–200 psi at 1650–1750° F for 30 min to 4 h.

During the forming operation, the metal sheet is being reduced uniformly in thickness; however, where the sheet makes contact with the die it sticks and no longer thins-out. Therefore, if the sheet is formed down over a male die, it will touch the top of the die first and this area will be the thickest. The thickness tapers down along the sides of the die to its thinnest point in the bottom corners which are formed last. For the same reason, when a sheet is formed into a female cavity, the first areas that make contact are the center of the bottom and the top of the sides. These areas are the thickest. The thickness tapers down the sides to the thinnest point in the bottom corners which again form last. To reduce these variations in thickness, overlay forming can be used.

In overlay forming, the sheet that will become the final part is cut smaller than the tool periphery. A sacrificial overlay sheet is then placed on top of it and clamped to the tool periphery. As gas is injected into the upper die cavity, the overlay sheet forms down over the lower die, forming the part blank simultaneously. While overlay forming does help to minimize thickness variations, it requires a sacrificial sheet for each run that is discarded. Dies for titanium SPF are high temperature steels such as ESCO 49C (Fe-22Cr-4Ni-9Mn-5Co) lubricated with either boron nitride or yttria.

For titanium alloys, superplastic forming (SPF) can be combined with diffusion bonding (DB), a processes known as superplastic forming/diffusion bonding (SPF/DB) to form one piece unitized structure. Titanium is very amenable to DB because the thin protective oxide layer (TiO_2) dissolves into the titanium above 1150° F leaving a clean surface. Several processes have been developed including two-, three-, and four-sheet processes.

In the two-sheet process, shown in Fig. 4.19, two sheets are welded around the periphery to form a closed envelope. The sheets can be welded by either resistance seam welding or laser welding; what is important is that the weld joints are vacuum tight and capable of resisting up to 200 psi gas pressure. The welded pack is placed in the die and heated to the forming temperature. Argon gas is used to form the lower sheet into the corrugated die cavity. Once the lower sheet is formed, the bladder below the lower die is inflated to push the lower skin up against the upper skin and the two are diffusion bonded together. To prevent the two sheets from sticking together prior to forming of the lower sheet, stop-off agents such as boron nitride or yttria suspended in an acrylic binder can be used, or a slight positive pressure can be used initially to keep the two sheets separate. The inner moldline of a two-sheet door is shown in Fig. 4.20, illustrating the stiffeners that were formed in the lower die during the process. This SPF/DB door replaced a built-up structure that consisted of two skins, seven formers, six intercostals, ten miscellaneous pieces, and several hundred fasteners.

The three-sheet process is shown in Fig. 4.21. The three sheets are welded around the periphery; however, in this process selected areas of the center sheet

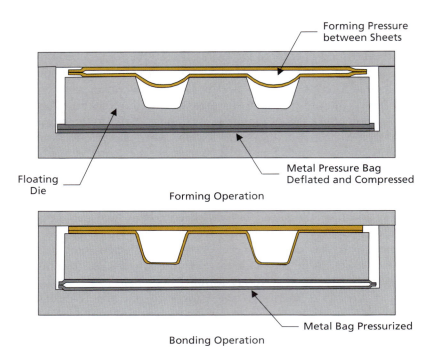

Fig. 4.19. Two-Sheet SPF/DB

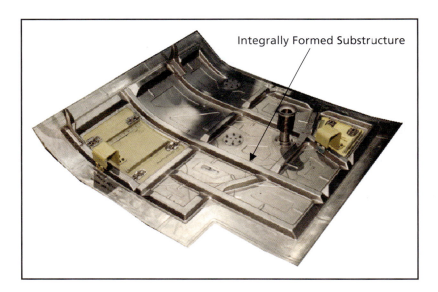

Fig. 4.20. Two-Sheet SPF/DB Aircraft Door
Source: The Boeing Company

Titanium

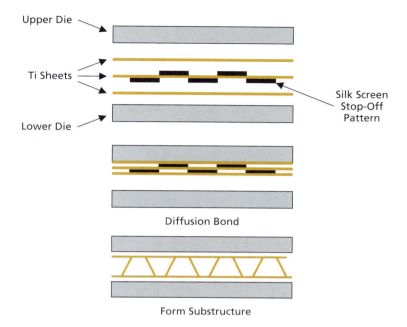

Fig. 4.21. Three-Sheet SPF/DB Process

are masked with a stop-off material to prevent bonding. The pack is then placed between the dies and the sheets are diffusion bonded together except at the locations that have been masked. After DB, gas pressure is applied to each side of the center sheet to expand the substructure. The areas of the center sheet that were masked stretch between the top and bottom sheets to form the stiffening ribs.

The four-sheet process,[19] shown in Fig. 4.22, utilizes two sheets that form the skin (skin pack) and two sheets that form the substructure (core pack). The two core pack skins are first selectively welded together in a pattern that will form the substructure. The two skins for the skin pack are then welded to the periphery of the core pack. Although either resistance seam welding or laser welding can be used, one advantage of laser welding is that an automated laser welder can be programmed to make weld patterns and resulting core geometries other than rectangular, possibly resulting in more structurally efficient substructure designs. The total pack is then placed in the die, heated to the SPF temperature, the face sheets are expanded against the tool to form the skins, and then the core pack is expanded to form the substructure. The process is not limited to four sheets; experimental structures with five and more sheets have been designed and fabricated.[20] A distinct advantage of both the three- and four-sheet processes is that the tooling is much simpler; instead of having cavities for substructure as

149

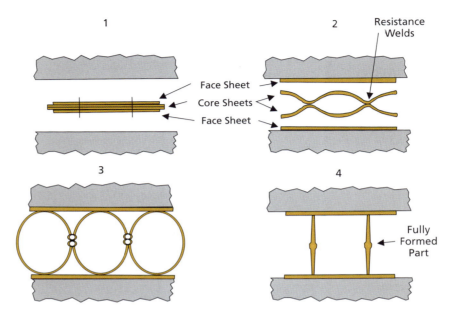

Fig. 4.22. Four-Sheet SPF/DB

in the two-sheet process, the tools have the more gentle curvature of the inner and outer part moldlines. The completed parts are pickled to remove any alpha case that may have formed at elevated temperature.

Superplastic forming and SPF/DB of titanium has been successfully used on a number of aircraft programs. An example is the F-15E, shown in Fig. 4.23, in which these processes allowed the elimination of over 726 detail parts and over 10 000 fasteners.

4.8 Heat Treating

Heat treatments for titanium alloys include stress relieving, annealing, and solution treating and aging (STA). All titanium alloys can be stress relieved and annealed, but only the alpha–beta and beta alloys can be STA to increase their strength. Since alpha and near-alpha alloys do not undergo a phase change during processing, they cannot be strengthened by STA. Since the response to STA is determined by the amount of beta phase, the beta alloys, with higher percentages of beta phase, are heat treatable to higher strength levels than the alpha–beta alloys. In addition, the beta alloys can be through-hardened to thicker sections than the alpha–beta alloys.

One great disadvantage of heat treating titanium alloys is the lack of a nondestructive test method to determine the actual response to heat treatment. While

Titanium

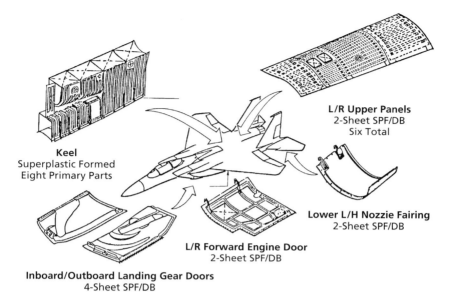

Fig. 4.23. SPF and SPF/DB Applications on F-15E
Source: The Boeing Company

hardness may be used with steel alloys and a combination of hardness and conductivity for aluminum alloys, there are no equivalent methods for titanium. Therefore, if verification is required, actual mechanical property tests have to be conducted to determine heat treat response.

4.8.1 Stress Relief

Stress relief is used to remove residual stresses that result from mechanical working, welding, cooling of castings, machining, and heat treatment. Stress relief cycles may be omitted if the part is going to be annealed or solution treated. All titanium alloys can be stress relieved without affecting their strength or ductility. Like most thermal processes, combinations of time and temperature may be used with higher temperatures requiring shorter times and lower temperatures requiring longer times. Thicker sections require longer times to insure uniform temperatures. Either furnace or air cooling is usually acceptable. While the cooling rate is not critical, uniformity of cooling is important, especially through the 600–900° F range. The stress relief temperature for the near-alpha and alpha–beta alloys is in the range of 850–1500° F. Ti-6-4 is normally stress relieved at 1000–1200° F with the time being dependent on the temperature used and the percent stress relief desired. At 8 h, 1000, 1100, and 1200° F will result in 55, 75 and

100% stress relief respectively. During stress relieving of STA parts, care has to be taken to prevent overaging to lower than desired strength levels.

4.8.2 Annealing

Annealing is similar to stress relief but is done at higher temperatures to remove almost all residual stresses and the affects of cold work. Common types of annealing operations for titanium alloys include mill annealing, duplex annealing, recrystallization annealing, and beta annealing.

As the name implies, mill annealing is conducted at the mill. It is not a full anneal and may leave traces of cold or warm working in the microstructure of heavily worked products, particularly sheet. For near-alpha and alpha–beta alloys, this is the heat treatment normally supplied by the manufacturer. Mill annealing of Ti-6-4 can be achieved by heating to 1300–1440° F and holding for a minimum of 1 h. Beta alloys are not supplied in the mill annealed condition because this condition is not stable at elevated temperatures and can lead to the precipitation of embrittling phases.

Duplex annealing can be used to provide better creep resistance for high temperature alloys such as Ti-6242S. It is a two-stage annealing process that starts with an anneal high in the alpha + beta field followed by air cooling. The second anneal is conducted at lower temperature to provide thermal stability, again followed by air cooling.

Recrystallization anneals are used to improve the fracture toughness. The part is heated into the upper range of the alpha + beta field, held for a period of time and then slowly cooled.

Beta annealing is conducted by annealing at temperatures above the beta transus followed by air cooling or water quenching to avoid the formation of grain boundary alpha. This treatment maximizes fracture toughness at the expense of a substantial decrease in fatigue strength.

A summary of these different annealing procedures and their effects on properties are given in Table 4.3. With the appropriate use of constraint fixtures, operations such as straightening, sizing, and flattening can be combined with annealing. The elevated temperature stability of alpha–beta alloys is improved by annealing because the beta phase is stabilized.

4.8.3 Solution Treating and Aging

The purpose of the solution treatment is to transform a portion of the alpha phase into beta and then to cool rapidly enough to retain the beta phase at room temperature. During aging, alpha precipitates from the retained beta. STA is used with both alpha–beta and beta alloys to achieve higher strength levels than can be obtained by annealing. Solution treating consists of heating the part to high in the two-phase alpha + beta field followed by quenching. Solution treatments for alpha–beta alloys are conducted by heating to slightly below

Table 4.3 Effects of Different Anneal Cycles on Titanium Properties

	Mill Anneal	Recrystallization Anneal	Duplex Anneal	Beta Anneal
Ultimate Tensile Strength	High	Low	Low	Low
Ductility	High	High	High	Lower
Fatigue Strength	Intermediate	Intermediate	Lower	Lower
Fracture Toughness	Lowest	High	Intermediate	Highest
Fatigue Crack Growth Rate	Lowest	Intermediate	Intermediate	Highest
Creep Resistance	Lowest	Lowest	Intermediate	Highest

Mill Anneal – Roughly 300–450° F below beta transus, air cool.
Recrystallization Anneal – Roughly 50–100° F below beta transus, slow cool.
Duplex Anneal – Roughly 50–100° F below beta transus, air cool followed by mill anneal.
Beta Anneal – Usually 50–100° F above beta transus, air cool.

the beta transus. To obtain the maximum strength with adequate ductility, it is necessary to solution treat within about 50–150° F of the beta transus. When an alpha–beta alloy is solution treated, the ratio of beta phase to alpha phase increases and is maintained during quenching. The effect of solution treating temperature on the strength and ductility of Ti-6-4 sheet is shown in Fig. 4.24. During aging, the unstable retained beta transforms into fine alpha phase which increases the strength.

The cooling rate after solution heat treating has an important effect on the strength of alpha–beta alloys. For most alpha–beta alloys, quenching in water or an equivalent quenchant is required to develop the desired strength levels. The time between removing from the furnace and the initiation of the quench is usually about 7 s for alpha–beta alloys and as long as 20 s for beta alloys. For alloys with appreciable beta stabilizing elements and moderate section thickness, air or fan cooling is usually adequate. Essentially, the amount and type of beta stabilizers in the alloy will determine the depth of hardening. Unless an alloy contains appreciable amounts of beta stabilizers, it will not harden through thick sections and will exhibit lower properties in the center where the cooling rates are lower.

Aging consists of reheating the solution treated part in the range of 800–1200° F. A typical STA cycle for Ti-6-4 would be to solution treat at 1660–1700° F followed by water quenching. Aging would then be conducted at 1000° F for 4 h followed by air cooling. Ti-6-4 is sometimes solution treated and overaged (STOA) to achieve modest decreases in strength while obtaining improved fracture toughness and good dimensional stability.

The solution treatment for beta alloys is carried out above the beta transus. Commercial beta alloys are usually supplied in the solution treated condition with a 100% beta structure to provide maximum formability and only need to be aged to achieve high strength levels. After forming, the part is aged to provide

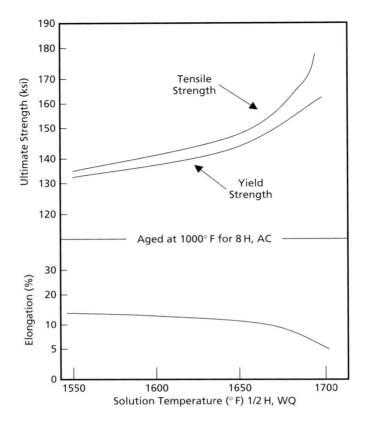

Fig. 4.24. Effect of Solution Temperature on Ti-6Al-4V Sheet

maximum strength. Beta processed alloys have improved fracture toughness, better creep strength, and more resistance to stress corrosion cracking; however, there is a considerable loss in ductility and fatigue strength. Beta alloys are usually air cooled from the solution treating temperature.

Although there are coatings that can be used to protect titanium alloys during heat treatment from oxygen, hydrogen, and nitrogen, the best practice is to conduct the heat treatments in a vacuum furnace. Prior to heat treating, it is important that the surfaces are clean and free of all organic contaminants, including finger prints. After heat treatment, any alpha case must be removed from the surface either by machining or chemical milling.

4.9 Investment Casting

Titanium alloys are difficult to cast due to their high reactivity; they will react with both the atmosphere and the casting mold. However, investment casting

Titanium

Fig. 4.25. Titanium Investment Casting
Source: The Boeing Company

procedures are now available, that in combination with HIP after casting, can produce aerospace quality near net shaped parts that can offer cost savings over forgings and built-up structure.[21] An example of the complexity that can be achieved with investment casting is shown in Fig. 4.25. This casting replaced 22 parts resulting in a significant cost savings.

Although a large number of titanium alloys have been successfully cast, by far the majority of titanium casting is done with Ti-6-4 (about 90% of all castings), with CP titanium representing the majority of the other 10%. It should be noted that there are no specific titanium casting alloys, and the same compositions that are used for wrought products are also used for castings. This is due to the lack of problems, such as fluidity or lower mechanical properties, which have been encountered with the wrought compositions for other metals. The advantages of investment castings over wrought titanium are lower costs for near net shaped complex parts, shorter lead times, and the ability to prototype new parts at reasonable costs. In general, the more complex the part, the better are the economics of using a casting. Rapid prototyping, using processes such as stereolithography, can be used to generate patterns for casting from CAD files.

Cast titanium parts approach the mechanical properties of wrought product forms. Static strength is usually the same while ductility is somewhat lower. Fracture toughness and fatigue crack growth rate are often as good, or better, than for wrought material. While fatigue strength is lower, HIP processing is used to close internal porosity and improve fatigue performance as shown in Fig. 4.26.

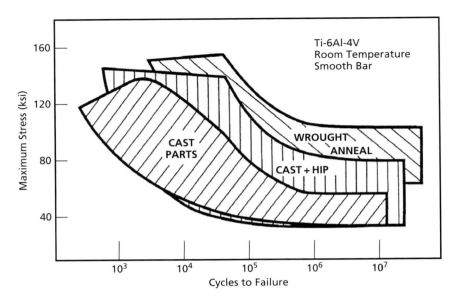

Fig. 4.26. Comparison of Cast and Wrought Fatigue Strength

In the investment casting process,[22] shown in Fig. 4.27, a pattern of the part is produced from wax. The waxes are formulated to give smooth, defect-free surfaces, be stable, maintain tolerances, and have a relatively long shelf life. The wax patterns are then robotically dipped in a fine ceramic slurry that contains refractories such as silica or alumina. The coated patterns are then stuccoed with dry coarser particles of the same material to make the slurry dry faster and insure adhesion between the layers. The dipping and stuccoing process is repeated until the desired thickness is obtained, usually 6–8 times. Once the mold is completely dry, it is placed in an oven and the wax is melted out. The ceramic mold is then fired at about 1800° F. The titanium melt is produced by vacuum arc remelting titanium in a water cooled copper crucible before pouring it into the mold. Sufficient preheat of the melt and preheating of the molds is used to maximize flow to achieve complete mold filling. As-cast titanium has a microstructure typical of titanium alloys worked in the beta field, which has lower ductility and fatigue strength than equiaxed structures. Due to the slow cooling rate from the HIP temperature, titanium castings are often heat treated to refine the microstructure and eliminate grain boundary alpha, large alpha plate colonies, and individual alpha plates.

One of the problems with investment castings has been shell inclusions (Fig. 4.28), which are small pieces of the ceramic shell that flake off during casting and can cause contamination that adversely affects fatigue strength. Very extensive and expensive non-destructive inspection procedures have been

Titanium

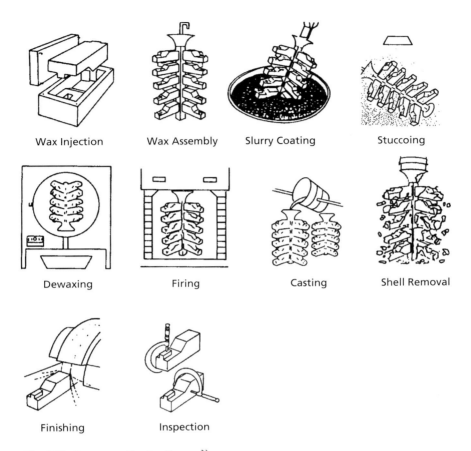

Fig. 4.27. *Investment Casting Process*[23]

developed to detect this defect and others in castings. To eliminate surface contamination, castings are usually chemical milled after casting. Unfortunately, titanium investment castings have a tendency not to completely fill the mold during casting, particularly for large and/or complex castings. Weld repair of surface defects must be carefully done to avoid oxygen and hydrogen pickup. Repairs are usually done using GTAW with filler wire. ELI filler wire is often used to help minimize the potential for oxygen contamination. Weld repaired castings must be stress relieved after welding. Fortunately, test programs have shown that the fatigue properties of castings with properly conducted weld repairs are not degraded.

To improve the fatigue properties, all aerospace grade titanium castings are processed by HIP to close off any internal porosity. Typical HIP processing is done under argon pressure of 15 ksi at 1750° F for 2 h. It should be noted

Fig. 4.28. Shell Inclusion Cross-Section

that HIP will collapse internal porosity but will not close off surface connected porosity, hence the need to repair those areas by welding prior to HIP.

While most Ti-6-4 titanium castings are supplied in the mill annealed condition, some are supplied in the beta-SOTA condition. In this heat treatment, a beta solution heat treatment at 1875° F in a vacuum for 1 h is followed by rapid cooling to room temperature. The castings are then aged at 1550° F for 2 h followed by air cooling. While some aerospace specifications require this heat treatment, a round robin test program conducted by the major aerospace companies in the U.S. showed that it offered no major advantages over a standard mill anneal, i.e. heating to around 1550° F for 2 h followed by air cooling,[24] except in high cycle fatigue (10^8 cycles) as experienced in some helicopter applications.

4.10 Machining[25,26]

> *Use low cutting speeds. . . . maintain high feed rates. . . . use copious amounts of cutting fluid. . . . use sharp tools and replace them at the first signs of wear. . . . moving parts of the machine tool should be free from backlash or torsional vibrations. . . .* Titanium Machining Techniques, Timet Titanium Engineering Bulletin No. 7, 1960s.

Unfortunately, not much has changed in machining titanium since the mid-1960s. Titanium was a difficult-to-machine metal then and remains a difficult-to-machine metal. Titanium is difficult to machine for several reasons:

1. Titanium is very reactive and the chips tend to weld to the tool tip leading to premature tool failure due to edge chipping. Almost all tool materials tend to react chemically with titanium when the temperature exceeds 950° F.[27]

2. Titanium's low thermal conductivity causes heat to build-up at the tool-workpiece interface. High temperatures at the cutting edge is the principal reason for rapid tool wear. When machining Ti-6-4, about 80% of the heat generated is conducted into the tool due to titanium's low thermal conductivity. The thermal conductivity of titanium is about 1/6 that of steel.[27] This should be contrasted with high speed machining of aluminum in which almost all of the heat of machining is ejected with the chip.
3. Titanium's relatively low modulus causes excessive workpiece deflection when machining thin walls, i.e. there is a bouncing action as the cutting edge enters the cut. The low modulus of titanium is a principal cause of chatter during machining operations.
4. Titanium maintains its strength and hardness at elevated temperatures, contributing to cutting tool wear. Very high mechanical stresses occur in the immediate vicinity of the cutting edge when machining titanium. This is another difference between titanium and the high speed machining of aluminum; aluminum becomes very soft under high speed machining conditions.

In addition, improper machining procedures, especially grinding operations, can cause surface damage to the workpiece that will dramatically reduce fatigue life.

The following guidelines are well established for the successful machining of titanium:

1. Use slow cutting speeds. A slow cutting speed minimizes tool edge temperature and prolongs tool life. Tool life is extremely short at high cutting speeds. As speed is reduced, tool life increases.
2. Maintain high feed rates. The depth of cut should be greater than the work hardened layer resulting from the previous cut.
3. Use generous quantities of cutting fluid. Coolant helps in heat transfer, reduces cutting forces, and helps to wash chips away.
4. Maintain sharp tools. As the tool wears, metal builds-up on the cutting edge resulting in a poor surface finish and excessive workpiece deflection.
5. Never stop feeding while the tool and workpiece are in moving contact. Tool dwell causes rapid work hardening and promotes smearing, galling, and seizing.
6. Use rigid setups. Rigidity insures a controlled depth of cut and minimizes part deflection.

Rigid machine tools are required for machining of titanium. Sufficient horsepower must be available to insure that the desired speed can be maintained for given feed rate and depth of cut. Titanium requires about 0.8 horsepower per cubic inch of material removed per minute.[26] The base and frame should be massive enough to resist deflections, and the shafts, gears, bearings, and other moving parts should run smoothly with no backlash, unbalance, or torsional vibrations. Rigid spindles with larger taper holders are recommended; a Number 50 taper or equivalent provides stability and the mass to counter the axial and radial loads encountered when machining titanium.

Cutting tools used for machining titanium include cobalt-containing high speed tool steels, such as M33, M40, and M42, and the straight tungsten carbide grade C-2 (ISO K20). While carbides are more susceptible to chipping during interrupted cutting operations, they can achieve about a 60% improvement in metal removal rates compared to HSS. Ceramic cutting tools have not made inroads in titanium machining due to their reactivity with titanium, low fracture toughness, and poor thermal conductivity. It should be noted that although improvements in cutting tool materials and coatings have resulted in tremendous productivity improvements in machining for a number of materials (e.g., steels), none of these improvements have been successful with titanium.[27]

Cutting fluids are required to achieve adequate cutting tool life in most machining operations. Flood cooling is recommended to help remove heat and act as a lubricant to reduce the cutting forces between the tool and workpiece. A dilute solution of rust inhibitor and/or water soluble oil at 5–10% concentration can be used for higher speed cutting operations, while chlorinated or sulfurized oils can be used for slower speeds and heavier cuts to minimize frictional forces that cause galling and seizing. The use of chlorinated oils requires careful cleaning after machining to remove the possibility of stress corrosion cracking.

For the production of airframe parts, end milling and drilling are the two most important machining processes, while turning and drilling are the most important for jet engine components.[27] In turning operations, carbide tools are recommended for continuous cuts to increase productivity, but for heavy interrupted cutting operations, high speed tool steel tools are needed to resist edge chipping. Tools need to be kept sharp and should be replaced at a wearland of about 0.015 in. for carbide and 0.030 in. for HSS. Tool geometry is important, especially the rake angle. Negative rake angles should be used for rough turning with carbide, while positive rakes are best for semi-finishing and finishing cuts and for all operations using HSS. Typical end mill configurations are shown in Fig. 4.29. Note the improvements in metal removal rates for some of the newer configurations, as compared to standard HSS four flute end mill.

When milling titanium, climb milling, as shown in Fig. 4.30, rather than conventional milling, is recommended to minimize tool chipping, the predominate failure mode in interrupting cutting. In climb milling, the tooth cuts a minimum thickness of chip, minimizing the tendency of the chip to adhere to the tool as it leaves the workpiece. Slow speeds and uniform positive feeds help to reduce tool temperature and wear. Tools should not be allowed to dwell in the cut or rub across the workpiece. This will result in rapid work hardening of the titanium making it even more difficult to cut. Both carbide and HSS cutting tools can be used; however, carbide tools are more susceptible to chipping and may not perform as well in heavy interrupted cutting operations. Increased relief angles

Titanium

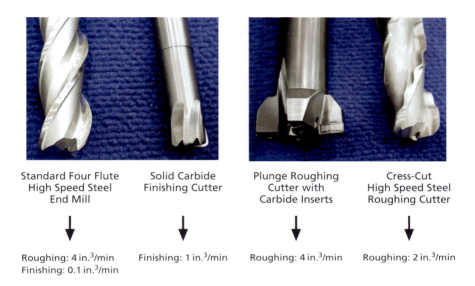

Fig. 4.29. Titanium Machining Improvements Due to Cutter Design

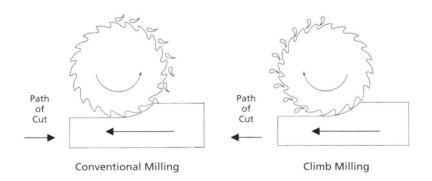

Fig. 4.30. Conventional and Climb Milling

increase tool life by reducing pressure and deflection. Rigid machine tools and set-ups are necessary along with flood cooling. Although few improvements have been realized for machining titanium, recent work[28] shows that productivity improvements can be realized if the effort is expended to scientifically analyze the process and cutting tools, as shown for the roughing and finishing cuts shown in Figures 4-31 and 4-32 respectively. In all, a 41% reduction in machining time was realized for this test bulkhead.

Many of the same guidelines for turning and milling also apply to drilling titanium. During drilling, positive power feed equipment is preferred over hand drilling to reduce operator fatigue and provide more consistent hole quality.

Manufacturing Technology for Aerospace Structural Materials

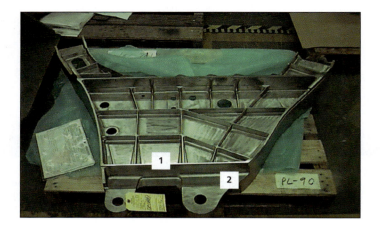

Location 1 Interior Roughening Cut		
	Old Process	New Process
Cutter	2 in. Cobalt Rougher	2 in. Indexable Carbide
RPM	96 RPM	210 RPM
Feed	2.3 IPM	6 IPM
ADOC	0.75 in.	0.50 in.
RDOC	2.0 in.	2.0 in.
MRR	3.45 in.^{3}/min	6.0 in.^{3}/min

Location 2 Exterior Roughing Cut		
	Old Process	New Process
Cutter	2 in. Cobalt Rougher	2 in. Indexable Carbide
RPM	96 RPM	210 RPM
Feed	2.3 IPM	3.0 IPM
ADOC	1.5 in.	3.0 in.
RDOC	1.0 in.	1.5 in.
MRR	3.45 in.^{3}/min	13.5 in.^{3}/min

Fig. 4.31. Comparison of Roughing Cuts[28]

Blunt point angles (140°) work best for small diameter holes (1/4 in. and less), while sharp points (90°) work better for larger diameter holes where higher pressures are needed to feed the drill. Relief angles also affect tool life; too small a relief angle will cause smearing and galling, while too large a relief angle results in edge chipping. When drilling holes deeper than one hole diameter, peck drilling will produce more consistent and higher quality holes. During peck drilling, the drill motor periodically extracts the bit from the hole to clear chips

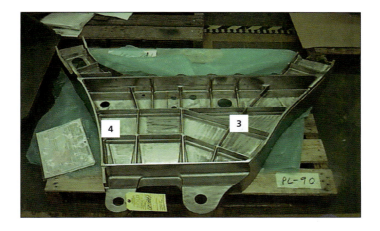

Location 3 Pocket Peripheral Finish Cut		
	Old Process	New Process
Cutter	3/4 in. Cobalt Finisher	3/4 in. Solid Carbide
RPM	250 RPM	2500 RPM
Feed	3.0 IPM	30 IPM
MRR	0.338/0.113 in.^{3}/min	1.125 in.^{3}/min

Location 4 Pocket Peripheral Finish Cut		
	Old Process	New Process
Cutter	3/4 in. Cobalt Finisher	3/4 in. Solid Carbide
RPM	250 RPM	2500 RPM
Feed	3.0 IPM	30 IPM
MRR	0.338/0.113 in.^{3}/min	1.5 in.^{3}/min

Fig. 4.32. Comparison of Finishing Cuts[28]

from the flutes and hole. Since it is not practical to use flood coolant during many assembly drilling operations, drills equipped with air blast cooling have been found to be effective in thin gage sheet.

In all machining operations, and especially grinding, the surface of titanium alloys can be damaged to the extent that it adversely affects fatigue life. Damage, including microcracks, built-up edges, plastic deformation, heat affected zones, and tensile residual stresses, can all result from improper machining operations. As shown in Fig. 4.33, machining operations that induce residual compressive stresses on the surface, such as gentle grinding, milling, and turning, are much less susceptible to fatigue life reductions. Note the devastating results for

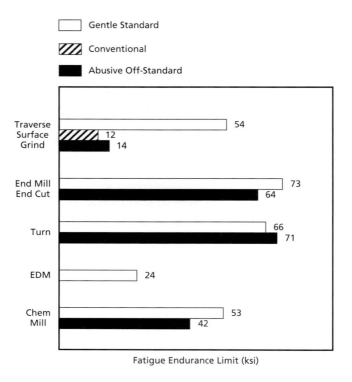

Fig. 4.33. Effects of Machining Conditions on Mill Annealed Ti-6Al-4V[4]

improper grinding procedures. Post-machining processes, such as grit blasting or shot peening, can also be used to induce residual compressive stresses at the surface and improve fatigue life as shown in Fig. 4.34.

Titanium is an extremely reactive metal and fine particles of titanium can ignite and burn; however, the use of flood coolant in most machining operations eliminates this danger to a large extent. Chips should be collected, placed in covered steel containers for recycling, and preferably stored outside of the building. In the event of a titanium fire, water should never be applied directly to the fire; it will immediately turn to steam and possibly cause an explosion. Instead, special extinguishers containing dry salt powders developed specifically for metal fires should be used.

Chemical milling is often used to machine pockets in skins in lower stressed areas to save weight. The use of maskants allows multiple step-cuts and tolerances as tight as 0.001 in. are possible. Chemical milling is conducted by masking the areas in which no milling is desired and then etching in a solution of nitric-hydrofluoric acid. The hydrofluoric acid removes the titanium by etching, while the nitric acid limits the pick-up of hydrogen. After milling, the maskant is stripped.

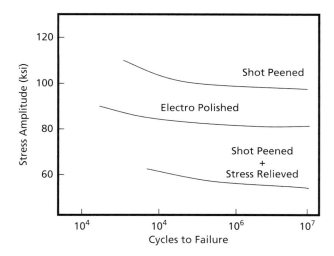

Fig. 4.34. Effects of Surface Residual Stress on Fatigue Strength[29]

4.11 Joining

Titanium can be joined by welding, diffusion bonding, brazing/soldering, adhesive bonding, and with mechanical fasteners. Adhesive bonding is covered in Chapter 8 on Adhesive Bonding and Integrated Cocured Structure, and mechanical fastening is covered in Chapter 11 on Structural Assembly.

4.12 Welding[30]

Titanium is considered highly weldable; however, attention to cleanliness is very important and inert gases are required to prevent contamination from the atmosphere. Several factors contribute to titanium's weldability: its low thermal conductivity prevents heat dissipation; its low coefficient of thermal expansion helps to reduce welding stresses; its relatively low modulus also contributes to low welding stresses; and its high electrical resistivity contributes to its ability to be resistance welded.

The CP grades, near-alpha alloys, and some of the alpha–beta alloys exhibit excellent weldability. The weldability of some of the higher strength alpha–beta alloys, such as Ti-6-6-2 and Ti-6246, are not as good as Ti-6-4 because they have a tendency to hot crack. Some of the beta stabilizing alloying elements added to these alloys lower the melting point and extend the freezing range. Thermal stresses introduced during cooling can cause the partially solidified weld to separate, a defect known as liquation cracking. Some of the highly beta-alloyed systems also have low ductility in both the fusion and the heat affected zones. While most titanium alloys require a post-weld stress relief, especially

Table 4.4 Relative Comparison of Titanium Welding Methods

	GTAW	GMAW	PAW	EB
Common Thickness Range (in.)	1/32–1/4	1/4–3+	1/8–3/8	Foil to 3
Ease of Welding	Good	Fair	Good	Excellent
Grooved Joint Required	Often	Always	No	No
Automatic or Manual	Both	Automatic	Both	Automatic
Mechanical Properties	Fair	Fair	Good	Excellent
Quality of Joint	Good	Fair–Good	Excellent	Excellent
Equipment Cost	Low	High	Moderate	Very High
Wire Required	Often	Always	Sometimes	No
Distortion	Very High	High	Moderate	Very Low

if the part is going to be subjected to fatigue loading, CP Grades 1, 2, and 3 do not require post-weld stress relief unless the part will be highly stressed in a reducing atmosphere.

Gas tungsten arc welding (GTAW), gas metal arc welding (GMAW), plasma arc welding (PAW), and electron beam (EB) welding are the well-established fusion welding methods for titanium, and laser welding is a rapidly emerging technology. The relative merits of GTAW, GMAW, PAW, and EB welding are given in Table 4.4. Many common welding processes are unsuitable for titanium because titanium tends to react with the fluxes and gases used, including gas welding, shielded metal arc, flux core, and submerged arc welding. In addition, titanium cannot be welded to many dissimilar metals due to the formation of intermetallic compounds. However, successful welds can be made to zirconium, tantalum, and niobium.

Since fusion welding results in melting and resolidification, there will exist a microstructural gradient from the as-cast nugget through the heat affected zone (HAZ) to the base metal. Pre-heating the pieces to be welded helps to control residual stress formation on cooling. It is also a common practice to use weld start and run-off tabs to improve weld quality. Fusion welding usually increases the strength and hardness of the joint material while decreasing its ductility.

Attention to cleanliness and the use of inert gas shielding, or vacuum, are critical to obtaining good fusion welds. Molten titanium weld metal must be totally protected from contamination by air. Since oxygen and nitrogen from the atmosphere will embrittle the joint, all fusion welding must be conducted either using a protective atmosphere (i.e., argon or helium) or in a vacuum chamber. Argon, helium, or a mixture of the two are used for shielding. Helium gases operate at higher temperatures than argon, allowing greater weld penetrations and faster speeds, but the hotter helium shielded arc is less stable, requiring better joint fit-up and more operator skill. Therefore, argon is usually the preferred shielding gas. All shielding gases should be free of water vapor; a dew point of $-50°$ F is recommended. The hot HAZs and root side of the welds must also be protected until the temperature drops below 800° F. If fusion welding is conducted in an open environment, local trailing and backing shields must be used to prevent weld contamination. The color of the welded joint is a fairly good way to assess atmospheric contamination. Welds that appear bright silver to straw indicate none to minimal contamination, while light blue to dark blue indicates unacceptable contamination.

Cleanliness is important because titanium readily reacts with moisture, grease, refractories, and most other metals to form brittle intermetallic compounds. Prior to welding, all grease and oil must be removed with a non-chlorinated solvent such as toluene or methyl ethyl ketone (MEK). Surface oxide layers can be removed by pickling or stainless steel wire brushing of the joint area. If the oxide layer is heavy, grit blasting or chemical descaling should be conducted prior to pickling. After pickling, the cleaned material should be wrapped in wax-free kraft paper and handled with clean white cotton gloves. Filler wire should be wrapped and stored in a clean dry location when not in use.

Gas tungsten arc welding is the most common fusion welding method for titanium. In GTAW, the welding heat is provided by an arc maintained between a non-consumable tungsten electrode and the workpiece. In GTAW, as shown in Fig. 4.35, the power supply is direct current with a negative electrode. The negative electrode is cooler than the positive weld joint, enabling a small electrode to carry a large current, resulting in a deep weld penetration with a narrow weld bead. The weld puddle and adjacent HAZ on the weld face are protected by the nozzle gas; trailing shields are used to protect the hot solidified metal and the HAZ behind the weld puddle; and back-up shielding protects the root of the weld and its adjacent HAZ. GTAW can be accomplished either manually or automatically in sheet up to about 0.125 in. in thickness without special joint preparation or filler wire. For thicker gages, grooved joints and filler wire additions are required. If high joint ductility is needed, unalloyed

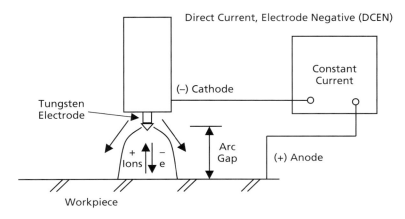

Fig. 4.35. Gas Tungsten Arc Welding (GTAW) Schematic

filler wire can be used at some sacrifice in joint strength. In manual welding, it is important that the operator make sure that the tungsten electrode does not make contact with the molten weld bead, as tungsten contamination can occur. Electrodes with ceria (2% cerium oxide) or lanthana (1–2% lanthanum oxide) are recommended because they produce better weld stability, superior arc starting characteristics, and operate cooler for a given current density than pure tungsten electrodes. Conventional GTAW equipment can be used but requires the addition of appropriate argon or helium shielding gases. Protection can be provided by either rigid chambers or collapsible plastic tents that have been thoroughly purged with argon. Other methods of local shielding have also successfully been used.

Plasma arc welding is similar to GTAW except that the plasma arc is constricted by a nozzle which increases the energy density and welding temperature. The higher energy density allows greater penetration than GTAW and faster welding speeds.

Gas metal arc welding uses a consumable electrode rather than a non-consumable electrode that is used in the GTAW process. In GMAW, as shown in Fig. 4.36, the power supply is direct current with a positive electrode. The positive electrode is hotter than the negative weld joint ensuring complete fusion of the wire in the weld joint. GMAW has the advantage of more weld metal deposit per unit time and unit of power consumption. For plates 0.5 in. and thicker, it is a more cost-effective process than GTAW. However, poor arc stability can cause appreciable spatter during welding which reduces its efficiency.

Electron beam (EB) welding uses a focused beam of high energy electrons resulting in a high depth of penetration and the ability to weld sections up to 3 in. thick. Other advantages of EB are a very narrow HAZ, low distortion,

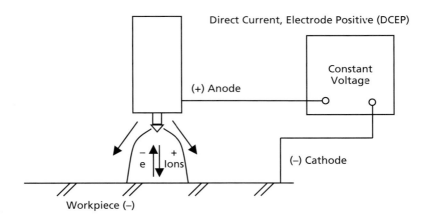

Fig. 4.36. *Gas Metal Arc Welding (GMAW) Schematic*

and clean welds since the welding is conducted in a vacuum chamber. The biggest disadvantage of EB is the high capital equipment cost. EB welding has been successfully used on two major fighter programs to make large unitized structures. In the late 1960s, Grumman Aerospace used EB welding on the F-14 aircraft to weld the folding wingbox. Since computer controls were not available at that time, almost every weld was made through constant thickness material with the beam perpendicular to the surface. In addition, a separate set-up was required for almost every weld. Nevertheless, Grumman successfully delivered over 700 production units. More recently, Boeing is using EB welding extensively on the aft fuselage of the F-22 aircraft. The four major assemblies being welded are the aft and forward booms (two each), which are fabricated from beta annealed Ti-6-4. This amounts to over 3000 linear inches of weldment which requires many less set-ups due to improvements in computer controls.[31] Weld thickness ranges from 1/4 in. to slightly more than 1 in. Gun-to-work distances vary from 10 to 25 in. with beam swings of more than 130°.[32] Both of these applications avoided the costly and error-prone process of installing thousands of mechanical fasteners.

Laser beam (LB) welding uses a high intensity coherent beam of light, which, like EB, results in a very narrow HAZ. However, LB welding is limited to sheet and plate up to about 0.50 in. One big advantage of LB is that it can be conducted in the open atmosphere with appropriate shielding while EB requires a vacuum chamber.

Titanium alloys can be readily resistance welded using both spot and seam welding. As with all titanium welding, cleanliness of the material to be welded is mandatory. Due to the rapid thermal cycles experienced with resistance welding, inert gas shielding is not required.

Diffusion bonding is a solid state joining process that relies on the simultaneous application of heat and pressure to facilitate a bond that can be as strong as the parent metal. DB, as shown in Fig. 4.37, occurs in four steps: (1) development of intimate physical contact through the deformation of surface

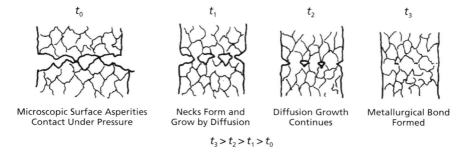

Fig. 4.37. Stages in Diffusion Bonding[33]

roughness through creep at high temperature and low pressure; (2) formation of a metallic bond; (3) diffusion across the faying surfaces; and (4) grain growth across the original interface. Clean and smooth faying surfaces are necessary to create high quality bonds. One advantage of titanium is that at temperatures exceeding 1150° F it dissolves its own surface oxide (TiO_2), leaving a surface that is very amenable to DB. Another advantage is that many titanium alloys display superplasticity at the bonding temperatures (1650–1750° F) which greatly facilitates intimate contact. Typical DB parameters for Ti-6-4 are 1650–1740° F at 200–2000 psi for 1–6 h.

4.13 Brazing

Brazing is often used to fabricate sandwich assemblies that contain titanium skins and welded titanium honeycomb core. Titanium brazing must be conducted in either vacuum or an inert gas atmosphere using either induction, furnace, or vacuum chambers. Fixture material selection is important; for example, nickel will react with titanium to form low melting point eutectics causing the part and fixture to fuse; coated graphite and carbon steel fixtures work well. Brazing can be achieved using several different classes of alloys: (1) aluminum alloys, (2) silver based alloys, (3) Ti-Cu or Ti-Cu-Ni/Zr alloys, (4) Cu-Ni or Cu-Ni-Ti alloys, and (5) Ti-Zr alloys. Braze materials for use in the 1600–1700° F range include Ag-10Pd, Ti-15Ni-15Cu, Ti-20Ni-20Cu, and 48Ti-48Zr-4Be, while lower temperature materials, such as 3003 Al and other aluminum compositions, have been used when the use temperature is in the range of 600° F. When corrosion resistance is a priority, the 48Ti-48Zr-4Be and 43Ti-43Zr-12Ni-2Be materials should be considered. Another alloy, Ag-9Pd-9Ga, which flows at 1650–1675° F, offers good filling when large gaps are encountered. The disadvantages of silver and aluminum based alloys is that they form brittle intermetallics, are corrosion prone, and are limited to lower temperature (<600° F) usage.[34]

Diffusion brazing, also known as liquid interface diffusion (LID) bonding and transient liquid phase (TLP) bonding, is a process similar yet somewhat different than conventional brazing.[35] Like brazing, a separate interfacial material is used that melts at a lower temperature than the base metals and forms a liquid at the interface. Interlayers in the form of foils or electroplated, evaporated, or sputtered coatings are used in the joint. In TLP bonding, the interlayer is heated to the bonding temperature and melted. Liquid film formation depends on the formation of a low melting point eutectic or peritectic at the joint interface. During an isothermal hold at the bonding temperature, diffusion changes the composition of the joint and the joint solidifies. Additional time at temperature allows time for further diffusion resulting in a dilution of the interlayer elements with minimal effect on the parent metal. All stages of TLP bonding depend on solute diffusion from the joint region into the base material. Structural titanium honeycomb panels have been bonded using a Ni-Ti eutectic alloy.[2]

Summary

Titanium is an attractive structural material due to its high strength, low density, and excellent corrosion resistance. However, due to its high melting point and extreme reactivity, the cost of titanium is high, including the mill operations (sponge production, ingot melting, and primary working) and many of the secondary operations conducted by the user.

Titanium alloys are classified according to the amount of alpha and beta retained in their structures at room temperature. Classifications include alpha, near-alpha, alpha–beta, and metastable beta. Since the near-alpha alloys retain their strength at elevated temperatures and have the best creep and oxidation resistance of the titanium alloys, they are often specified for high temperature applications. The formability of the near-alpha alloys is not as good as the alpha–beta or beta alloys. Because these are single-phase alloys containing only alpha, they cannot be strengthened by heat treatment.

The alpha–beta alloys have the best balance of mechanical properties and are the most widely used. The alpha–beta alloy Ti-6-4 is by far and away the most utilized alloy. The alpha–beta alloys can be heat treated to moderate to high strength levels, although their heat treat response is not as great as for the beta alloys. In general, the alpha–beta alloys have good strength at room temperature and for short times at elevated temperatures, although they are not noted for their creep resistance. The weldability of some of these alloys is poor due to the two-phase microstructure.

The beta alloys exhibit much better formability than the near alpha or alpha–beta alloys. Where the alpha and alpha–beta alloys would require hot forming operations, some of the beta alloys can be formed at room temperature. The beta alloys can be solution treated and aged to higher strength levels than the alpha–beta alloys while still retaining sufficient toughness. The biggest drawbacks of the beta alloys are higher densities and reduced ductilities when heat treated to peak strength levels, and some have limited weldability.

Titanium has a great affinity for interstitial elements such as oxygen and nitrogen and readily absorbs them at elevated temperature. Titanium absorbs oxygen at temperatures above 1300° F, which complicates the processing and increases the cost, since many hot working operations are conducted at temperatures exceeding 1300° F. Some alloys are available in an ELI grade that is used for applications requiring maximum ductility and fracture toughness.

Due to their high reactivity with refractory lined vacuum induction furnaces, titanium alloys are made by consumable vacuum arc melting. To insure cleanliness and homogeneity, either double or triple melting practices are used. Cold hearth melting is another melting practice that is now being used to produce even cleaner and more homogeneous ingots. Alloys produced by VAR, in combination with the cold hearth process, have proven to be essentially free of melt related inclusions. Primary hot working operations refine the grain size, produce a uniform microstructure, and reduce segregation. These initial hot working operations will significantly affect the properties of the final product.

Whether the part is forged above or below the beta transus temperature will have a pronounced effect on both the forging process and the resultant microstructure and mechanical properties. Alpha–beta forging is used to develop optimal combinations of strength and ductility and optimal fatigue strength, while beta forging is used to maximize fracture toughness and fatigue crack growth rate, with a significant reduction in fatigue strength.

Investment casting, in combination with HIP after casting, can produce aerospace quality titanium near net shaped parts that can offer cost savings over forgings and built-up structure.

Although some of the beta alloys can be cold formed, hot forming greatly improves formability, reduces springback, and eliminates the need for stress relief. However, the requirement to hot form increases the cost of titanium structures by increasing the cost by having to heat the material; the need for more expensive tools capable of withstanding the temperature; and the requirement to remove surface contamination (i.e., alpha case) after forming. Titanium is also very amendable to SPF, and when combined with diffusion bonding (SPF/DB) it is capable of producing complex unitized structures. In directed metal deposition, a focused laser or electron beam is used to melt titanium powder and deposit the melt in a predetermined path on a titanium substrate plate. The deposited preform is then machined to the final part shape. This near net process leads to savings in materials, machining costs, and cycle times over conventional forged or machined parts.

Titanium is very difficult to machine because of its high reactivity, low thermal conductivity, relatively low modulus, and high strength at elevated temperatures. In addition, improper machining procedures, especially grinding operations, can cause surface damage to the workpiece that will dramatically reduce fatigue life. In machining titanium, use slow speeds, maintain high feed rates, use flood cooling, maintain sharp tools, and use rigid setups.

Gas tungsten arc welding (GTAW), gas metal arc welding (GMAW), plasma arc welding (PAW), and electron beam (EB) are the well-established fusion welding methods for titanium, and laser welding is a rapidly emerging technology. All fusion welding operations must be conducted either in vacuum or inert gas shielding must be used to prevent weld embrittlement.

Recommended Reading

[1] Donachie, M.J., *Titanium: A Technical Guide*, ASM International, 2nd edition, 2000.
[2] "Titanium Alloys", Titanium Metals Corporation, 1997.
[3] Boyer, R., Collings, E.W., Welsch, G., *Materials Properties: Titanium Alloys*, ASM International, 1994.
[4] *ASM Handbook Vol. 2: Properties and Selection: Nonferrous Alloys and Special Purpose Materials*, ASM International, 1990.

[5]   *ASM Handbook Vol. 4: Heat Treating*, ASM International, 1991.
[6]   *ASM Handbook Vol. 6: Welding, Brazing, and Soldering*, ASM International, 1993.
[7]   *ASM Handbook Vol. 9: Metallography and Microstructures*, ASM International, 2004.
[8]   *ASM Handbook Vol. 14: Forming and Forging*, ASM International, 1988.
[9]   *ASM Handbook Vol. 15: Casting*, ASM International, 1988.
[10] *ASM Handbook Vol. 16: Machining*, ASM International, 1990.

References

[1]   Boyer, R.R., "An Overview of the Use of Titanium in the Aerospace Industry", *Materials Science and Engineering*, A213, 1996, pp. 103–114.
[2]   Cotton, J.D., Clark, L.P., Phelps, H.R., "Titanium Alloys on the F-22 Fighter Aircraft", *Advanced Materials & Processes*, May 2002, pp. 25–28.
[3]   Williams, J.C., Starke, E.A., "Progress in Structural Materials for Aerospace Systems", *Acta Materialia*, Vol. 51, 2003, pp. 5775–5799.
[4]   Donachie, M.J., *Titanium: A Technical Guide*, ASM International, 2nd edition, 2000.
[5]   Prasad, Y.V.R.K., Seshacharyulu, T., Medeiros, S.C., Frazier, W.G., Morgan, J.T., Malas, J.C., "Titanium Allow Processing", *Advanced Materials & Processes*, June 2000, pp. 85–89.
[6]   Allen, P., "Titanium Alloy Development", *Advanced Materials & Processes*, October 1996, pp. 35–37.
[7]   Gerdemann, S.J., "Titanium Process Technologies", *Advanced Materials & Processes*, July 2001, pp. 41–43.
[8]   Donachie, M.J., "Selection of Titanium Alloys for Design", in *Handbook of Materials Selection*, John Wiley & Sons, Inc., 2002, pp. 201–234.
[9]   Williams, J.C., "Titanium Alloys: Production, Behavior and Application", in *High Performance Materials in Aerospace*, Chapman & Hall, 1995, pp. 85–134.
[10] Mitchell, A., "Melting, Casting and Forging Problems in Titanium Alloys", *Materials Science and Engineering*, A243, 1998, pp. 257–262.
[11] Williams, J., "Thermo-Mechanical Processing of High-Performance Ti Alloys: Recent Progress and Future Needs", *Journal of Materials Processing Technology*, Vol. 117, 2001, pp. 370–373.
[12] Tamirisakandala, S., Medeiros, S.C., Frazier, W.G., Prasad, Y.V.R.K., "Strain-Induced Porosity During Cogging of Extra-Low Interstitial Grade Ti-6Al-4V", *Journal of Materials Engineering and Performance*, Vol. 10, 2001, pp. 125–130.
[13] Prasad, Y.V.R.K., Seshacharyulu, T., Medeiros, S.C., Frazier, W.G., "A Study of Beta Processing of Ti-6Al-4V: Is it Trivial?", *Journal of Engineering Materials and Technology*, Vol. 123, 2002, pp. 355–360.
[14] Semiatin, S.L., Seetharaman, V., Weiss, I., "Hot Workability of Titanium and Titanium Aluminide – An Overview", *Materials Science and Engineering*, A243, 1998, pp. 1–24.
[15] Shen, G., Furrer, D., "Manufacturing of Aerospace Forgings", *Journal of Materials Processing Technology*, Vol. 98, 2000, pp. 189–195.
[16] Arcella, F.G., Abbott, D.H., House, M.A., "Titanium Alloy Structures for Airframe Application by the Laser Forming Process", American Institute of Aeronautics and Astronautics, Inc., AIAA-2000-1465, 2000.
[17] Arcella, F.G., Froes, F.H., "Producing Titanium Aerospace Components from Powder Using Laser Forming", *Journal of Materials*, May 2000, pp. 28–30.

[18] Semiatin, S.L., Kobryn, P.A., Roush, E.D., Furrer, D.U., Howson, T.E., Boyer, R.R., Chellman, D.J., "Plastic Flow and Microstructure Evolution During Thermomechanical Processing of Laser-Deposited Ti-6Al-4V Preforms", *Metallurgical and Materials Transactions A*, 32A 2002, pp. 1801–1811.
[19] Brewer, W.D., Bird, R.K., Wallace, T.A., "Titanium Alloys and Processing for High Speed Aircraft", *Materials Science and Engineering*, A243, 1998, pp. 299–304.
[20] Hatakeyama, S.J., "SPF/DB Wing Structures Development for the High-Speed Civil Transport", Society of Automotive Engineers, SAE 942158, 1994.
[21] Veeck, S., Lee, D., Tom, T., "Titanium Investment Castings", *Advanced Materials & Processes*, January 2002, pp. 59–62.
[22] Marz, S.J., "Birth of An Engine Blade", *Machine Design*, July 1997, pp. 39–44.
[23] *Metals Handbook-Desk Edition*, ASM International, 2nd edition, 1998, p. 744.
[24] Lei, C.S.C., Davis, A., Lee, E.W., "Effect of BSTOA and Mill Anneal on the Mechanical Properties of Ti-6Al-4V Castings", *Advanced Materials & Processes*, May 2000, pp. 75–77.
[25] "Titanium Alloys", Titanium Metals Corporation, 1997.
[26] "Titanium Machining Techniques", *Titanium Engineering Bulletin No. 7*, Titanium Metals Corporation, 1960s.
[27] Ezugwu, E.O., Wang, Z.M., "Titanium Alloys and Their Machinability- A Review", *Journal of Materials Processing Technology*, Vol. 68, 1997, pp. 262–274.
[28] West, D., "High Performance Aluminum & Titanium Machining", High Speed Machining Seminar, Sponsored by CTA, September 2002.
[29] Yue, S., Durham, S., "Titanium", in *Handbook of Materials for Product Design*, McGraw-Hill, 2001, pp. 3.1–3.61.
[30] "Titanium Welding Techniques", *Titanium Engineering Bulletin No. 6*, Titanium Metals Corporation, 1964.
[31] Lee, J.H., "Applications and Development of EB Welding on the F-22", Society of Automotive Engineers, SAE 972201, 1997.
[32] Irving, R., "EB Welding Joins the Titanium Fuselage of Boeing's F-22 Fighter", *Welding Journal*, December 1994, pp. 31–36.
[33] Paez, C., Messler, R.W., "Design and Fabrication of Advanced Titanium Structures", American Institute of Aeronautics and Astronautics, AIAA-79-0757, 1979.
[34] Huang, X., Richards, N.L., "Activated Diffusion Brazing Technology for Manufacture of Titanium Honeycomb Structures – A Statistical Study", *Welding Journal*, March 2004, pp. 73–81.
[35] Zhou, Y., Gale, W.F., North, T.H., "Modelling of Transient Liquid Phase Bonding", *International Materials Reviews*, Vol. 40, No. 5, 1995, pp. 181–196.

Chapter 5

High Strength Steels

While high strength steels normally account for only about 5–15% of the airframe structural weight, they are often used for highly critical parts such as landing gear components, control surface hinges, and helicopter transmissions. The main advantages of high strength steels are their extremely high strengths and stiffness. Although high strength steels are often defined as those with a minimum yield strength of 200 ksi, there are steels capable of being heat treated to yield strengths exceeding 300 ksi. This can be extremely important in landing gear applications where it is critical to minimize the volume of the gear components, such as the one shown in Fig. 5.1. In addition, steel alloys have a modulus of elasticity of 28–29 msi, which allows landing gears to maintain their shape during hard landings. The disadvantages of high strength steels are primarily their high densities and susceptibility to brittle fracture. At a density of around 0.29 lb/in.^{3}, steel alloys are considerably heavier than other structural materials such as aluminum (0.1 lb/in.^{3}) and titanium (0.16 lb/in.^{3}). Also, as a result of their high strength levels, they are susceptible to hydrogen embrittlement and stress corrosion cracking, both of which can cause sudden brittle failures.

Four types of high strength steels will be covered in this chapter: medium carbon low alloy steels, high fracture toughness steels, maraging steels, and precipitation hardening stainless steels. The medium carbon low alloy steels and high fracture toughness steels are used primarily in aircraft landing gears. Landing gear steels are subject to some of the most severe loadings associated with the airframe environment. They experience high loads and the structure is often non-redundant, with restrictions on both weight and space. Landing gear steels usually account for 2.5–4% of the airframe weight and operate in the range of −70 to 210° F, with temperatures as high as 750° F for emergency braking operations.[1] The requirements for landing gear steels include high static and fatigue strength, high stiffness to resist deformation, resistance to stress corrosion, adequate toughness, and wear resistance where there are moving surfaces. A high ultimate-to-yield strength ratio is desired, along with a totally martensitic microstructure with no ferrite, bainite, or retained austenite. Fine grain sizes, fine carbides, and few non-metallic inclusions all improve the static and fatigue strength. The maraging and precipitation hardening stainless steels are also used for structural components where high strength is required. The compositions of a number of these high strength steels are given in Table 5.1. Since heat treatment is critical to the performance of these steels, heat treating procedures will be emphasized in this chapter.

5.1 Metallurgical Considerations

Steels are alloys of iron and carbon that contain the BCC crystalline structure at room temperature. Since steels contain the BCC structure, their formability is not as good as metals with an FCC structure but better than those with the HCP structure. In general, most forming operations can be conducted at room

High Strength Steels

Fig. 5.1. Mass and Complexity of F/A-18 Landing Gear

temperature as long as the material has not been heat treated to high strength and hardness levels.

As shown in the iron–carbon phase diagram of Fig. 5.2, when steel alloys are heated sufficiently above the A_3 temperature, they transform to the FCC austenite (γ) structure. On slow cooling, the structure transforms back into the BCC ferrite (α) structure, along with cementite (Fe_3C), to form a structure called pearlite. This transformation forms the basis for heat treatment by quenching and tempering. If the steel is austenitized at a temperature sufficiently above

Table 5.1 Compositions of Select High Strength Steels

Alloy	C	Cr	Mo	Ni	Other
Medium Carbon Low Alloy Steels					
4130	0.28–0.33	0.80–1.10	0.15–0.25	–	0.40–0.60 Mn, 0.20–0.35 Si
4140	0.38–0.43	0.80–1.10	0.15–0.25	–	0.75–1.00 Mn, 0.20–0.35 Si
4340	0.38–0.43	0.70–0.90	0.20–0.30	1.65–2.00	0.60–0.80 Mn, 0.20–0.35 Si
300M	0.40–0.46	0.70–0.95	0.30–0.45	1.65–2.00	0.65–0.90 Mn, 1.45–1.80 Si, 0.05 min V
High Fracture Toughness Steels					
AF1410	0.13–0.17	1.80–2.20	0.90–1.10	9.50–10.50	0.10 max Mn, 0.10 max Si, 13.50–14.50 Co
AerMet 100	0.23	3.1	1.2	11.1	13.4 Co
HP 9-4-30	0.29–0.34	0.90–1.10	0.90–1.0	7.0–8.0	0.10–0.35 Mn, 0.20 max Si, 4.25–4.75 Co, 0.06–0.12 V
Maraging Steels					
18Ni(200)	0.03 max	–	3.3	18	8.5 Co, 0.2 Ti, 0.1 Al
18Ni(250)	0.03 max	–	5.0	18	8.5 Co, 0.4 Ti, 0.1 Al
18Ni(300)	0.03 max	–	5.0	18	9.0 Co, 0.7 Ti, 0.1 Al
18Ni(350)	0.03 max	–	4.2	18	12.5 Co, 1.6 Ti, 0.1 Al
Semiaustenitic Precipitation Hardening Stainless Steels					
17-7PH	0.09	16.0–18.0	–	6.50–7.75	1.0 Mn, 1.0 Si, 0.75–1.50 Al
PH 15-7Mo	0.09	14.0–16.0	2.0–3.0	6.50–7.75	1.0 Mn, 1.0 Si, 0.75–1.50 Al
PH 14-8Mo	0.05	13.75–15.0	2.0–3.0	7.50–8.75	1.0 Mn, 1.0 Si, 0.75–1.50 Al
AM-350	0.07–0.11	16.0–17.0	2.50–3.25	4.0–5.0	0.50–1.25 Mn, 0.50 Si, 0.07–0.13 N
AM-355	0.10–0.15	15.0–16.0	2.50–3.25	4.0–5.0	0.50–1.25 Mn, 0.50 Si, 0.07–0.13 N
Martensitic Precipitation Hardening Stainless Steels					
17-4PH	0.07	15.0–17.5	–	3.0–5.0	1.0 Mn, 3.0–5.0 Cu, 1.00 Si, 0.15–0.45 Nb
15-5PH	0.07	14.0–15.5	–	3.5–5.5	1.0 Mn, 2.5–4.5 Cu, 1.00 Si, 0.15–0.45 Nb
PH 13-8Mo	0.05	12.3–13.3	2.0–2.5	7.5–8.5	0.10 Mn, 0.90–1.35 Al, 0.10 Si
Custom 450	0.05	14.0–16.0	0.5–1.0	5.0–7.0	1.0 Mn, 1.25–1.75 Cu, 1.00 Si, 8×%C min Nb
Custom 455	0.05	11.0–12.5	0.5	7.5–9.5	0.50 Mn, 1.5–2.5 Cu, 0.50 Si, 0.8–1.4 Ti, 0.1–0.50 Nb
Custom 465	0.02	11.0–12.5	0.75–1.25	10.75–11.25	0.25 Mn, 0.25 Si, 1.50–1.80 Ti

1333°F for some period of time and then quenched to room temperature, it does not convert to the normal BCC structure. Instead, it converts to a body centered tetragonal (BCT) structure called martensite. This transformation is shown in the isothermal transformation diagram for 4340 steel in Fig. 5.3. The BCT martensite structure is essentially a BCC structure distorted by interstitial carbon atoms into a tetragonal structure (Fig. 5.4). The distortion severely strains the crystalline lattice and dramatically increases the strength and hardness.

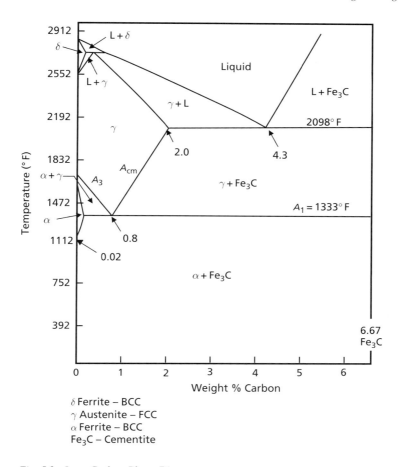

Fig. 5.2. Iron–Carbon Phase Diagram

Unfortunately, it also makes the steel extremely brittle; therefore, the steel is reheated, or tempered, at intermediate temperatures as shown in the schematic of Fig. 5.5 to restore some ductility and toughness, although the strength also decreases as the tempering temperature is increased.

A key variable in heat treating alloy steels is the cooling rate during quenching. The quench rate that will provide the desired hardness for a given thickness is determined primarily by alloying additions. Some alloys require a water quench, others an oil quench, and some are so highly alloyed that they can be air cooled to room temperature to form a martensitic structure. Since the cooling rate is also dependent on section size, the quench may have to be changed as the thickness increases, for example, a steel that could be through-hardened with an oil quench at a thickness of 1/2 in. may have to be water quenched when the thickness is increased to an inch.

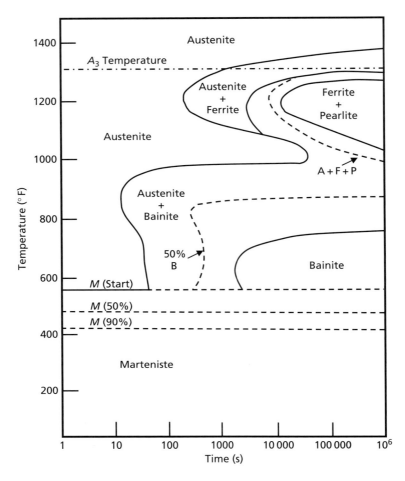

Fig. 5.3. Isothermal Transformation Diagram for 4340 Steel

While some steels are hardened by the conversion of austenite to martensite through a quench and temper process, others, such as the maraging and precipitation hardening steels, are strengthened by precipitation hardening. The maraging steels, with nominal nickel contents of 18% and carbon contents of only 0.03%, will form martensite on air cooling from the austenitizing temperature. Even very slow cooling of heavy sections produces a fully martensitic structure. However, this low carbon martensite is not the high strength martensite that forms in the higher carbon alloy steels. The influence of carbon on the strength and hardness of steel is shown in Fig. 5.6. The low carbon martensite that is formed is a tough and ductile iron–nickel martensite. The strength in the maraging steels results during age hardening at 850–950° F to form precipitates of Ni_3Mo and Ni_3Ti.[2] Since the carbon content is extremely low, maraging steels

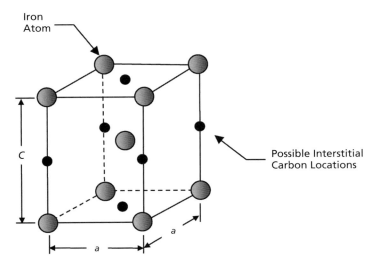

Fig. 5.4. Body Centered Tetragonal Structure of Martensite

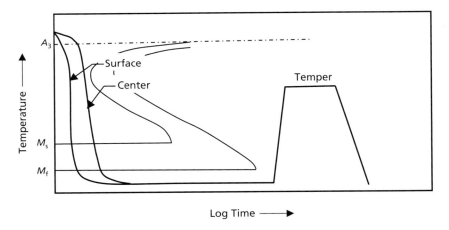

Fig. 5.5. Heat Treatment for Medium Carbon Low Alloy Steels

are characterized by a combination of high strength, ductility, and excellent toughness.

The precipitation hardening (PH) stainless steels are classified as either semi-austenitic or martensitic. The semiaustenitic grades contain an austenitic structure in the annealed or solution treated condition. After fabrication operations are complete, they can be transformed to martensite by a simple conditioning treatment followed by precipitation hardening. The conditioning treatment consists of heating the alloy to a high enough temperature to remove carbon from solid

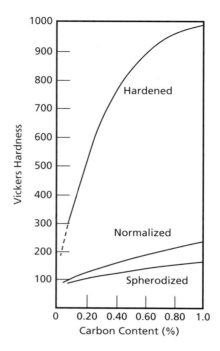

Fig. 5.6. Effect of Carbon Content on Hardness of Steel

solution and precipitate it in the form of chromium carbide $(Cr_{23}C_6)$.[3] Removing carbon from solid solution makes the austenite unstable, and it starts transforming to martensite on cooling to the martensite start (M_s) temperature. Depending on the specific alloy composition and the conditioning temperature, some of these grades must be cooled to subzero temperatures to cause the conversion from austenite to martensite. The final step is precipitation hardening in which the alloy is reheated to 900–1200° F. During this aging treatment, aluminum in the martensite combines with nickel to produce precipitates of NiAl and Ni_3Al.[3]

The martensitic grades of precipitation hardening stainless steels are essentially martensite after solution treating. They are therefore harder and stronger than the semiaustenitic grades before precipitation hardening. Since they are already martensitic, a conditioning treatment is not required; they only require the final precipitation hardening treatment to attain maximum strength and hardness.

5.2 Medium Carbon Low Alloy Steels

Medium carbon low alloy steels contain carbon in the range of 0.30–0.50%, with alloying elements added to provide deeper hardening and higher strength and toughness. Typical alloying elements include manganese, silicon, nickel, chromium, molybdenum, vanadium, and boron. Medium carbon low alloy steels are identified by the American Iron and Steel Institute (AISI) four digit system of numbers. The first two digits identify the specific alloy group while the last two digits give the approximate carbon content in hundredths of a percent. For example, the designation 4130 would indicate that the steel is from the 41XX series of steels that are chromium–molybdenum steels, and the 30 would indicate that the carbon content is 0.30%.

Medium carbon low alloy steels are normally hardened to develop the desired strength. Some are also carburized and then heat treated to produce a combination of high surface hardness and good core toughness. The medium carbon low alloy steels include a number of important steels such as 4340 and vanadium modified 4340 (4340V). Several modifications to the basic 4340 have been developed, such as the silicon modified 300M. Typical airframe applications include landing gear components, shafts, gears, and other parts requiring high strength, through-hardening, or toughness.

The 41XX series of alloys are classified as chromium–molybdenum steels containing 0.5–0.95% chromium and 0.13–0.20% molybdenum. Chromium is added to increase hardenability and strength; however, the addition of chromium can also make this series susceptible to temper embrittlement. Due to its low-to-intermediate hardenability, 4130 must be water quenched. It has good tensile, fatigue, and impact properties up to about 700° F; however, the impact properties at cryogenic temperatures are low. The 4140 is similar to 4130 except for a higher carbon content, which results in higher strengths with some sacrifice in formability and weldability. When heat treated to high strength levels, 4140 is susceptible to hydrogen embrittlement that can result from acid pickling or from electroplating with cadmium or chromium. After pickling or electroplating, it is baked for 2–4 h at 375° F to remove any absorbed hydrogen.

Nickel is added along with chromium and molybdenum to form the 43XX class of alloys. Their composition is about 0.5–0.8% Cr, 0.20% Mo, and 1.8% Ni. Nickel in combination with chromium improves strength and provides greater hardenability, higher impact strength, and better fatigue resistance. The addition of 0.2% molybdenum further increases hardenability and minimizes the susceptibility to temper embrittlement. The 4340 is the benchmark by which other high strength steels are judged. It combines deep hardenability with high strength, ductility, and toughness. It also has good fatigue and creep resistance. It is often used where high strength in thick sections is required. The 4340 can be oil quenched to full hardness in sections up to 3 in. in diameter with thicker sections requiring water quenching. However, water quenching significantly increases the danger of cracking during quenching. It is immune to temper embrittlement. Parts exposed to hydrogen during pickling or plating operations should be baked to remove any hydrogen. The effects of various baking cycles on the notched bar strength of hydrogen-charged 4340 is shown in Fig. 5.7. Unfortunately, it is very susceptible to stress corrosion cracking when heat treated to the highest

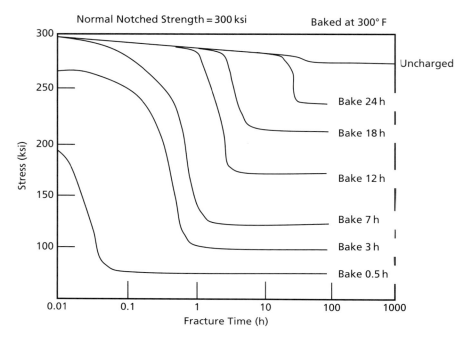

Fig. 5.7. Static Fatigue of 4340 Notched Bars[4]

strength levels (220–280 ksi). The 4340 can be modified with vanadium (4340V) that forms a stable high melting point carbide that helps in pinning grain boundaries and preventing grain growth during hot working operations. The vanadium addition also serves as a grain refiner that increases toughness.

When the silicon content of 4340 is increased to 2%, the strength and toughness increases in the manner shown in Fig. 5.8. The increased silicon content provides deeper hardenability, increases solid solution strengthening, and provides better higher temperature resistance. The increase in toughness is attributed to silicon retarding the precipitation of cementite from retained austenite during tempering and to the stabilization of carbides. Silicon added to the basic 4340 composition forms the alloy 300M, which nominally contains 1.6% silicon. Vanadium is added for grain refinement, and the sulfur and phosphorus levels are kept very low to reduce temper embrittlement and increase toughness and transverse ductility. The 300M is also vacuum arc remelted to lower the hydrogen and oxygen contents. The lower oxygen content minimizes the formation of oxide inclusions and increases toughness. However, due to 300Ms high silicon and molybdenum contents, it is extremely prone to decarburization during heat treatment, and when heat treated to strength levels above 200 ksi, it is also susceptible to hydrogen embrittlement.

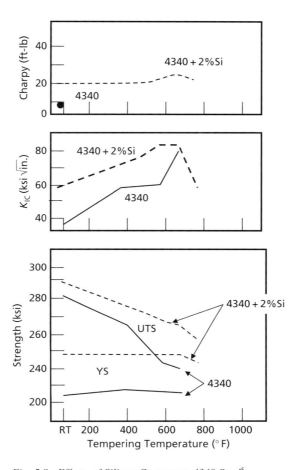

Fig. 5.8. Effects of Silicon Content on 4340 Steel[5]

High strength steels are available in a variety of quality levels depending on the type of melting practice used. While many of these steels were originally air melted, the trend has been to move to more advanced melting techniques such as vacuum degassing, electroslag remelting (ESR), VAR, and double vacuum melting (vacuum induction melting followed by vacuum are remelting (VIM-VAR)) for improved cleanliness and higher quality. These methods reduce both the quantity of dissolved gases (hydrogen, oxygen, and nitrogen) and the non-metallic inclusions. As the high strength steels have evolved since the mid-1970s, improvements in melting process control and inspection have steadily increased fracture toughness, ductility, and fatigue resistance. A comparison of air and vacuum melted 300M, shown in Fig. 5.9, illustrates the property advantages imparted by vacuum processing. Both VAR and ESR are acceptable

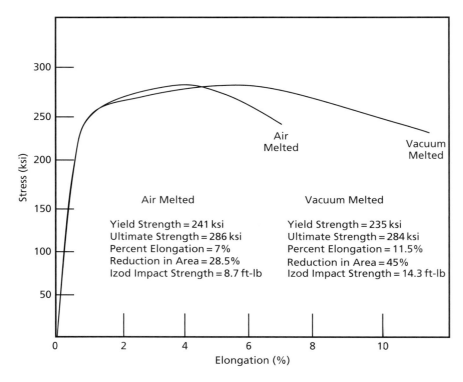

Fig. 5.9. Comparison of Air and Vacuum Melted 300M Steel[6]

melting methods since the mechanical properties are essentially equivalent for both methods.

5.3 Fabrication of Medium Carbon Low Alloy Steels

Forging[7] is often selected for critical components because it creates a grain flow (Fig. 5.10) that increases ductility, impact strength, and fatigue strength. Forging breaks up casting segregation, reduces the as-cast grain structure, heals porosity, and helps to homogenize the structure. The improvement in transverse ductility as a result of forging for 4340 is shown in Fig. 5.11.

Medium carbon low alloy steels are only slightly more difficult to forge than carbon steels. They are supplied from the mill in either the annealed or the normalized and tempered condition. The selection of the forging temperature is based on carbon content, alloy composition, temperature range for optimum plasticity, and the amount of reduction required to forge the workpiece. In general, the forging temperature decreases with increases in carbon content and alloying elements. Maximum recommended forging temperatures are generally about 50°F lower than those used for plain carbon steels of the same carbon content. Medium carbon

High Strength Steels

Fig. 5.10. Etched 4140 Forging Showing Grain Flow[7]

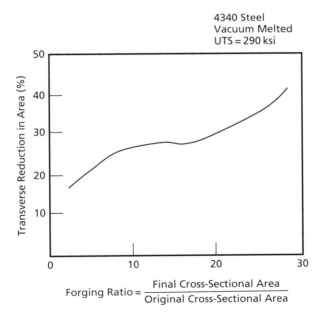

Fig. 5.11. Influence of Forging Ratio on 4340 Steel[8]

low alloy steels are hot forged at temperatures ranging from 1950 to 2250° F. To avoid stress cracks that can result from air cooling, the forged part should be slowly cooled in a furnace or embedded in an insulating material. The section thickness, complexity, and size are often limited by the cooling that occurs when the part comes in contact with the cold dies; therefore, hammer forging with its short contact times is often used for forging intricate shapes. However, large landing gear components (Fig. 5.12) are usually forged in hydraulic presses due to their large sizes and the slow controlled strain rates that can be achieved in hydraulic presses. For steel forgings that are to be heat treated above a tensile strength of 150 ksi, the normal practice is to normalize before heat treatment to produce a uniform grain size and minimize internal residual stresses.

Medium carbon low alloy steels are usually formed in the annealed condition.[9] Their formability depends mainly on the carbon content and is generally slightly less than for unalloyed steels of the same carbon content. They can be cut, sheared, punched, and cold formed in the annealed condition and then heat treated to the desired hardness. Because of their high strength and limited ductility, forming operations are not conducted in the quenched and tempered condition.

During machining, the higher hardness of these steels requires lower speeds and feeds than those used for the plain carbon steels. The machinability ratings[10] of 4130, 4140, 4340, and 300M compared to cold rolled 1212 steel (100%) are

Fig. 5.12. Large Landing Gear Forging
Source: Schultz Steel

70, 65, 50, and 50% respectively, indicating that these materials are considerably harder to machine and require lower speeds and feeds. To improve machinability, medium carbon low alloy steels are normalized at 1600–1700° F and then tempered at 1200–1250° F prior to machining to produce a partially spherodized microstructure. Highly alloyed steels should be machined before hardening to martensite. If it is necessary to conduct machining operations after hardening, then a two-step process should be used in which the majority of the machining is done before heat treatment, and the machining conducted after hardening is essentially a light finish machining to provide dimensional accuracy. Finish machining, because of the relatively high hardness of the material, necessitates the use of sharp, well-designed carbide cutting tools with proper feeds, speeds, and a generous supply of coolant.

Due to their extreme strength and hardness, high strength steels are often finished by grinding to provide precise dimensions, remove any nicks or scratches, and provide smooth surfaces. However, great care must be taken when grinding these ultrahigh strength steels. Improper or abusive grinding can result in grinding burns in which the surface is heated above the austenitizing temperature and the austenite formed converts to untempered martensite on cooling, as shown in Fig. 5.13. This untempered martensitic surface layer is brittle and susceptible to forming a network of fine cracks that can reduce the fatigue strength by as much as 30%. Even if the grinding temperature does not produce austenite, it can result in overtempered martensite on the surface that is lower in hardness and strength.

Medium carbon low alloy grades are welded or brazed by a number of techniques.[11] Alloy welding rods, comparable in strength to the base metal, are used and moderate preheating (200–600° F) is usually necessary. At higher carbon levels, higher preheating temperatures and post-weld stress relieving are often required. To avoid brittleness and cracking, preheating and interpass heating are used, and complex structures should be stress relieved or hardened and tempered immediately after welding. Since the weld joint cannot usually develop the high strength levels required in the as-welded or stress relieved condition, medium carbon low alloy steels are usually reaustenitized and then quenched and tempered after welding. The best condition for welding is either the normalized or the annealed condition.

Typical welding processes include inert gas tungsten arc, shielded metal arc, inert gas metal arc, and, in some instances, electron beam welding. The 4340 is weldable by a number of methods including both gas and arc welding. Welding rods of the same composition should be used. Since 4340 develops considerable hardness on air cooling, the welded parts should be either annealed or normalized and tempered shortly after welding. The medium carbon low alloy steels are also susceptible to hydrogen-induced cracking; therefore, every effort must be exerted to minimize any possible absorption of hydrogen gas. For some steels, such as 300M, welding is not recommended. Other common welding processes

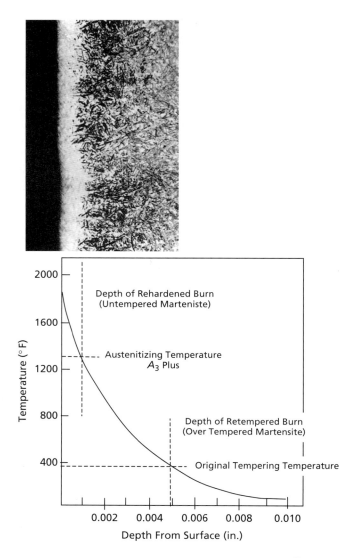

Fig. 5.13. *Effects of Grinding Burns on High Strength Steel*[12]

for high strength steels include friction welding and pressure welding. Post-weld tempering is necessary for many processes to prevent cracking.

Many high strength steels are susceptible to stress corrosion cracking when placed in-service. The plane strain stress corrosion cracking fracture toughness of many high strength steels is only about one half that of the non-exposed fracture toughness ($K_{Icscc} \sim 1/2\ K_{Ic}$). Surface coatings, such as cadmium or chromium plating, are normally used to prevent access to the environment and

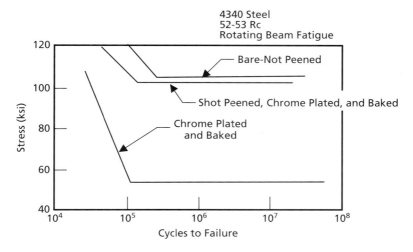

Fig. 5.14. *Fatigue Improvement Due to Shot Peening*

sacrificially corrode instead of the base metal. Hard chrome plating is often used where wear surfaces are involved. However, both of these plating methods can cause hydrogen embrittlement so it is important to stress relieve before plating and then bake immediately after plating to remove any hydrogen. If the steel is subject to possible hydrogen pickup due to pickling or electroplating operations and it has been heat treated to 200 ksi, or higher, it should be immediately baked at 365–385° F for at least 8 h or for 24 h if it is thicker than 1.5 in. Shot peening before chrome plating can significantly improve the fatigue life, as shown in Fig. 5.14. Shot peening induces a residual stress pattern (Fig. 5.15) near the surface that helps prevent plating cracks from propagating into the base metal.

5.4 Heat Treatment of Medium Carbon Low Alloy Steels

Typical heat treatments for several medium carbon low alloy steels are given in Table 5.2, and the effects of different tempering temperatures are shown in Table 5.3.[13] Annealing produces maximum softness by heating to 1350–1600° F followed by slow furnace cooling. Normalizing is somewhat similar except that the steel is air cooled from the austenization range to produce a structure of pearlite (ferrite + cementite) and partially spherodized carbides. Medium carbon low alloy steel forgings are generally normalized before hardening to produce a uniform grain structure and to minimize residual stresses. For the optimum machinability, it is best to normalize at 1500–1700° F and then temper at 1200–1250° F to produce a partially spherodized structure. Although an even softer structure can be achieved by annealing, it is not as good as a structure for machining as the slightly harder normalized and tempered structure. A completely spherodized carbide structure can be produced by spherodization,

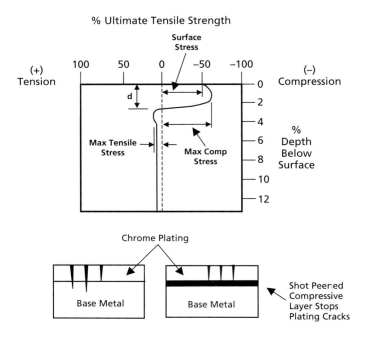

Fig. 5.15. Example of Residual Stress Profile Created by Shot Peening

in which the steel is held just below the austenization temperature for 4–12 h; however, this treatment is generally not used because it is lengthy and therefore more costly.

Stress relieving is used to reduce residual stresses that result from welding or machining. If stress relief is done before hardening, it can be conducted at 1200–1250° F, or at about 50° F below the tempering temperatures given in Table 5.2, if it is done after hardening. Weldments, especially if the part is complex, should be stress relieved immediately after welding. Welded parts are stress relieved and sometimes normalized prior to hardening and tempering.

Medium carbon low alloy steels are hardened by austenitizing, water or oil quenching, and then tempering to their final hardness. Since the composition of these steels varies quite a bit, it is important to: (1) understand the maximum section thickness that can be hardened in a specific quench media (e.g., water or oil) and (2) realize the large variations in final strength and ductility that can be obtained by tempering at different temperatures. For example, the effects of different tempering temperatures on the mechanical properties of 4340 are shown in Fig. 5.16.

All of these steels are very susceptible to decarburization during heat treatment. Decarburization is controlled by either leaving sufficient excess stock that will be removed after heat treating, copper plating, controlled atmosphere heat treating, or by salt bath heat treating.

High Strength Steels

Table 5.2 Heat Treatments for Select Medium Carbon Low Alloy Steels

Steel	Normalizing	Annealing	Hardening	Tempering
4130	Heat to 1600–1700° F; hold for a minimum of 1 h or 15–20 min per 1 in. of max section thickness; air cool.	Heat to 1525–1600° F and hold for a period that depends on section thickness or furnace load; furnace cool at a rate of about 30° F/h to 900° F and then air cool.	Heat to 1550–1600° F and hold, and then water quench, or heat to 1575–1625° F and hold, and then oil quench. Holding time depends on section thickness and is typically 1 h minimum or 15–20 min per 1 in. of max section thickness.	Temper at least 2 h at 400–1300° F; air cool or water quench. Tempering temperature and time depends on desired hardness and strength.
4140	Heat to 1500–1700° F; hold for a minimum of 1 h or 15–20 min per 1 in. of max section thickness; air cool.	Heat to 1525–1600° F and hold for a period that depends on section thickness or furnace load; furnace cool at a rate of about 30° F/h to 900° F and then air cool.	Heat to 1525–1600° F and hold, and then oil quench, or heat to 1575–1625° F and hold, and then oil quench. Holding time depends on section thickness and is typically 1 h minimum or 15–20 min per 1 in. of max section thickness.	Temper at least 1/2–2 h at 350–450° F for "ultra-high" strength, or 725–1300° F for yield strength below 200 ksi; air cool or water quench. Tempering temperature and time depends on desired hardness.
4340V	Heat to 1600–1700° F for a minimum of 1 h per 1 in. of max thickness, and then air cool.	Heat to 1525–1600° F for 1 h per 1 in. of max thickness, furnace cool (approximately 30° F/h) to 900° F, and then air cool.	Heat to 1550–1660° F for 1 h minimum or about 15 min per 1 in. of max thickness, quench in warm oil at 75–140° F, and air cool.	Temper at least 2 h at 500–1100° F depending on desired tensile strength. For ultimate tensile strength of 220–240 ksi, temper between 500 and 700° F.
300M	Heat to 1675–1725° F and hold for 15 to 20 min per 1 in. of section thickness; air cool. For enhanced machinability, temper at 1200–1250° F.	Heat to temperature no higher than 1350° F and hold for a period that depends on section thickness or furnace load. Cool to 1200° F at a rate no faster than 10° F/h, cool to 900° F no faster than 20° F/h, and finally air cool to room temperature.	Heat to 1575–1625° F, oil quench to 160° F; or quench in salt at 390–410° F, hold 10 min and then air cool to 160° F or below.	Double temper for 2–4 h at 575 ± 25° F.

Table 5.3 Typical Properties of Medium Carbon Low Alloy Steels[13]

Alloy	Temper Temperature (°F)	Ultimate Tensile Strength (ksi)	Yield Strength (ksi)	Elongation (%) 2 in.	Reduction in Area (%)
4130	400	256	220	10.0	33.0
	600	228	195	13.0	41.0
	800	200	179	16.5	49.0
	1200	140	120	22.0	63.0
4140	400	285	252	11.0	42.0
	600	250	228	11.5	46.0
	800	210	195	15.0	50.0
	1200	130	114	21.0	61.0
4340	400	287	270	11.0	39.0
	600	255	235	12.0	44.0
	800	217	198	14.0	48.0
	1200	148	125	20.0	60.0
300M	200	340	280	6.0	10.0
	400	310	240	8.0	27.0
	600	289	245	9.5	34.0
	800	260	215	8.5	23.0

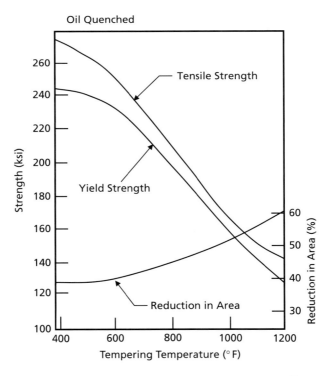

Fig. 5.16. Effects of Tempering Temperature on 4340 Steel[14]

High Strength Steels

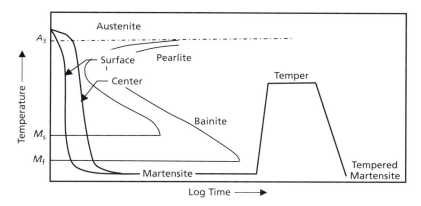

Fig. 5.17. Heat Treatment for Medium Carbon Low Alloy Steels

The procedure for hardening is illustrated in Fig. 5.17 and consists of the following:[15]

(1) *Austenitizing*. During austenization, the steel is heated into the austenite (γ) field and held for a sufficient period of time to dissolve many of the carbides and put them into solution. As shown in Fig. 5.18, the

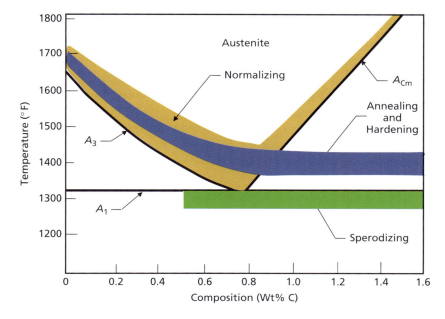

Fig. 5.18. Steel Heat Treating Ranges[15].

temperature required for austenization is a function of the carbon content; with increasing carbon contents the temperature decreases along the A_3 line to a minimum value A_1 at the eutectoid composition (0.8% carbon) and then increases again along the A_{Cm} line. The first stage in the formation of austenite is the nucleation and growth of austenite from pearlite (ferrite $+ Fe_3C$). Even after the complete disappearance of pearlite, some carbides will remain in the austenite. To minimize the time for austenization, the temperature used is about 100° F above the minimum temperature for 100% austenite, and the time is about one hour per inch of thickness. However, it is also important to keep the austenization temperature as low as possible to reduce the tendency toward cracking and distortion, minimize oxidation and decarburization, and minimize grain growth. The addition of alloying elements, such as manganese and nickel, helps to reduce the temperature necessary for austenite formation.

(2) *Quenching*. While the FCC austenite that forms during austenization is capable of dissolving as much as 2% carbon, only a small fraction of carbon can be retained in the lower temperature BCC ferrite. If the steel is slowly cooled from the austenization temperature, carbon atoms are rejected as the FCC austenite transforms to the BCC ferrite, and alternating layers of ferrite and cementite form pearlite through a nucleation and growth process. However, if the steel is rapidly cooled from the austenization temperature (quenched), the carbon does not have time to diffuse out of the austenite structure as it transforms to the BCC ferrite, and the BCC structure becomes distorted into a BCT tetragonal structure called martensite. The objective of the quenching process is to cool at a sufficient rate to form martensite. The distortion of the BCT structure results in high strength and hardness of the quenched steel. As previously shown in Fig. 5.17, the steel must be cooled past the nose of the isothermal transformation diagram to form 100% martensite. Martensite does not form until it reaches the martensite start temperature (M_s) and is complete after it is cooled below the martensite finish temperature (M_f). The addition of alloying elements increases the hardenability of steels by moving the nose of the isothermal transformation diagram to the right, allowing slower cooling rates for alloy steels to form martensite. However, alloying elements also often depress the M_s and M_f temperatures so that some highly alloyed steels must be cooled to below room temperature to obtain fully martensitic structures. Medium carbon low alloy steels are quenched in either water or oil to produce adequate cooling rates. While water quenching produces the fastest cooling rates, it also produces the highest residual stresses and can often cause warpage and distortion; therefore, higher alloy grades are often used so that oil quenching can be used.

(3) *Tempering.* While as-quenched steel is extremely hard and strong, it is also very brittle. Tempering, in which the steel is reheated to an intermediate temperature, is used to increase the ductility and toughness with some loss of strength and hardness. During tempering, the highly strained BCT structure starts losing carbon to transformation products which reduces lattice strains producing an increase in ductility and a reduction in strength. One of the advantages of medium carbon low alloy steels is the large range of strength values that can be obtained by varying the tempering temperature, as previously shown in Fig. 5.16 for 4340 steel.

Medium carbon low alloy steels can be susceptible to two types of temper embrittlement during heat treating: (1) one-step temper embrittlement and (2) two-step temper embrittlement.[2] Commercial heat treatments and compositional controls to eliminate unwanted impurities are designed to avoid these embrittlement mechanisms.

One-step temper embrittlement. This type of embrittlement, also known as 660° F embrittlement, occurs in high strength low alloy steels that have quenched and tempered martensitic structures. When the alloy is austenitized, quenched, and then tempered for a short time (about 1 h) between 480 and 660° F, it can cause a decrease in notched toughness and impact strength. The failure mode is intergranular and is thought to be due to the impurity elements phosphorous, nitrogen, and possibly sulfur, since tests on high purity alloys do not exhibit embrittlement. Manganese may play an indirect role by helping the impurity elements segregate to the grain boundaries.

Two-step temper embrittlement. This type of embrittlement causes a decrease in notch toughness when tempered alloy steels are isothermally aged in the temperature range of 700–1040° F, or are slowly cooled after tempering. Two-step embrittlement results in intergranular failure modes and is also attributed to the presence of impurities that segregate to the grain boundaries. The ductile-to-brittle transition is directly dependent on the grain boundary concentration of impurities. The relative effect of these impurities has been found to be tin > antimony > phosphorous. Alloying elements can also cosegregate to the grain boundaries along with the impurities, i.e. nickel cosegregates with antimony. The rate and amount of impurity segregation, and hence the resultant intergranular embrittlement, depends on the total composition of the system. Nickel, chromium, and manganese increase two-step temper embrittlement caused by antimony, tin, phosphorous, or arsenic. Additions of molybdenum retard temper embrittlement since molybdenum inhibits the segregation of impurities, i.e. molybdenum readily precipitates as phosphides in the matrix and inhibits segregation.

5.5 High Fracture Toughness Steels

The three high fracture toughness steels, HP-9-4-30, AF1410, and AerMet 100 have somewhat lower carbon contents than the medium carbon low alloy steels 4340 and 300M. The lower carbon content significantly contributes to their better ductilities and higher fracture toughness. In addition, these alloys have high nickel contents that provides deep hardening and toughness and cobalt that helps to prevent retained austenite. To obtain the desired fracture toughness, all of these steels are vacuum melted. The recommended heat treatments for HP-9-4-30, AF1410, and AerMet 100 are given in Table 5.4 and some of the mechanical properties are shown in Table 5.5. These alloys are not corrosion resistant and parts must be protected with a corrosion resistant coating.

The 9Ni-4Co family of steels[16] were developed as high fracture toughness steels capable of being heat treated to high strength levels in thick sections. The highest strength of these alloys, HP 9-4-30, contains nominally 0.30% carbon, 9% nickel, and 4% cobalt. It is capable of being hardened in sections up to 6 in. thick to an ultimate tensile strength level of 220–240 ksi while maintaining a fracture toughness $K_{Ic} = 100$ ksi $\sqrt{in}$. Double tempering is normally employed to prevent retained austenite. HP-9-4-30 is available as billet, bar, rod, plate, sheet, and strip. It can be formed by bending, rolling, or shear spinning. Heat treated HP-9-4-30 can be welded using GTAW without preheating or post-heating. Welded parts should be stress relieved at 1000° F for 24 h.

The AF1410 was developed specifically to have high strength, excellent fracture toughness, and excellent weldability when heat treated to 235–255 ksi ultimate tensile strength.[16] The nominal composition is 14% cobalt, 10% nickel, 2% chromium, 1% molybdenum, and 0.15% carbon. AF1410 maintains good toughness at cryogenic temperatures and has high strength and stability at temperatures up to 800° F. The general corrosion resistance is similar to the maraging steels. The alloy is highly resistant to stress corrosion cracking compared to other high strength steels. AF1410 is produced by VIM followed by VAR. It is available as billet, bar, plate, and die forgings. AF1410 has good weldability using GTAW provided high purity welding wire is used and oxygen contamination is avoided. Preheating prior to welding is not required.

AerMet 100 is a nickel–cobalt high strength steel that can be heat treated to 280–300 ksi or to 290–310 ksi tensile strength while exhibiting excellent fracture toughness and high resistance to stress corrosion cracking.[17] AerMet 100 is replacing older steels such as 4340, 300M, HP 9-4-30, and AF1410 in many applications due to its good combination of strength (UTS = 285 ksi) and toughness ($K_{Ic} = 100$ ksi $\sqrt{in}$.). Other advantages include good toughness at cryogenic temperatures, a critical flaw length of nearly 0.25 in., and an operating temperature up to 750° F.[18] It is highly resistant to stress corrosion cracking compared to other high strength steels of the same strength level. AerMet 100, produced by VIM followed by VAR, is available as billet, bar, sheet, strip, plate, wire, and die forgings. Impurity concentrations and inclusions are kept

High Strength Steels

Table 5.4 Heat Treatments for High Fracture Toughness Steels[16,17]

Steel	Normalizing	Annealing	Hardening	Tempering
HP-9-4-30	Heat to 1625–1675° F and hold 1 h for each 1 in. of thickness (1 h min.); air cool.	Heat to 1150° F and hold 24 h; air cool.	Austenitize at 1475–1575° F and hold 1 h for each 1 in. of thickness (1 h min.); water or oil quench. Complete martensitic transformation by refrigerating at least 1 h at −125 to −75° F; let warm to RT.	Temper at 400–1100° F, depending on desired strength; double tempering preferred. Most widely used tempering temperature is double tempering (2 h or more at temperature) from 1000 to 1075° F.
AF1410	Normalize and overage by heating to 1620–1675° F; hold 1 h for each 1 in. of thickness; air cool and overage at 1250° F for 5 h min.	Normalizing and overaging are used to soften and stress relieve the product. A stress relief of 1250° F can be applied to relieve mechanical stresses.	Renormalize and austenitize at 1475–1575° F and hold 1 h for 1 in. of thickness; oil, water, or air cool depending on section size. An alternative method is double austenitizing, first at 1600–1650° F with holding times of 1 h for 1 in. of section thickness. Then cool with oil, water, or air, and reaustenitize as with the single austenitizing treatment Refrigeration treatment of −100° F optional.	Age at 900–950° F for 5–8 h and air cool.
AerMet 100	Heat to 1650° F; hold 1 h for each 1 in. of thickness; air cool. Optimum machinability: normalize followed by 16 h at 1250° F age.	Anneal at 1250° F for 16 h to obtain hardness of 40 HRC.	Heat to 1625 ± 25° F for 1 h, cool to 150° F in 1–2 h using either air cooling or oil quenching, cool to −100° F and hold for 1 h.	Age at 900 ± 10° F for 5 h and air cool. Never age at temperatures below 875° F.

to a minimum by double vacuum processing. Unlike conventional steels, the manganese and silicon concentrations are also kept close to zero because both reduce austenite grain boundary cohesion. AerMet 100 has good weldability and does not require preheating prior to welding.

Table 5.5 Typical Properties of High Fracture Toughness Steels[16,17]

Alloy	Heat Treatment	Ultimate Tensile Strength (ksi)	Yield Strength (ksi)	Elongation (%) 4D Gage	Reduction in Area (%)	Fracture Toughness K_{Ic} (ksi $\sqrt{in.}$)
HP-9-4-30	1550° F, Oil Quench, Cool to −100° F, Double Temper at 400° F	240–260	200–210	8–12	25–35	60–90
	1550° F, Oil Quench, Cool to −100° F, Double Temper at 1025° F	220–240	190–200	12–16	35–50	90–105
AF1410	1650° F, Air Cool, 1525° F, Air Cool, Cool to −100° F, Temper at 950° F	244	214	16	69	158
AerMet 100	1625° F, Air Cool, Cool to −100° F, Aged at 900° F	285	250	13–14	55–65	110–126

5.6 Maraging Steels

Maraging steels are a class of high strength steels with very low carbon contents (0.030% maximum) and additions of substitutional alloying elements that produce age hardening of iron–nickel martensites.[19] The term "maraging" was derived from the combination of the words martensite and age hardening. Maraging steels have high hardenability and high strength combined with good toughness. The maraging steels have a nominal composition of 18% nickel, 7–9% cobalt, 3–5% molybdenum, less than 1% titanium, and very low carbon contents. Carbon is actually viewed as an impurity and kept to as low a level as possible to minimize the formation of titanium carbide (TiC), which can adversely impact strength, ductility, and toughness. During air cooling from the annealing or hot working temperature, maraging steels transform to a relatively soft martensite (30–35 HRC) which can be easily machined or formed. They are then aged to high strength levels at 850–950° F for times ranging from 3 to 9 h.

The commercial maraging steels 18Ni(200), 18Ni(250), 18Ni(300), and 18Ni(350) have nominal yield strengths after heat treatment of 200, 250, 300, and 350 ksi respectively. Typical properties of maraging steels are shown in Table 5.6. Due to their extremely low carbon content, the fracture toughness of the maraging steels is considerably higher than that of conventional high strength steels. Maraging steels can be used for prolonged service at temperatures up to 750° F. Maraging steels are also more resistant to hydrogen embrittlement than the medium carbon low alloy steels. Although they are susceptible to stress corrosion cracking, they

Table 5.6 Typical Properties of Maraging Steels[20]

Alloy	Heat Treatment	Ultimate Tensile Strength (ksi)	Yield Strength (ksi)	Elongation (%) 2 in.	Reduction in Area (%)	Fracture Toughness K_{Ic} (ksi $\sqrt{in.}$)
18Ni(200)	1500° F, Air Cool,	218	203	10	60	140–220
18Ni(250)	Age 3 h at 900° F	260	247	8	55	110
18Ni(300)		297	290	7	40	73
18Ni(350)	1500° F, Air Cool, Age 12 h at 900° F	358	355	6	25	32–45

are more resistant than the medium carbon low alloy steels. Processing techniques that improve the fracture toughness, such as vacuum melting, proper hot working, and keeping residual impurities low, also improve the resistance to stress corrosion cracking. The alloys are available in the form of sheet, plate, bar, and die forgings. Most applications use bar or forgings.

Maraging steels are either air melted followed by VAR or vacuum induction melted followed by VAR. Aerospace grades are tripled melted using air, vacuum induction and vacuum arc remelting, to minimize the residual elements carbon, manganese, sulfur, and phosphorous and the gases oxygen, nitrogen, and hydrogen. Carbon and sulfur are the most deleterious impurities because they tend to form brittle carbide, sulfide, carbonitride, and carbosulfide inclusions that can crack when the metal is strained, lowering the fracture toughness and ductility.

The maraging steels are readily hot worked by conventional rolling and forging operations. As the titanium content increases, hot working becomes more difficult due to increased hot strength and either higher loads or higher temperatures are required. The precipitation of TiC at the grain boundaries, which can form during slow cooling through the temperature range of 2000–1380° F, must be avoided. They also have good cold forming characteristics in spite of relatively high hardness in the annealed condition. Their machinability is similar to 4330 at equivalent hardness. As a result of their low carbon contents, weldability is excellent. The maraging steels can be readily welded by GTAW in either the annealed or the aged conditions. During welding, avoid prolonged dwell times at high temperatures, avoid preheat, keep interpass temperatures low (250° F), avoid slow cooling, and keep welds as clean as possible. Welding of aged material should be followed by aging at 900° F to strengthen the weld area.

Heat treatment consists of solution annealing, air cooling, and then aging. Solution annealing is usually conducted at 1500° F for 1 h. Since the nickel content is so high, austenite transforms to martensite on cooling from the austenitic temperature. The martensite start temperature (M_s) is about 310° F and the martensite finish temperature (M_f) is about 210° F. The formation of martensite is not affected by cooling rate and thick sections can be air cooled and

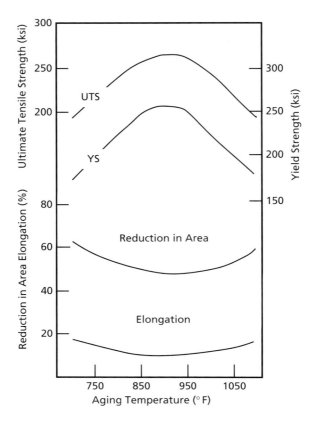

Fig. 5.19. Effect of Aging Temperature on 18Ni(250) Maraging Steel[20]

still be fully martensitic. Since the martensitic transformation involves only an austenite-to-martensite transformation of iron–nickel and does not involve carbon to any considerable extent, the martensite formed is relatively ductile. Before aging, maraging steels have yield strengths in the range of 95–120 ksi. The effect of aging temperature on 18Ni(250) is shown in Fig. 5.19.

5.7 Precipitation Hardening Stainless Steels

Important properties of the precipitation hardening (PH) stainless steels are ease of fabrication, high strength, good ductility, and excellent corrosion resistance. There are two main types of PH stainless steels: semiaustenitic and martensitic. The semiaustenitic grades are essentially austenitic in the solution annealed condition. After fabrication operations are completed, they can be transformed to martensite by an austenite conditioning heat treatment that converts the austenite to martensite followed by precipitation hardening. The martensitic types are already martensitic in the solution annealed condition and only require precipitation hardening after fabrication.

Stainless steels are more difficult to forge than alloy steels because they maintain greater strength at elevated temperatures, and they have to be forged at lower temperatures to avoid microstructural damage, and the PH grades are the most difficult of the stainless steels to forge.[21] For example, the PH grades require about 30–50% greater pressures than 4340; therefore, heavier duty equipment is required. However, the PH stainless steels are much less prone to decarburization during forging than alloy steels. Power drop hammers are used for open die forgings and mechanical presses for small forgings. Hydraulic presses are often used for the final forging operations because there is less chance of overheating due to the slower forging action than with hammers. However, die life is shorter for hydraulic presses due to the longer contact times with the hot metal. Both gas-fired and electrically heated furnaces can be used for preheating. Oil-fired furnaces are avoided due to the potential for sulfur contamination. The maximum forging temperatures for the PH grades are in the range of 2150–2250° F. Due to their low thermal conductivities, it takes longer to heat the PH steels to the forging temperature; however, once the forging temperature is reached, they should not be soaked but immediately forged. The steel must be reheated if it falls below 1800° F. To prevent cracking due to thermal gradients during heating, the PH grades can be preheated in the 1200–1700° F range. The forging dies should be heated to at least 300° F and lubricated before each blow. The forging flash must be hot trimmed to prevent flash-line cracks, which can penetrate into the forging. Hot trimming is done immediately after forging before the forging falls below a red heat. The cooling rate after forging for the martensitic grades is important because they convert directly to martensite on cooling from the austenite range, and too fast a cooling rate can result in cracking as a result of residual stresses. Slow furnace cooling, or an equalization temperature hold (e.g., 1900° F for 30 min), during cooling are used to prevent cracking. Although stainless steels do not scale as badly as carbon or alloy steels, they form a hard and abrasive scale that should be removed prior to machining or cutting tool life will be adversely affected.

The semiaustenitic alloys are generally supplied from the mill in the solution annealed condition (Condition A). In Condition A, these alloys can be formed almost as easily as if they were true austenitic stainless steels.[22] 17-7PH has approximately the same chromium and nickel contents as austenitic type 301 stainless but also contains 1.2% aluminum for precipitation hardening. After fabrication in the soft condition, the austenite is conditioned to allow transformation to martensite. Because of their relatively high hardness in the solution annealed condition, the martensitic types are used principally in the form of bar, rod, wire, and heavy forgings, and only to a minimum extent the form of sheet. The martensitic precipitation hardening steels, before aging, are similar to the chromium martensitic stainless steels (e.g., 410 or 431) in their general fabrication characteristics.

Stainless steels are difficult to machine because of their high strength, high ductility and toughness, high work hardening rates, and low thermal conductivities. They are often characterized as being "gummy" during machining, producing long stringy chips which seize or form built-up edges on the cutting tool, resulting in reduced tool life and degraded surface finishes. The chips removed during machining exert high pressures on the nose of the tool, and these pressures, when combined with the high temperature at the chip-tool interface, cause pressure welding of portions of the chip to the tool. In addition, their low thermal conductivities contribute to a continuing heat build-up. In general, more power is required to machine stainless steels than carbon steels; cutting speeds are usually lower; a positive feed must be maintained; tooling and fixtures must be rigid; and flood cooling should be used to provide lubrication and temperature control. The martensitic PH grades are somewhat easier to machine in the solution annealed condition than the semiaustenitic grades because they are harder and cut cleaner. The following guidelines should be used when machining stainless steels:[23]

- Because more power is generally required to machine stainless steels, equipment should be used only up to about 75% of the rating for carbon steels,
- To avoid chatter, tooling and fixtures must be as rigid as possible. Overhang of either the workpiece or the tool must be minimized,
- To avoid glazed work hardened surfaces, particularly with the semiaustenitic alloys, a positive feed must be maintained. In some cases, increasing the feed and reducing the speed may be necessary. Dwelling, interrupted cuts, or a succession of thin cuts should be avoided,
- Lower cutting speeds may be necessary. Excessive cutting speeds result in tool wear or tool failure and shutdown for frequent tool replacement. Slower speeds with longer tool life are often a better answer to higher output and lower costs,
- Both high speed steel and carbide cutting tools must be kept sharp with a fine surface finish to minimize friction with the chip. A sharp cutting edge produces the best surface finish and provides the longest tool life, and
- Cutting fluids must be used to provide proper lubrication and heat removal. Fluids must be carefully directed to the cutting area at a sufficient flow rate to prevent overheating.

These alloys can be welded by the conventional methods used for the austenitic stainless steels. Inert gas shielded welding is recommended to prevent the loss of titanium or aluminum. Post-weld annealing is recommended for some grades.

The conditioning treatment for the semiaustenitic alloys consists of heating to a high enough temperature to remove carbon from solid solution and precipitate

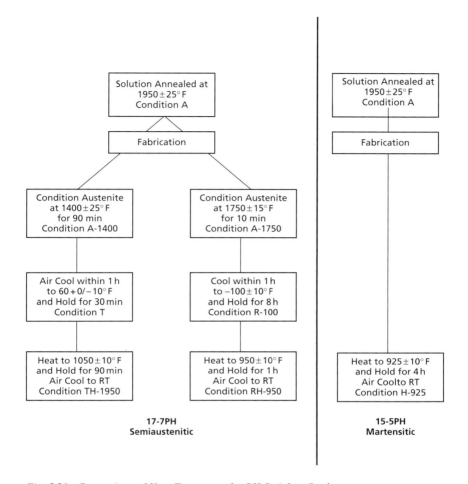

Fig. 5.20. Comparison of Heat Treatments for PH Stainless Steels

it as chromium carbide ($Cr_{23}C_6$). Removing carbon and some chromium from the austenite matrix makes the austenite unstable and on cooling to the M_s temperature, the austenite transforms to martensite. As shown in Fig. 5.20, 17-7PH is conditioned at 1400° F and then cooled to 60° F to produce the T condition. If the conditioning is done at a higher temperature (1750° F), fewer carbides are precipitated and the steel must be cooled to a lower temperature (−110° F) to transform the austenite to martensite. The final step is PH, which is carried out in the 900–1200° F range. During PH, aluminum in the martensite combines with some of the nickel to produce precipitates of NiAl and Ni_3Al.

Since the martensitic PH grades are martensitic after solution annealing, they do not require conditioning but only a precipitation hardening treatment. As shown in the right side of Fig. 5.20, 15-5PH is solution annealed at 1950° F

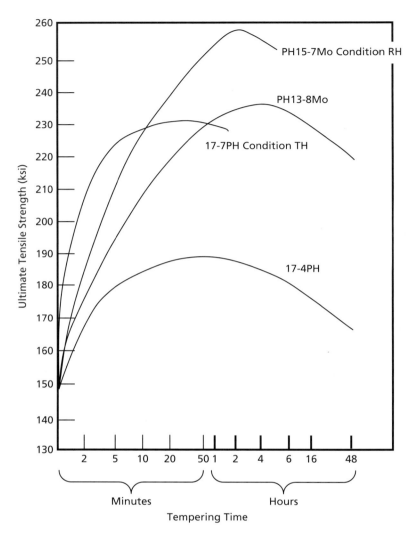

Fig. 5.21. Tempering Response of Several PH Stainless Steels[24]

followed by air cooling. PH during aging at 925° F produces the H-925 condition. The tempering response of a number of PH stainless steels is shown in Fig. 5.21. If stress corrosion is a concern, the PH steels should be aged at the highest temperature that will maintain an adequate strength level.

Some dimensional change are experienced during the heat treatment of the semiaustenitic steels. A dimensional expansion of approximately 0.0045 in./in. occurs during the transformation from the austenitic to the martensitic condition, and, during aging, a contraction of about 0.0005 in./in. takes place. Vapor blasting of scaled parts after final heat treatment is recommended because of the hazards of intergranular corrosion in inadequately controlled acid pickling operations.

Summary

Although high strength steels normally account for only about 5–15% of the airframe structural weight, they are often used for highly critical parts such as landing gear components because of their capability to be heat treated to high strength levels (250–300 ksi) and because of their high stiffness. The disadvantages of high strength steels are primarily their high densities and susceptibility to brittle fracture. As a result of their high strength levels, they are susceptible to both hydrogen embrittlement and stress corrosion cracking.

Medium carbon low alloy steels contain carbon in the range of 0.30–0.50% with alloying elements added to provide deeper hardening and higher strength and toughness. The medium carbon low alloy steels are hardened by austenitizing, quenching in water or oil, and then tempering to their final hardness.

The three high fracture toughness steels, HP-9-4-30, AF1410, and AerMet 100 have somewhat lower carbon contents than the high strength medium carbon low alloy steels 4340 and 300M. The lower carbon content significantly contributes to their better ductilities and higher fracture toughness. In addition, these alloys have high nickel contents that provides deep hardening and toughness, and cobalt that helps to prevent retained austenite.

Maraging steels are a class of high strength steels with very low carbon contents (0.030% maximum) and additions of substitutional alloying elements that produce age hardening of iron–nickel martensites. The maraging steels are characterized by high hardenability and high strength combined with good toughness.

Important properties of the precipitation hardening (PH) stainless steels are ease of fabrication, high strength, good ductility, and excellent corrosion resistance. There are two main types of PH stainless steels: semiaustenitic and martensitic. The semiaustenitic alloys are essentially austenitic in the solution annealed condition. After fabrication operations are completed, they can be transformed to martensite by an austenite conditioning heat treatment that converts the austenite to martensite followed by precipitation hardening. The martensitic types are already martensitic in the solution annealed condition and only require precipitation hardening after fabrication.

Recommended Reading

[1] Metallic Materials Properties Development and Standardization, U.S. Department of Transportation, Federal Aviation Agency, DOT/FAA/AR-MMPDS-01, January 2004.
[2] *ASM Handbook Vol. 1: Properties and Selection: Irons, Steels, and High-Performance Alloys*, ASM International, 1990.

[3] *ASM Handbook Vol. 4: Heat Treating*, ASM International, 1991.
[4] *ASM Handbook Vol. 6: Welding, Brazing, and Soldering*, ASM International, 1993.
[5] *ASM Handbook Vol. 14: Forming and Forging*, ASM International, 1988.
[6] *ASM Handbook Vol. 16: Machining*, ASM International, 1990.

References

[1] Davis, D.P., "Structural Steels", in *High Performance Materials in Aerospace*, Chapman & Hall, 1995, pp. 151–181.
[2] Smith, W.F., "Alloy Steels", in *Structure and Properties of Engineering Alloys*, McGraw-Hill, Inc., 2nd edition, 1993, pp. 125–173.
[3] Smith, W.F., "Stainless Steels", in *Structure and Properties of Engineering Alloys*, McGraw-Hill, Inc., 2nd edition, 1993, pp. 288–332.
[4] Morlett, J.O., Johnson, H., Troiano, A., *Journal of Iron and Steel Institute*, Vol. 189, 1958, p. 37.
[5] Parker, E.R., *Metallurgical Transactions*, 8A, 1977, p. 1025.
[6] Imrie, W.M., *Royal Society London Philosophical Transactions*, A282, 1976, p. 91.
[7] "Forging of Carbon and Alloy Steels", in *ASM Handbook Vol. 14: Forming and Forging*, ASM International, 1988.
[8] Sprague, L.E., "The Effects of Vacuum Melting on the Fabrication and Mechanical Properties of Forging", *Steel Improvement and Forge Company*, 1960.
[9] Philip, T.V., McCaffrey, T.J., "Ultrahigh Strength Steels", in *ASM Handbook Vol. 1: Properties and Selection: Irons, Steels, and High-Performance Alloys*, ASM International, 1990.
[10] Davis, J.R., "Carbon and Alloy Steels", in *Alloying: Understanding The Basics*, ASM International, 2001, p. 190.
[11] Somers, B.R., "Introduction to the Selection of Carbon and Low-Alloy Steels", in *ASM Handbook Vol. 6: Welding, Brazing, and Soldering*, ASM International, 1993.
[12] Totten, G.E., Narazaki, M., Blackwood, R.R., Jarvis, L.M., "Factors Relating to Heat Treating Operations", in *ASM Handbook Vol. 11: Failure Analysis and Prevention*, ASM International, 2002.
[13] Becherer, B.A., Withford, T.J., "Heat Treating of Ultrahigh-Strength Steels", in *ASM Handbook Vol. 4: Heat Treating*, ASM International, 1991.
[14] Callister, W.D., "Phase Transformations", in *Fundamentals of Materials Science and Engineering*, John Wiley & Sons, Inc., 5th edition, 2001, p. 345.
[15] Guy, A.G., "Heat Treatment of Steel", in *Elements of Physical Metallurgy*, Addison-Wesley, 2nd edition, 1959, pp. 465–494.
[16] Philip, T.V., McCafferty, T.J., "High Fracture Toughness Steels", in *ASM Handbook Vol. 1: Properties and Selection: Irons, Steels, and High-Performance Alloys*, ASM International, 1990.
[17] "AerMet 100 Alloy" datasheet, Carpenter Technology Corporation, 1995.
[18] Dahl, J.M., "Ferrous-Base Aerospace Alloys", *Advanced Materials & Processes*, May 2000, pp. 33–36.
[19] Rohrbach, K., Schmidt, M., "Maraging Steels", in *ASM Handbook Vol. 1: Properties and Selection: Irons, Steels, and High-Performance Alloys*, ASM International, 1990.
[20] Schmidt, M., Rohrbach, K., "Heat Treatment of Maraging Steels", in *ASM Handbook Vol. 4: Heat Treating*, ASM International, 1991.
[21] Harris, T., Priebe, E., "Forging of Stainless Steels", in *ASM Handbook Vol. 14: Forming and Forging*, ASM International, 1988.

[22] Washko, S.D., Aggen, G., "Wrought Stainless Steels: Fabrication Characteristics", in *ASM Handbook Vol. 1: Properties and Selection: Irons, Steels, and High-Performance Alloys*, ASM International, 1990.

[23] Kosa, T., Ney, R.P., "Machining of Stainless Steels", in *ASM Handbook Vol. 16: Machining*, ASM International, 1990.

[24] Pollard, B., "Selection of Wrought Precipitation-Hardening Stainless Steels", in *ASM Handbook Vol. 6: Welding, Brazing, and Soldering*, ASM International, 1993.

Chapter 6

Superalloys

Superalloys are heat resistant alloys of nickel, iron–nickel and cobalt that frequently operate at temperatures exceeding 1000° F. However, some superalloys are capable of being used in load bearing applications in excess of 85% of their incipient melting temperatures. They are required to exhibit combinations of high strength, good fatigue and creep resistance, good corrosion resistance, and the ability to operate at elevated temperatures for extended periods of time (i.e., metallurgical stability).[1] Their combination of elevated temperature strength and resistance to surface degradation is unmatched by other metallic materials.

Superalloys are the primary materials used in the hot portions of jet turbine engines, such as the blades, vanes and combustion chambers, constituting over 50% of the engine weight. Typical applications are shown in Fig. 6.1. Superalloys are also used in other industrial applications where their high temperature strength and/or corrosion resistance is required. These applications include rocket engines, steam turbine power plants, reciprocating engines, metal processing equipment, heat treating equipment, chemical and petrochemical plants, pollution control equipment, coal gasification and liquification systems, and medical applications.[2]

In general, the nickel-based alloys are used for the highest temperature applications, followed by the cobalt-based alloys and then the iron–nickel alloys.

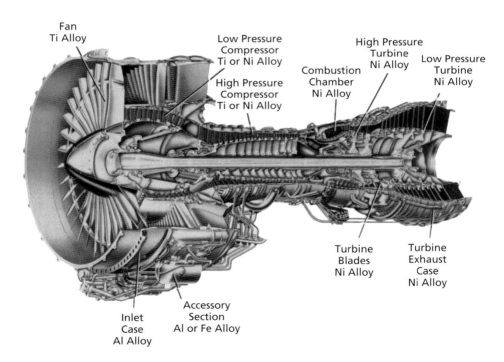

Fig. 6.1. Typical Material Distribution in Jet Engine

Superalloys

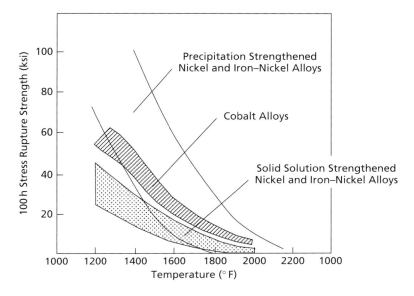

Fig. 6.2. Stress Rupture Comparison of Wrought Superalloys[3]

A relative comparison of their stress rupture properties is shown in Fig. 6.2. Superalloys are produced as wrought, cast and powder metallurgy product forms. Some superalloys are strengthened by precipitation hardening mechanisms, while others are strengthened by solid solution hardening. For jet engine applications, large grain cast alloys are preferred for creep and stress rupture limited turbine blade applications, while small grain forged alloys are preferred for strength and fatigue limited turbine disk applications. The large grain sizes help in preventing creep, while the smaller grain sizes enhance strength and fatigue resistance.

6.1 Metallurgical Considerations[4-6]

Nickel has an FCC crystalline structure, a density of 0.322 lb/in.^{3}, and a melting point of 2650° F. While iron has a BCC structure at room temperature and cobalt a HCP structure at room temperature, both iron- and cobalt-based superalloys are so highly alloyed that they have an austenitic γ FCC structure at room temperature. Therefore, the superalloys display many of the fabrication advantages of the FCC structure.

Nickel and iron–nickel based superalloys are strengthened by a combination of solid solution hardening, precipitation hardening and the presence of carbides at the grain boundaries.[1] The FCC nickel matrix, which is designated as austenite (γ), contains a large percentage of solid solution elements such as iron, chromium, cobalt, molybdenum, tungsten, titanium and aluminum. Aluminum

213

and titanium, in addition to being potent solid solution hardeners, are also precipitation strengtheners. At temperatures above $0.6\,T_m$, which is in the temperature range for diffusion controlled creep, the slowly diffusing elements molybdenum and tungsten are beneficial in reducing high temperature creep.

The most important precipitate in nickel and iron–nickel based superalloys is γ' FCC ordered $Ni_3(Al,Ti)$ in the form of either Ni_3Al or Ni_3Ti. The γ' phase is precipitated by precipitation hardening heat treatments: solution heat treating followed by aging. The γ' precipitate is an A_3B type compound where A is composed of the relatively electronegative elements nickel, cobalt and iron, and B of the electropositive elements aluminum, titanium or niobium. Typically, in the nickel based alloys, γ' is of the form $Ni_3(Al,Ti)$, but if cobalt is added, it can substitute for some nickel as $(Ni,Co)_3(Al,Ti)$. The precipitate γ' has only about an 0.1% mismatch with the γ matrix; therefore, γ' precipitates homogeneously with a low surface energy and has extraordinary long-term stability. The coherency between γ' and γ is maintained to high temperatures and has a very slow coarsening rate, so that the alloy overages extremely slowly even as high as $0.7\,T_m$. Since the degree of order in $Ni_3(Al,Ti)$ increases with temperature, alloys with a high volume of γ' actually exhibit an increase in strength as the temperature is increased up to about 1300°F. The γ/γ' mismatch determines the γ' precipitate morphology, with small mismatches ($\sim 0.05\%$) producing spherical precipitates and larger mismatches producing cubical precipitates as shown in Fig. 6.3.

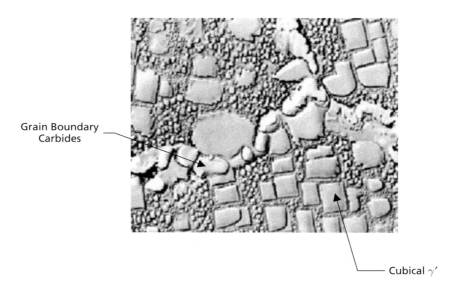

Fig. 6.3. Microstructure of Precipitation-Strengthened Nickel Based Superalloy

If appreciable niobium is present, the body centered tetragonal ordered γ'' (Ni_3Nb) precipitate can form. This is an important strengthening precipitate in some of the iron–nickel based superalloys, and forms the basis for strengthening the important alloy Inconel 718. Other less frequent precipitates include the hexagonal ordered η (Ni_3Ti) and the orthorhombic δ (Ni_3Nb) phases which help to control the structure of wrought alloys during processing.

The compositions of commercial superalloys are complex (some contain as many as a dozen alloying elements), with the roles of various alloying elements shown in Table 6.1. Chromium and aluminum additions help in providing oxidation resistance. Chromium forms Cr_2O_3 on the surface, and when aluminum is present, the even more stable Al_2O_3 is formed. A chromium content of 5–30% is usually found in superalloys. Solid solution strengthening is provided by molybdenum, tantalum, tungsten and rhenium. Rhenium also helps to retard the coarsening rate of γ' and is used in some of the latest cast nickel based alloys. Cobalt helps to increase the volume percentage of helpful precipitates that form with additions of aluminum and titanium (γ') and niobium (γ''). Unfortunately, increasing the aluminum and titanium content lowers the melting point, thereby narrowing the forging range, which makes processing more difficult. Small additions of boron, zirconium and hafnium improve the mechanical properties of nickel and iron–nickel alloys. A number of alloying elements, although added for their favourable characteristics, can also form the undesirable σ, μ and Laves phases which can cause in-service embrittlement if the composition and processing are not carefully controlled.

The carbon content of nickel based superalloys varies from about 0.02 to 0.2% for wrought alloys and up to about 0.6% for some cast alloys. Metallic carbides can form in both the matrix and at the grain boundaries. In high temperature service, the properties of the grain boundaries are as important as the strengthening by γ' within the grains. Grain boundary strengthening is

Table 6.1 Role of Alloying Elements in Superalloys[7]

Alloy Additions	Solid Solution Strengtheners	γ' Formers	Carbide Formers	Grain Boundary Strengtheners	Oxide Scale Formers
Chromium	X		X		X
Aluminum		X			X
Titanium		X	X		
Molybdenum	X		X		
Tungsten	X		X		
Boron				X	
Zirconium				X	
Carbon				X	
Niobium		X	X		
Hafnium			X	X	
Tantalum		X	X	X	

produced mainly by precipitation of chromium and refractory metal carbides. Small additions of zirconium and boron improve the morphology and stability of these carbides.

Carbides in superalloys perform three functions. First, grain boundary carbides, when properly formed, strengthen the grain boundaries, prevent or retard grain boundary sliding and permit stress relaxation along the grain boundaries. Second, a fine distribution of carbides precipitated within the grains increases strength; this is especially important in the cobalt based alloys that cannot be strengthened by γ' precipitates. Third, carbides can tie-up certain elements that would otherwise promote phase instability during service. Since carbides are harder and more brittle than the alloy matrix, their distribution along the grain boundaries will affect the high temperature strength, ductility and creep performance of nickel based alloys. There is an optimum amount and distribution of carbides along the grain boundaries. If there are no carbides along the grain boundaries, voids will form and contribute to excessive grain boundary sliding. However, if a continuous film of carbides is present along the grain boundaries, continuous fracture paths will result in brittleness. Thus, the optimum distribution is a discontinuous chain of carbides along the grain boundaries, since the carbides will then hinder grain boundary sliding without adversely affecting ductility. Multistage heat treatments are often used to obtain the desired grain boundary distribution, along with a mix of both small and large γ' precipitates, for the best combination of strength at intermediate and high temperatures.

Some of the important carbides are MC, $M_{23}C_6$, M_6C and M_7C_3. In nickel based alloys, M stands for titanium, tantalum, niobium or tungsten. MC carbides usually form just below the solidification temperature during ingot casting and are usually large and blocky, have a random distribution and are generally not desirable.[4] However, MC carbides tend to decompose during heat treatment into other more stable carbides, such as $M_{23}C_6$ and/or M_6C. In the $M_{23}C_6$ carbides, M is usually chromium but can be replaced by iron and to a smaller extent by tungsten, molybdenum or cobalt, depending on the alloy composition. $M_{23}C_6$ carbides can form either during heat treatment or during service at temperatures between 1400 and 1800° F. They can form from either the degeneration of MC carbides or from soluble carbon in the matrix. $M_{23}C_6$ carbides tend to precipitate along the grain boundaries and enhance stress rupture properties. M_6C carbides form at temperatures in the range of 1500–1800° F. They are similar to the $M_{23}C_6$ carbides and have a tendency to form when the molybdenum and tungsten contents are high. Although not nearly as prevalent as the other carbides, M_7C_3 carbides also precipitate on the grain boundaries, and are beneficial if they are discrete particles but detrimental if they form a grain boundary film.

Over the years, a number of undesirable topologically closed-packed (TCP) phases have appeared either during heat treatment or in-service, the most important being σ, μ and Laves. These phases, which usually form as thin plates or needles, can lead to lower stress rupture strengths and a loss in ductility.

Superalloys

They also remove useful strengthening elements from the matrix, such as the refractory elements molybdenum, chromium and tungsten, which reduces both solid solution strengthening and the γ/γ' mismatch. Modern computer modeling programs (e.g., Phacomp) are capable of predicting their occurrence so they can be avoided in alloy design. As with most high performance alloys, hydrogen, oxygen and nitrogen are considered detrimental and are held to very low levels.

Superalloys are used in the cast, rolled, extruded, forged and powder produced forms as shown in the overall process flow in Fig. 6.4. Wrought alloys are generally more uniform with finer grain sizes and superior tensile and fatigue properties, while cast alloys have more alloy segregation and coarser grain sizes

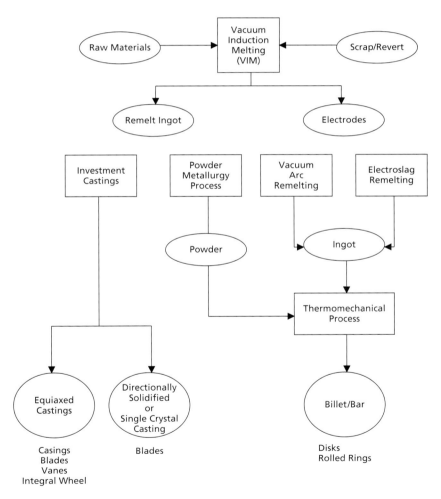

Fig. 6.4. Process Flow for Superalloy Components[8]

217

but better creep and stress rupture properties. Accordingly, wrought alloys are used where tensile strength and fatigue resistance are important, such as disks, and cast alloys are used where creep and stress rupture are important, such as turbine blades. It should also be noted that the amount of alloying is so high in some superalloys that they cannot be produced as wrought products (which are produced from large diameter ingots); they must be produced as either castings or by powder metallurgy methods. In general, the more heat resistant the alloy, the more likely it is to be prone to segregation and brittleness, and therefore formable only by casting or by using powder metallurgy techniques.

Two processes are used for the production of forged superalloys, ingot metallurgy and powder metallurgy. Ingot metallurgy[9] often involves triple melt technology which includes three melting steps followed by homogenizing and hot working to achieve the desired compositional control and grain size. The three melting steps include (1) VIM to prepare the desired alloy composition, (2) ESR to remove oxygen containing inclusions and (3) VAR to reduce compositional segregation that occurs during ESR solidification. Melting is followed by homogenizing and hot working to achieve the desired ingot homogeneity and grain size.

Powder metallurgy (PM)[9] is often required for high volume fraction γ' strengthened alloys, such as René 95 and Inconel 100, which cannot be made by conventional ingot metallurgy and forging without cracking. The PM process (Fig. 6.5) includes (1) VIM to prepare the desired alloy composition, (2) remelting and atomizing to produce powder, (3) sieving to remove large particles and inclusions (>50–100 μm in diameter), (4) canning to place the powder in a container suitable for consolidation, (5) vacuum degassing and sealing to remove the atmosphere and (6) hot isostatic pressing (HIP) or extrusion to consolidate the alloy to a billet. During HIP, small (<20 μm) pores containing argon can be collapsed and tend to stay closed if subsequent processing temperatures are not significantly above the original HIP temperature. Billets are then subsequently forged to final part shape.

Creep failures are known to initiate at transverse grain boundaries, and it is possible to eliminate them in cast turbine blades to obtain further improvements in creep and stress rupture resistance. This can be achieved by directional solidified (DS) castings with columnar grains aligned along the growth direction, with no grain boundaries normal to the high stress direction. Further, by incorporating a geometric constriction in the mold or by the use of a seed crystal, it is possible to eliminate grain boundaries entirely and grow the blade as one single crystal (SX). The elimination of grain boundaries also removes the necessity for adding grain boundary strengthening elements, such as carbon, boron, zirconium and hafnium.[10] The removal of these elements raises the melting point and allows higher solution heat treatment temperatures with improvements in chemical homogeneity and a more uniform distribution of γ' precipitates.[7]

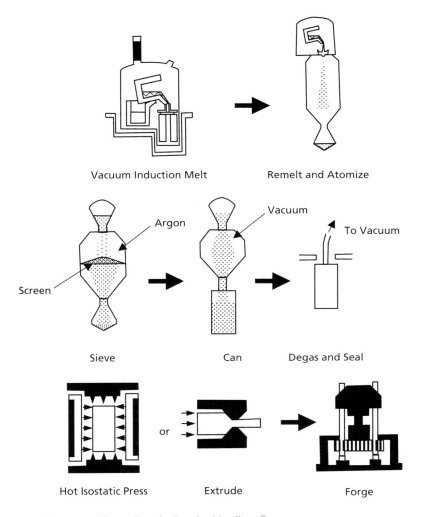

Fig. 6.5. *Forged Superalloys by Powder Metallurgy*[9]

6.2 Commercial Superalloys

The number of superalloys that have been developed and used over the years is large; a small number of these are listed in Tables 6.2 (wrought) and 6-3 (cast). Note that the designation in Table 6.2 breaks them into either solid solution or precipitation hardened. In reality, the solid solution strengthened alloys are strengthened both by solid solution hardening and by the presence of carbides, while the precipitation hardened alloys are strengthened by the combination of precipitates, solid solution hardening and the presence of carbides.

Table 6.2 Nominal Compositions of Select Wrought Superalloys

Alloy	Cr	Ni	Co	Mo	W	Nb	Ti	Al	Fe	C	Other
Nickel Based											
Solid Solution Hardened											
Hastelloy X	22.0	49.0	<1.5	9.0	0.6	–	–	2.0	15.8	0.15	–
Inconel 625	21.5	61.0	–	9.0	–	3.6	–	0.2	2.5	0.05	–
Nimonic 75	19.5	75.0	–	–	–	–	0.4	0.15	2.5	0.12	<0.25 Cu
Precipitation Hardened											
Astroloy	15.0	56.5	15.0	5.25	–	–	3.5	4.4	<0.3	0.06	0.03 B, 0.06 Zr
Inconel 100	10.0	60.0	15.0	3.0	–	–	4.7	5.5	<0.6	0.15	1.0 V, 0.06 Zr, 0.015 B
Inconel 706	16.0	41.5	–	–	–	–	1.75	0.2	37.5	<0.08	2.9 (Nb+Ta), <0.15 Cu
Nimonic 95	19.5	53.5	18.0	–	–	–	2.9	2.0	<5.0	<0.15	+B+Zr
René 95	14.0	61.0	8.0	3.5	3.5	3.5	2.5	3.5	<0.3	0.16	0.01 B, 0.5 Zr
Waspaloy	19.5	57.0	13.5	4.3	–	–	3.0	1.4	<2.0	0.07	0.006 B, 0.09 Zr
Iron–Nickel Based											
Solid Solution Hardened											
19-9 DL	19.0	9.0	–	1.25	1.25	0.4	0.3	–	66.8	0.30	1.10 Mn, 0.60 Si
Haynes 556	22.0	21.0	20.0	3.0	2.5	0.1	0.1	0.3	29.0	0.10	0.50 Ta, 0.02 La, 0.002 Zr
Incoloy 802	21.0	32.5	–	–	–	–	–	0.58	44.8	0.36	–
Precipitation Hardened											
A-286	15.0	26.0	–	1.25	–	–	2.0	0.2	55.2	0.04	0.005 B, 0.3 V
Inconel 718	19.0	52.5	–	3.0	–	5.1	0.91	0.5	18.5	<0.08	<0.15 Cu
Incoloy 903	<0.1	38.0	15.0	0.1	–	3.0	1.4	0.7	41.0	0.04	–
Cobalt Based											
Solid Solution Hardened											
Haynes 25 (L605)	20.0	10.0	50.0	–	15.0	–	–	–	3.0	0.10	1.5 Mn
Haynes 188	22.0	22.0	37.0	–	14.5	–	–	–	<3.0	0.10	0.90 La
MP35-N	20.0	35.0	35.0	10.0	–	–	–	–	–	–	–

The iron–nickel based alloys, which are an extension of stainless steel technology, are generally wrought, whereas the cobalt and nickel based alloys are both wrought and cast.[1] The iron–nickel based alloys have high strengths below 1200° F and are more easily processed and welded than the nickel based alloys. Cobalt based superalloys have high melting points and high temperature capability at moderate stress levels, excellent hot salt corrosion resistance and better weldability than the nickel based alloys.[11] However, they cannot compete with the nickel based alloys at high temperatures and high stress levels.

The most commercially important superalloy, Inconel 718, is listed as an iron–nickel based alloy even though it contains more nickel than iron. This classification fits with the traditional classification for this alloy, although many newer works list it as a nickel based alloy. The significance of Inconel 718 is shown in the Fig. 6.6 material distribution, where it accounts for 34% of the General Electric CF6 engine. Also note that for the cobalt based alloys, there are none listed as being precipitation hardened, because unfortunately, these alloys do not precipitation harden like the nickel and iron–nickel alloys. Also note that the composition of the cast alloys in Table 6.3 is generally more complex than for the wrought alloys.

6.2.1 Nickel Based Superalloys

Nickel based superalloys are the most complex of the superalloys and are used in the hottest parts of aircraft engines, constituting over 50% of the engine weight. They are either solid solution hardened for lower temperature use or precipitation hardened for higher temperature use. The nickel based alloys contain at least 50% nickel and are characterized by the high phase stability of the FCC austenitic (γ) matrix. Many nickel based alloys contain 10–20% chromium, up to about 8% aluminum and titanium combined, 5–15% cobalt, and small amounts of boron, zirconium, hafnium and carbon. Other common alloying additions are molybdenum, niobium, tantalum, tungsten and rhenium. Chromium and aluminum are important in providing oxidation resistance by forming the oxides Cr_2O_3 and Al_2O_3 respectively.

The most important precipitate in the nickel based alloys is γ' in the austenitic γ nickel matrix. An example of a γ' strengthened alloy is the wrought alloy Waspaloy and the cast alloys René 80 and Inconel 713C. In general, strengthening increases with increasing amounts of γ', which is a function of the combined aluminum and titanium content as shown in Figure 6.7. When the volume fraction of γ' is less than about 25%, the precipitate particles are spherical, changing to a cubical shape at volume percentages greater than 35%. Most wrought nickel based alloys contain between 20 and 45% γ', while cast alloys can contain as much as 60% γ'. As the amount of γ' increases, the elevated temperature resistance increases.

Alloys containing niobium are strengthened primarily by γ''. Alloys can also contain both niobium and titanium and/or aluminum and be strengthened by

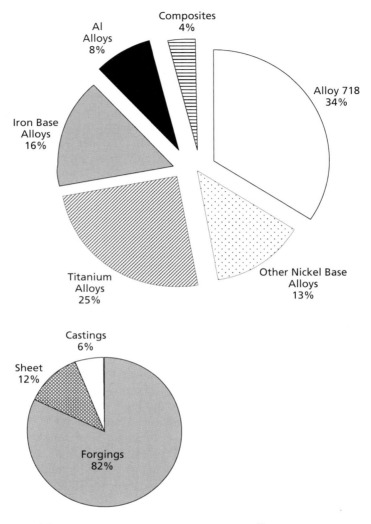

Fig. 6.6. Materials Distribution for GE CF6 Engine[11]

a combination γ'' and γ', such as Inconel 706. Some nickel based alloys, such as Hasteloy X and Inconel 625, are predominately hardened by solid solution strengthening and are therefore used at lower temperatures, predominately being specified for corrosive environments. There are also oxide dispersion hardened alloys that are strengthened by a dispersion of inert particles, such as yttria (Y_2O_3), including MA-754 and MA-6000, where the MA stands for the mechanical alloying process used to prepare metal powders for consolidation.

Table 6.3 Nominal Compositions of Select Cast Superalloys

Alloy	Composition												
	C	Ni	Cr	Co	Mo	Fe	Al	B	Ti	Ta	W	Zr	Other
Nickel Based													
CMSX-2	–	66.2	8	4.6	0.6	–	56	–	1	6	8	6	–
Inconel 713C	0.12	74	12.5	–	4.2	–	6	0.012	0.8	1.75	–	0.1	0.9 Nb
Inconel 738	0.17	61.5	16	8.5	1.75	–	3.4	0.01	3.4	–	2.6	0.1	2 Nb
MAR-M-247	0.15	59	8.25	10	0.7	0.5	5.5	0.015	1	3	10	0.05	1.5 Hf
PWA 1480	–	Bal	10	5.0	–	–	5.0	–	1.5	12	4.0	–	–
René 41	0.09	55	19	11.0	10.0	–	1.5	0.01	3.1	–	–	–	–
René 80	0.17	60	14	9.5	4	–	3	0.015	5	–	4	0.03	–
René 80 Hf	0.08	60	14	9.5	4	–	3	0.015	4.8	–	4	0.02	0.75 Hf
René N4	0.06	62	9.8	7.5	1.5	–	4.2	0.004	3.5	4.8	6	–	0.5 Nb, 0.15 Hf
Udimet 700	0.1	53.5	15	18.5	5.25	–	4.25	0.03	3.5	–	–	–	–
Waspaloy	0.07	57.5	19.5	13.5	4.2	1	1.2	0.005	3	–	–	0.09	–
Iron–Nickel Based													
Inconel 718	0.04	53	19	–	3	18	0.5	–	0.9	–	–	–	0.1 Cu, 5 Nb
Cobalt Based													
AirResist 215	0.35	0.5	19	63	–	0.5	4.3	–	–	7.5	4.5	0.1	0.1 Y
FSX-414	0.25	10	29	52.5	–	1	–	0.010	–	–	7.5	–	–
Haynes 25	0.1	10	20	54	–	1	–	–	–	–	15	–	–
MAR-M 918	0.05	20	20	52	–	–	–	–	0.2	7.5	–	0.5	–
X-40	0.50	10	22	57.5	–	1.5	–	–	–	–	7.5	–	0.5 Mn, 0.5 Si

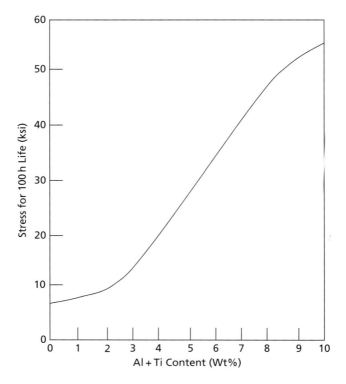

Fig. 6.7. Effect of Al + Ti *Content on Strength of Nickel Based Alloys at* $1600°$ F[12]

6.2.2 Iron–Nickel Based Superalloys

The iron–nickel based superalloys originally evolved from austenitic stainless steels with at least 25% nickel required to stabilize the FCC austenitic matrix. Most of them contain 25–45% nickel and 15–60% iron. Chromium from 15 to 28% is added for oxidation resistance at elevated temperature, while 1–6% molybdenum is added for solid solution strengthening. Titanium, aluminum and niobium are added for precipitation hardening. In the iron–nickel based alloys,[1] the most common precipitate is γ' as in the alloy A-286, which contains 26% nickel. In alloys containing niobium, such as Inconel 718, the γ'' precipitate Ni_3Nb is the predominate strengthener. Due to the lower temperature stability of the γ'' precipitate compared to the γ' precipitate, the maximum use temperature of Inconel 718 is about $1200°$ F; however, Inconel 718 is still the most widely used of all of the superalloys. It is one of the strongest at low temperatures but rapidly loses strength in the $1200–1500°$ F range. The iron–nickel based family also contains some low expansion alloys, such as Incoloy 903, which are important in applications requiring closely controlled clearances between rotating and static components. Some iron–nickel alloys are strengthened primarily by solid solution hardening, such as 19-9DL, which is essentially 18–8 stainless steel with slight chromium and nickel adjustments. As a class, the iron–nickel based alloys have useful strengths to about 1200° F.

6.2.3 Cobalt Based Superalloys

Unalloyed cobalt has a hexagonal close-packed matrix at temperatures below 780° F that then transforms to an FCC structure at higher temperatures; however, nickel alloying additions are used to stabilize the FCC austenitic structure between room temperature and the melting point. As a class, cobalt based superalloys are much simpler than the nickel based alloys. Cast cobalt alloys contain about 50–60% cobalt, 20–30% chromium, 5–10% tungsten and 0.1–1.0% carbon. Wrought alloys contain about 40% cobalt and high nickel contents ($\sim 20\%$) for increased workability. Unfortunately, no precipitates that result in a large strength increase have been found for the cobalt based alloys, and therefore they must rely on a combination of solid solution and carbide strengthening, which limits their use in many applications. However, a fine dispersion of carbides contributes significantly to the strength of these alloys. In general, there are three main classes of carbides: MC, $M_{23}C_6$ and M_6C. In MC carbides, M stands for the reactive metals titanium, tantalum, niobium and zirconium. In the $M_{23}C_6$ carbides, M is mostly chromium but can also be molybdenum or tungsten. When the molybdenum or tungsten content exceeds about 5 atomic percent, M_6C carbides can form. The cobalt alloys display good stress rupture properties at temperatures higher than 1830° F but cannot compete with the nickel based alloys for highly stressed parts, so are used for low stress long lived static parts. They also have superior hot corrosion resistance at elevated temperature, probably due to their quite high chromium contents. Examples of important cobalt based alloys are the wrought alloys Haynes 25 (L605) and Haynes 188 and the cast alloy X-40.

6.3 Melting and Primary Fabrication[6,13]

There are three major types of nickel deposits: nickel–copper–iron sulfides, nickel silicates, and nickel laterites and serpentines.[5] The sulfide deposits, which are located in Canada, provide most of the Western world's supply. The nickel–copper–iron sulfide ore is crushed and ground and the iron sulfide is separated magnetically. The remaining nickel and copper ores are then separated by flotation. The nickel concentrate is roasted, smelted in a reverberatory furnace and converted to a Bessemer matte which consists mainly of nickel and copper sulfides. The copper–nickel matte is cooled under controlled conditions so that discrete crystals of nickel and copper sulfides and a nickel–copper metallic alloy are formed. After the cooled matte is crushed and ground, the metallic alloy is separated magnetically and treated. The remaining copper and nickel sulfides are separated by froth flotation. The nickel sulfide is roasted to produce various grades of nickel oxides which are then converted to pure nickel and nickel alloys.

Vacuum induction melting (VIM), vacuum arc remelting (VAR) and electroslag remelting (ESR) are used in the production of nickel and iron–nickel based superalloy ingots. In the VIM process, liquid metal is processed under vacuum in an induction heated crucible. VIM is used as the initial melting method to reduce interstitial gases to low levels, enable higher and more controllable levels of the reactive strengthening elements aluminum and titanium, and eliminate the slag or dross problem inherent in air melting. After initial VIM processing, the alloys are then remelted using either VAR or ESR.

Vacuum induction melting is used to produce the desired alloy composition. Feedstock for VIM includes pure elements, master alloys and recycled scrap. VIM is used to remove dissolved gases (oxygen, nitrogen and hydrogen) and other impurities before the addition of reactive alloying elements to the melt. Ceramic filters are also used to remove large oxide and nitride inclusions during the final pour. The VIM process may be the only melting process if the ingot is going to be remelted for casting; however, if the ingot is going to be hot worked, it is must be secondarily melted, because VIM ingots generally have coarse and non-uniform grain sizes, shrinkage and alloying element segregation that restricts hot workability during forging.

Vacuum arc melting and electroslag remelting are used to further refine the ingot after initial VIM processing. In the VAR process, an arc is struck between the end of the ingot electrode and the bottom of a water cooled copper crucible. The arc generates the heat to melt the end of the electrode which drips down into the crucible. In the ESR process, remelting does not occur by striking an arc; instead, the ingot is built-up in a water-cooled crucible by melting a consumable electrode that is immersed in a slag that is heated by resistance heating. The ESR process does not require a vacuum since the molten metal is protected from the atmosphere by the slag covering.

In addition to refining the composition, VAR and ESR refine the solidification structure of the resulting ingot. The ESR process is inherently capable of producing cleaner metal than VAR, while VAR is capable of producing larger ingots with fewer segregation defects. Therefore, a triple melting process (VIM–ESR–VAR) is used for producing large ingots for forging stock for gas turbine components,[13] as shown in Fig. 6.8. Other triple melt processing options include double VIM followed by VAR or ESR and VIM–VAR–VAR. In some highly alloyed nickel based alloys, containing a high volume fraction of γ', even VIM–VAR or VIM–ESR does not provide an ingot structure satisfactory for hot working. These alloys are normally processed by using powder metallurgy methods.

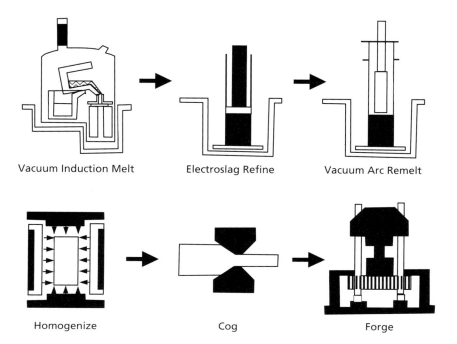

Fig. 6.8. Forged Superalloys by Ingot Metallurgy[9]

Cobalt based alloys generally do not require vacuum melting and are melted in air induction melting furnaces; however, these alloys are quite frequently remelted using either VAR or ESR. Cobalt based alloys containing aluminum, titanium, zirconium and tantalum must be vacuum melted using VIM. Vacuum melting of the other cobalt based alloys can be used to enhance strength and ductility as a result of improved cleanliness and compositional control.

Initial ingot breakdown is conducted well above the γ' solvus to allow chemical homogenization and microstructural refinement; however, care must be taken to make sure that the temperature does not exceed the incipient melting temperature. Cast ingots are initially refined by cogging (Fig. 6.9), a forging operation that uses open dies and hydraulic presses. New cogging equipment and tighter, more reproducible controls are being used to produce more uniform billets. Automated equipment is used to manipulate the billet and carry out the deformation producing uniform deformation along the length of the billet. Continued deformation is conducted at successively lower temperatures to give greater microstructural refinement. Producing a fine, uniform grained billet is important for at least three reasons: (1) fine grained billets lead to fine grained forgings which improves fatigue life; (2) fine grained billets yield lower flow stresses and uniform reductions during forging; and (3) fine grained billets are easier to ultrasonically inspect assuring increased levels of cleanliness.[14]

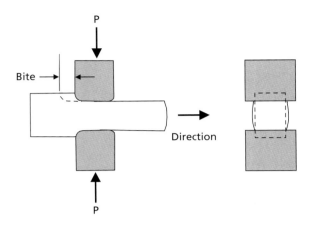

Fig. 6.9. Cogging

The important alloy Inconel 718 is processed by billet breakdown at temperatures in the range of 1940–2010° F region, followed by finishing temperatures in the range of 1865–1900° F.[15]

6.4 Powder Metallurgy[6,16]

While PM product forms do not currently play an important role in aluminum or titanium alloys, powder metallurgy is important for some of the superalloys. The majority of PM superalloy parts fall into one of two categories. In the first, spherical superalloy powders are produced, screened to remove oversize particles, blended to provide a uniform distribution, loaded into container, vacuum degassed and sealed, and then hot isostatic pressed (HIP) to provide a near net shape. This process is used with highly alloyed compositions that would crack during hot working if they were made using conventional ingot metallurgy. In the second, mechanically alloyed powders strengthened by ultrafine dispersions of yttria (Y_2O_3) are consolidated by hot extrusion and hot rolling. In this case, oxide dispersion strengthened (ODS) alloys are being fabricated for improved creep resistance.

6.4.1 Powder Metallurgy Forged Alloys[16]

As the alloy contents of conventional wrought forged superalloys increased, the forging of these alloys became more and more difficult. The use of PM techniques permits the attainment of fine grain sizes that are often superplastic during hot forging operations. In addition, since PM allows near net shape capabilities, fewer processing steps and better material utilization are obtainable. The main advantages of using PM for superalloys are: the ability to add more alloying elements for greater high temperature capability, more uniform composition and phase distribution, finer grain sizes, reduced carbide segregation and, in some cases, higher material yields. However, potential problems with the powder metallurgy process include increased residual gas contents, carbon contamination, ceramic inclusions and the formation of prior particle boundary oxides and/or carbon films.

Superalloy powders are made by gas atomization processes that produce spherical powders. The most common method is inert gas atomization (Fig. 6.10). A molten metal is poured through a refractory orifice. A high pressure argon gas stream breaks up the molten metal into liquid droplets which solidify on the fly at cooling rates of around $10^{2°}$ K/s. The spherical powder is collected at the outlet of the atomization chamber. The maximum particle diameter produced depends on the surface tension, viscosity and density of the melt, as well as the velocity

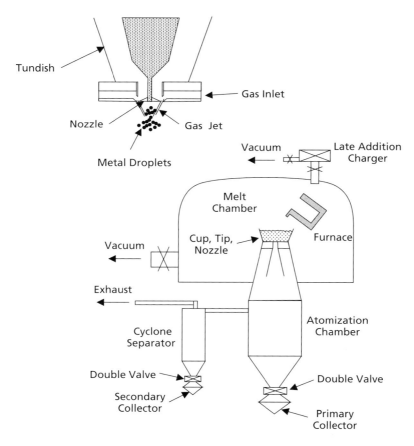

Fig. 6.10. Gas Atomization System for Superalloy Powder Production[17]

of the gas stream. Oxygen contents are on the order of 100 ppm, depending on particle size. Finer particle sizes are obtained by screening out oversize particles. Other powder production processes include soluble gas atomization, centrifugal atomization and the plasma rotating electrode process.

Powder consolidation is usually conducted by HIP, or by extrusion, followed by isothermal forging. For HIP, the powder is packed in a sheet metal container, evacuated at either room or elevated temperature, sealed and placed in the HIP chamber and consolidated at 15 ksi argon pressure and 2000–2200° F. HIP consolidated billets can then be processed by isothermal forging or hot die forging. In isothermal forging, the workpiece and the die are heated to the same temperature, while in hot die forging, the die temperature is higher than that for conventional forging but lower than that for isothermal forging. The use of hot tooling and slow strain rates allows more precise shapes. Compared to conventional forging, these processes offer the following advantages: reduced flow stresses, enhanced workability, greater microstructural control and greater dimensional accuracy. The low strain rates, in combination with fine grain sizes, allows production of near net shapes which reduces material and machining cost. HIP followed by forging can be more economical than conventional cast or wrought products, due to improved material utilization, fewer forging steps and reduced final machining. A comparison between conventional ingot metallurgy and forging with the powder metallurgy approach is shown in Fig. 6.11. The biggest disadvantages are high die costs (e.g., TZM molybdenum) and the requirement for forging in vacuum or in an inert atmosphere.

Superalloy powders can also be consolidated by extrusion. Powder, normally containerized, is hot extruded at a reduction ratio in the range of 13–1 to achieve a fully consolidated billet. The extruded billet is then hot worked by conventional press forging, isothermal forging or hot die forging. More recently, as-HIP PM products (no forging), which require only minimal machining, are emerging as an even lower cost route.

6.4.2 Mechanical Alloying[1]

Mechanical alloying (MA) is a process used to make ODS alloys such as MA6000. Some of these alloys combine γ' strengthening along with oxide dispersion so that their stress rupture properties at 1830° F (and higher) are superior to γ' strengthening alone.

Mechanical alloying[1] is a high energy dry ball milling process used to make composite metallic powders with a controlled fine microstructure. As shown in Fig. 6.12, powders of the desired composition are blended and then placed in a high energy ball mill. The intensive milling process repeatedly fractures and then rewelds the powder particles. During each collision with the grinding balls, the particles are plastically deformed to the extent that the surface oxides are broken, exposing clean metal surfaces.

Superalloys

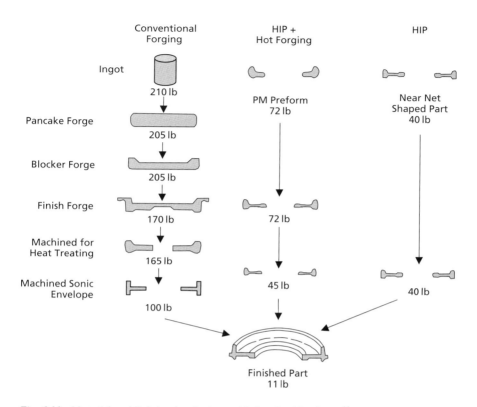

Fig. 6.11. *Material and Fabrication Savings with Powder Metallurgy*[16]

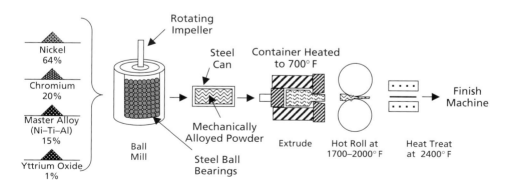

Fig. 6.12. *Mechanical Alloying Powder Metallurgy Process*[1]

On subsequent impacts, the clean surfaces are welded together. This cold welding process increases the size of the particles, while at the same time additional impacts are fracturing particles and reducing their size. As the process continues, the microstructure of the particles is continually being refined. Very fine yttria (25 nm), which is added to the mixture, becomes entrapped between fragments of the composite metal powders. Milling is continued until every powder particle contains the same composition as the starting mix.

Powders for mechanical alloying include pure metals, master alloys and refractory compounds. Pure metals include nickel, iron, chromium, cobalt, tungsten, molybdenum and niobium, while master alloys are nickel based alloys with large concentrations of the more reactive metals aluminum, titanium, zirconium and hafnium. About 2% volume of fine yttria (Y_2O_3) is added to form the dispersoid. After milling, a uniform interparticle spacing of about 0.5 μm is achieved.

After mechanical alloying, the powder is packed into steel tubes and extruded at 1950–2050° F with reduction ratios of 13 to 1. The consolidated extruded stock is then hot rolled into mill shapes at 1750–1950° F to produce product forms such as bar, plate, sheet and wire. Since most applications are for creep resistant structure, a grain coarsening anneal at 2400° F is applied after fabrication.

6.5 Forging[18]

Because of their strength retention at elevated temperatures, superalloys are more difficult to forge than most metals. In reality, forgeability varies widely depending on the type of superalloy and its exact composition. For example, some of the iron–nickel based alloys, such as A-286, are similar to the austenitic stainless steels. At the other extreme, some superalloy compositions are intrinsically so strong at elevated temperatures that they can only be processed by powder metallurgy or by casting. In general, as the alloying content has been increased to obtain even greater elevated temperature strength, the forgeability has been degraded, i.e. the γ' strengthened alloys are much more difficult to forge than the solid solution strengthened alloys. A comparison of the specific energy required to upset three alloys is shown in Fig. 6.13. Note that A-286 is only marginally harder to forge than 4340 steel, while René 41, which contains a larger amount of γ' than A-286, is significantly more difficult to forge.

Forging of superalloys has evolved from the simple process of making a specific shape to a very sophisticated one that not only fabricates the correct shape, but also imparts a great degree of microstructural control for enhanced properties. Computer modeling programs, such as ALPID (Analysis of Large Plastic Incremental Deformation), are used to design dies and processes that predict the metal flow during forging.[14] It should be noted that a great deal of the specifics of forging superalloys is proprietary information to the individual forging suppliers.

A number of different forging methods are used in superalloy part fabrication,[19] including open die forging, closed die forging, upset forging,

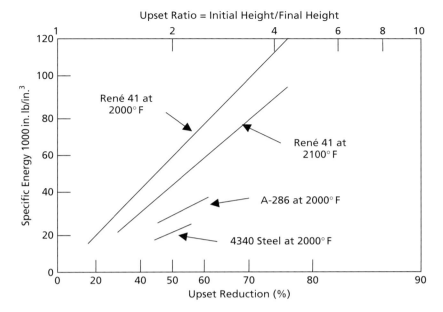

Fig. 6.13. Effect of Upset Reduction and Forging Temperature on Specific Energy Required[18]

extrusion, roll forging, ring rolling and hot die and isothermal forging. Open die forging is often used for making preforms for relatively large parts, such as gas turbine disks. Open die preforms are usually finished by close die forging, the most common forging process for superalloys. Preforms made by open die, upsetting, extrusion and roll forging are often finished by closed die forging. Upset forging and extrusion are commonly used for finished forgings, but more commonly for producing preforms for closed die finishing. In upset forging, the maximum unsupported length is usually about two diameters. Roll forging can also be used to produce preforms for closed die forging. Roll forging is attractive because it can save material and reduce the number of closed die operations. Ring rolling is another method to save material when producing ring-like parts from hollow billets. Isothermal and hot die forging offers the advantages of closer tolerances or near net shape parts that can significantly reduce machining costs.

Superalloy billets are usually furnace heated for hot forging. Although nickel base alloys have a greater resistance to scaling at hot working temperatures than steels, they are more susceptible to attack by sulfur during heating. Cleanliness is extremely important. All potential contaminants, such as lubricants and paint markers, must be removed before heating. Low sulfur fuels must also be used. Gas fuels, such as natural gas, butane and propane, are the best fuels. Oil is also a satisfactory fuel provided it has low sulfur content.

Dies are lubricated to facilitate removal of the workpiece after forging. Again, sulfur-free lubricants are necessary. Colloidal graphite lubricants give good results and are usually applied by spraying. Preheating of all tools and dies to above 500° F is recommended to avoid chilling the metal during working. Dies heated in the range of 1200–1600° F allow better control of workpiece temperature during forging. However, since the dies are then made out of superalloys, the die cost is higher than for steels, which are limited to around 800° F.

Hot working can be defined as plastic deformation performed at temperatures sufficiently high for recovery and recrystallization to counteract strain hardening.[20] Typical hot working temperature ranges for a number of superalloys are given in Fig. 6.14. The lower limit for hot working is usually determined by the increase in flow stress as the temperature is lowered, while the upper limit is determined by the incipient melting temperature or by excessive grain growth. Initial forging operations are generally conducted well above the γ' solvus temperature to allow both chemical homogenization and some microstructural refinement.[15] During this step, it is important that the forging temperature does not exceed low liquation temperatures associated with localized segregation.

Slow strain rates are used during the initial closed die reductions to avoid cracking. Subsequent deformation can be completed below the γ' solvus to give greater degrees of microstructural refinement, but still at high enough temperatures to avoid excessive warm working and an unrecrystallized microstructure. Approximately 80% of the reduction is done above the recrystallization temperature, with the remaining 20% conducted at lower temperatures to introduce some warm work for improved mechanical properties.[19] Typically, forgings for strength and fatigue resistance applications will be finished at temperatures at or below the γ' solvus to produce fine grain sizes, while forgings for creep and stress rupture applications will be finished above the γ' solvus to produce larger recrystallized grain structures. However, it should be pointed out that most applications for forgings are strength and fatigue critical. Therefore, the objectives of forging are normally uniform fine grained structures, controlled grain flow and structurally sound parts.

Recrystallization must be achieved during each hot working operation to obtain the desired grain size and flow characteristics, and to reduce the effects of continuous grain boundary networks that can develop during heating and cooling. Continuous grain boundary networks contribute more to low mechanical properties and other problems than any other single factor.[21] Poor weldability, low cycle fatigue and stress rupture problems are often associated with continuous grain boundary carbide networks. In order to achieve uniform mechanical properties, all portions of a part must receive some hot work after the final heating operation. In general, the precipitation hardening alloys should be cooled in air after forging. Water quenching is not recommended, because of the possibility of thermal cracking, which can occur during subsequent heating for further forging or heat treating.

Superalloys

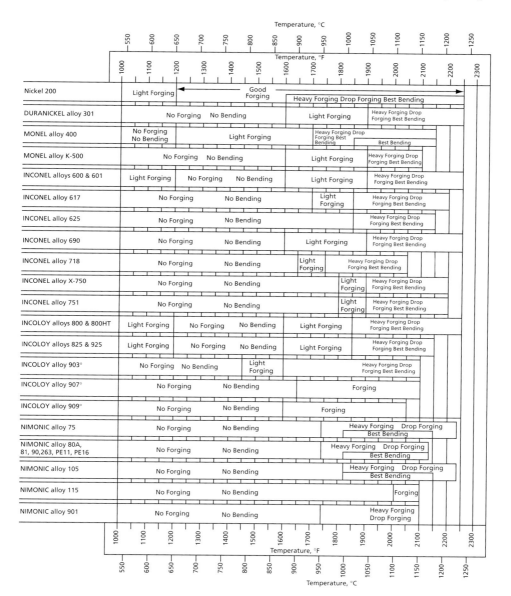

Fig. 6.14. Hot Forming Temperature Ranges[22]
*INCONEL, INCOLOY and NIMONIC are registered trademarks of the Special Metals Corporation group of companies

Isothermal forging is the most common forging method for powder metallurgy preforms. In isothermal forging, the dies are maintained at the same temperature as the workpiece. Since this temperature is often of the order of 1800–2000° F, the dies are usually made of TZM molybdenum, for elevated temperature strength. In addition, isothermal forging is conducted in a vacuum or inert atmosphere in order to protect the die materials from oxidation.

Hot isostatic pressed or extruded PM billets have an extremely fine grain size and a fine distribution of carbides that makes them superplastic or near-superplastic during forging. The key to isothermal forging is to start with a fine grain size and then maintain it during the forging process. A large amount of γ' precipitate is useful in preventing grain growth. Alloys that contain large amounts of γ', such as Inconel 100, René 95 and Astroloy, develop superplastic properties during isothermal forging. In contrast, Waspaloy, which contains less than 25 vol. % γ' at isothermal forging temperatures, is only marginally superplastic. Iron–nickel base alloys, such as Inconel 718, have even lower volume fractions of precipitate and are therefore even less frequently isothermally forged.

Compared to conventional forging, isothermal forging deformation rates are slow; hydraulic press speeds of approximately 0.1 in./min are typical. The slow strain rates help to control both the shape of the forged part and the microstructure.[11] Accurate temperature and strain rate can be controlled by sensors, in conjunction with the press equipment to give a closed loop control system.[23] The slower production rate is largely offset by the ability to forge complex shapes to closer tolerances, which leads to less machining and substantial material savings. In addition, a large amount of deformation is accomplished in one operation. Pressures are low and uniform microstructures are achieved. For example, the as-forged weight of a finish machined 150 lb Astroloy disk is about 245 lb for a conventional forging versus 160 lb for a corresponding isothermal forging.[21]

Forging quality is controlled by a combination of ultrasonic inspection, microstructural quality and mechanical property tests.

6.6 Forming

Wrought superalloys can often be formed using techniques similar to stainless steels, although forming is more difficult. Like stainless steels, superalloys work harden rapidly during forming. The work hardening rate usually increases with alloy complexity, i.e. the precipitation hardening alloys have higher work hardening rates than the solid solution strengthened alloys.[22] The cobalt based alloys require greater forces than the nickel or iron based alloys. Alloys that contain substantial amounts of molybdenum or tungsten are also harder to form. Forming presses are the same as those used for forming steel; however, because of their higher strength, more power is needed, usually 50–100% more power.

In cold forming operations, the rate of work hardening in relation to the amount of cold deformation achieved determines the frequency of intermediate anneals. For example, some alloys, such as A-286, Nimonic 75 and Hastelloy X, can be reduced as much as 90% before annealing, while others, such as René 41 and Udimet 500,

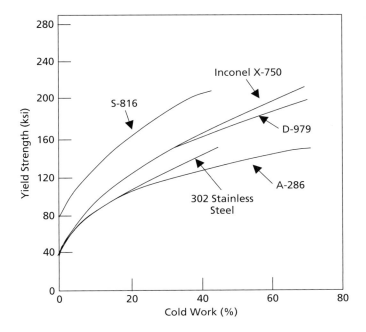

Fig. 6.15. Work Hardening for Several Superalloys[24]

can only be reduced 35–40% before an intermediate anneal is required.[24] The rate of work hardening for four superalloys is compared to type 302 stainless in Fig. 6.15. The iron–nickel base alloy A-286 work hardens similar to type 302, while the others work harden at higher rates, with the cobalt base alloy S-816 displaying the highest work hardening rate.

Since the window for hot forming is narrow (1700–2300° F), cold forming with intermediate anneals is preferred for thin sheets. To preserve a fine grain structure during cold forming operations, a critical amount of cold work (8–10%) must be achieved prior to intermediate anneals. Annealing material with less than sufficient amounts of cold working can result in abnormal grain growth, which can cause undesirable orange peel effects during subsequent forming, and also adversely affects fatigue strength. Cold working can accelerate the aging process in precipitation hardened alloys; in other words, overaging can occur at otherwise normal aging temperatures.[24] The aging temperature, or time, may have to be adjusted in such situations. If annealing causes distortion, the part can be formed to within 10% of its final shape, annealed and then give a light forming operation to bring it into tolerance.

The annealed condition is usually soft enough to permit mild forming. If the annealed alloy is not soft enough for the forming operation, a higher temperature solution annealing treatment can be used that will remove all of the effects of cold work, but will dissolve the age hardening and other secondary phases. Some grain size control is sacrificed, but if cooling from the solution annealing temperature is very rapid, the age hardening elements will be retained in solution. Several process anneals may be required in severe forming, but the high temperature solution anneal need not be repeated. Annealing should be performed at a temperature that produces optimal ductility for the specific metal. Typical minimum bend diameters for annealed material are in the range of 1–2 X thickness.

Superalloys often exhibit galling between the workpiece and tool. The oxide film formed due to chromium addition does not help in reducing galling, since it is easily fractured during forming. Therefore, steel dies, punches and mandrels are often chrome plated to minimize adherence. Cast iron tools are often used for low production runs while steel is used for longer runs. Areas which experience high wear rates are often locally hardened. Lubrication is required for drawing, stretch forming or spinning. Although lubrication is seldom needed for the press brake forming of V-bends, its use will improve the quality. Mild forming operations (<10% reduction) can be conducted with unpigmented mineral oils and greases. For more severe forming, lubricants include chlorinated, sulfochlorinated, or sulfurized oils or waxes. After forming, it is important to thoroughly clean the material before heating it to elevated temperatures. Due to the tendency to rapidly work harden and gall, forming operations are usually performed at relatively slow speeds.

6.7 Investment casting[25–27]

Since the mid-1960s, the hottest parts of the engine, the blades (rotating) and vanes (non-rotating), have been manufactured by investment casting. As the alloy content of nickel based superalloys was continually increased to obtain better creep and stress rupture capability, the alloys became increasingly difficult to forge. To allow even higher contents of alloying elements, it became necessary to change the fabrication process to casting. Investment casting became the process of choice because it is amendable to the fabrication of hollow blades with intricate cooling passages, which allows higher operating temperatures. Since the mid-1980s, turbine inlet temperatures have increased by 500° F. About half of this increase is due to more efficient designs, while the other half is due to improved superalloys and casting processes. The introduction of directional solidification allowed about a 50° F increase in operating temperature, while the single crystal process produced another 50° F increase.[11] The reader may want to refer back to Chapter 4 on Titanium for the detailed steps involved in investment casting.

The original cast blades and vanes were fine grained polycrystalline structures made using conventional investment casting procedures. These blades were then heat treated to coarsen the grain structures for enhanced creep resistance. Eventually it became possible to produce directionally solidified (DS) structures with columnar grains oriented along the longitudinal axis of the blade. The columnar grain structure enhances the elevated temperature ductility by eliminating the grain boundaries as failure initiation sites. The DS process also creates a preferred low modulus texture, or orientation, parallel to the solidification direction that helps in preventing thermal fatigue failures. An extension of the DS process was the development of the single crystal (SX) process in which a single crystal grows to form the entire blade. Since there are no grain boundaries in the SX process, this allowed the removal of grain boundary control alloying elements that were detrimental to high temperature creep performance, allowing even higher operating temperatures. However, the SX process is significantly more expensive than the DS process because yields are lower; therefore, both processes are used extensively for fabricating blades and vanes. The macrostructures of polycrystalline, DS and SX blades are shown in Fig. 6.16, and the dramatic improvements in creep resistance are shown in Fig. 6.17.

6.7.1 Polycrystalline Casting[28]

Although DS and SX castings have replaced polycrystalline castings for applications like blades and vanes, polycrystalline castings are still used for structures such as compressor housings, diffuser cases, exhaust cases and engine frames. Some of these parts are very large (e.g., 60 in. in diameter) and the castings can weigh up to 1500 lb. The casting of large structural parts allows reduced manufacturing costs with more unitized structure by eliminating the casting of many smaller parts and then welding them together. Since many of these parts are strength and fatigue critical, innovations have been developed to produce castings with finer and more uniform grain sizes. Examples include mold agitation

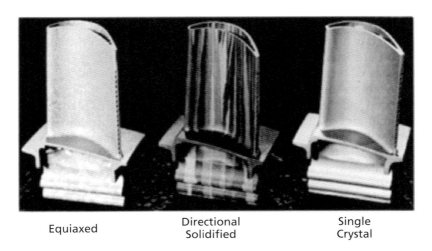

Equiaxed   Directional Solidified   Single Crystal

Fig. 6.16. Cast Turbine Blades[25]

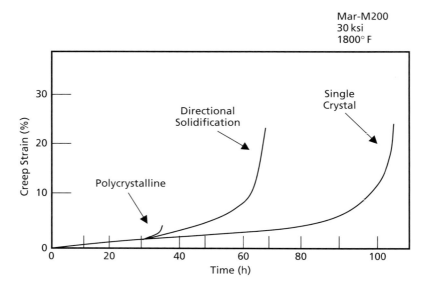

Fig. 6.17. Creep Comparison for Different Casting Procedures[12]

during solidification (Grainex process) and the use of low superheat during pouring (Microcast-X process).

6.7.2 Directional Solidification (DS) Casting[29]

To develop a directionally solidified structure, it is necessary for the dendrites (grains) to grow from one end of the casting to the other. This is accomplished by creating a sharp temperature gradient, by removing the majority of the heat from one end of the casting. As shown in Fig. 6.18, a water chilled mold is slowly withdrawn from the furnace, setting up a strong temperature gradient in the freezing metal. A thin wall investment casting mold, that is open at the bottom, is placed on a water-cooled copper chill plate. The mold is then heated to the casting temperature. The alloy is heated under vacuum in an upper chamber of the furnace and then poured into the heated mold. After a couple of minutes to allow the grains to nucleate on the chill plate, the mold is slowly withdrawn from the hot zone and moved to the cold zone. The chill plate insures that there is a good nucleation of grains to start the process. Although the grains nucleate with random orientations, those with the preferred growth direction normal to the chill surface grow and crowd out the other grains.

The thermal gradient is established in the zone between the liquidus and solidus temperatures, and is passed from one end of the casting to the other at a slow rate that maintains the steady growth of the grains. Temperature control and extraction rate are critical. If the progression of the thermal gradient is too fast, grains will nucleate ahead of the solid/liquid interface, while if the

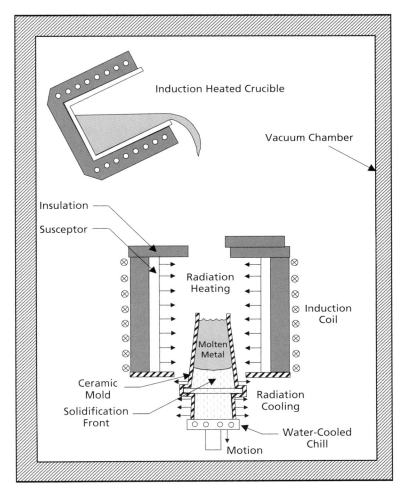

Fig. 6.18. Directional Solidification Process for Superalloy Castings[30]

movement is too slow, excessive macrosegregation will occur along with the formation of freckles, which are defects of equiaxed grains of interdendritic composition. Typical temperature gradients are 170–330° F/in. with withdrawal rates of around 12 in./h. In general, higher temperature gradients produce better quality castings.

Since a very high degree of process control is necessary to achieve proper grain growth, the entire process is automated. One reason why automation is necessary is that the withdrawal rate is not necessarily constant. For example, large differences in section size change the solidification rate, and the withdrawal rate must be changed to compensate. In addition, the effect of cooling rates due to the ceramic filled cores must be included since the cores lengthen the time to preheat the mold and slow the withdrawal rate.

Typical defects in DS castings include equiaxed grains, misoriented grains, grain boundary cracking, excessive shrinkage, microporosity, and mold or core distortion. Increasing the thermal gradient during casting is used to control the formation of equiaxed grains, misoriented grains and microporosity. Hafnium, in the amount of 0.8–2.0%, can be added to the alloy composition to avoid grain boundary cracking. Shrinkage on the upper surfaces (the last surfaces to solidify) is sometimes encountered, because risers are not used since they interfere with the radiation heat transfer. Inverting the casting can sometimes help to eliminate this problem. Due to the thin mold cross-sections used and the long exposure times at high temperatures, ceramic mold or core distortion can also be a problem. Careful control of their compositions, uniformity and firing conditions are required to prevent mold/core distortion problems.

6.7.3 Single Crystal (SC) Casting[29]

After the development of the DS casting process, it was recognized that if all but one of the growing columnar grains could be suppressed, it would be possible to cast parts with only a single grain, thereby eliminating all of the grain boundaries. In addition, alloying elements that were necessary to prevent grain boundary cracking, but were detrimental to creep strength, namely boron, hafnium, zirconium and carbon, could be eliminated.

Single crystal castings are produced in a manner similar to DS castings with one important difference; a method of selecting a single properly oriented grain is used. In the most prevalent method, a helical section of mold (Fig. 6.19) is placed between the chill plate and casting mold. This helix, or spiral grain selector, acts as a filter and allows only a single grain to pass through it. Because superalloys solidify by dendritic growth that can occur in only three mutually orthogonal directions, the continually changing direction of the helix gradually

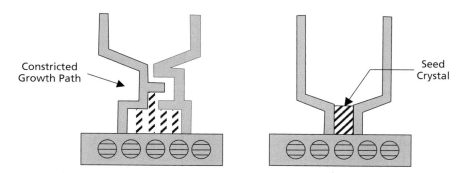

Fig. 6.19. Single Crystal Initiation

constricts all but one grain, resulting in a single crystal emerging from the helix into the mold. Seeding is another method of creating a single grain. Instead of the helical section to control grain growth, a single crystal of the alloy is placed on the chill plate. The crystal is oriented so that its orientation will be repeated in the alloy that fills the mold. The temperature of the seed is controlled so that the seed does not completely melt, allowing the molten metal to solidify with the same orientation as the seed crystal. One advantage of the seed crystal method is that the orientation of the metal that grows from the crystal can be controlled, to a degree, by the orientation of the crystal that is placed in the mold.

With the exception of grain boundary cracking, SX castings can experience the same defects as those found in DS castings. In addition, SX castings can form low angle grain boundaries. Although the misorientation is usually less than 15°, they can act as crack initiation sites. The misorientation allowed depends on the specific application, but boundaries above 10° are usually not permitted. Single crystal castings are usually inspected by X-ray diffraction to check for crystallographic orientation.

6.8 Heat Treatment[31–33]

Heat treatments that are used for superalloys include annealing and precipitation hardening (solution heat treatment followed by aging). The precipitation hardening treatments are varied and often have multiple aging cycles to optimize properties for different applications. Although stress relieving can be done at temperatures lower than annealing, it is often not advisable due to the possibility of undesirable carbide precipitation.

6.8.1 Solution Strengthened Superalloys

Solid solution strengthened wrought superalloys of nickel, iron–nickel and cobalt are normally supplied in the solution heat treated condition with virtually all of the secondary carbides dissolved in solution, i.e. the microstructure consists of the primary carbides dispersed in a single phase matrix, with the grain boundaries relatively free of carbides. During in-service exposure to elevated temperatures, the secondary carbides precipitate at the grain boundaries, providing the desired elevated temperature strength properties.

Full annealing is used to cause complete recrystallization and to obtain maximum softness for continued cold or hot working operations, to relieve stresses after welding, and to homogenize cast ingots. Typical annealing heat treatments are conducted at minimum temperatures of 1750–2050° F for 5–20 min followed by rapid cooling through the range of 1200–1600° F. Rapid cooling is necessary to prevent premature secondary carbide precipitation. The temperatures are selected so that a recrystallized structure will develop from a cold or warm worked condition to produce a low enough yield strength and sufficient ductility for subsequent cold forming operations. While most wrought superalloys can be cold formed, they may require one or more intermediate annealing operations to prevent cracking. Even during hot working operations, they will store energy and need to be reheated for subsequent deformation processes, as for example during forging operations. Annealing can also be used for finished parts where a finer grain size is desired, and where strength and fatigue are more important than creep resistance.

Stress relief treatments at lower temperatures (1200–1600° F) are sometimes used, but normally result in excessive carbide precipitation. Therefore, full annealing is often a better choice. Stress relief treatments conducted at temperatures less than 1200° F avoid the carbide precipitation problem but are less effective in reducing the residual stresses.

Solution heat treating is conducted at 1925–2250° F for 10–30 min, followed by rapid cooling. Rapid heating to the solution temperature is used to minimize carbide precipitation and preserve the cold or warm work required to provide recrystallization and/or grain growth during the solutioning treatment. Since the temperatures are higher than for mill annealing, the carbon is in an even greater state of supersaturation, and it is even more important to rapidly cool to prevent premature carbide precipitation. The response to solution heat treatment is very dependent on the material condition. If the material contains sufficient stored energy from working, then recrystallization will occur, and either a fine or coarse grained structure can be produced, by selecting either a lower or higher solution treating temperature, respectively. A problem can occur if the amount of stored energy is low ($\sim$ <10%). Then, the response to heat treatment can be non-uniform and abnormal grain growth can be a problem. For these low levels of cold work, the problem can be minimized by: (1) solution treating at the low end of the allowable temperature range, (2) mill annealing during fabrication, and (3) stress relieving just prior to final solution treating.

Types of surface attack and contamination which can occur during heat treatment can include oxidation, carbon pickup, alloy depletion and contamination. Vacuum atmospheres below 2×10^{-3} torr, or inert atmospheres, such as dry argon, should be used to prevent oxidation during solution treating or annealing at temperatures above 1500° F.

6.8.2 Precipitation Strengthened Nickel Base Superalloys

Wrought precipitation strengthened nickel based superalloys are solution heat treated and then aged to produce the desired properties. Solution treating temperatures range from about 1800–2250° F, or even up to 2400° F for some single crystal alloys. Long exposure times at solution treatment temperatures can result in partial dissolution of primary carbides with subsequent grain growth. The solution treatment may be above or below the γ' solvus depending on the desired microstructure and application. A higher temperature is used to develop coarser grain sizes for creep and stress rupture critical applications, while a lower solution temperature will produce a finer grain size for enhanced tensile and fatigue properties. To retain the supersaturated solution obtained during solution treating, the part should be rapidly cooled to room temperature using either gas cooling or water or oil quenching.

Aging treatments are used to strengthen precipitation strengthened alloys by precipitating one or more phases (γ' or γ''). Aging treatments vary from as low as 1150° F to as high as 1900° F. Double aging treatments are also used to produce different sizes and distributions of precipitates. A principal reason for double aging treatments, in addition to γ' and γ'' control, is the need to precipitate or control grain boundary carbide morphology. Aging heat treatments usually range from 1600–1800° F, with times of about 4–32 h. Either single or multiple aging treatments are then used to precipitate γ'. Like the solution temperature, the aging temperatures and times are selected depending on the intended application. Higher aging temperatures will produce coarse γ' precipitates desirable for creep and stress rupture applications, while lower aging temperatures produce finer γ' precipitates for applications requiring strength and fatigue resistance.

Waspaloy is an example of a wrought nickel based superalloy in which the heat treatment can be varied depending on the application. For applications where creep and stress rupture properties are the most important, such as turbine blades, the heat treatment might consist of:[33]

- Solution treat at 1975° F for 4 h followed by air cooling,
- Age at 1550° F for 24 h followed by air cooling, and
- Age at 1400° F for 16 h followed by air cooling.

The high solution heat treating temperature (1975° F), above the γ' solvus, dissolves the maximum amount of alloying elements in solution but also promotes grain growth. The first aging treatment at 1550° F for 24 h stabilizes the alloy for high temperature service by forcing MC carbides to react with the matrix alloying elements to form $M_{23}C_6$ grain boundary carbides and γ'. The second aging treatment at 1400° F completes the precipitation of a finer γ' precipitate that is advantageous to creep, stress rupture, and tensile properties.

For applications requiring finer grain sizes for enhanced tensile and fatigue properties, such as disks, the heat treatment might consist of:[33]

- Solution treat at 1825–1900° F for 4 h followed by air cooling,
- Age at 1550° F for 4 h followed by air cooling, and
- Age at 1400° F for 16 h followed by air cooling.

The lower solution treating temperature (1825–1900° F), below the γ' solvus, is only a partial solution treatment; however, the lower temperature results in a finer grain size and starts the carbide precipitation process, so the first age at 1550° F requires only 4 h as compared to 24 h for the previous heat treatment. A comparison of these two heat treatments on mechanical properties is shown in

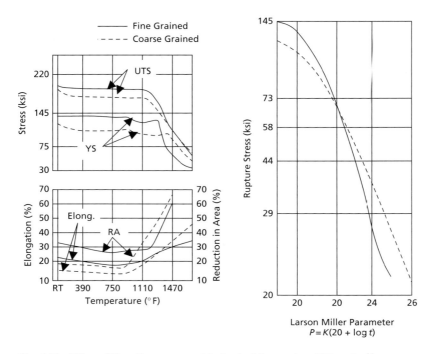

Fig. 6.20. Effect of Heat Treatment on Mechanical Properties of Waspaloy[34]

Fig. 6.20. Note that the blade heat treatment (coarse grain size) results in better rupture strength at high temperatures, while the disk heat treatment (fine grain size) gives better tensile and stress rupture properties at lower temperatures.

6.8.3 Precipitation Strengthened Iron–Nickel Base Superalloys

Although some of the wrought iron–nickel based superalloys are strengthened by γ', Inconel 718, the most widely used wrought superalloy, accounting for 45% of all wrought alloy production, is strengthened by γ'' (Ni_3Nb). The γ'' phase precipitates in the range of 1300–1650°F and has a solvus temperature of about 1675°F. In Inconel 718, the δ phase, which is present above the γ'' solvus, can be used for grain size control. The δ phase (Ni_3Nb) precipitates between 1600 and 1850°F and has a solvus temperature of about 1850°F. For applications requiring a good combination of tensile/fatigue and stress rupture properties, Inconel 718 can be heat treated below the δ solvus as follows:[33]

- Solution treat at 1700–1850°F for 1–2 h followed by air cooling,
- Age at 1325°F for 8 h followed by furnace cooling to 1150°F, and
- Age at 1150°F for a total aging time of 18 h followed by air cooling.

Higher strength levels can be obtained by forging below the δ solvus, quenching after forging and conducting a double aging treatment without a solution treatment. The precipitation of the δ phase is helpful in achieving a fine grain size during hot working and avoiding undesirable grain growth.[35]

In general, lower solution treating temperatures produce better strength, while higher solution treating temperatures provide better creep and stress rupture properties.

6.8.4 Cast Superalloy Heat Treatment

Heat treatment of cast superalloys includes homogenization, solution heat treatment, and aging heat treatments. Stress relief treatments are also used to reduce residual stresses resulting from casting, weld repair, and machining. Cobalt based castings, which are usually given stress relief or aging treatments, can be heat treated in air while nickel base alloys are always heat treated using inert atmospheres or vacuum. Cobalt based casting aging treatments are conducted at approximately 1400° F to promote formation of discrete $Cr_{21}C_6$ carbides. Some superalloy castings are put into service in the as-cast condition. However, stress relieving of non-precipitation hardening alloys may be used if the casting is extremely complex, or when dimensional tolerances are stringent. Weld repaired castings are always stress relieved after welding. When heating to high solution treating temperatures, slow heat-up rates are often used for castings to allow some homogenization to minimize the danger of incipient melting due to alloy segregation.

To obtain optimum properties after casting, precipitation-strengthened superalloys must be homogenized at temperatures between the γ' solvus and the solidus. At these temperatures, although the γ' phase dissolves relatively rapidly in the γ matrix, a substantial time (many hours) is required to obtain a uniform distribution of alloying elements through diffusion.[36] Residual stresses also develop during casting due to non-uniform expansion and contraction rates at different points within the casting. Residual stresses are a function of the superheat prior to casting (higher superheat creates higher residual stresses), casting complexity (complex castings produce higher residual stresses) and is also dependent on alloy composition.[37]

Many conventionally cast polycrystalline nickel based alloys are not solution heat treated but all DS (directionally solidified) and SX (single crystal) alloys are solution heat treated. Because polycrystalline alloys respond differently to solution heat treatment, some are only given an aging treatment. Solution treatments are performed at temperatures above or near the γ' solvus for times ranging from 2 to 6 h, with the objective of dissolving all of the phases into the as-cast structure. Since precipitation occurs during cooling from the solution treating temperatures, fast cooling rates (e.g., gas furnace quenching) are used to prevent coarsening of the γ' precipitate which can degrade the mechanical

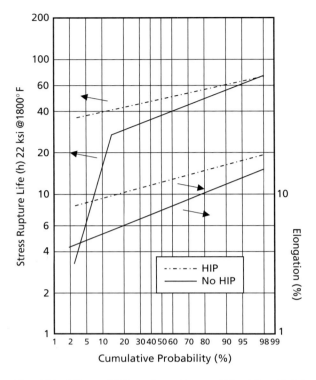

Fig. 6.21. *Effects of HIP on Stress Rupture Life of Cast Nickel Based Alloy*[31]

properties. If a vacuum atmosphere is used for solution treating, a partial pressure of inert gas, rather than a hard vacuum, is used to prevent surface depletion of chromium and aluminum. Slow heating rates to the solution temperature are used to allow time for casting homogenization. Cooling rates in excess of 100° F/min can be achieved by backfilling with cold argon and rapidly circulating the gas past the parts being cooled.[10]

Hot isostatic pressing can be used to improve quality by healing internal casting porosity. A typical HIP cycle would be 2175° F for 4 h at 15 ksi argon pressure. The beneficial effects of HIP on the stress rupture properties of a polycrystalline cast nickel based alloy (Inconel 738) are shown in Fig. 6.21.

6.9 Machining[38]

Superalloys are difficult to machine, perhaps second only to titanium in machining difficulty, although those who machine superalloys would probably maintain that superalloys are the most difficult to machine of the aerospace metals.

Superalloys

However, a relative ranking of aerospace metallic alloys, in order of decreasing machinability is:[39]

- Magnesium
- Aluminum
- Low Alloy Steels
- Stainless Steels
- Hardened and High Alloy Steels
- Nickel Based Superalloys
- Titanium Alloys.

Many of the same characteristics that make superalloys good high temperature materials also make them difficult to machine, namely:

- Retention of high strength levels at elevated temperature
- Rapid work hardening during machining
- Presence of hard abrasive carbide particles
- Generally low thermal conductivities and
- Tendency of chips to weld to cutting edges and form built-up edges.

However, it should also be noted that superalloys cover a wide range of alloys that have somewhat different machining characteristics. For example, the iron–nickel based alloys, which resemble stainless steels, are easier to machine than the nickel and cobalt based alloys. In general, as the amount of alloying elements increases for higher temperature service, the alloy becomes more difficult to machine.

These high temperature characteristics place cutting tools under tremendous heat, pressure, and abrasion, leading to rapid flank wear, crater wear, and tool notching at the tool nose and/or depth of cut region, as illustrated in Fig. 6.22.

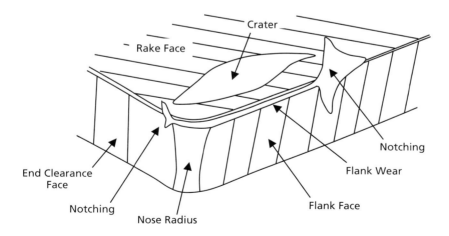

Fig. 6.22. Typical Tool Wear Features[40]

Due to their high temperature strengths, superalloys remain hard and stiff at the cutting temperature, resulting in high cutting forces that promote chipping or deformation of the tool cutting edge. In addition, since superalloys retain a large percentage of their strength at elevated temperatures, more heat is generated in the shear zone resulting in greater tool wear than with most metals. Since the forces required to cut superalloys are about twice those required for alloy steels, tool geometry, tool strength, and rigidity are all important variables.

Their low thermal conductivities cause high temperatures during machining. The combination of high strength, toughness, and ductility impairs chip segmentation, while the presence of abrasive carbide particles accelerates tool wear. Superalloys also have a tendency to rapidly work harden which can create a hardened surface layer that degrades the surface integrity and can lead to lower fatigue life.

General guidelines for machining superalloys are very similar to those for titanium alloys:[41]

- Conduct the majority of machining in the softest state possible,
- Use positive rake angles,
- Use sharp cutting tools,
- Use strong geometries,
- Use rigid set-ups,
- Prevent part deflection,
- Use high lead angles, and
- When more than one pass is required, vary the depth of cut.

To make machining somewhat easier, precipitation hardened alloys are often solution heat treated, rough machined, aged, and then finish machined. This approach results in superior surface finishes with minimal distortion. Positive rake angles minimize work hardening and built-up edges, by shearing the chip away from the workpiece. Sharp cutting edges also help to minimize built-up edges and provide better surface finishes. Dull or improperly ground cutting tools cause higher machining forces leading to built-up edges, metal tearing, and excessive part deflection. However, since sharp edges are more fragile and susceptible to chipping, honed edges are frequently used for roughing operations and sharp edges for finishing.

A large nose radius, where possible, also results in a stronger, less chip prone tool. Rigid set-ups with minimal part deflection reduce vibration and chatter that can cause tool failures and deteriorate tolerances and surface finishes. Large leading edge angles, which engage more of the cutting tool in the workpiece and spread the wear over a larger distance, help to prevent localized notching. In turning operations using NC equipment, the depth of cut can be varied from one pass to another to spread the wear along the entire insert and thus prolong tool life.

6.9.1 Turning

Standard grades of cemented carbide are usually used for uninterrupted turning operations with C-2 grade used for roughing and C-3 grade used for finishing. Although high speed steel cutting tools wear much faster than carbides, they are sometimes required if heavy interrupted cuts are being made due to their greater resistance to chipping and breakage. The cobalt grades, such as M33 and M42, are recommended because of their greater heat resistance than the standard high speed grades. A comparison of the cutting speeds in surface feet per minute (sfm) illustrates the productivity advantages when carbides can be used.

High speed steel:

- Roughing: 0.250 in. depth of cut, 12–18 sfm, 0.010 ipr
- Finishing: 0.050 in. depth of cut, 15–20 sfm, 0.008 ipr

Cemented Carbide:

- Roughing: 0.250 in. depth of cut, 30–40 sfm, 0.010 ipr
- Finishing: 0.050 in. depth of cut, 40–50 sfm, 0.008 ipr

Even further productivity gains, through higher cutting speeds, can often be achieved with cubic boron nitride (CBN) and ceramic tools, but the setup needs to be extremely rigid to preclude tool chipping.[42] CBN tools are often used when turning the harder nickel base (wrought and cast) and cobalt base cast alloys. Ceramic cutting tools allow speeds of 500–700 sfm with feeds of about 80% those of carbide. Typical ceramic tools include SiAlON and SiC whisker reinforced Al_2O_3.

The recommended geometries for single point turning tools are shown in Fig. 6.23. As mentioned earlier, it is important to use positive rake angles to insure that the metal is cut, rather than pushed. Positive rake angles also help to guide the chip away from the workpiece surface. The side and end relief angles, which provide clearance between the tool flank and the workpiece, are usually a compromise: too small of an angle will cause rubbing, while too large of an angle will provide insufficient cutting edge strength.

The side cutting angle affects the load on the cutting edge and provides thickness and directional control to the chips. The nose radius provides strength to the tool nose and helps to dissipate heat generated by the cut. A scallop produced by a tool with a nose radius (Fig. 6.24) gives a better surface finish, shallower scratches, and a stronger workpiece, with less tendency to crack at sharp corners than the notch effect produced by a sharp tool.

Water soluble oils in mixtures (1 part oil with 20–40 parts water) are frequently used in turning. Flood cooling is used to minimize excessive heat build-up. If sulfurized or chlorinated oil is used as a cutting fluid, the workpieces must be thoroughly cleaned before heat treatment or high temperature service. Serious damage to workpieces during heating cycles can result if any residue remains.

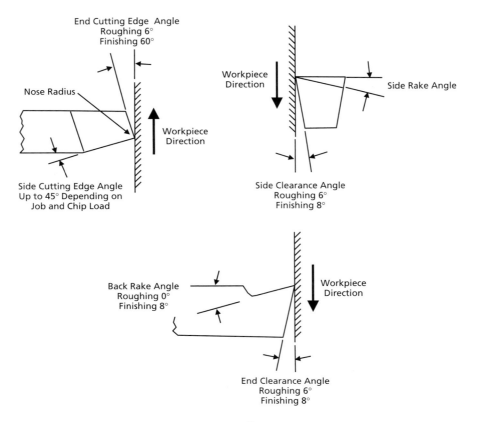

Fig. 6.23. Single Point Turning Tool Geometries[43]

6.9.2 Milling

To provide dimensional accuracy and a smooth surface finish during milling operations, it is important to use sharp tools along with a rigid machine and setup. Since most milling operations involve interrupted cutting, high speed steel cutting tools are often used.

Low cutting speeds (10–20 sfm) and light chip loads are required. In some applications, it is necessary to mill nickel and cobalt base alloys at speeds as low as 5 sfm for acceptable tool life. Typical milling parameters for roughing operations are 10–20 sfm with feed rates of 0.001–0.004 in. per tooth, while finishing is conducted in the range of 5–15 sfm at feeds of 0.001–0.004 in. per tooth. Too light a feed, approximating rubbing, will cause an excessive work hardened layer. Because this rubbing action at the beginning of the cut is avoided in climb milling, it is preferred to conventional or up-milling. In addition, the downward motion of the cut assists rigidity and diminishes chatter. However, climb milling requires a very rigid setup and a machine equipped with a backlash

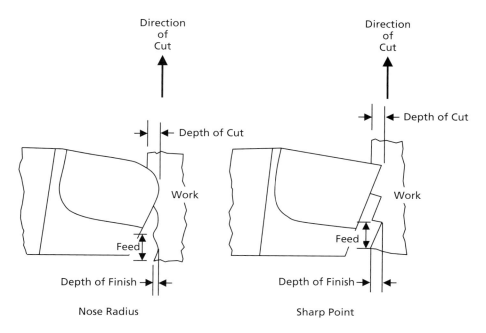

Fig. 6.24. *Effect of Turning Tool Nose Radius on Workpiece*[43]

eliminator. Cuts deeper than 0.060 in. are seldom attempted with climb milling, because it is virtually impossible to obtain the required rigidity. Face milling is preferable to slab milling, because it reduces work hardening and chatter.

Two cutter design principles require special consideration for milling cutters: (1) tooth strength must be greater than that required for milling steel or cast iron, and (2) relief angles must be large enough to prevent rubbing and subsequent work hardening. Milling cutters should be designed so that the teeth have positive rake and helix angles. Inserted blades are used on nearly all but the smallest cutters, because even under the most favorable machining conditions, the life of the cutting edges is short. Mechanical methods of securing the blades in the cutter body are preferred because replacement of chipped or broken blades is easier. Because of the interrupted cutting action, high speed steel is used for cutters in most applications. However, carbide is frequently more economical than high speed steel when milling the more difficult to machine alloys, such as the highly alloyed precipitation hardening grades.

Sulfochlorinated oil introduced in copious amounts at the exhaust side of the cutter is the preferred condition for milling superalloys. Soluble oil emulsions are often used, and they provide better cooling for the tools and workpieces than straight oils. However, some sacrifice in surface finish and tool life occurs with the use of soluble oil emulsions compared to sulfochlorinated oils. The latter are often diluted with mineral oil (up to 50%) to obtain fluidity with no large sacrifice in the ability to promote cutting and achieve a good surface finish. Workpieces milled with sulfochlorinated or other chemically active oils must be thoroughly cleaned before being heated to elevated temperature.

6.9.3 Grinding[44]

Grinding is often used to machine highly alloyed heat treated cast turbine blades. When the machining operation calls for a significant amount of material removal, a rough grind followed by a finish grind can be used to produce a defect-free surface. Because high temperature superalloys are sensitive to the level of energy used during grinding, metallurgical alterations and microcracking may occur at the surface. If an extremely accurate surface is required, the work should be allowed to cool to room temperature after the final roughing grind. This allows a redistribution of internal stresses, and the resulting distortion, if any, can be corrected in the final grinding operation.

Aluminum oxide wheels work best for superalloys; however, CBN is also used for some precision grinding applications. Grinding pressures should be great enough to cause slight wheel breakdown. Coarse wheels (46–60 grit) produce the best finishes during surface grinding. Low pressures help prevent distortion during grinding, especially with annealed material. Using moderate wheel speeds (e.g., 4000 sfm) reduces grinding heat and the probability of the workpiece cracking. Reciprocating tables are preferred to rotary tables, because they have reduced wheel contact, generate less heat, and cause less distortion of the workpiece.

Flood cooling is used during grinding to prevent heat checking (cracking) of the workpiece surface. Due to the low thermal conductivity of superalloys, copious amounts of grinding fluid are important. Highly sufurized water based soluble oils provide the best heat removal. Chlorinated oils and chlorinated water soluble oils with about 1% chlorine are often used for wet dressing of form grinding wheels to a tolerance of 0.002 in. or less. However, residual or entrapped fluid will react with the alloy during high temperature service and are often avoided for this reason.

Creep feed grinding is frequently used for turbine components, such as cast single crystal blades. A comparison of creep feed grinding with conventional grinding is shown in Fig. 6.25. In conventional reciprocating surface grinding, the grinding wheel makes rapid traversals removing only a small amount of material for each pass. In creep feed grinding, the workpiece is traversed at a very low table speed with a large depth of material removed per pass. In addition, the entire width of the wheel is used in creep feed grinding. Since creep feed grinding uses a low workpiece speed and a large depth of cut, it requires a larger total force, and therefore more power, than conventional surface grinding. However, creep feed grinding allows much higher metal removal rates. A typical grinding cycle would be to take the majority of the stock in one or two roughing passes, dress the wheel, and then take a finishing pass.

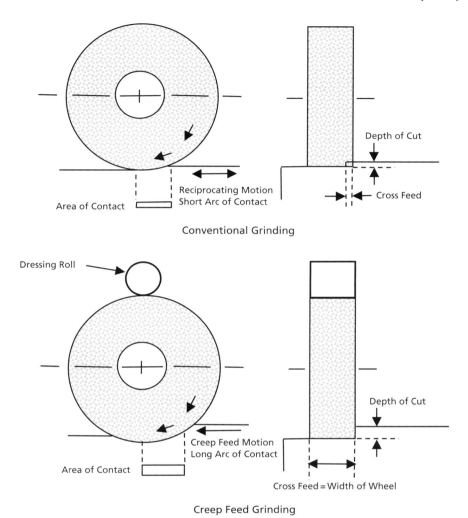

Fig. 6.25. *Conventional and Creep Feed Grinding*[44]

An extension of creep feed grinding is the Very Impressive Performance Extreme Removal (VIPER) process developed by Rolls Royce in the UK. In the VIPER process, small diameter (<8 in.) aluminum oxide wheels are combined with high pressure coolant (1000 psi) that is precisely injected into the grinding wheel ahead of the grind. Centrifugal force moves the coolant out of the wheel during the grind, both cleaning the wheel and cooling the workpiece. Productivity improvements in the range of 5–10 times have been reported, depending on whether or not the wheel is intermittently dressed or continuously dressed during the grind operation.

6.10 Joining

Superalloys can be joined by welding, diffusion bonding, brazing/soldering, adhesive bonding, and with mechanical fasteners. Although adhesive bonding is rarely used for superalloys, adhesive bonding is covered in Chapter 8 on Adhesive Bonding and Integrally Cocured Structure. Mechanical fastening is covered in Chapter 11 on Structural Assembly.

6.10.1 Welding[45,46]

The solid solution strengthened iron–nickel, nickel, and cobalt based superalloys are readily welded by arc welding processes such as GTAW and GMAW. However, the γ' strengthened iron–nickel and nickel based alloys are susceptible to hot cracking during welding, or may crack after welding (delayed cracking). The susceptibility to hot cracking is a function of the aluminum and titanium contents, which forms γ', as shown in Fig. 6.26. Cracking usually occurs in the HAZ and welding is usually restricted to wrought alloys with about 0.35 or less volume fraction of γ'. Casting alloys with high aluminum and titanium contents, like Inconel 713C and Inconel 100, are considered unweldable because they will

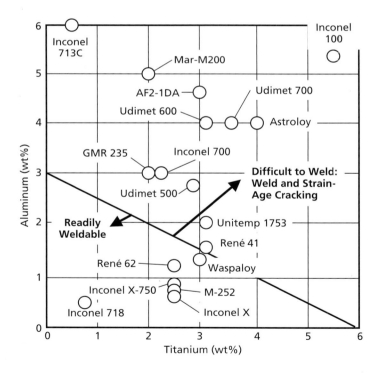

Fig. 6.26. Influence of Titanium and Aluminum on Superalloy Weldability[1]

Superalloys

usually hot crack during the welding operation. Borderline alloys, such as René 41 and Waspaloy, can usually survive the welding process but may crack later.

Alloys that are hardened with γ'' (Ni_3Nb), such as Inconel 718, are not subject to strain age cracking, because γ'' precipitates at a much slower rate than γ'. This allows them to be heated into the solution temperature range without suffering aging and resultant cracking. The delayed precipitation reaction enables the alloys to be welded and directly aged with less possibility of cracking. Inconel 718 is readily weldable in the solution treated condition, followed by a combined stress relieving and aging treatment. The weldability of Inconel 718 is yet another reason for its high usage.

If the surface of the material to be welded is not free of oxide, superalloys, due to the same oxides that give them good oxidation resistance, can experience trapped oxides in the weld metal and lack-of-fusion defects at the weld metal/parent metal interface. The oxide cannot be removed by simple wire brushing; an abrasive grinding operation is needed to positively remove the oxide. Ineffective inert gas protection can allow the reformation of an oxide film on the surface of the deposited weld metal. Care needs to be taken to remove this oxide before multipass welding to avoid the problems of entrapped oxide and lack-of-fusion defects.

Typical joint designs are shown in Fig. 6.27. Beveling is not required for material that is 0.1 in. or thinner. Thicker material should be beveled using either

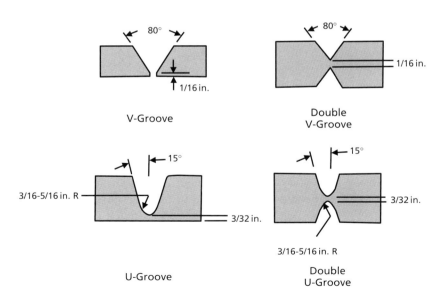

Fig. 6.27. Common Joint Designs[45]

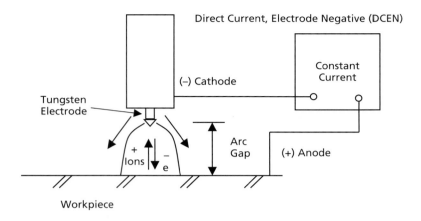

Fig. 6.28. Gas Tungsten Arc Welding (GTAW) Schematic

a V or U groove. For material thicker than 0.375 in., a double V or U groove joint design is recommended.

In GTAW, the welding heat is provided by an arc maintained between a non-consumable tungsten electrode and the workpiece. In GTAW, as shown in Fig. 6.28, the power supply is direct current with a negative electrode (DCEN). The negative electrode is cooler than the positive weld joint, enabling a small electrode to carry a large current, resulting in a deep weld penetration with a narrow weld bead. A high frequency circuit for assistance in starting the arc, and a current decay unit for slowly stopping the arc, should also be used. The high frequency circuit eliminates the need to contact the work with the electrode to start the arc. Contact starting can damage the electrode tip and also result in tungsten inclusions in the weld metal. A current decay unit gradually lowers the current before the arc is broken, to reduce the puddle size and end the bead smoothly.

The weld puddle and adjacent HAZ on the weld face are protected by the nozzle gas. Trailing shields are used to protect the hot solidified metal and the HAZ behind the weld puddle. Back-up shielding protects the root of the weld and its adjacent HAZ. Recommended shielding gases are helium, argon, or a mixture of the two. For welding thin material without the addition of filler metal, helium has the advantages over argon of reduced porosity and increased welding speed. Welding speeds can be increased as much as 40% over those achieved with argon; however, the arc voltage for a given arc length is about 40% greater with helium, and the heat input is therefore greater. Since welding speed is a function of heat input, the hotter arc permits higher speeds. The arc is more difficult to start and maintain in helium when the welding current is below about 60 amps. When low currents are required for joining small parts or thin material, either argon shielding gas should be used, or a high frequency current arc starting system should be added. Shielding gas flow rate is critical. Low rates will not protect the weld while high rates can cause turbulence and aspirate air, destroying the gas shield.

Tungsten electrodes, or those alloyed with thorium, are normally used for GTAW. A 2% thoria electrode will give good results for most welding applications; however, thoria tungsten electrodes are mildly radioactive, so alternate electrodes with ceria (2% cerium oxide) or lanthana (1–2% lanthanum oxide) are available. Although the initial cost of alloyed electrodes is greater, their longer life resulting from lower vaporization and cooler operation, along with their greater current carrying capacity, makes them more cost effective in the long term. The electrode will become contaminated if it contacts the weld metal or the base metal surface during the welding operation. If this occurs, the electrode should be cleaned and reshaped by grinding. In addition, the welder should stop and grind out the electrode debris from the weld metal.

Gas metal arc welding uses a consumable electrode, rather than a nonconsumable electrode, as used in the GTAW process. In GMAW, as shown in Fig. 6.29, the power supply is direct current with a positive electrode (DCEP). The positive electrode is hotter than the negative weld joint ensuring complete fusion of the wire in the weld joint. GMAW has the advantage of more weld metal deposited per unit time and unit of power consumption. For plates 0.5 in. and thicker, it is a more cost-effective process than GTAW. However, poor arc stability can cause appreciable spatter during welding, which reduces its efficiency.

Filler metals are normally similar in chemical composition to the base metals with which they are used. Because of high arc currents and high puddle temperatures, filler metals often contain small additions of alloying elements to deoxidize the weldment, and thus help to prevent solidification cracking and hot cracking.

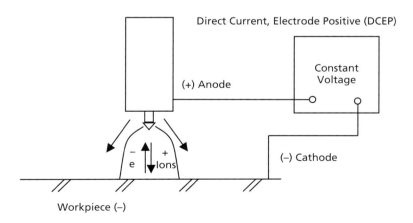

Fig. 6.29. Gas Metal Arc Welding (GMAW) Schematic

Heat input during welding should be kept as low as practical. For multiple bead or multiple layer welds, several small beads should be used instead of a few large heavy beads. Aluminum and titanium readily form high melting oxide compounds. Even under ideal welding conditions, aluminum and titanium oxides form and float to the top of the molten weld puddle. When depositing multiple pass weldments, the oxide particles can accumulate to the point that a scale is formed that can inhibit proper fusion. Also, flakes of oxide scale can become entrapped in the weldment. The flakes can act as mechanical stress raisers and significantly reduce joint efficiency and service life. The oxide film should be removed by abrasive blasting, or by grinding, when it becomes heavy enough to be visually apparent on the weld surface. Wire brushing is not recommended, because it does not actually remove the oxide film but only polishes it, hiding it from sight.

Another welding process used is linear friction welding, in which blades are attached to a hub to produce an integrated one piece engine rotor. In this method, one part is oscillated back and forth in a straight line, with a force applied normal to the plane of oscillation. The oscillatory motion, combined with the pressure, creates the heat needed to soften and upset the metal, forming a solid state joint in which the original surfaces are expelled as flash. Although machining this flash is an exacting process, in some situations, this still produces a significant cost savings as compared to machining the entire assembly from a large forging.

6.10.2 Brazing[45,47]

Since the aging cycle(s) used for the precipitation hardening alloys can introduce residual stresses that can cause joint cracking, the precipitation hardened alloys are brazed in the solution treated condition and then aged after brazing. Nickel brazing alloys produce brazed joints with strength and oxidation resistance for service at temperatures up to 2000° F. While the joints have good strength, often approaching that of the base metal (Fig. 6.30), they generally exhibit only moderate ductility.

Nickel brazing alloys (Table 6.4) contain 70–95% nickel. Generally, these alloys contain boron and/or silicon, which are melting point depressants and act as oxide reducing agents. In many commercial brazing filler metals, the levels are 2–3.5% boron and 3–10% silicon. Chromium, in amounts up to 20%, is often present to provide oxidation and corrosion resistance; however, higher amounts tend to lower the joint strength. Braze alloys are often produced in the form of powder (200 mesh or finer). They are also available as powder impregnated sheets that can be cut or formed into rings, washers, and other shapes. Powder can be applied directly to the joints; however, it is usually mixed with a binder and applied as a slurry by brushing, spraying, or dipping. Acrylic resins are often used as binders, because they do not leave a residue that could contaminate the joint. It is important that the slurry be allowed to thoroughly dry before brazing.

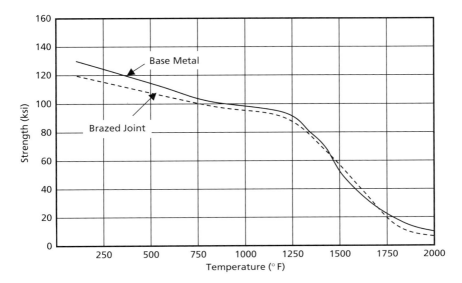

Fig. 6.30. Joint Strength for Brazed Superalloy[45]

Four factors that are important in achieving high strength brazed joints include: (1) joint design, (2) close control of joint clearances, (3) elimination of flux inclusions and unfused areas, and (4) effective wetting of the base material by the brazing alloy.

- Lap joints which load the brazed joint in shear are much stronger than butt joints that load the joint in tension.
- For high strength joints, joint clearances should be small. Clearances should be between 0.0005 and 0.005 in. Slightly larger clearances, along with some defects, such as flux inclusions or incomplete brazed areas, can be tolerated if the lap joint is designed with an increased overlap (i.e., $3X$ the thickness). The best approach is to design parts that are self-locating; however, if this is not possible, then mechanical fixturing, riveting, bolting, or spot welding can be used.
- In all brazing operations, it is essential that the parts are thoroughly cleaned and that all oxides have been removed from the surfaces. Grinding is frequently used to insure the removal of oxide films. To improve wetting, fluoride based fluxes can be used, or the surfaces can be plated with nickel. It is imperative that all flux residue be removed after brazing to avoid corrosive attack of the base metal. Flux residue is generally glass like and quite tenacious, requiring grinding, chipping, or abrasive blasting for removal. Fluxes should not be used in a vacuum environment.
- Precipitation hardening alloys present several difficulties not normally encountered with solid solution alloys. Precipitation hardening alloys often

Table 6.4 Nickel Based Brazing Alloys[45]

	Nominal Composition						Solidus (°F)	Liquidus (°F)	Brazing Range (°F)	AMS Spec.	Remarks
Ni	Cr	B	Si	Fe	C	P					
73.25	14	3.5	4	4.5	<0.75	–	1790	1900	1950–2200	4775	High strength and heat resistance
82.4	7	3.1	4.5	3	–	–	1780	1830	1850–2150	4777	Similar to 4775. Lower brazing temp.
90.9	–	3.1	4.5	1.5	–	–	1800	1900	1850–2150	4778	Good flow
92.5	–	1.5	3.5	1.5	–	–	1800	1950	1850–2150	4779	Forms large ductile fillets
70.9	19	–	10.1	–	–	–	1975	2075	2100–2200	4782	Can be used in Nuclear apps.
89	–	–	–	–	–	11	1610	1610	1700–1875	–	Extremely free flowing
77	13	–	–	–	–	10	1630	1630	1700–1900	–	Heat resistance for thin joints
70	16.5	3.75	4	4	0.95	–	1740	1950	–	4575A	Good corrosion resistance
70	16.5	3.75	4	4	<0.6	–	1740	1950	–	4776	Low C version of 4575A

contain appreciable (>1%) quantities of aluminum and titanium. The oxides of these elements are almost impossible to reduce in a controlled atmosphere (e.g., vacuum or hydrogen). Therefore, nickel plating or the use of a flux is necessary to obtain a surface that allows wetting by the filler metal. Because it is difficult to obtain wetting on alloys containing more than about 0.5% aluminum and/or titanium, precipitation hardened nickel base alloys are often nickel plated prior to brazing. The thickness of the plating required is dependent upon the brazing time and temperature. In general, a plating thickness of 0.0005–0.0015 in. is sufficient.

While nickel brazing is often performed in a furnace with highly reducing atmospheres, brazing of superalloys is best done in a vacuum furnace or an inert atmosphere. After evacuation, the chamber is sometimes backfilled with dry argon.

Brazing of cobalt based alloys is conducted using the same techniques used for nickel based alloys. These alloys are usually brazed in either a hydrogen or vacuum atmosphere. Filler metals are usually nickel, cobalt based alloys, or gold–palladium compositions. In addition to boron and silicon, these alloys usually contain chromium, nickel, and tungsten to provide corrosion and oxidation resistance and improve strength. Copper-containing filler metals should be avoided, as they can cause liquid metal embrittlement. Although cobalt base superalloys do not contain appreciable amounts of aluminum or titanium, an electroplate or flash coating of nickel is often used to promote better wetting.

6.10.3 Transient Liquid Phase (TLP) Bonding[48]

Transient liquid phase bonding, also known as diffusion brazing, is closely related to brazing; however, in brazing, the braze alloy melts and wets the metal with only a minimal amount of intermixing with the base metal. In transient liquid phase bonding, the filler metal melts and wets the metal, and the two intermix to form a higher temperature melting joint. An isothermal hold that results in continued diffusion is used to raise the melting point of the joint, to produce joints that match the base metal properties. The diffusion process often results in the total loss of identity of the original joint.

Transient liquid phase bonding can require extended diffusion cycles, ranging from 30 min up to 80 h, or even longer. Once the liquid has been distributed throughout the joint area, the assembly is held at the bonding temperature long enough for the filler metal elements to adequately diffuse into the base metal. In this way, the mechanical properties of the bonded joint area become essentially identical to those of either or both of the base metals. The amount of diffusion that occurs will be a function of the bonding temperature, the length of time the part is held at temperature, the quantity of filler metal available for diffusion, and the mutual solubility of filler metal and the base metals.

A fully diffused joint will lose all joint identity because of base metal grain growth across the entire joint. As an example, a cobalt based alloy was bonded at 2050° F in a vacuum furnace using a nickel–chromium–boron filler material in sheet form. The assembly was then held at 2050° F for 8 h. Even though the solidus of the filler metal was only 1780° F, the bonded joint completely lost its identity through diffusion, and the resulting remelt temperature was above 2500° F.

6.11 Coating Technology[49]

Superalloy engine components are often coated to prevent environmental degradation and more recently to provide thermal barriers which allow higher operating temperatures. Environmental coatings are used to prevent environmental attack (i.e., oxidation) of the substrate for the maximum possible time with the maximum degree of reliability. It should be noted that these coatings are not inert but react with the atmosphere to provide protection by forming a dense, tightly adherent oxide scale (Cr_2O_3 and Al_2O_3) through the interaction of chromium and aluminum with oxygen in the atmosphere. All oxidation resistant coatings will eventually fail, due to interdiffusion between the coating and substrate, i.e. the coating chemistry changes with time so that it is no longer protective. It is also important that coatings are compatible with the superalloy substrate, to prevent cracking and spalling of the coating.

At temperatures below about 1800° F, chromium in the form of Cr_2O_3 provides adequate protection, while at temperatures above 1800° F, it is aluminum in the form of Al_2O_3 that provides the most protection. Since chromium degrades the high temperature strength of γ', there has been a strong incentive to minimize the amount of chromium in the newer high temperature superalloys. The level of chromium has been reduced from about 20% down to a little as 9% in some of the newer alloys, thus decreasing oxidation resistance.

Superalloy coatings are divided into two main categories: (1) diffusion coatings are coatings that diffuse into the surface and react with alloying elements to form the protective coating, and (2) overlay coatings that are deposited on the surface but only react with the substrate to the extent that an adherent bond is formed.

6.11.1 Diffusion Coatings

Diffusion coatings are often applied by a pack cementation process. Aluminum powder, a chemical activator, and an inert filler of Al_2O_3 are sealed in a metallic pack along with the part. The pack is then heated in an inert atmosphere so that the reactants form a vapor that diffuses into the surface of the part enriching it with aluminum. The aluminum reacts with the nickel at the surface to form the intermetallic compounds Ni_3Al, $NiAl$ and Ni_2Al_3. If the substrate is a cobalt based alloy, the compounds are CoAl and FeAl.

Diffusion coatings are classified as either inward or outward diffusion coatings. In inward diffusion coatings, the aluminum in the coating diffuses faster inward than the nickel diffuses outward. These coatings are formed at a relatively low reaction temperature (1300–1450° F), by aluminum diffusing into the Ni_3Al phase. Since this produces a brittle coating with a low melting point, the part is then heated to higher temperatures (1900–2000° F) to convert the Ni_3Al to the more refractory NiAl phase. For outward diffusion coatings, the initial reaction is conducted at higher temperatures (1800–2000° F), in which the nickel diffuses outward through the aluminide layer.

For aircraft turbine blades fabricated from moderately corrosion resistant nickel alloys containing 12–15% chromium, simple inward and outward diffusion coatings provide adequate oxidation resistance. When the chromium content is lower than 7–10%, the coatings are normally modified by chromizing prior to aluminizing. For low temperature hot corrosion protection that occurs in the temperature range of 1650–1920° F, silicon or platinum is usually added to the pack to provide protection. Platinum-modified coatings on nickel based alloys form the intermetallic compounds $PtAl_2$, Pt_2Al_3, and PtAl to provide additional protection.

Chemically vapor-deposited coatings offer certain advantages over coatings applied by pack cementation. Since the coating is a vapor, it is much more effective in coating the internal passages of film-cooled turbine blades. The vapor is pumped through the passages to provide a relatively uniform coating on both the external and internal surfaces.

6.11.2 Overlay Coatings

Overlay coatings differ from diffusion coatings in that diffusion of the coating into the substrate is not required to obtain the desired coating composition and structure. Although some elemental diffusion at the interface between the coating and substrate is necessary for coating adherence, overlay coatings do not rely on reactions with the substrate for their formation. Instead, the prealloyed powder contains all of the necessary constituents to form the desired composition. Overlay coatings for nickel based superalloys are generally of the MCrAlY type, where M represents either nickel, cobalt, iron, or some combination of these three elements. MCrAlY coatings may contain 15–25% chromium, 10–15% aluminum, and 0.2–0.5% yttrium. Aluminum provides the primary oxidation resistance through the formation of Al_2O_3, while chromium is effective in combating hot corrosion and also increases the effective aluminum chemical activity. A small amount of yttrium helps to improve the adherence of the oxidation product.

Overlay coatings are applied by electron beam physical vapor deposition (EB-PVD) or by plasma spraying. The EB-PVD method usually produces superior coatings, but the plasma spray process is more cost effective. In EB-PVD, an electron beam gun is used to vaporize the coating alloy, which condenses out on the preheated part to form equilibrium or metastable phases. The coating grows as columnar grains normal to the substrate. Plasma-sprayed coatings are produced by introducing the coating powder into a high temperature plasma stream. The molten particles solidify on contact with the surface to form the coating. The process is generally carried out in a low pressure vacuum chamber, termed "low pressure plasma spray" (LPPS), to minimize the formation of oxide-related defects.

6.11.3 Thermal Barrier Coatings

While diffusion and overlay coatings are applied to provide environmental resistance to oxidation and hot corrosion, thermal barrier coatings (TBC) are applied to allow the turbine blades to operate at even higher temperatures. The TBC must be sufficiently thick; have a low thermal conductivity; have a high resistance to thermal shock; and contain a certain percentage of voids to provide even more thermal insulation. The temperature difference between the outer and inner surfaces of a TBC can be as much as 300° F.

A TBC consists of a ceramic outer layer (top coat) and a metallic inner layer (bond coat). A schematic of a typical coating system is shown in Fig. 6.31. The outer ceramic layer is typically zirconium oxide (ZrO_2) with 6–8% yttrium oxide added to partially stabilize the zirconia tetragonal phase. A dense interfacial ZrO_2 film provides chemical bonding between the columnar zirconia top coat and the inner oxidation resistant bond coat.

Although TBCs can be applied by either plasma spraying or EB-PVD, EB-PVD is the preferred method because it produces a columnar grain morphology, in which the grains are strongly bonded at their base but have a weak bond between the grains. This helps to reduce residual stress build-up within the coating, by allowing the grains to expand or contract into the gaps between the grains. The inner metallic bond coat (MCrAlY) aids in adhesion of the ceramic top coat, provides resistance to oxidation and hot corrosion, and helps in handling the thermal mismatch between the ceramic and the superalloy substrate.

Summary

Superalloys are a family of heat resistant alloys of nickel, iron–nickel, and cobalt that normally operate at temperatures exceeding 1000° F. They are required to exhibit combinations of high strength; good fatigue and creep resistance; good corrosion resistance; and the ability to operate at elevated temperatures for extended periods of time (i.e., metallurgical stability). Their combination of high temperature strength and resistance to surface degradation is unmatched by other metallic materials.

Strengthening of the nickel and iron–nickel based superalloys is due to the combination of solid solution hardening, precipitation hardening, and the

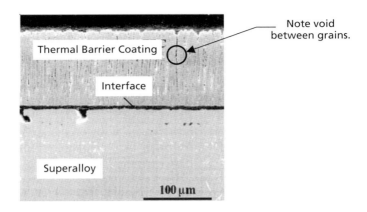

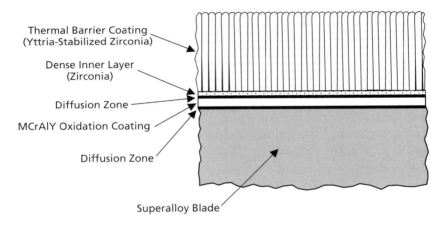

Fig. 6.31. Thermal Barrier Coating[49,50]

presence of carbides. Metallic carbides form in both the matrix, and at the grain boundaries, to help in providing high temperature stability. Cobalt based alloys are not precipitation strengthened; they are strengthened by a combination of solid solution strengthening and by carbides. The most important precipitate in nickel, and some iron–nickel based superalloys, is γ' $Ni_3(Al, Ti)$ in the form of either Ni_3Al or Ni_3Ti. The γ' phase is precipitated by heat treatments. The important strengthening precipitate in the iron–nickel based alloy Inconel 718 is γ'' (Ni_3Nb).

Superalloys are used in cast, rolled, extruded, forged, and powder-produced forms. Wrought alloys are generally more uniform with finer grain sizes and superior tensile and fatigue properties, while cast alloys have more alloy segregation and coarser grain sizes but better creep and stress rupture properties. Accordingly, wrought alloys are used where tensile strength and fatigue resistance are important, such as disks, while cast alloys are used where creep and stress rupture are important, such as turbine blades. The amount of alloying is so high in some superalloys that they cannot be produced as wrought products; they must be produced as either castings or by PM methods. In general, the more heat resistant the alloy, the more likely it is to be prone to segregation and brittleness, and therefore producible only by casting or by using powder metallurgy techniques.

Two processes are used for the production of forged superalloys, ingot metallurgy, and PM. Ingot metallurgy often involves triple melt technology, which includes three melting steps, followed by annealing and hot working to achieve the desired compositional control and grain size. The three melting steps include (1) VIM to prepare the desired alloy composition; (2) ESR to remove oxygen containing inclusions; and (3) VAR to reduce compositional segregation that occurs during ESR solidification. Melting is followed by homogenizing and hot working to achieve the desired homogeneity and grain size.

Powder metallurgy is often required for high volume fraction γ' strengthened alloys, such as René 95 and Inconel 100, which cannot be made by conventional ingot metallurgy and forging without cracking. The PM process includes (1) VIM to prepare the desired alloy composition; (2) remelting and atomizing to produce powder; (3) sieving to remove large particles and inclusions; (4) canning to place the powder in a container suitable for consolidation; (5) vacuum degassing and sealing to remove the atmosphere; and (6) HIP or extrusion to consolidate the alloy to a billet. Billets are then subsequently forged to final part shape.

Because of their strength retention at elevated temperatures, superalloys are more difficult to forge than most metals. The forgeability varies widely depending on the type of superalloy and its exact composition. For example, some of the iron–nickel based alloys, such as A-286, are similar to the austenitic stainless steels. At the other extreme, some superalloy compositions are intrinsically so strong at elevated temperatures that they can only be processed by casting or PM. In general, as the alloying content has been increased to obtain even greater elevated temperature strength, the forgeability has been degraded, i.e. the γ' strengthened alloys are much more difficult to forge than the solid solution strengthened alloys.

Wrought superalloys can often be formed using techniques similar to stainless steels, although forming is more difficult. Like stainless steels, superalloys work harden rapidly during forming. The cobalt based alloys require greater forces than the nickel or iron based alloys. Forming presses are the same as those used for forming steel; however, because of their higher strength, more power is needed, usually 50–100% more power.

The hottest parts of the engine, the blades and vanes, are manufactured by investment casting. As the alloy content of nickel based superalloys was continually increased to obtain better creep and stress rupture capability, the alloys became increasingly difficult to forge. To allow even higher contents of alloying elements, it became necessary to change the fabrication process to casting. The original cast blades and vanes were fine grained polycrystalline structures made using conventional investment casting procedures. These blades were then heat treated to coarsen the grain structures to enhance creep resistance. Eventually it became possible to produce DS structures with columnar grains oriented along the longitudinal axis of the blade. The columnar grain structure enhances the elevated temperature ductility by eliminating the grain boundaries as failure initiation sites. The DS process also creates a preferred low modulus texture or orientation parallel to the solidification direction that helps in preventing thermal fatigue failures. An extension of the DS process is the development of the single crystal (SX) process, in which a single crystal grows to form the entire blade. The elimination of grain boundaries also removes the necessity for adding grain boundary strengthening elements, namely boron, hafnium, zirconium, and carbon. The removal of these elements raises the melting point, and allows a higher solution heat treatment temperature, with a consequent improvement in chemical homogeneity and more uniform distribution of γ' precipitates.

Solution heat treating and aging is used to precipitation harden a great many of the nickel and iron–nickel based alloys. Solution treating temperatures range from about 1800–2250° F, or even up to 2400° F for single crystal alloys. Aging treatments are then used to strengthen precipitation-strengthened alloys by precipitating one or more phases (γ' or γ''). Aging treatments vary from as low as 1150° F to as high as 1900° F. Double aging treatments are used to produce different sizes and distributions of precipitates. A principal reason for double aging treatments, in addition to γ' and γ'' control, is the need to control grain boundary carbide morphology. Aging heat treatments normally range from 1600 to 1800° F with times from 4 to 32 h.

Superalloys are difficult to machine, perhaps second only to titanium in machining difficulty. Many of the same characteristics that make superalloys good high temperature materials also make them difficult to machine. General guidelines for machining superalloys are: conduct majority of machining in the softest state possible; use positive rake angles; use sharp cutting tools; use strong geometries; use rigid set-ups; prevent part deflection; use large lead angles; and when more than one pass is required, vary the depth of cut.

Because of the large amounts of γ' strengthening in nickel and iron–nickel superalloys, they are considerably less weldable than the cobalt alloys. The γ' strengthened alloys are susceptible to hot cracking during welding, or may crack after welding (delayed cracking). The susceptibility to hot cracking is a function of their aluminum and titanium contents, which forms γ'. Cracking usually occurs in the HAZ and welding is usually restricted to wrought alloys with about 0.35 or less volume fraction of γ'. Casting alloys with high aluminum and titanium contents are considered unweldable, because they will usually hot crack during the welding operation. Borderline alloys, such as René 41 and

Waspaloy, can usually survive the welding process but may crack later. One big advantage of Inconel 718 is its ability to be successfully fusion welded. Nickel brazing alloys produce joints with strength and oxidation resistance for service at temperatures up to 2000° F.

Superalloy engine blades are often coated to prevent environmental degradation, and more recently, to provide thermal barriers which allow even higher operating temperatures. Superalloy coatings are divided into two main categories: diffusion coatings are coatings that diffuse into the surface and react with alloying elements to form the protective coating, and overlay coatings that are deposited on the surface but only react with the substrate to the extent that an adherent bond is formed. While the diffusion and overlay coatings are applied to provide environmental resistance to oxidation and hot corrosion, ceramic TBC are applied to allow the turbine blades to operate at even higher temperatures. The TBC must be sufficiently thick, have a low thermal conductivity, have a high resistance to thermal shock, and contain a certain percentage of voids to provide more thermal insulation.

Recommended Reading

[1] Davis, J.R., *Heat-Resistant Materials*, ASM International, 1997.
[2] Donachie, M.J., Donachie, S.J., *Superalloys: A Technical Guide*, 2nd edition, ASM International, 2002.
[3] Tien, J.K., Caulfield, T., *Superalloys, Supercomposites and Superceramics*, Academic Press, Inc., 1989.
[4] *ASM Handbook Vol. 1: Properties and Selection: Irons, Steels, and High-Performance Alloys*, ASM International, 1990.
[5] *ASM Handbook Vol. 4: Heat Treating*, ASM International, 1991.
[6] *ASM Handbook Vol. 6: Welding, Brazing, and Soldering*, ASM International, 1993.
[7] *ASM Handbook Vol. 14: Forming and Forging*, ASM International, 1988.
[8] *ASM Handbook Vol. 15: Casting*, ASM International, 1988.
[9] *ASM Handbook Vol. 16: Machining*, ASM International, 1990.

References

[1] Davis, J.R., "Superalloys", in *Alloying: Understanding The Basics*, ASM International, 2001, pp. 290–307.
[2] Choudhury, I.A., El-Baradie, M.A., "Machinability of Nickel-Base Super Alloys: A General Review", *Journal of Materials Processing Technology*, Vol. 77, 1998, pp. 278–284.
[3] Davis, J.R., "Metallurgy, Processing, and Properties of Superalloys", in *Heat-Resistant Materials*, ASM International, 1997, pp. 221–254.
[4] Mankins, W.L., Lamb, S., "Physical Metallurgy of Nickel and Nickel Alloys", in *ASM Handbook Vol. 1: Properties and Selection: Irons, Steels, and High-Performance Alloys*, ASM International, 1990.
[5] Smith, W.F., "Nickel and Cobalt Alloys", in *Structure and Properties of Engineering Alloys*, 2nd edition, McGraw-Hill, Inc., 1993, pp. 487–536.
[6] Stoloff, N.S., "Wrought and P/M Superalloys", in *ASM Handbook Vol. 1: Properties and Selection: Irons, Steels, and High-Performance Alloys*, ASM International, 1990.

[7] Smallman, R.E., Bishop, R.J., "Modern Alloy Developments", in *Modern Physical Metallurgy and Materials Engineering*, 6th edition, Butterworth-Heinemann, 1999, pp. 305–308.
[8] Maurer, G.E., "Primary and Secondary Melt Processing- Superalloys", in *Superalloys, Supercomposites, and Superceramcis*, Academic Press, 1989, pp. 64–96.
[9] Benz, M.G., "Preparation of Clean Superalloys", GE Technical Information Series No. 98CRD128, General Electric Research & Development Center, 1998.
[10] Duhl, D.N., "Single Crystal Superalloys", in *Superalloys, Supercomposites and Superceramics*, Academic Press, 1989, pp. 149–182.
[11] Schafrik, R., Sprague, R., "Gas Turbine Materials", in *Advanced Material & Processes*, March/June 2004.
[12] Bradley, E.F., "Microstructure", in *Superalloys: A Technical Guide*, ASM International, 1988, pp. 31–51.
[13] Donachie, M.J., "Superalloy Processing", in *ASM Handbook Vol. 1: Properties and Selection: Irons, Steels, and High-Performance Alloys*, ASM International, 1990.
[14] Howson, T.E., Couts, W.H. "Thermomechanical Processing of Superalloys", in *Superalloys, Supercomposites and Superceramics*, Academic Press, Inc., 1989, pp. 183–213.
[15] Brooks, J.W., "Forging of Superalloys", *Materials & Design*, Vol. 21, 2000, pp. 297–303.
[16] Davis, J.R., "Powder Metallurgy Superalloys", in *Heat-Resistant Materials*, ASM International, 1997, pp. 272–289.
[17] Reichman, S., Chang, D.S., in *Superalloys II*, John Wiley & Sons, 1987, p. 459.
[18] Bradley, E.F., "Forging", in *Superalloys: A Technical Guide*, ASM International, 1988, pp. 133–141.
[19] Srivastava, S.K., "Forging of Heat-Resistant Alloys", in *ASM Handbook Vol. 14: Forming and Forging*, ASM International, 1988.
[20] Forbes Jones, R.M., Jackmann, L.A., "The Structural Evolution of Superalloy Ingots During Hot Working", *Journal of Metals*, Vol. 51, No. 1, 1999, pp. 27–31.
[21] Ruble, H.H., "Forging of Nickel Alloys", in *ASM Handbook Vol. 14: Forming and Forging*, ASM International, 1988.
[22] "Fabrication the Special Metals Corporation Alloys", Special Metals Corporation.
[23] Shen, G., Furrer, D., "Manufacturing of Aerospace Forgings", *Journal of Materials Processing Technology*, Vol. 98, 2000, pp. 189–195.
[24] Bradley, E.F., "Heat Treating", in *Superalloys: A Technical Guide*, ASM International, 1988, pp. 163–183.
[25] Davis, J.R., "Directionally Solidified and Single-Crystal Superalloys", in *Heat-Resistant Materials*, ASM International, 1997, pp. 255–271.
[26] Erickson, G.L., "Polycrystalline Cast Superalloys", in *ASM Handbook Vol. 1: Properties and Selection: Irons, Steels, and High-Performance Alloys*, ASM International, 1990.
[27] Harris, K., Erickson, G.L., Schwer, R.E., "Directionally Solidified and Single-Crystal Superalloys", in *ASM Handbook Vol. 1: Properties and Selection: Irons, Steels, and High-Performance Alloys*, ASM International, 1990.
[28] Bouse, G.K., Mihalsin, J.R., "Metallurgy of Investment Cast Superalloy Components", in *Superalloys, Supercomposites and Superceramics*, Academic Press, 1989, pp. 99–148.
[29] Piwonka, T.S., "Directional and Monocrystal Solidification", in *ASM Handbook Vol. 15: Casting*, ASM International, 1988.
[30] Mraz, S.J., "Birth of an Engine Blade", *Machine Design*, June 24, 1997, pp. 39–44.
[31] Davis, J.R., "Effect of Heat Treating on Superalloy Properties", in *Heat-Resistant Materials*, ASM International, 1997, pp. 290–308.
[32] Donachie, M.J., Donachie, S.J., "Heat Treating", in *Superalloys: A Technical Guide*, 2nd edition, ASM International, 2002, pp. 135–147.

[33] DeAntonio, D.A., Duhl, D., Howson, T., Rothman, M.F., "Heat Treating of Superalloys", in *ASM Handbook Vol. 4: Heat Treating*, ASM International, 1991.
[34] Schubert, F., "Temperature and Time Dependent Transformation: Application to Heat Treatment of High Temperature Alloys", in *Superalloy Source Book*, ASM International, 1989, p. 88.
[35] Medeiros, S.C., Prasad, Y.V.R.K., Frazier, W.G., Srinivasan, R., "Microstructural Modeling of Metadynamic Recrystallization in Hot Working of IN 718 Superalloy", *Materials Science and Engineering*, A293, 2000, pp. 198–207.
[36] Semiatin, S.L., Kramb, R.C., Turner, R.E., Zhang, F., Antony, M.M., "Analysis of the Homogenization of a Nickel-Base Alloy", *Scripta Materialia*, Vol. 51, 2004, pp. 491–495.
[37] Farhngi, H., Horouzi, S., Nili-Ahmadabadi, M., "Effects of Casting Process Variables on the Residual Stress in Ni-Base Superalloys", *Journal of Materials Processing Technology*, Vol. 153–154, 2004, pp. 209–212.
[38] "Machining of Heat-Resistant Alloys", in *ASM Handbook Vol. 16: Machining*, ASM International, 1990.
[39] "Machining", in *Metals Handbook Desk Edition*, 2nd edition, ASM International, 1989, p. 893.
[40] Ezugwu, E.O., "High Speed Machining of Aero-Engine Alloys", *Journal of Brazilian Society of Mechanical Science & Engineering*, Vol. XXVI, No. 1, 2004, pp. 1–11.
[41] Graham, D., "Turning Difficult-To-Machine Alloys", Modern Machine Shop, July 2002.
[42] Ezugwu, E.O., Wang, Z.M., Machado, A.R., "The Machinability of Nickel-Based Alloys: A Review", *Journal of Materials Processing Technology*, Vol. 86, 1999, pp. 1–16.
[43] "Machining", Special Metals Corporation.
[44] "Final Shaping and Surface Finishing", in *Engineered Materials Handbook Desk Edition*, ASM International, 1995, pp. 830–834.
[45] "Joining", Publication SMC-055, Special Metals Corporation, 2003, pp. 1–48.
[46] Bradley, E.F., "Welding", in *Superalloys: A Technical Guide*, ASM International, 1988, pp. 197–220.
[47] Bradley, E.F., "Brazing", in *Superalloys: A Technical Guide*, ASM International, 1988, pp. 221–232.
[48] Kay, W.D., "Diffusion Brazing", in *ASM Handbook Vol. 6: Welding, Brazing, and Soldering*, ASM International, 1993.
[49] Davis, J.R., "Protective Coatings for Superalloys", in *Heat-Resistant Materials*, ASM International, 1997, pp. 335–344.
[50] Clarke, D.R., Phillpot, S.R., "Thermal Barrier Coating Materials", *Materials Today*, June 2005, p. 23.

Chapter 7
Polymer Matrix Composites

The advantages of high performance composites are many, including lighter weight; the ability to tailor lay-ups for optimum strength and stiffness; improved fatigue strength; corrosion resistance; and with good design practice, reduced assembly costs due to fewer detail parts and fasteners. The specific strength (strength/density) and specific modulus (modulus/density) of high strength fiber composites, especially carbon, are higher than other comparable aerospace metallic alloys. This translates into greater weight savings resulting in improved performance, greater payloads, longer range, and fuel savings. A comparison of the overall structural efficiency of carbon/epoxy, Ti-6Al-4V, and 7075-T6 aluminum is given in Fig. 7.1.

Composites do not corrode and their fatigue resistance is outstanding. Corrosion of aluminum alloys is a major cost, and a constant maintenance problem, for both commercial and military aircraft. The corrosion resistance of composites can result in major savings in supportability costs. The superior fatigue resistance of composites, compared to high strength metals, is shown in Fig. 7.2. As long as reasonable design strain levels are used, fatigue of carbon fiber composites should not be a problem.

Assembly costs usually account for about 50% of the cost of an airframe. Composites offer the opportunity to significantly reduce the amount of assembly labor and fasteners. Detail parts can be combined into a single cured assembly, either during initial cure or by secondarily adhesive bonding.

Disadvantages of composites include high raw material costs and high fabrication and assembly costs; composites are adversely affected by both temperature and moisture; composites are weak in the out-of-plane direction, where the matrix carries the primary load; composites are susceptible to impact damage

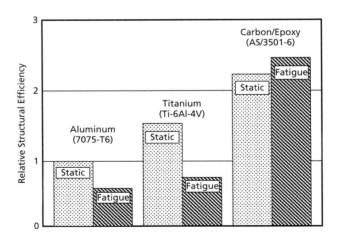

Fig. 7.1. *Relative Structural Efficiency of Aircraft Materials*[1]

Polymer Matrix Composites

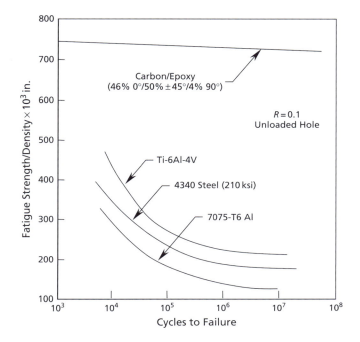

Fig. 7.2. *Fatigue Properties of Aerospace Materials*[1]

and delaminations or ply separations can occur; and composites are more difficult to repair than metallic structure.

The major cost driver in fabrication for a conventional hand layed-up composite part is the cost of laying-up, or collating, the plies. This cost (Fig. 7.3)

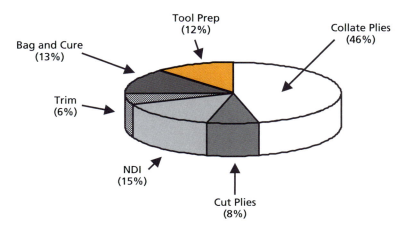

Fig. 7.3. *Cost Drivers for Composite Fabrication*[2]

275

generally consists of 40–60% of the fabrication cost depending on part complexity. Assembly cost is another major cost driver, accounting for about 50% of the total part cost. As previously stated, one of the potential advantages of composites is to cure or bond a number of detail parts together to reduce assembly costs and the number of required fasteners.

7.1 Materials

Continuous fiber composites are laminated materials (Fig. 7.4), in which the individual layers, plies, or lamina are oriented in directions that will enhance the strength in the primary load direction. Unidirectional (0°) laminates are extremely strong and stiff in the 0° direction but are also very weak in the 90° direction because the load must be carried by the much weaker polymeric

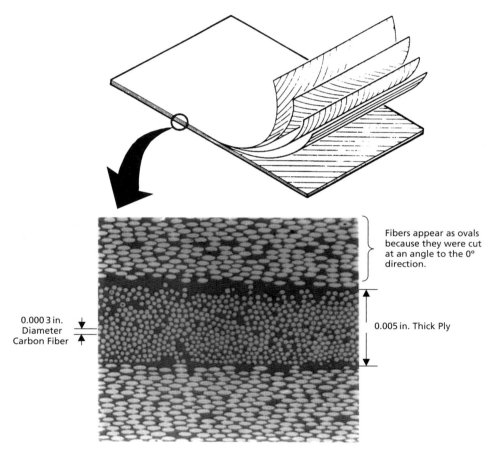

Fig. 7.4. Laminate Construction[1]

matrix. While a high strength fiber can have a tensile strength of 500 ksi or more, a typical polymeric matrix normally has a tensile strength of only 5–10 ksi. The longitudinal tension and compression loads are carried by the fibers, while the matrix distributes the loads between the fibers in tension, and stabilizes and prevents the fibers from buckling in compression. The matrix is also the primary load carrier for interlaminar shear (i.e., shear between the layers) and transverse (90°) tension. Since the fiber orientation directly impacts the mechanical properties, it would seem logical to orient as many of the layers as possible in the main load carrying direction. While this approach may work for some structures, it is almost always necessary to balance the load-carrying capability in a number of different directions, such as the 0°, +45°, −45°, and 90° directions.

7.1.1 Fibers

The primary role of the fibers is to provide strength and stiffness; however, as a class, high strength fibers are brittle; posses linear stress–strain behavior with little or no evidence of yielding; have low strain to failures (1–2% for carbon); and exhibit larger strength variations than metals. A summary of the major composite reinforcing fibers is given in Table 7.1.

Due to their good balance of mechanical properties and low cost, glass fibers are the most widely used reinforcement. E-glass, or "electrical" glass, is the most common glass fiber and is used extensively in commercial composite products. E-glass is a low cost, high density, low modulus fiber that has good corrosion resistance and good handling characteristics. S-2 glass, or "structural" glass, was developed in response to the need for a higher strength fiber for filament wound pressure vessels and solid rocket motor casings. It has a density, performance, and cost between E-glass and carbon. Quartz fiber is used in some electrical applications due to its low dielectric constant; however, it is expensive.

Table 7.1 Properties of Typical High Strength Fibers[1]

Fiber	Density $lb/in.^3$	Tensile Strength (ksi)	Elastic Modulus (msi)	Strain to Failure (%)	Diameter (mil)	Thermal Expansion Coefficient $10^{-6} in./in./°F$
E-glass	0.090	500	11.0	4.8	0.36	2.8
S-glass	0.092	650	12.6	5.6	0.36	1.3
Quartz	0.079	490	10.0	5.0	0.35	1.0
Aramid (Kevlar 49)	0.052	550	19.0	2.8	0.47	−1.1
Spectra 1000	0.035	450	25.0	0.7	1.00	−1.0
Carbon (AS4)	0.065	530	33.0	1.5	0.32	−0.2
Carbon (IM-7)	0.064	730	41.0	1.8	0.20	−0.2
Graphite (P-100)	0.078	350	107	0.3	0.43	−0.3
Boron	0.093	520	58.0	0.9	4.00	2.5

Aramid fiber (e.g., Kevlar) is an organic fiber that has a low density and is extremely tough, exhibiting excellent damage tolerance. Although it has a high tensile strength, it performs poorly in compression. It is also sensitive to ultraviolet light and should be limited to long-term service at temperatures less than 350° F. Another organic fiber is made from Ultra-High Molecular Weight Polyethylene (UHMWPE), e.g., Spectra. It has a low density with excellent radar transparency and a low dielectric constant. Due to its low density, it exhibits a very high specific strength and modulus at room temperature. However, being polyethylene, it is limited in temperature capability to 290° F or lower. Like aramid, UHMWPE has excellent impact resistance; however, poor adhesion to the matrix is a problem. However, plasma treatments have been developed to improve the adhesion.

Carbon fiber contains the best combination of properties but is also more expensive than either glass or aramid. Carbon fiber has a low density, a low coefficient of thermal expansion (CTE), and is conductive. It is structurally very efficient and exhibits excellent fatigue resistance. It is also brittle (strain-to-failure less than 2%) and exhibits low impact resistance. Being conductive, it will cause galvanic corrosion if placed in direct contact with aluminum. Carbon fiber is available in a wide range of strength and stiffness, with strengths ranging from 300 to 1000 ksi and moduli ranging from 30 to 145 msi. With this wide range of properties, carbon fiber (Table 7.2) is frequently classified either as: (1) high strength, (2) intermediate modulus, or (3) high modulus.

The terms "carbon" and "graphite" are often used to describe the same material. However, carbon fibers contain ~95% carbon and are carbonized at 1800–2700° F, while graphite fibers contain ~99% carbon and are first carbonized, followed by graphitizing at temperatures between 3600 and 5500° F. In general, the graphitization process results in a fiber with a higher modulus.

Table 7.2 Properties of PAN Based Carbon Fibers[3]

Property	Commercial High Strength	Aerospace		
		High Strength	Intermediate Modulus	High modulus
Tensile Modulus (msi)	33	32–35	40–43	50–65
Tensile Strength (ksi)	550	500–700	600–900	600–800
Elongation at Failure (%)	1.6	1.5–2.2	1.3–2.0	0.7–1.0
Electrical Resistivity ($\mu\Omega$ – cm)	1650	1650	1450	900
Thermal Conductivity (Btu/ft-h-° F)	11.6	11.6	11.6	29–46
Coefficient of Thermal Expansion Axial Direction (10^{-6} K)	−0.4	−0.4	−0.55	−0.75
Density (lb/in.^{3})	0.065	0.065	0.065	0.069
Carbon Content (%)	95	95	95	+99
Filament Diameter (μm)	6–8	6–8	5–6	5–8

Polymer Matrix Composites

Carbon and graphite fibers are made from rayon, PAN (polyacrylonitrile), or petroleum based pitch with PAN based fibers producing the best combination of properties. Rayon was developed as a precursor prior to PAN but is rarely used today due to its higher cost and lower yield. Petroleum based pitch based fibers were developed as a lower cost alternative to PAN but are mainly used to produce high and ultra-high modulus graphite fibers. A comparison of the PAN and pitch manufacturing processes is shown in Fig. 7.5. Both carbon and graphite fibers are produced as untwisted bundles called tows. Common tow sizes are 1k, 3k, 6k, 12k, and 24k, where k = 1000 fibers. Immediately after fabrication, carbon and graphite fibers are normally surface treated to improve their adhesion to the polymeric matrix. Sizings, often epoxies without a curing agent, are frequently applied as thin films (1% or less) to improve handleability and protect the fibers during weaving or other handling operations.

Several other fibers are occasionally used for polymeric composites. Before carbon was developed, boron fiber was the original high performance fiber. Boron is a large diameter fiber that is made by pulling a fine tungsten wire through a long slender reactor where it is chemically vapor deposited with boron. Since it is made one fiber at a time, rather than thousands of fibers at a time,

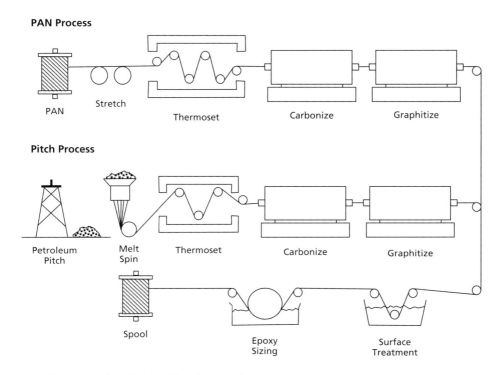

Fig. 7.5. PAN and Pitch Fiber Manufacturing Processes

it is very expensive. However, due to its large diameter and high modulus, it exhibits outstanding stiffness and compression properties. On the negative side, it does not conform well to complicated shapes and is very difficult to machine. High temperature ceramic fibers, such as silicon carbide (Nicalon), aluminum oxide, and alumina–boria–silica (Nextel), are frequently used in ceramic matrix composites, but rarely in polymeric composites.

7.1.2 Matrices

The matrix holds the fibers in their proper position; protects the fibers from abrasion; transfers loads between fibers; and provides interlaminar shear strength. A properly chosen matrix will also provide resistance to heat, chemicals, and moisture; have a high strain-to-failure; cure at as low a temperature as possible and yet have a long pot or out-time life, and not be toxic. The most prevalent thermoset resins used for composite matrices (Table 7.3) are polyesters, vinyl esters, epoxies, bismaleimides, polyimides, and phenolics.

Matrices for polymeric composites can be either thermosets or thermoplastics. Thermoset resins usually consist of a resin (e.g., epoxy) and a compatible curing agent. When the two are initially mixed, they form a low viscosity liquid that cures as a result of either internally generated (exothermic) or externally applied heat. The curing reaction, as shown schematically in Fig. 7.6, forms a series of crosslinks between the molecular chains so that one large molecular network is formed, resulting in an intractable solid that cannot be reprocessed on reheating. On the other hand, thermoplastics start as fully reacted, high viscosity materials that do not crosslink on heating. On heating to a high enough temperature, they either soften or melt, so they can be reprocessed a number of times. Although a lot of research and development has been conducted on thermoplastic composites, thermoset resins are by far the most widely used resin systems for current high performance composite applications.

Table 7.3 Relative Characteristics of Composite Resin Matrices[1]

Polyesters	Used extensively in commercial applications. Relatively inexpensive with processing flexibility. Used for continuous and discontinuous composites.
Vinyl Esters	Similar to polyesters but are tougher and have better moisture resistance.
Epoxies	High performance matrix systems for primarily continuous fiber composites. Can be used at temperatures up to 250–275° F. Better high temperature performance than polyesters and vinyl esters.
Bismaleimides	High temperature resin matrices for use in the temperature range of 275–350° F with epoxy-like processing. Requires elevated temperature post-cure.
Polyimides	Very high temperature resin systems for use at 550–600° F. Very difficult to process.
Phenolics	High temperature resin systems with good smoke and fire resistance. Used extensively for aircraft interiors. Can be difficult to process.

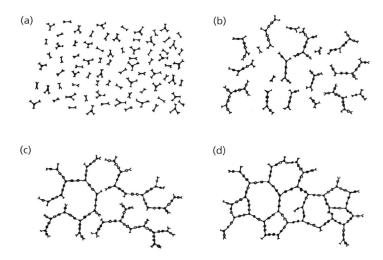

(a) Polymer and curing agent prior to reaction
(b) Curing initiated with size of molecules increasing
(c) Gellation with full network formed
(d) Full cured and crosslinked

Fig. 7.6. Crosslinking Reaction for Thermoset Resin[4]

The first consideration in selecting a resin system is the service temperature required for the part. The glass transition temperature T_g is a good indicator of the temperature capability of the matrix. The glass transition temperature (T_g) is the temperature at which a polymer changes from a rigid glassy solid into a softer, semi-flexible material. At this point, the polymer structure is still intact but the crosslinks are no longer locked in position. A resin should never be used above its T_g unless the service life is very short (e.g., a missile body). A good rule of thumb is to select a resin in which the T_g is 50° F higher than the maximum service temperature. Since most polymeric resins absorb moisture that lowers the T_g, it is not unusual to require that the T_g be as much as 100° F higher than the service temperature. It should be noted that different resins absorb moisture at different rates, and the saturation levels can be different. Therefore, the specific resin candidate must be evaluated for environmental performance. Most thermoset resins are fairly resistant to solvents and chemicals.

Although fiber selection usually dominates the mechanical properties of the composite, matrix selection can also influence performance. Some resins wet-out and adhere to fibers better than others, forming a chemical and/or mechanical bond that can improve the fiber-to-matrix load transfer capability. The matrix can also microcrack during cure or in-service. Resin rich pockets and brittle resin systems are susceptible to microcracking, especially when the processing temperatures are high and the use temperatures are low (e.g., −65° F), since this condition creates a very large difference in thermal expansion between the fibers and the matrix. Toughened resins can help in preventing microcracking but often at the expense of elevated temperature performance. The selection of a matrix material has the largest effect on the fabrication and processing conditions.

7.1.3 Product Forms

There are a multitude of material product forms used in composite structures. The fibers can be continuous or discontinuous. They can be oriented or disoriented (random). They can be furnished as dry fibers or preimpregnated with resin (prepreg). Since the market drives availability, not all fiber or matrix combinations are available in all product forms. In general, the more operations required by the supplier, the higher the cost. For example, prepreg cloth is more expensive than dry woven cloth. While complex dry preforms may be expensive, they can translate into lower fabrication costs by reducing or eliminating hand lay-up costs. If structural efficiency and weight are important design parameters, then continuous reinforced product forms are normally used because discontinuous fibers yield lower mechanical properties.

Rovings, tows, and yarns are collections of continuous fiber. This is the basic material form that can be chopped, woven, stitched, or prepregged into other product forms. It is the least expensive product form and available in all fiber types. Rovings and tows are supplied with no twist, while yarns have a slight twist to improve their handleability. Some processes, such as wet filament winding and pultrusion, use rovings as their primary product form.

Continuous thermoset prepreg materials are available in many fiber and matrix combinations. A prepreg is a fiber form that has a predetermined amount of uncured resin impregnated on the fiber by the material supplier. Prepreg rovings and tapes are usually used in automated processes, such as filament winding and automated tape laying, while unidirectional tape and prepreg fabrics are used for hand-lay up. Unidirectional prepreg tapes (Fig. 7.7) offer better structural performance than woven prepregs, due to absence of fiber crimp and the ability to more easily tailor the designs. However, woven prepregs offer increased drapeablility. With the exception of predominantly unidirectional designs, unidirectional tapes require placement of more individual plies during lay-up. For example, with cloth, for every 0° ply in the lay-up, a 90° reinforcement is also included. With unidirectional tape, a separate 0° ply and a separate 90° ply must be placed onto the tool.

Prepregs are supplied with either a net resin (prepreg resin content $\approx$ final part resin content) or excess resin (prepreg resin content > final part resin content). The excess resin approach relies on the matrix flowing through the plies and removing entrapped air, while the extra resin is removed by impregnating bleeder plies on top of the lay-up. The amount of bleeder used in the lay-up will dictate

Polymer Matrix Composites

Fig. 7.7. Unidirectional Prepreg Tape

the final fiber and resin content. To insure proper final physical properties, accurate calculations of the number, and areal weight, of bleeder plies for a specific prepreg are required. Since the net resin approach contains the final resin content weight in the fabric, no resin removal is necessary. This is an advantage because the fiber and resin volumes can more easily be controlled.

Woven fabric, shown in Fig. 7.8, consisting of interlaced warp and fill yarns, is the most common continuous dry material form. The warp is the 0° direction as the fabric comes off the roll and the fill, or weft, is the 90° fiber. Typically, woven fabrics are more drapeable than stitched materials; however, the specific weave pattern will affect their drapeablility characteristics. The weave pattern will also affect the handleability and structural properties of the woven fabric. Many weave patterns are available. All weaves have their advantages and disadvantages, and consideration of the part configuration is necessary during fabric selection. Two of the more widely used weave patterns are shown in Fig. 7.9. The plain weave has the advantage that it has good stability and resists distortion, while the satin weaves have higher mechanical properties, due to less fiber crimp, and are more drapeable. Most fibers are available in woven fabric form; however, it can be very difficult to weave some high modulus fibers due to their inherent brittleness. Advantages of woven fabric include drapeablility, ability to achieve high fiber volumes, structural efficiency, and market availability. A disadvantage of woven fabric is the crimp that is introduced to the warp or fill fiber during weaving. Finishes or sizings are typically put on the fibers to aid in the weaving process and minimize fiber damage. However, it is important to insure that the finish is compatible with the matrix selection when specifying a fabric.

Manufacturing Technology for Aerospace Structural Materials

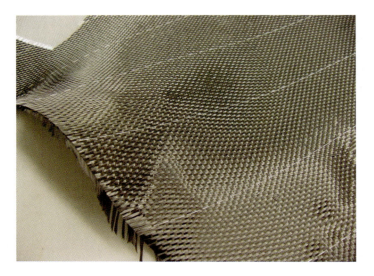

Fig. 7.8. Dry Woven Carbon Cloth

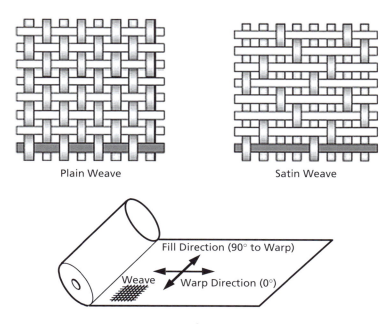

Fig. 7.9. Plain and Satin Weave Cloth[5]

A stitched fabric consists of unidirectional fibers oriented in specified directions that are then stitched together to form a fabric. A common stitched design includes 0°, +45°, 90°, and −45° plies in one multi-directional fabric. Advantages include: (1) the ability to incorporate off-axis orientations as the fabric is removed from the roll. Off-axis cutting is not needed for a multi-directional stitched fabric and, when compared to conventional woven materials, can reduce scrap rates (up to 25%); (2) labor costs are also reduced when using multi-ply stitched materials, because fewer plies are required to be cut and handled during lay-up; and (3) due to the z-axis stitch threads, ply orientation remains intact during handling. A disadvantage is the availability of specific stitched ply set designs. Typically, a special order is required due to the tailoring requested by the customer, such as fiber selection, fiber volume, and stitching requirements. In addition, not as many companies stitch as weave and drapeablility characteristics are reduced. However, this can be an advantage for parts with large simple curvature. Careful selection of the stitching thread is necessary to insure compatibility with the matrix and process temperatures.

Hybrids are material forms that make use of two or more fiber types. Common hybrids include glass/carbon, glass/aramid, and aramid/carbon fibers. Hybrids are used to take advantage of properties or features of each reinforcement type. In a sense, a hybrid is a "trade-off" reinforcement that allows increased design flexibility. Hybrids can be interply (two alternating layers), intraply (present in one layer), or in selected areas. Hybridization in selected areas is usually done to locally strengthen or stiffen a part.

A preform is a pre-shaped fibrous reinforcement that has been formed into shape, on a mandrel or in a tool, before being placed into a mold. As shown in Fig. 7.10, the shape of the preform closely resembles the final part configuration. A simple multi-ply stitched fabric is not a preform unless it is shaped to near its final configuration. The preform is the most expensive dry, continuous, oriented fiber form; however, using preforms can significantly reduce fabrication labor. A preform can be made using rovings, chopped, woven, stitched, or

Fig. 7.10. Fiberglass Preform

unidirectional material forms. These reinforcements are formed and held in place by stitching, binders or tactifiers, braiding, or three-dimensional (3-D) weaving. Advantages include reduced labor costs; minimal material scrap; reduced fiber fraying of woven or stitched materials; improved damage tolerance for 3-D stitched or woven preforms; and the desired fiber orientations are locked in place. Disadvantages include high preform costs; fiber wetability concerns for complex shapes; tackifier or binder compatibility concerns with the matrix; and limited flexibility if design changes are required.

7.2 Fabrication Processes

Automated ply cutting, manual ply collation, or lay-up, and autoclave curing is the most widely used process for high performance composites in the aerospace industry. While manual ply collation is expensive, this process is capable of making complex and high quality parts. Since cost has become a major driver, a number of other processes such as automated tape laying, filament winding, and fiber placement are used for certain classes of parts. In addition, other low cost processes, such as liquid molding and pultrusion, are either in limited production or are emerging as production ready processes.

7.3 Cure Tooling

The purpose of the bond tool is to transfer the autoclave heat and pressure during cure to yield a dimensionally accurate part. Tooling for composite fabrication is a major up-front non-recurring cost. It is not unusual for a large bond tool to cost as much as $500 000–$1 000 000. Unfortunately, if the tool is not designed and fabricated correctly, it can become a recurring headache, requiring continual maintenance and modifications, and, in the worst-case scenario, replacement. Tooling for composite structures is a complex discipline in its own right, largely built on years of experience. It should be pointed out that there is no single correct way to tool a part. There are usually several different approaches that will work, with the final decision based largely on experience of what has worked in the past and what did not work.

7.3.1 Tooling Considerations

There are many requirements a tool designer must consider before selecting a tooling material and fabrication process for a given application. However, the number of parts to be made on the tool and the part configuration are often the overriding factors in the selection process. It would not make good economic sense to build an inexpensive prototype tool that would only last for several parts when the application calls for a long production run, or vice versa. Part configuration or complexity will also drive the tooling decision process. For example, while welded steel tools are often used for large flat pasts, such as wing skins, it would not be cost effective to use steel for a highly contoured fuselage section, due to the high fabrication cost and complexity.

One of the first choices that must be made is which side of the part should be tooled, i.e. the inside or outside surface as shown in Fig. 7.11. Tooling a skin to the outside surface or outer moldline (OML) surface provides the opportunity to produce a part with an extremely smooth outside surface finish. However, if the part is going to be assembled to substructure, for example with mechanical fasteners, tooling to the inside or inner moldline (IML) surface will provide better fit with fewer gaps and less shimming required. Ease of part fabrication is another concern. It would certainly be easier to collate, or lay-up, the plies on a male tool than down inside the cavity of a female tool.

Selection of the material used to make the tool is another important consideration. Several of the key properties of various tooling materials are given in Table 7.4. Normally, reinforced polymers can be used for low-to-intermediate temperatures, metals for low-to-high temperatures, and monolithic graphite or

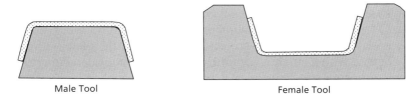

Fig. 7.11. *Male and Female Tooling*[1]

Table 7.4 Properties of Typical Tooling Materials[1]

Material	Max Service Temp. (°F)	CTE × 10^{-6}/(°F)	Density (lb/in.³)	Thermal Conductivity (Btu/h-ft-°F)
Steel	1500	6.3–7.3	0.29	30
Aluminum	500	12.5–13.5	0.10	104–116
Electroformed Nickel	550	7.4–7.5	0.32	42–45
Invar/Nilo	1500	0.8–2.9	0.29	6–9
Carbon/Epoxy 350° F	350	2.0–5.0	0.058	2–3.5
Carbon/Epoxy RT/350° F	350	2.0–5.0	0.058	2–3.5
Glass/Epoxy 350° F	350	8.0–11.0	0.067	1.8–2.5
Glass/Epoxy RT/350° F	350	8.0–11.0	0.067	1.8–2.5
Monolithic Graphite	800	1.0–2.0	0.060	13–18
Mass Cast Ceramic	1650	0.40–0.45	0.093	0.5
Silicone	550	45–200	0.046	0.1
Isobutyl Rubber	350	≈ 90	0.040	0.1
Fluoroelastomer	450	≈ 80–90	0.065	0.1

Note: For reference only. Check with material supplier for exact values.

ceramics for very high temperatures. Traditionally, tools for autoclave curing have generally been made of either steel or aluminum. Electroformed nickel became popular in the early-1980s, followed by the introduction of carbon/epoxy and carbon/bismaleimide composite tools in the mid-1980s. Finally, in the early-1990s, a series of low expansion iron–nickel alloys was introduced under the trade names Invar and Nilo.

Steel has the attributes of being a fairly cheap material with exceptional durability. It is readily castable and weldable. It has been known to withstand over 1500 autoclave cure cycles and still be capable of making good parts. However, steel is heavy, has a higher coefficient of thermal expansion than the carbon/epoxy parts usually built on it, and, for large massive tooling, can experience slow heat-up rates in an autoclave. When a steel tool fails in-service, it is usually due to a cracked weld.

On the other hand, aluminum is much lighter and has a much higher coefficient of thermal conductivity. It is also much easier to machine than steel, but is more difficult to produce pressure tight castings and welds. The two biggest drawbacks of aluminum are, being a soft material, it is rather susceptible to scratches, nicks, and dents, and it has a very high coefficient of thermal expansion. Due to its lightweight and ease of machinability, aluminum is often used for what are called "form block" tools. A number of aluminum form block tools can be placed on a large flat aluminum project plate and then the plate with all of the parts is covered with a single vacuum bag for cure, a considerable cost savings compared to bagging each individual part. Another application for aluminum tools is matched-die tooling, where all surfaces are tooled as in the example shown in Fig. 7.12. The attractiveness of aluminum for matched-die tooling is that on heating, it expands to help consolidate the part, while on cooling, it contracts, making part removal easier.

Electroformed nickel has the advantages that it can be made into complex contours and does not require a thick faceplate. When backed with an open tubular type substructure, this type of tool experiences excellent heat-up rates in an autoclave. However, to make an electroformed nickel tool requires a plating mandrel be fabricated to the exact contour of the final tool.

Carbon/epoxy, or glass/epoxy, tools also require a master or mandrel for lay-up during tool fabrication. A distinct advantage of carbon/epoxy tools is that their CTE can be tailored to match that of the carbon/epoxy parts they build. In addition, composite tools are relatively light, exhibit good heat-up rates during autoclave curing, and a single master can be used to fabricate duplicate tools. On the downside, there has been a lot of negative experience with composite tools that are subjected to 350° F autoclave cure cycles. The matrix has a tendency to crack and, with repeated thermal cycles, develop leaks. An additional consideration is that composite tools will absorb moisture if not in continual use. It may be necessary, after prolonged storage, to slowly dry tools in an oven to allow the moisture to diffuse out. A moisture saturated tool

Polymer Matrix Composites

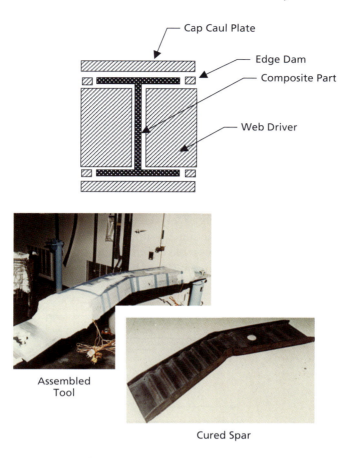

Fig. 7.12. Example of Matched-Die Tool
Source: The Boeing Company

placed directly in an autoclave and heated to 350°F could very easily develop blisters and internal delaminations due to the absorbed moisture.

Invar and the Nilo series of alloys were introduced in the early 1990s as the answer for composite tooling. Being low expansion alloys, they very closely match the CTE of the carbon/epoxy parts. Their biggest disadvantages are cost and weight that produces slow heat-up rates. The material itself is very expensive and it is more difficult to work with than even steel. It can be cast, machined, and welded. It is used for premium tooling applications such as wing skins.

Since many common tooling materials, such as aluminum and steel, expand at greater rates than the carbon/epoxy part being cured on them, it is necessary to correct their size, or compensate for the differences in thermal expansion. As the tool heats-up during cure, it grows, or expands, more than the composite laminate. During cool-down, the tool contracts more than the cured laminate. If not handled correctly, both of these conditions can cause problems, ranging from incorrect part size to cracked and damaged laminates. Thermal expansion is normally handled by shrinking the tool at room temperature using the calculation method shown in Fig. 7.13. For example, an aluminum tool producing a part 120.0 in. in length might actually be made as 119.7 in. long, assuming it will be cured at 350° F.

Another correction required for tooling for parts with geometric complexity is spring-in. When sheet metal is formed at room temperature, it normally springs-back, or opens up, after forming. To correct for springback, sheet metal parts are over formed to compensate for the springback. The opposite phenomenon occurs in composite parts. They tend to spring-in, or close up, during the cure process. Therefore, it is necessary to compensate angled parts by opening the angles on the tool, as shown in Fig. 7.14. The degree of compensation required is somewhat dependent on the actual lay-up orientation and thickness of the laminate. A great deal of progress has been made in calculating the degree of spring-in using finite element analysis, but it still usually requires some

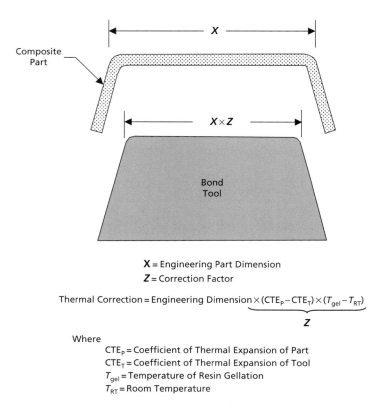

X = Engineering Part Dimension
Z = Correction Factor

$$\text{Thermal Correction} = \text{Engineering Dimension} \times \underbrace{(CTE_P - CTE_T) \times (T_{gel} - T_{RT})}_{Z}$$

Where
CTE_P = Coefficient of Thermal Expansion of Part
CTE_T = Coefficient of Thermal Expansion of Tool
T_{gel} = Temperature of Resin Gellation
T_{RT} = Room Temperature

Fig. 7.13. Thermal Expansion Correction Factors for Tooling[1]

Polymer Matrix Composites

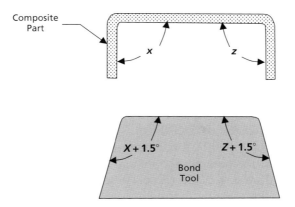

Note: 1.5° shown. Typical values range from 0 to 5° depending on tool material used.

Fig. 7.14. Spring-in Correction Factors[1]

experimental data for the particular material system, cure conditions, orientation, and thickness to establish tool design guidelines.

7.4 Ply Collation

Cutting and manual ply collation are the major cost drivers in composite part fabrication, normally accounting for 40–60% of the cost, depending on part size and complexity. Ply collation can be accomplished by hand, automated tape laying, filament winding, or fiber placement. Hand lay-up is generally the most labor intensive method, but may be the most economical, if the number of parts to be built is limited, the part size is small, or the part configuration is too complex to automate. Automated tape laying is advantageous for flat or mildly contoured skins, such as large thick wing skins. Filament winding is a high rate process that is used primarily for bodies of revolution or near bodies of revolution. Fiber placement is a hybrid process that possesses some of the characteristics of automated tape laying and filament winding. It was developed to allow the automated fabrication of large parts that could not be fabricated by either tape laying or filament winding.

7.4.1 Manual Lay-up

Manual hand collation is conducted using either prepreg tape (24 in. maximum width) or broadgoods (60 in. maximum width). Prior to actual lay-up, the plies are usually precut and kitted into ply packs for the part. The cutting operations are normally automated unless the number of parts to be built does not justify the cost of programming an automated ply cutter. However, if hand cutting is selected, templates to facilitate the cutting operation may have to be fabricated. If the lay-up has any contour of the plies, the contour will also have to be factored into the templates.

Automated ply cutting of broadgoods, usually 48–60 in. wide material, is the most prevalent method used today. Both reciprocating knives and ultrasonically driven ply-cutting methods are currently used in the composites industry. The reciprocating knife concept originated in the garment industry. In this process, a carbide blade reciprocates up and down, similar to a saber saw, while the lateral movement is controlled by a computer-controlled driven head. To allow the blade to penetrate the prepreg, the bed supporting the prepreg consists of nylon bristles that allow the blade to penetrate during the cutting operation. With a reciprocating knife cutter, normally one to five plies can be cut during a single pass.

The ultrasonic ply cutter operates in a similar manner; however, the mechanism is a chopping rather than a cutting action. Instead of a bristle bed that allows the cutter to penetrate, a hard plastic bed is used with the ultrasonic method. A typical ultrasonic ply cutter, shown in Fig. 7.15, can cut at speeds approaching 2400 fpm, while holding accuracies of ± 0.003 in. One of the primary advantages of any of the automated methods is that they can be programmed off-line and nesting routines are used to maximize material utilization. In addition, many of these systems have automated ply-labeling systems in which the ply identification label is placed directly on the prepreg release paper. A typical ply

Fig. 7.15. Large Ultrasonic Ply Cutter
Source: The Boeing Company

Polymer Matrix Composites

label will contain both the part number and the ply identification number. This makes sorting and kitting operations much simpler after the cutting operations are completed. Modern automated ply-cutting equipment is fast and produces high quality cuts.

Prior to ply collation, the tool should have either been coated with a liquid mold release agent, or covered with a release film. If the surface is going to be painted or adhesively bonded after cure, some lay-ups also require a peel ply on the tool surface. Peel plies are normally nylon, polyester, or fiberglass fabrics. Some are coated with release agents and some are not. It is important to thoroughly characterize any peel ply material that is bonded to a composite surface, particularly if that surface is going to be structurally adhesively bonded in a subsequent operation.

The plies are placed on the tool in the location and orientation as specified by the engineering drawing or shop work order. Prior to placing a ply onto the lay-up, the operator should make sure that all of the release paper is removed and that there are no foreign objects on the surface. Large Mylar (clear polyester film) templates are often used to define ply location and orientation. However, these are quite bulky and difficult to use and are rapidly been displaced by laser projection units. These units, shown schematically in Fig. 7.16, use low intensity laser beams to project the ply periphery on the lay-up. They are programmed off-line using CAD data for each ply and, with advanced software, are capable

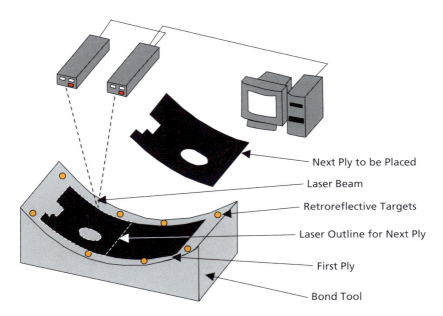

Fig. 7.16. Principle of Laser Ply Projection
Source: Virtek

of projecting ply locations on both flat and highly contoured lay-ups. The accuracy is generally in the ±0.015 to 0.040 in. range depending on the projection distance required for the part.[6] Ply location accuracy requirements are normally specified on the engineering drawing or applicable process specification. For unidirectional material, gaps between plies are normally restricted to 0.030 in. and overlaps and butt splices are not permitted. For woven cloth, butt splices are usually permitted but require an overlap of 0.5–1.0 in. The engineering drawing should also control the number of ply drop-offs at any one location in the lay-up.

The lay-up should be vacuum debulked every 3–5 plies, or more often if the shape is complex. Vacuum debulking consists of covering the lay-up with a layer of porous release material, applying several layers of breather material, applying a temporary vacuum bag and pulling a vacuum for a few minutes. This helps to compact the laminate and remove entrapped air from between the plies. For some complex parts, hot debulking, or prebleeding, in an oven under a vacuum or autoclave pressure at approximately 150–200° F can be useful for reducing the bulk factor. Prebleeding is similar to hot debulking except that in prebleeding some of the resin is intentionally removed with the addition of bleeder cloth, while in hot debulking no resin is intentionally removed.

7.4.2 Flat Ply Collation and Vacuum Forming

To lower the cost of ply-by-ply hand collation directly to the contour of the tool, a method called flat ply collation was developed in the early 1980s. This method, shown schematically in Fig. 7.17, consists of manually collating the

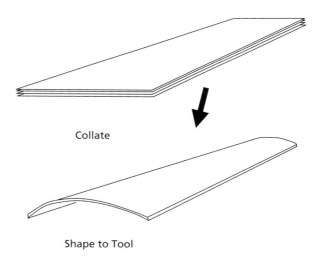

Fig. 7.17. Flat Ply Collation[1]

laminate in the flat condition, and then using a vacuum bag to form it to the contour of the tool. If the laminate is thick, this process may have to be done in several steps to prevent wrinkling and buckling of the ply packs. Heat ($<150°$ F) can be used to soften the resin to aid in forming if the contour is severe.

This process has also been used to successfully make substructure parts, such as C-channels. Normally woven cloth is used and the parts are again flat ply collated, placed on a form block tool, covered with a release film, and then vacuum formed to shape using a silicone rubber vacuum bag. Note that it is important to keep the fibers in tension during the forming process. If compression occurs, the fibers will wrinkle and buckle. To maintain uniform tension during the forming operation, a double diaphragm forming technique[7] can be used, in which the plies are sandwiched between two thin flexible diaphragms pulled together with a vacuum. Again, the application of low heat to soften the resin and aid in forming is quite prevalent. After cure, these long parts can be trimmed into shorter lengths, thus saving the cost of laying-up each individual part on its individual tool.

7.5 Automated Tape Laying

Automated tape laying (ATL) is a process that is very amenable to large flat parts, such as wing skins. Tape layers usually lay-down either 3, 6, or 12 in. wide unidirectional tape, depending on whether the application is for flat structure or mildly contoured structure. Automated tape layers are normally gantry style machines (Fig. 7.18) which can contain up to 10 axes of movement.[8] Normally, 5 axes of movement are associated with the gantry itself and the other 5 axes with the delivery head movement. A typical tape layer consists of a large floor-mounted gantry with parallel rails, a cross-feed bar that moves on precision ground ways, a ram bar that raises and lowers the delivery head, and the delivery head that is attached to the lower end of the ram bar. Commercial tape layers can be configured to lay either flat or mildly contoured parts. Flat tape laying machines (FTLM) are either fixed bed machines or open bay gantries, while contour tape laying machines (CTLM) are normally open bay gantries. The tool is rolled into the working envelope of the gantry, secured to the floor, and the delivery head is initialized onto the working surface.

The delivery heads (Fig. 7.19) for both FTLM and CTLM are basically the same configuration and will normally accept 3, 6, or 12 in. wide unidirectional tape. To facilitate the tape laying process, the unidirectional tape purchased for ATL applications is closely controlled for width and tack. FLTM uses either 6 or 12 in. wide tape to maximize material deposition rates for flat parts, while most CTLMs are restricted to 3 or 6 in. wide tape to minimize tracking errors (gaps and overlaps) when laying contoured parts. The term "CTLM" currently applies to mild contours that rise and fall at angles up to about 15%. More highly contoured parts normally are made by processes such as hand lay-up, filament

Manufacturing Technology for Aerospace Structural Materials

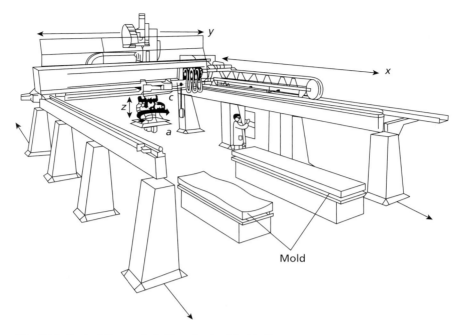

Fig. 7.18. *Typical Gantry Style Tape Laying Machine*[8]

winding, or fiber placement, depending on the geometry and complexity of the part. Material for ATL comes in large diameter spools, some containing almost 3000 lineal ft of material. The tape contains a backing paper that must be removed during the tape laying operation.

The spool of material is loaded onto the delivery head supply reel (reels as large as 25 in. in diameter are used) and threaded through the upper tape guide shoot and past the cutters. The material then passes through the lower tape guides, under the compaction shoe and onto a backing paper take-up reel. The backing paper is separated from the prepreg and wound onto a take-up roller. The compaction shoe makes contact with the tool surface and the material is laid onto the tool with compaction pressure. To insure uniform compaction pressure, the compaction shoe is segmented so that it follows the contour of the lay-up. The segmented compaction shoe is a series of plates that are air pressurized and conform to lay-up surface deviations, maintaining a uniform compaction pressure. The machine lays the tape according to the previously generated NC program, cuts the material at the correct length and angle when a length (course) is completed, lays out tail, lifts off the tool, retracts to the course start position, and begins laying the next course.[9]

Modern tape laying heads have optical sensors that will detect flaws during the tape laying process and send a signal to the operator. In addition, machine suppliers now offer a laser boundary trace in which the boundary of a ply can be

Polymer Matrix Composites

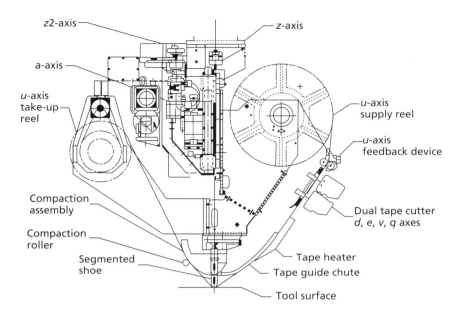

Fig. 7.19. Composite Tape Layer Delivery Head
Source: The Boeing Company

traced by the operator to verify the correct position. Modern tape laying heads also contain a hot air heating system that will preheat the tape (80–110° F) to improve the tack and tape-to-tape adhesion. Computer controlled valves maintain the temperature in proportion to the machine speed, i.e. if the head stops, the system diverts hot air flow to prevent overheating the material.

Software to drive modern tape layers has improved dramatically in the last 10 years. All modern machines are programmed off-line with systems that

297

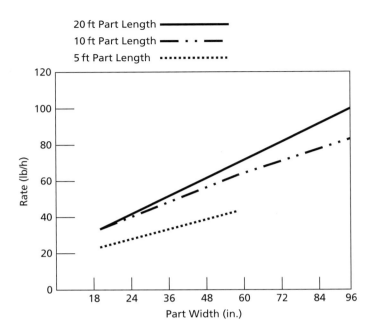

Fig. 7.20. Tape Laying Efficiency vs. Part Size[9]

automatically compute the "natural path" for tape laying over a contoured surface. As each ply is generated, the software updates the surface geometry, eliminating the need for the designer to redefine the surface for each new ply. The software can also display detailed information about the fiber orientation of each course and the predicted gaps between adjacent courses. Once the part has been programmed, the software will generate NC programs that optimize the maximum quantity of composite tape laid per hour.

Part size and design are key drivers for composite tape layer efficiency. As a general rule of thumb, bigger parts and simpler lay-ups are more efficient. This is illustrated in Fig. 7.20 for a FTLM.[9] If the design is highly sculptured (lots of ply drop-offs), or the part size is small, the machine will spend a significant amount of time slowing down, cutting, and then accelerating back to full speed.

7.6 Filament Winding

Filament winding is a high rate process in which a continuous fiber band is placed on a rotating mandrel. Lay-down rates as high as 100–400 lb/h are not uncommon. It is also a highly repeatable process that can fabricate large and thick-walled structure. Filament winding is a mature process, having been in continuous use since the mid-1940s. It can be used to fabricate almost any body of revolution, such as cylinders, shafts, spheres, and cones. Filament winding can also fabricate a large range of part sizes; parts smaller than 1 in. in diameter (e.g., golf club shafts) and as large as 20 ft in diameter have been wound. The major restriction on geometry is that concave contours cannot be wound, because the fibers are under tension and will bridge across the cavity. Typical applications for filament winding are cylinders, pressure vessels, rocket motor cases, and engine cowlings. End fittings are often wound into the structure producing strong and efficient joints.

A typical filament winding process is shown in Fig. 7.21. Dry tows are drawn through a bath containing liquid resin, collimated into a band, and then wound on a rotating mandrel.[10] To deliver the fiber tows from the spool to the part requires that the band pass through a series of guides, redirects, and spreader bars. During the entire delivery process, tension on the tows is minimized to preferably 1 lb or less. Low tension helps to reduce abrasion to the fibers, minimizes the possibility of tow breakage, and helps to spread the band as it passes over the spreader bar. Many modern filament winding machines are equipped with automatic tensioning devices to help control the amount of tension during the winding process.

Actually, there are three main variants of the filament winding process:[11] (1) wet winding, in which the dry reinforcement is impregnated with a liquid

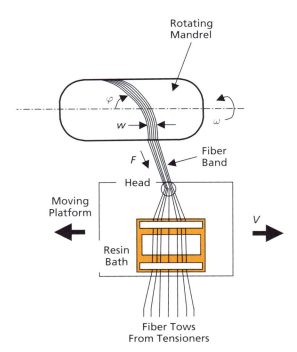

Fig. 7.21. Filament Winding Process[10]

resin just prior to winding; (2) wet rolled prepreg winding, in which the dry reinforcement is impregnated with the liquid resin and then rewound prior to filament winding; and (3) towpreg winding, in which a commercially impregnated tow is purchased from a material supplier.

Filament winding equipment costs can be low, moderate, or high, depending on part size, the type of winder, and the sophistication of the control system (mechanical or NC control). For high rate applications, some winders have been designed with multiple pay-out systems and multiple mandrels so that several parts can be fabricated simultaneously. Filament wound parts are usually cured in ovens rather than autoclaves, the compaction pressure being provided by mandrel expansion and fiber tension on the lay-up. Circular windings, separated from the part by caul plates or separator sheets, can also be used to provide compaction during cure. Forced air convection ovens are the most common curing equipment. Others, such as microwave curing are faster but result in higher equipment costs. Mandrel costs can be moderate to high, depending on part size and complexity. The mandrel must be able to be removed from the part. This is often accomplished by shrinkage of the mandrel during cool-down, incorporation of a slight draft or taper, wash-out mandrels, plaster break-out mandrels, inflatable mandrels, or for complex parts, segmented mandrels that can be taken out of the inside of the part in sections. While the inner surface (mandrel side) of the part is usually smooth, the outer surface can be quite rough. If this presents a problem, it is possible to wind sacrificial layers on the outer surface and then grind, or machine, the outer surface smooth after cure.

Fiber orientation can be a problem for some filament wound designs, i.e. the minimum fiber angle that can usually be wound is 10–15° due to slippage of the fiber bands at the mandrel ends. However, schemes such as temporary pins inserted in the mandrel ends during winding can sometimes be used to overcome this limitation.

Helical, polar, and hoop are the three dominant winding patterns used in filament winding. Helical winding (Fig. 7.22) is a very versatile process that can produce almost any combination of length and diameter. In helical winding the mandrel rotates, while the fiber carriage traverses back and forth at the speed necessary to generate the desired helical angle (θ).[12] As the band is wound, the circuits are not adjacent and additional circuits must be applied before the surface begins to be covered with the first layer. This winding pattern produces band cross-overs at periodic locations along the part, which can be somewhat controlled by the newer NC winding machines. Due to this cross-over winding pattern, a layer is made up of a two-ply balanced laminate. If the end dome openings are the same size, a geodesic wind pattern may be used. This pattern produces the shortest band path possible and results in uniform tension in the filaments throughout their length. An additional advantage of the geodesic pattern is that it produces a no-slip condition, i.e. there is no tendency for the bands to slip or shift on the mandrel surface.

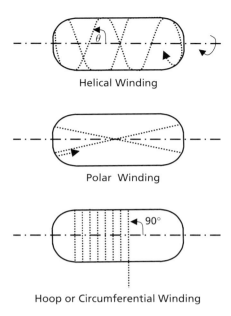

Fig. 7.22. Filament Winding Patterns[12]

Polar winding (Fig. 7.22) is somewhat simpler than helical winding in that: (1) a constant winding speed can be used; (2) it is not necessary to reverse the carriage during winding; and (3) the bandwidths are laid adjacent to each other as the part is wound. This is an excellent method for fabricating spherical shapes. In this process, the bands pass tangentially to the opening at one end of the part, reverse direction, and then pass tangentially to the opening at the opposite end of the part. The lay-down is planar, with the bandwidths adjacent to each other due to the winding arm, generating a great circle during each pass. Simple polar winders have only two axes of motion, the mandrel and the winding arm. Polar winding machines are generally much simpler than helical winding machines, but they are also somewhat limited in their capabilities. The length-to-diameter ratio must be less than 2.0. They are frequently used to wind spherical shapes by utilizing a continuous step-out pattern. A variation of the polar winder is the tumble winder, in which the mandrel is mounted at an inclined axis and tumbles in a polar path, while the roving strands remain stationary. While tumble winders are very efficient for spherical shapes, they are usually limited to diameters of 20 in. or less.

Hoop winding, also known a circumferential or circ winding, is the simplest winding process. This winding action, shown in Fig. 7.22, is similar to a lathe, where the mandrel speed is much greater than the carriage travel. Each full rotation of the mandrel advances the carriage one full bandwidth, so that the bands are wound adjacent to each other. During part fabrication, hoop winding is often combined with longitudinal (helical or polar) winding to provide adequate part strength and stiffness. The hoop windings can be applied to the cylindrical portion of the part, while the longitudinal windings are applied to both the cylindrical and domed portions of the part. It again should be pointed out that the minimum wind angle for longitudinal winding is generally about 10–15°, to preclude slippage of the bands at the ends of the mandrel.

The majority of filament winding fabricators formulate their own resin systems for both wet winding and wet rolled prepreg.[11] If a prepreg product form is specified, then they will normally purchase the preimpregnated tow from one of the major prepreg suppliers. Epoxies are the most prevalent matrix resins used for high performance filament wound parts, while polyesters and vinyl esters are often used for less demanding commercial applications. However, many different types of resins have been successfully used for filament winding, including cyanate esters, phenolics, bismaleimides, polyimides, and others.

Viscosity and pot life are two of the main factors in selecting a resin for wet winding. Low viscosity, generally around 2000 Centipoise (cP), is desirable to help wet the fibers, spread the band, and lower the friction over the guides during the winding process. Pot life is primarily a function of the time it will take to wind the part, i.e. large and thicker parts will require a longer pot life than smaller thinner parts. A number of premixed wet winding resin systems are also available from material suppliers. While preimpregnated tow (towpreg) is more expensive than wet winding resin systems, it does offer several important advantages: (1) a qualified fiber and resin system can often be prepregged onto a tow; (2) it allows the best control of resin content; (3) it allows the highest winding speeds because there is no wet resin that will be thrown-off during winding; and (4) the tack can be adjusted to allow less slippage when winding shallow angles.

Wet winding is accomplished by either pulling the dry tows through a resin bath or directly over a roller that contains a metered volume of resin controlled by a doctor blade. The resin content of wet wound parts is difficult to control, being affected by the resin reactivity, the resin viscosity, the winding tension, the pressure at the mandrel interface, and the mandrel diameter. For example, too low a viscosity resin will impregnate the strands thoroughly but will tend to squeeze out during the pressure of the winding operation, resulting in an excessively high fiber content. At the other extreme, too high a viscosity will not sufficiently impregnate the strands and there will be a tendency for the cured part to contain excessive porosity. Due to the generally low viscosity of wet winding resins, it is not uncommon to have parts with higher fiber volume percentages (70% and sometimes higher) than are normally found in composite parts fabricated with higher viscosity prepreg resins (60 volume percent).

To circumvent some of the problems with controlling a direct wet winding process, wet rolled prepreg is sometimes manufactured by wet impregnating the strands in the normal manner, and then respooling them prior to winding.

There are two main advantages to this process: (1) the fabricator can conduct off-line quality assurance on the wet wound prepreg prior to use, and (2) they can somewhat control the viscosity and tack by room temperature staging. Staging at room or slightly elevated temperature is commonly called B-staging. The objective is to advance the resin to increase the viscosity and tack. On the negative side, wet wound prepreg has to be packaged and refrigerated for storage, unless it is immediately used for winding.

Commercially supplied prepregs offer the best control of resin content, uniformity, and band width control, but are also the most expensive of the product forms, usually 1.5–2 times the cost of wet winding materials. While prepreg tows are the predominant prepreg form using in filament winding, some aerospace manufacturers specify slit prepreg tape to insure extremely tight control of the bandwidth, and the resultant gaps, during the fiber placement of flight critical hardware. Prepreg tows for filament winding generally (1) have the longest pot lives; (2) allow higher winding speeds because there is less chance of "resin throw" during the winding process; and (3) allow winding angles closer to longitudinal (0°), because they contain higher tack than most wet winding systems and will not tend to slip as much at the ends.

The choice of a mandrel material and design is to a great extent a function of the design and size of the part to be built. A large number of materials have been used for filament winding mandrels. Dissolvable mandrels are often used for parts with only small openings. This type of mandrel includes water soluble sand, water soluble or breakout plaster, low temperature eutectic salts, and occasionally low melting point metals. After cure, the disposable mandrel is dissolved out with hot water, melted, or broken into small pieces for removal. An alternate to these approaches would be to use an inflatable mandrel that can be either left inside the part as a liner or extracted through an opening. Reusable mandrels can either be segmented or non-segmented. Segmented mandrels are required when the part geometry does not allow the part to be removed by simply sliding the part off the mandrel after cure. Segmented mandrels are generally more expensive to fabricate and use than non-segmented mandrels. Non-segmented mandrels usually have a slight draft or taper to ease part removal after cure.

After the winding operation is complete, wet wound parts are often B-staged prior to final cure to remove excess resin by heating the part to a slightly elevated temperature but below the resin gel temperature. Frequently, the part is heated with heat lamps and the excess resin is removed as the part rotates. The great majority of filament wound parts are cured in an oven (electric, gas fired or microwave) without a vacuum bag or any other supplemental method of applying pressure. As the part heats-up to the cure temperature, the mandrel expands but is constrained by the fibers in the wound part. This creates pressure that helps to compact the laminate and reduce the amount of voids and porosity. Since the majority of filament wound parts are cured in ovens rather than autoclaves, filament winding is capable of making very large structures, limited only by the size of the winder and the curing oven available.

Autoclave curing may also be used to further reduce the amount of porosity; however, the compaction pressure applied by an autoclave can also induce fiber buckling, and even wrinkles in the part. The use of thin caul plates that are allowed to slip over the surface may help to alleviate some wrinkling on cylindrical surfaces, but these are prone to leaving mark-off on the part surface where they terminate. Caul plates with circumferential windings over the outside of the caul plates have also been used in oven cured parts to improve compaction and provide a smoother surface finish. Occasionally, the part will be wrapped with shrink tape to provide compaction pressure, a common method employed in manufacturing carbon fiber golf club shafts.

7.7 Fiber Placement

In the late 1970s, Hercules Aerospace Co. (now Alliant Techsystems) developed the fiber placement process. Shown conceptually in Fig. 7.23, it is a hybrid between filament winding and tape laying. A fiber placement, or tow placement, machine allows individual tows of prepreg to be placed by the head. The tension on the individual tows normally ranges from 0 up to about 2 lb. Therefore, true 0° (longitudinal) plies pose no problems. In addition, a typical fiber placement machine (Fig. 7.24) contains either 12, 24, or 32 individual tows that may be individually cut and then added back in during the placement process. Since the tow width normally ranges from 0.125 to 0.182 in., bands as wide as 1.50–5.824 in. can be applied depending on whether a 12 or 32 tow head

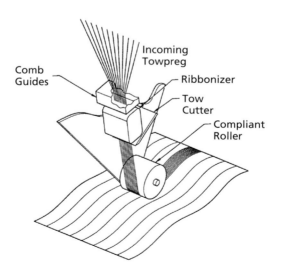

Fig. 7.23. Fiber Placement Process[1]

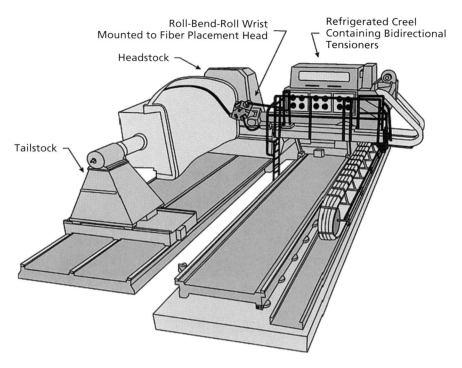

Fig. 7.24. Fiber Placement Machine[13]

is used. The adjustable tension employed during this process also allows the machine to lay tows into concave contours, limited only by the diameter of the roller mechanism. This allows complicated ply shapes, similar to those that can be obtained by hand lay-up. In addition, the head (Fig. 7.25) contains a compliant compaction roller that applies pressure in the range of 10–400 lb during the process, effectively debulking the laminate during lay-up. Advanced fiber placement heads also contain heating and cooling capability. Cooling is used to decrease the towpreg tack during cutting, clamping, and restarting processes, while heating can be used to increase the tack and compaction during lay-down. For the current generation of fiber placement heads, a minimum convex radius of approximately 0.124 in. and a minimum concave radius of 2 in. are obtainable. One limitation of the fiber placement process is that there is a minimum course (or ply) length, normally about 4 in. This is a result of the cut-and-add process. A ply that is cut or added must then pass under the compliant roller, resulting in a minimum length that is dependent on the roller diameter. Fiber-placed parts are usually autoclave cured on carbon/epoxy, steel, or low-expansion invar tools to provide dimensionally accurate parts. Typical applications for fiber placement are engine cowls, inlet ducts, fuselage sections, pressure tanks, nozzle cones,

Fig. 7.25. Fiber Placement Head
Source: The Boeing Company

tapered casings, fan blades, and C-channel spars. The aft section of a V-22, in which the skin is fiber placed over cocured stiffeners, is shown in Fig. 7.26.

Extensive testing has shown that the mechanical properties of fiber-placed parts can be essentially equivalent to hand layed-up parts.[14] Like hand layed-up parts, gaps and overlaps are typically controlled to 0.030 in. or less. One difference between fiber placed and hand layed-up plies are the "stair-step" ply terminations obtained with fiber placement, since each tow is cut perpendicular to the fiber direction. Again, this stair-step ply termination has been shown to be equivalent in properties to the smooth transition you obtain with manual lay-up. In fact, some parts have been designed so that either fiber placement or manual hand lay-up may be used for fabrication. Since the tows are added-in and taken-out as they are needed, there is very little wasted material; scrap rates of only 2–5% are common in fiber placement. In addition, since the head can "steer" the fiber tows, there is the potential for the design of highly efficient load-bearing structure.

The software required to program and control a fiber placement machine is even more complex than that required for an automated tape layer or modern filament winder. The software translates CAD part and tooling data into 7-axis commands, developing the paths and tool rotations for applying the composite tows to the part's curved and geometric features, while keeping the compaction roller normal to the surface. A simulator module confirms the part program with 3-D animation, while integrated collision avoidance post-processing of the NC program automatically detects interferences.

Polymer Matrix Composites

Fig. 7.26. V-22 Aft Fuselage
Source: The Boeing Company

Modern fiber placement machines are extremely complex and can be very large installations. Most machines contain seven axes of motion (cross-feed, carriage traverse, arm tilt, mandrel rotation, and wrist yaw, pitch, and roll). The larger machines are capable of handling parts up to 20 ft in diameter and 70 ft long, with mandrel weights up to 80 000 lb They typically contain refrigerated creels for the towpreg spools, towpreg delivery systems, redirect mechanisms to minimize twist, and tow sensors to sense the presence or absence of a tow during placement.

Although complex part geometries and lay-ups can be fabricated using fiber placement, the biggest disadvantages are that the current machines are very expensive, complex, and the lay-down rates are slow compared to most conventional filament winding operations.

7.8 Vacuum Bagging

After ply collation, the laminate is sealed in a vacuum bag for curing. A typical bagging schematic is shown in Fig. 7.27. To prevent resin from escaping from the edges of the laminate, dams are placed around the periphery of the lay-up. Typically, cork, silicone rubber, or metal dams are used. The dams should be butted up against the edge of the lay-up to prevent resin pools from forming between the laminate and dams. The dams are held in place with either double-sided tape or Teflon pins.

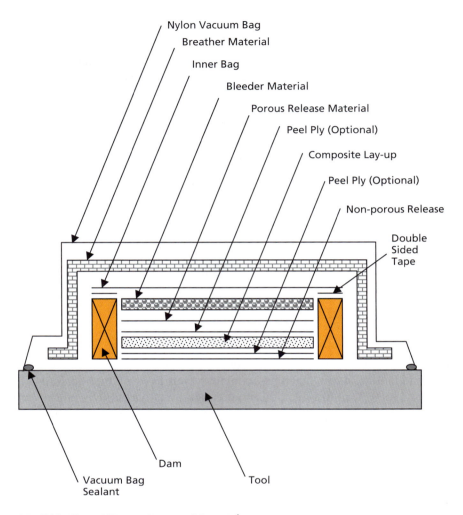

Fig. 7.27. Typical Vacuum Bagging Schematic[1]

A peel ply may be applied directly to the laminate surface if the surface is going to be subsequently bonded or painted. Then, a layer of porous release material, usually a layer of porous glass cloth coated with Teflon, is placed over the lay-up. This layer allows resin and air to pass through the layer without having the bleeder material bond to the laminate surface. The bleeder material can be a synthetic material (e.g., polyester mat) or dry fiberglass cloth, such as 120 or 7781 style glass. The amount of bleeder material depends on the laminate thickness and the desired amount of resin to be removed. For the newer net resin content prepregs, bleeder cloth is not required since it is not necessary to remove any excess resin.

After the bleeder is placed on the lay-up, an inner bag made of Mylar (polyester), Tedlar (PVF), or Teflon (TFE) is placed over the lay-up. The purpose of the inner bag is to let air escape while containing the resin within the bleeder pack. The inner bag is sealed to the edge dams with double-sided tape and then perforated with a few small holes to allow air to escape into the breather system. The breather material is similar to the bleeder material, either a synthetic mat material or a dry glass cloth can be used. If dry glass cloth is used, the last layer next to the vacuum bag should be no coarser than 7781 glass. Heavy glass fabrics, such as style 1000, have been known to cause vacuum bag ruptures during cure. The nylon bagging material can be pushed down into the coarse weave of the fabric and rupture. The purpose of the breather is to allow air and volatiles to evacuate out of the lay-up during cure. It is important to place the breather over the entire lay-up and extend it pass the vacuum ports.

The vacuum bag, which provides the membrane pressure to the laminate during autoclave cure, is normally a 3–5 mil thick layer of nylon-6 or nylon-66. It is sealed to the periphery of the tool with a butyl rubber or chromate rubber sealing compound. Nylon vacuum bags can be used at temperatures up to 375° F. If the cure temperature is higher than 375° F, a polyimide material called Kapton can be used to approximately 650° F, along with a silicone bag sealant. Higher temperatures usually require the use of a metallic bag (e.g., aluminum foil) and a mechanical sealing system. It should be noted that Kapton bagging films are stiffer and harder to work with than nylon. Some manufacturers have invested in reusable silicone rubber vacuum bags to reduce cost, and reduce the chance of a leak or bag rupture during cure. These normally require some type of mechanical seal to the tool. Also, if the part is large, reusable rubber vacuum bags become heavy and difficult to handle, so they may require a handling system to facilitate their installation and removal, as shown for the extremely large bag in Fig. 7.28. There are suppliers who sell both the materials to make silicone rubber vacuum bags, or will provide a complete bag and sealing system ready for use.

Caul plates, or pressure plates, can be used to provide a smoother part surface on the bag side. Caul plates are frequently made of mold release coated aluminum, steel, fiberglass, or glass reinforced silicone rubber. They range in thickness from as thin as 0.060 in. up to about 0.125 in. The design of a caul plate and its location in the lay-up are important considerations in achieving the desired surface finish. The caul plate may be placed above the bleeder pack or within the bleeder pack, but close to the laminate surface to provide a smooth surface. However, it will then require a series of small holes (e.g., 0.060–0.090 in. diameter) to allow resin to pass through the caul plate into the top portion of the bleeder pack. It should be noted that a caul plate containing holes is usually not placed next to the laminate surface, because the holes will mark-off on the laminate surface. In general, the further the caul plate is from the laminate surface, the less effective it is in producing a smooth surface, due to the cushioning effect of more and more bleeder material.

Fig. 7.28. Large Reusable Silicone Rubber Vacuum Bag
Source: The Boeing Company

The current trend in the composites industry is toward net or near-net resin systems (32–35% resin by weight), which require little or no bleeding, in contrast to the more traditional 40–42% resin systems. Since the labor and cost of the bleeder material is eliminated, this simplifies the bagging system. However, when using this type of material, it is even more important to properly seal the inner bag system, to prevent resin loss during cure, or resin-starved laminates may result. The edges are a particularly critical area, where excessive gaps or leaks in the dams can result in excessive resin loss and thinner-than-desired edges. In addition to the elimination of the need for a bleeder pack, net resin content prepregs produce laminates with a more uniform thickness and resin content. The problem with the traditional 40–42% resin content prepregs is that as the laminate gets thicker (i.e., larger number of plies) the ability to bleed resin through the thickness decreases. As more and more bleeder is added, the plies closest to the surface become overbled, while those in the middle and on the tool side of the laminate are underbled.

Once the vacuum bag has been successfully leak checked and the thermocouples applied, it is ready for autoclave cure. A slight vacuum should be maintained on the bag while it is waiting for autoclave cure to make sure that nothing shifts, or that wrinkles will not form, when the full vacuum is applied in the autoclave. If the lay-up contains honeycomb core, the maximum vacuum

Polymer Matrix Composites

that should be applied during leak checking, or cure, is 8–10 in. of mercury vacuum. Higher vacuums have been known to cause core migration, and even crushing, due to the differential pressure that can develop in the core cells.

7.9 Curing

Autoclave curing is the most widely used method of producing high quality laminates in the aerospace industry. An autoclave works on the principle of differential gas pressure, as illustrated in Fig. 7.29. The vacuum bag is evacuated to remove the air, and the autoclave supplies gas pressure to the part. Autoclaves are extremely versatile pieces of equipment. Since the gas pressure is applied isostatically to the part, almost any shape can be cured in an autoclave. The only limitation is the size of the autoclave, and the large initial capital investment to purchase and install an autoclave. A typical autoclave system, shown in Fig. 7.30, consists of a pressure vessel, a control system, an electrical system, a gas generation system, and a vacuum system. Autoclaves lend considerable versatility to the manufacturing process. They can accommodate a single large composite part, such as a large wing skin, or numerous smaller parts loaded onto racks and cured as a batch. While autoclave processing is not the most significant cost driver in total part cost, it does represent a culmination of all the previously performed manufacturing operations, because final part quality (per ply thickness, degree of crosslinking, and void and porosity content) is often determined during this operation.

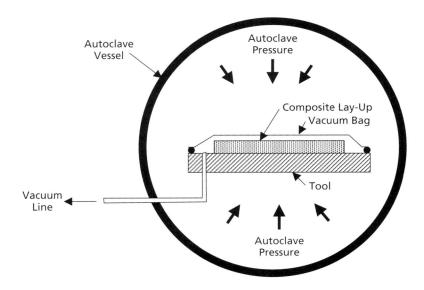

Fig. 7.29. Principle of Autoclave Curing

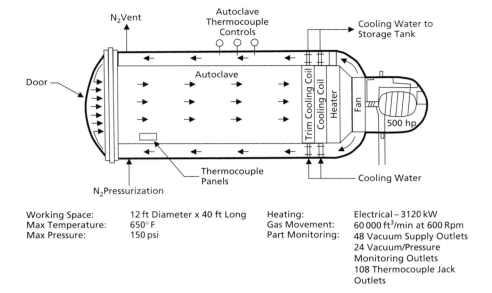

Fig. 7.30. Typical Production Autoclave Schematic[1]

Autoclaves are normally pressurized with inert gas, usually nitrogen or carbon dioxide. Air can be used, but it increases the danger of a fire within the autoclave during the heated cure cycle. The gas is circulated by a large fan at the rear of the vessel and passes down the walls next to a shroud containing the heater banks, usually electrical heaters. The heated gas strikes the front door and then flows back down the center of the vessel to heat the part. There is considerable turbulence in the gas flow near the door,[15] which produces higher velocities that stabilize as the gas flows toward the rear. The practical effect of this flow field is that you can often encounter higher heating rates for parts placed close to the door; however, the flow fields are dependent on the actual design of the autoclave and its gas flow characteristics. Another problem that can be encountered is blockage, in which large parts can block the flow of gas to smaller parts located behind them. Manufacturers typically use large racks to insure uniform heat flow and also maximize the number of parts that can be loaded for cure.

Composite parts can also be cured in presses or ovens. The main advantage of a heated platen press is that much higher pressures (e.g., 500–1000 psi) can be used to consolidate the plies and minimize void formation and growth. Presses are often used with polyimides that give off water, alcohols, or high boiling point solvents, such as NMP (N-methylpyrrolidone). On the other hand, presses usually require matched metal tools for each part configuration, and are limited by platen size to the number of parts that can be processed at one time. Ovens, usually heated by convective forced air, can also be used to cure composite structures. However, since pressure is provided by only a vacuum bag (⩽14.7 psia), the void contents of the cured parts are normally much higher (e.g., 5–10%) than those of autoclave cured parts (<1%).

7.9.1 Curing of Epoxy Composites

A typical cure cycle for a 350° F curing thermoset epoxy part is shown schematically in Fig. 7.31. It contains two ramps and two isothermal holds. The first ramp and isothermal hold, usually in the range of 240–280° F, is used to allow the resin to flow (bleed) and volatiles to escape. The imposed viscosity curve on the figure shows that the semi-solid resin matrix melts on heating and experiences a dramatic drop in viscosity. The second ramp and hold is the polymerization portion of the cure cycle. During this portion, the resin viscosity initially drops slightly due to the application of additional heat, and then rises dramatically, as the kinetics of the resin start the crosslinking process. The resin gels into a solid and the crosslinking process continues during the second isothermal hold, usually at 340–370° F for epoxy resin systems. The resin is held at this cure temperature for normally 4–6 h, allowing time for the crosslinking process to be completed. It should be noted that as the industry has moved toward net resin content systems, the use of the first isothermal hold, which allows time for resin bleeding, has been eliminated by many manufacturers, resulting in a straight ramp-up to the cure temperature.

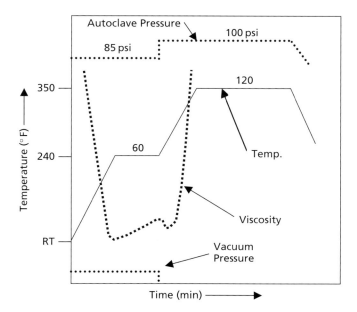

Fig. 7.31. Typical Autoclave Cure Cycle[1]

High pressures (i.e., 100 psi) are commonly used during autoclave processing to provide ply compaction and suppress void formation. Autoclave gas pressure is transferred to the laminate due to the pressure differential between the autoclave environment and the vacuum bag interior. Translation of the autoclave pressure to the resin depends on several factors, including the fiber content, laminate configuration, and the amount of bleeder used. Even though a relatively high autoclave pressure (e.g., 100 psi) may be used during the cure cycle, the actual pressure on the resin, the hydrostatic resin pressure, can be significantly less.

7.9.2 Theory of Void Formation

Porosity and voids have been one of the major problems in composite part fabrication. As shown in Fig. 7.32, voids and porosity can occur at either the ply interfaces (interlaminar) or within the individual plies (intralaminar). The terms "voids" and "porosity" are used fairly interchangeably in industry; however, the term "void" usually implies a large pore whereas "porosity" implies a series of small pores. Void formation and growth in addition curing composite laminates is primarily due to entrapped volatiles.[16] Higher temperatures result in higher volatile pressures. Void growth will potentially occur if the void pressure (i.e., the volatile vapor pressure) exceeds the actual pressure on the resin (i.e., the hydrostatic resin pressure), while the resin is a liquid (Fig. 7.33). The prevailing relationship is:

$$\text{If } P_{\text{Void}} > P_{\text{Hydrostatic}} \rightarrow \text{ then void formation and growth}$$

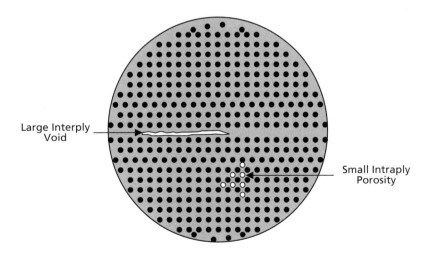

Fig. 7.32. Interply and Intraply Voids and Porosity[1]

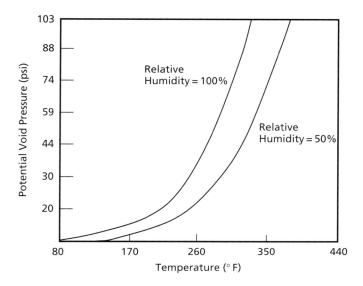

Fig. 7.33. Potential for Void Formation[16]

When the liquid resin viscosity dramatically increases, or gellation occurs, the voids are locked into the resin matrix. Note that the applied pressure on the laminate is not necessarily a factor. As will be shown later, the hydrostatic resin pressure can be low, even though the applied autoclave pressure is high, leading to void formation and growth.

Composite prepregs, like most organic materials, absorb moisture from the atmosphere. The amount of moisture absorbed is dependent on the relative humidity of the surrounding environment, while the rate of moisture absorption is dependent on the ambient temperature. While the carbon fibers themselves absorb minimal moisture, epoxy resins readily absorb moisture. Thus, the final prepreg moisture content is a function of the relative humidity, ambient temperature, and prepreg resin content.

Since moisture is typically the most predominate volatile present in hot melt addition curing prepregs, the amount of absorbed moisture in the prepreg determines the resultant vapor pressure of volatiles generated during the cure cycle. An examination of Fig. 7.33 explains why composite fabricators control the lay-up room environment, i.e. higher moisture contents result in higher vapor pressures, increasing the propensity for void formation and growth. Even though we are dealing with relatively small amounts of moisture (e.g., 1%), even these small percentages can generate large gas volumes and pressures, when heated. It should also be pointed out that other types of volatiles, such as solvents used in a solvent impregnation process or volatiles resulting from condensation curing reactions, can greatly complicate this problem, leading to much higher vapor pressures and a much greater propensity for void formation.

To fully understand void growth, an appreciation of the importance of hydrostatic resin pressure must be developed. Because of the load-carrying capability of the fiber bed in a composite lay-up, the hydrostatic resin pressure needed to suppress void formation and growth is typically only a fraction of the applied autoclave pressure.[16] The hydrostatic resin pressure is critical, because it is this pressure that helps to keep volatiles dissolved in solution. If the resin pressure drops below the volatile vapor pressure, then volatiles will come out of solution and form voids. To develop a better understanding of the interactions of the resin flow process, hydrostatic resin pressure and the load-carrying capability of the fiber bed, a mechanical analogy[17] is presented in Fig. 7.34. In this analogy, a laminate undergoing cure is simulated as a piston-spring-valve setup. The spring represents the fiber bed and is assumed to have a load carrying capability. Just like a spring, the fiber bed will support larger and larger loads as it undergoes compression. The liquid contained in the piston represents the ungelled resin. Finally, the valve is the means by which the liquid resin can leave the system, i.e. it could be representative of bleeder, a poorly dammed part, or any other system leak.

A description of each step in this simplified model is given below:

Step 1   Initially, there is no load on the system. The liquid hydrostatic pressure and the load carried by the fiber bed is zero.

Step 2   A 100 lb load is applied to the system, but no liquid has escaped (closed valve). The liquid carries the entire load and the load on the fiber bed is zero. Note that the downward force (in this case 100 lb) is equal to the upward force (again 100 lb). This upward force is the sum of the load carried by the liquid (100 lb) and the spring-like fiber bed (0 lb).

Step 3   The valve is now opened, allowing resin to escape (e.g., resin bleeding). However, at this point the resin still carries the entire 100 lb load.

Step 4   Liquid continues to escape, but at a decreasing rate due to a portion of the load now being carried by the spring (25 lb). This is analogous to bleeding in a laminate occurring rapidly until the fiber bed starts supporting a portion of the applied autoclave pressure.

Steps 5 and 6   Liquid continues to escape, but at an ever decreasing rate as a greater portion of the load is being carried by the spring. In an actual laminate, the rate of bleeding would be retarded both by the increasing fiber bed load-carrying capability and the reduced permeability of the fiber bed as it is compacted.

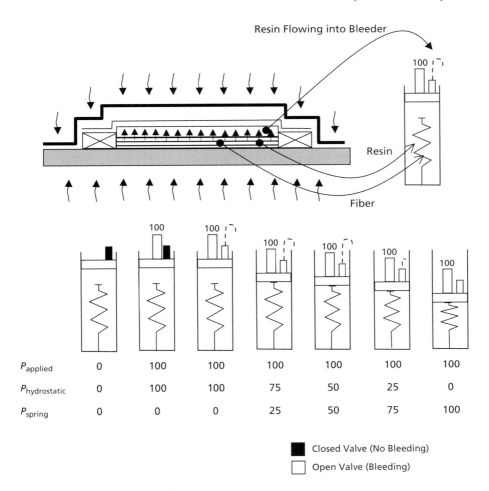

Fig. 7.34. Resin Flow Analogy[17]

Step 7   No further bleeding occurs because the pressure on the resin has now dropped to zero, and the entire load (100 lb) is being carried by the spring. If this condition occurs during actual autoclave processing before the resin gels (solidifies), it would be quite easy for dissolved volatiles to vaporize out of solution and form voids.

Although this analogy greatly simplifies the composite flow process, it does illustrate several key points. In the early stages of the cure cycle, the hydrostatic resin pressure should be equal to the applied autoclave pressure. As resin flow occurs, the resin pressure drops. If a laminate is severely overbled, the resin pressure could drop to a low enough value to allow void formation; therefore, the hydrostatic resin pressure is directly dependent on the amount of resin bleeding that occurs. As the amount of bleeding increases, the fiber volume increases, resulting in an increase in the load-carrying capability of the fiber bed. It should be noted that resin flow and bleeding can be intentional or unintentional. Intentional bleeding is, of course, the bleeder cloth used to remove the excess resin from the prepreg during cure. Examples of unintentional bleeding are excessive gaps between the dams and laminate, tears in the inner bag that allow resin to flow into the breather material, and mismatched tooling details that allow escape paths for the liquid resin. Therefore, the hydrostatic resin pressure is directly dependent on the amount of resin bleeding that occurs. As the amount of resin bleeding increases, the fiber volume increases, resulting in increase in the load-carrying capability of the fiber bed and a decrease in the hydrostatic resin pressure. Referring back to the cure cycle shown in Fig. 7.31, the second ramp portion of this cure cycle is critical from a void nucleation and growth standpoint. During this ramp, the temperature is high, the resin pressure can be near its minimum, and the volatile vapor pressure is high and rising with the temperature. These are the ideal conditions for void formation and growth.

Unfortunately, the void problem cannot be resolved simply by maintaining the hydrostatic resin pressure above the potential void pressure of the volatiles (although this is a good start). During collation or ply lay-up, air can become entrapped between the prepreg plies. The amount of air entrapped depends on many variables: the prepreg tack, the resin viscosity at room temperature, the degree of impregnation of the prepreg and its surface smoothness, the number of intermediate debulk cycles used during collation, and geometrical factors such as ply drop-offs, radii, and so forth. An obvious place where entrapped air pockets form is at the terminations of internal ply drop-offs. In addition, air can be entrained in the resin itself during the mixing and prepregging operations. This entrained air can also lead to voids, or at least serve as nucleation sites. To summarize, the void formation and growth process is complex and yet to be fully understood. However, a number of basic principles are fairly well understood and have been investigated through several research studies. Several of these studies are summarized in the next section.

7.9.3 Hydrostatic Resin Pressure

Considerable research[17] has been conducted to develop a better understanding of resin pressure and the many variables that influence resin pressure. The majority of this work has been done with unidirectional carbon/epoxy; however, woven and unidirectional carbon/bismaleimide have also been studied. The experimental setup used for these studies is shown in Fig. 7.35. To measure the hydrostatic resin pressure, a transducer was recessed into the tool surface and filled with an uncatalyzed liquid resin. To assure that the laminate did not deflect under pressure and contact the transducer, a stiff wire screen was placed over the transducer recess.

Polymer Matrix Composites

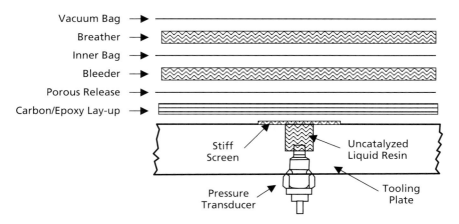

Fig. 7.35. *Resin Pressure Setup*[17]

Studies were initially conducted on unidirectional carbon/epoxy composites to evaluate the effects of (1) the type of epoxy resin (Hexcel's 3501-6 and 3502), (2) laminate bleeding (normal and overbleed), and (3) pressure application (normal applied autoclave and internally pressurized bag). In all cases, the lay-ups were cross-plied containing 0°, 90°, and ±45° plies.

These tests showed that:

(1) A high flow resin system experiences a larger pressure drop than a low flow system.

A comparison of the resin pressures for 3501-6 and 3502 (Fig. 7.36) shows that the higher flow 3502 resin system experiences a larger pressure drop than the lower flow 3501-6 system. The 3501-6 system is a lower flow system because it contains a boron trifluoride (BF_3) catalyst that significantly alters the cure behavior, resulting in a lower flow system that gels at a lower temperature. Since high flow resin systems are more prone to bleeding, additional care must be taken when they are tooled or bagged. High flow laminates should be tightly bagged and sealed to eliminate leak paths. Since high flow resin systems also typically have high gel temperatures, additional care must be taken to insure that the potential void pressure does not exceed the hydrostatic resin pressure prior to resin gellation.

(2) Overbleeding a laminate causes a large drop in resin pressure.

A comparison of normal bleeding with overbleeding is shown in Fig. 7.37. Overbleeding was accomplished by using three times the normal amount of glass bleeder and by removing the inner bag to create a free-bleeding condition. While this would normally not be done in composite

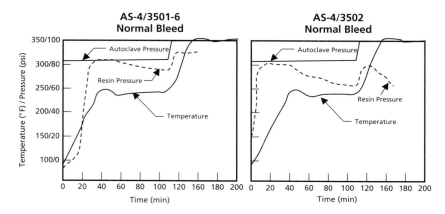

Fig. 7.36. Resin Flow Comparison[17]

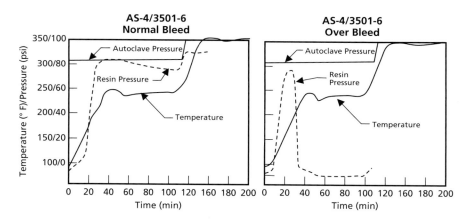

Fig. 7.37. Overbleeding Causes Pressure Loss[17]

part fabrication, overbleeding can, and does, occur because of leaky damming systems or leaky matched die molds. The analysis of these cured laminates included non-destructive testing (NDT), thickness measurements, resin content determinations, and materiallographic cross-sections. The resin contents and thickness measurements showed the dramatic effect of overbleeding. The resin contents and thickness values of the overbled laminate were significantly lower than those for standard bled laminate. Since the resin pressure actually dropped to below 0 psi due to the vacuum pulled underneath the bag, little resistance to void growth existed. As expected, the ultrasonic NDT results and materiallographic cross-sections revealed gross voids and porosity in the overbled laminate.

(3) Internal bag pressure can be used to maintain hydrostatic resin pressure and reduce resin flow.

Internally Pressurized Bag (IPB) curing was originally developed during the Reference 18 program. In this process, two separate pressure sources are used: (1) a normal external applied autoclave pressure, which provides the ply compaction, and (2) a somewhat lesser internal bag pressure, which applies hydrostatic pressure directly to the liquid resin to keep volatiles in solution and thereby prevent void nucleation and growth. An autoclave setup for IPB curing is shown in (Fig. 7.38). In the cure cycle used for this experiment (Fig. 7.39), an external applied autoclave pressure of 100 psi was used along with an internal bag pressure of 70 psi. This results in a compaction, or membrane, pressure of 30 psi (applied autoclave pressure−internal bag pressure = 100 psi − 70 psi) on the plies and a hydrostatic pressure of 70 psi minimum on the resin. There is nothing magical about these pressure selections. If desired, the compaction pressure could be raised back up to 100 psi by simply increasing the applied autoclave pressure to 170 psi. The only restriction is that the applied autoclave pressure must be greater than the internal bag pressure, to prevent blowing the bag off of the tool. Of course, the internal bag pressure needs to be high enough to keep the volatiles in solution to prevent void nucleation and growth. To test a worst-case condition, the IPB laminate was bagged in the same manner as the previous overbleed laminate in which the resin pressure had dropped to zero. Even though this

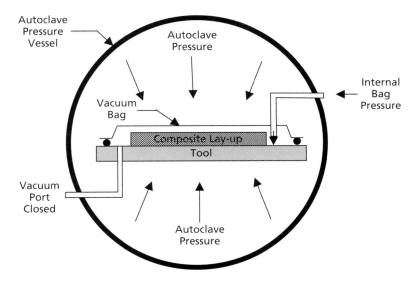

Fig. 7.38. Schematic of Internally Pressurized Bag Curing[17]

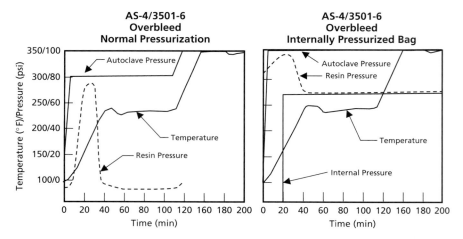

Fig. 7.39. IPB Maintains Resin Pressure[17]

bagging procedure resulted in a severely overbled and porous laminate in the previous test, the addition of internal bag pressure prevented both overbleeding and porosity. The absence of overbleeding was a result of the lower membrane pressure (30 psi for the IPB cure vs. 100 psi for the normal cure).

7.9.4 Resin and Prepreg Variables

The resin mixing and prepregging operations can also influence the processability of the final prepreg. During normal mixing operations, air can easily be mixed into the resin. This entrained air can later serve as nucleation sites for voids and porosity. However, some mixing vessels are equipped with seals that allow vacuum degassing during the mixing operation, a practice that has been found to be effective in removing entrained air and may be beneficial in producing superior quality laminates.

Prepreg physical properties can also influence final laminate quality. Prepreg tack is one such property. Prepreg tack is a measure of the stickiness, or self-adhesive nature, of the prepreg plies. Many times, prepregs with a high tack level have resulted in laminates with severe voids and porosity. This could be due to the potential difficulty of removing entrapped air pockets during collation with tacky prepreg. Again, moisture can be a factor. Prepregs with a high moisture content have been found to be inherently tackier than low moisture content material. Previous work[18] has indicated a possible correlation between prepreg tack and resin viscosity, i.e. prepregs that are extremely tacky also have high initial resin viscosities. Resins with such high viscosities will be less likely to cold flow and eliminate voids at ply terminations.

Prepreg physical quality can greatly influence final laminate quality. Ironically, prepreg that appears "good" (i.e., smooth and well impregnated) may not necessarily produce the best laminates. Several material suppliers have determined that only partially impregnating the fibers during prepregging results in a prepreg that consistently yields high quality parts, whereas a "good" (i.e., smooth and well impregnated) prepreg can result in laminates with voids and porosity. Partially impregnated prepregs have the same resin content and fiber areal weight as the fully impregnated material. The only difference is the placement of the resin with respect to the fibers. The partial impregnation process provides an evacuation path for air, and low temperature volatiles, entrapped in the lay-up. As the resin melts and flows, full impregnation occurs during cure. A closely related phenomenon is the surface condition of the prepreg. A fully impregnated prepreg will not cause a problem if a surface has the impressions of the fibers (sometimes called a corduroy texture), again providing an evacuation path. These three prepreg conditions are summarized in the highly idealized schematic shown in Fig. 7.40.

7.9.5 Condensation Curing Systems

The chemical composition of a thermoset resin system can dramatically affect volatile evolution, resin flow, and reaction kinetics. Addition curing polymers, in which no reaction by-products are given off during crosslinking, are, in general, much easier to process than condensation systems. Because of the reaction by-products and solvents that evolve during processing, condensation systems, such as phenolics and polyimides, are extremely difficult to process without voids and porosity.

Condensation curing systems, such a polyimides and phenolics, give off water and alcohols as part of their chemical crosslinking reactions. In addition, to allow prepregging, the polymer reactants are often dissolved in high temperature boiling point solvents, such as DMF (dimethylformamide), DMAC (dimethylactamide), NMP (N-methylpyrrolidone), or DMSO (dimethylsufoxide). Even the addition curing polyimide PMR-15 uses methanol as a solvent for prepregging.

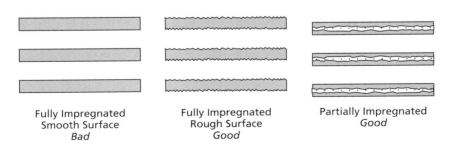

Fig. 7.40. Effects of Prepreg Physical Quality[17]

The eventual evolution of these volatiles during cure creates a major volatiles management problem, which can result in high void and porosity percentages in the cured part. Unless a heated platen press or a hydroclave with extremely high pressures (e.g., 1000 psi) is used to keep the volatiles in solution until gellation, they must be removed either before the cure cycle, or during cure heat-up when the resin viscosity is low. In addition, since these materials boil or condensate at different temperatures during heat-up, it is important to know the point(s) during the cycle when the different species will evolve.

There are three strategies for volatile management: (1) use a press or a hydroclave with an applied hydrostatic resin pressure greater than the volatile vapor pressure to keep the volatiles in solution until the resin gels; (2) remove the volatiles by laying-up only a few plies at a time and hot debulking under vacuum bag pressure at a temperature higher than the volatile boiling point; or (3) use slow heat-up rates and vacuum pressure during cure, with intermediate holds, to remove the volatiles before resin gellation. It should be noted that more than one of these strategies can be used at the same time. The advantage of a heated platen press or hydroclave is that high pressures can be used to suppress volatile evolution. However, the tooling must be designed to withstand the higher pressures, and special damming systems must be incorporated to prevent excessive resin squeeze out. The second method, intermediate hot debulks under vacuum pressure, is effective but is very costly and labor intensive, since the ply collation operation has to be interrupted every several plies; the part bagged and moved to an oven; hot debulked; and then cooled before further collation. The last method, as shown in Fig. 7.41 for a typical autoclave cure cycle for PMR-15, incorporates multiple holds under vacuum during heat-up to evacuate the volatiles during various points in the cure cycle. It should be noted that some manufacturers use a 600° F cure and post-cure rather than the 575° F shown in the figure. In addition, some use only a partial vacuum during the early stages of cure, and apply a full vacuum in the latter stages. Although the final cure of PMR-15 is an addition reaction, it undergoes condensation reactions early in the cure cycle during the imidization stage that creates a volatile management problem. The tricky part to this approach is determining the optimum times and temperatures for the hold periods, and the heat-up rates to use. Physiochemical test methods can be used in helping to design these cure cycles. To obtain full crosslinking, polyimides often require extended post-cure cycles. Note that even the post-cure cycle incorporates multiple hold periods during heat-up to help minimize residual stress build-up and thus reduce the likelihood of matrix microcracking.

7.9.6 Residual Curing Stresses

Residual stresses develop during the elevated temperature cure of composite parts. They can result either in physical warpage, or distortion, of the part

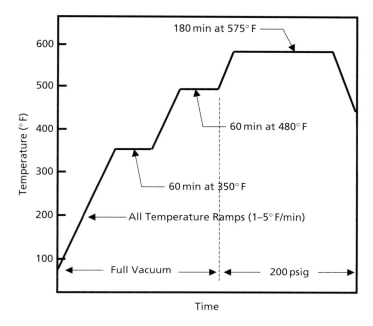

Fig. 7.41. Typical PMR-15 Cure Cycle[19]

(particularly thin parts) or in matrix microcracking either immediately after cure or during service. Distortion and warpage causes problems during assembly and is more troublesome for composite parts than metallic ones. While the distortion in thin sheet metal parts can often be pulled-out during assembly, composite parts run the danger of cracking, and even delamination, if they are stressed during assembly. Microcracking is known to result in degradation of the mechanical properties of the laminate, including the moduli, Poisson's ratio, and the CTE.[20] Microcracking (Fig. 7.42) can also induce secondary forms of damage, such as delaminations, fiber breakage, and the creation of pathways for the ingression of moisture and other fluids. Such damage modes have been known to result in premature laminate failure.[21]

The major cause of residual stresses in composite parts is due to the thermal mismatch between the fibers and the resin matrix. Recalling that the residual stress on a simple constrained bar is:

$$\sigma = \alpha E \Delta T$$

where

σ = Residual stress
α = Coefficient of thermal expansion (CTE)

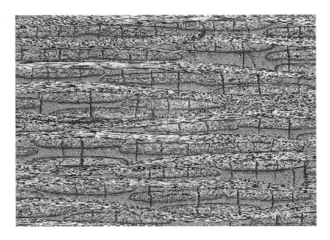

Fig. 7.42. Matrix Microcracking

E = Modulus of elasticity
ΔT = Temperature change

A rather simplified analogy for a composite part is that the CTE difference between the fibers (≈ 0 for carbon fiber) and the resin is large (≈ 20–$35 \times 10^{-6}/°$F for thermoset resins). The modulus difference between the fibers (30–140 msi) and the resin (0.5 msi) is also large. The temperature difference (ΔT) is the difference from when the resin becomes a solid gel during cure and the use temperature. The so-called "stress free temperature" is somewhere between the gel temperature and the final cure temperature, as the crosslinking structure develops strength and rigidity. The use temperature for epoxy composites usually ranges anywhere from -67 to $250°$F.

There are several observations we can make from this simplified analogy. High modulus carbon, graphite, and aramid fibers have negative CTEs. Normally, the higher the fiber modulus, the more negative the CTE becomes, which leads to increases in residual stresses and helps to explain why more matrix microcracking is observed with high modulus graphite fibers than with high strength carbon fibers. Carbon/epoxy resin systems are usually cured at either 250 or 350°F. Since there will be a smaller ΔT for the systems cured at 250°F, they should experience less microcracking than the systems cured at 350°F. Very high temperature polyimides, and many thermoplastic resins, that are often cured or processed at temperatures in the range of 600–700°F develop very high residual stresses and are very susceptible to microcracking. Since the ΔT differential becomes larger when the use temperature is lowered, for example, when the temperature is -40 to $-67°$F for a cruising airliner at 30 000–40 000 ft, more microcracking is normally observed after cold exposures than elevated temperature exposures. The analogy presented above greatly oversimplifies the residual stress problem in composite structures. In fact, analysis of residual stresses in composites is probably one of the most complex problems analysts have tried to address. There is quite a bit of conflicting data in the literature over the various causes of residual stresses and the effects of material, lay-up, tooling, and processing variables on residual stresses.

While residual stresses in composites are extremely complicated and there is considerable conflicting data on the effects of different variables, the following guidelines are offered for minimizing their effects:

- Use only balanced and symmetric laminates. Minimize ply lay-up misorientation or distortion whenever possible.
- Design tools with compensation factors to account for thermal growth and angular spring-in. The use of low CTE tools will probably help to minimize residual stresses when curing carbon fiber composites.
- The use of lower modulus fibers and tougher resin systems helps to minimize residual stresses and microcracking.
- Slow heat-up rates during cure with intermediate holds and lower curing temperatures probably helps in minimizing residual stresses by balancing the rate of chemical resin shrinkage with the rate of thermal expansion. Likewise, there is some evidence that slow cool-down rates help.

7.10 Liquid Molding

Liquid molding is a composite fabrication process that is capable of fabricating extremely complex and accurate dimensionally parts. One of the main advantages of liquid molding is part count reductions, in which a number of parts that would normally be made individually, and either fastened or bonded together, are integrated into a single molded part. Another advantage is the ability to incorporate molded-in features, such as a sandwich core section in the interior of a liquid molded part.

Resin transfer molding (RTM), the most widely used of the liquid molding processes, is a matched mold process that is well suited to fabricating three-dimensional structures requiring tight dimensional tolerances on several surfaces. Excellent surface finishes are possible, mirroring the surface finish of the tool. The major limitation of the RTM process is the relatively high initial investment in the matched-die tooling. Sufficient part quantities, usually in the 100–5000 range, are necessary to justify the high non-recurring cost of the tooling. A summary of the advantages and disadvantages of the RTM process are given in Table 7.5.

The RTM process consists of fabricating a dry fiber preform which is placed in a closed mold, impregnated with a resin, and then cured in the mold. The

Table 7.5 RTM Process Advantages and Disadvantages[22]

Advantages	Disadvantages
• Best tolerance control-tooling controls dimensions • Class A surface finish possible • Surfaces may be gel coated for better surface finish • Cycle times can be very short • Molded-in inserts, fittings, ribs, bosses, and reinforcements possible • Low pressure operation (usually less than 100 psi) • Prototype tooling costs relatively low • Volatile emissions (e.g., styrene) controlled by close mold process • Lower labor intensity and skill levels • Considerable design flexibility: reinforcements, lay-up sequence, core materials, and mixed materials • Mechanical properties comparable to autoclave parts (void content < 1%) • Part size range and complexity makes RTM appealing • Smooth finish on both surfaces • Near net molded parts.	• Mold and tool design critical to part quality • Tooling costs can be high for large production runs • Mold filling permeability based on limited permeability data base • Mold filling software still in development stages • Preform and reinforcement alignment in mold is critical • Production quantities typically range from 100–5000 parts • Requires matched, leakproof molds.

basic resin transfer molding process, shown in Fig. 7.43, consists of the following steps:

- Fabricate a dry composite preform.
- Place the preform in a closed mold.
- Inject the preform with a low viscosity liquid resin under pressure.
- Cure the part at elevated temperature in the closed mold under pressure.
- Demold and clean up the cured part.

Over the past several years, there have been many variations developed for this process, including RFI (Resin Film Infusion), VARTM (Vacuum Assisted Resin Transfer Molding) and SCRIMP (Seeman's Composite Resin Infusion Molding Process), to name a few. The objective of all of these processes is to fabricate near net molded composite parts at low cost.

7.11 Preform Technology

The most important types of preforms for liquid molding processes are (1) woven, (2) knitted, (3) stitched, and (4) braided. In many cases, conventional textile machinery has been modified to handle the high modulus fibers needed in structural applications and to reduce costs through automation. In addition, to

Polymer Matrix Composites

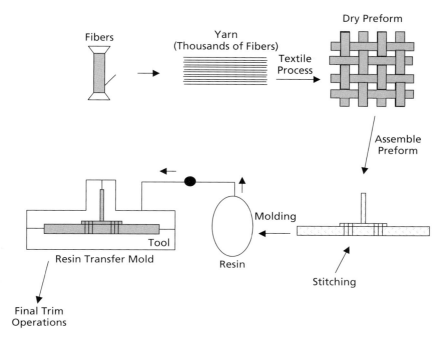

Fig. 7.43. *Process Flow for Resin Transfer Molding (RTM)*[23]

meet the growing demand for three-dimensional reinforced preforms, manufacturers have developed specialized machinery. NASA Langley Research Center has led much of the development of textile perform technology for aerospace applications. An excellent overview of their work is available.[24,25]

7.11.1 Fibers

Textile machines have been adapted to handle most of the fibers commonly used in structural composites, including glass, quartz, aramid, and carbon. The main limitation is that most textile processes subject yarns to bending and abrasion. Although machines have been modified to minimize fiber damage, in many processes, exceptionally brittle or stiff fibers will suffer significant strength degradation. In general, the higher the modulus of the fiber, the harder it will be to process, and the more prone it is to damage. Strength reductions can vary, depending upon the property being measured, and the textile process used to fabricate the preform. Polymeric sizings are usually applied to fibers to improve their handling characteristics and minimize strength degradation during processing. The sizings may be removed after processing, or left on the fibers for the lamination process. If the sizing remains on the fibers, it is important that it be compatible with the matrix resin. A surface treatment is also generally used to improve the adhesion between the fibers and the matrix.

In traditional textile processes, yarns are usually twisted to improve handling, structural integrity, and their ability to hold shape. However, twist reduces the axial strength and stiffness of the fibers, which is paramount in structural applications. Therefore, yarns with minimal or nominally zero twist (strands and tows) are preferred. Different processes and weaves require different strand or tow sizes. In general, the smaller the tow size, the more expensive the material will be on a per pound basis, particularly for carbon fiber.

7.11.2 Woven Fabrics

Woven fabrics are available as two-dimensional (2-D) reinforcements (x- and y-directions) or three-dimensional (3-D) reinforcements (x-, y-, and z-directions). When high in-plane stiffness and strength are required, 2-D woven reinforcements are used. As pointed out earlier, 2-D woven products can be supplied as either a prepreg or as a dry cloth for either hand lay-up, preforming, or repair applications. Two-dimensional weaves have the following advantages: (1) they can be accurately cut using automated ply cutters; (2) complicated lay-ups with ply drop-offs are possible; (3) there are a wide variety of fibers, tow sizes and weaves that are commercially available; and (4) 2-D weaves are more amendable to thinner structures than 3-D weaves.

Three-dimensional reinforced fabrics are normally used to (1) improve the handleability of the preform, (2) improve the delamination resistance of the composite structure, or (3) carry a significant portion of the load in the composite structure, such as a composite fitting that would be subject to complex load paths and major out-of-plane loading. If improved handlablity is the objective, usually z-direction fiber volumes as low as 1–2% will suffice. Improvements in the delamination resistance of composite structures can be obtained with as little as 3–5% z-direction fiber; however, as the amount of z-directional fiber is increased, the delamination resistance and durability increases.[26] If the application calls for major out-of-plane loading, as much as 33% z-direction fiber reinforcement may be required. The fibers will then be arranged with roughly equal load-bearing capacity along all three axes of a Cartesian coordinate system. However, it should be recognized that when the volume of reinforcement in the z-direction is increased, the volume percentages in the x- and y-directions will be decreased.

Historically, composite designs have been restricted to structure that experiences primarily in-plane loading, such as fuselages or wing skins. One of the key reasons for the lack of composite structures in complex substructure, such as bulkheads or fittings, is the inability of 2-D composites to handle complex, out-of-plane loads effectively. The planar load-carrying capability of composites is primarily a fiber-dominated property, requiring stiff and well-defined load paths. Unfortunately, the planar loads must ultimately be transferred through a 3-D joint into adjacent structure (e.g., a skin attached to a bulkhead). These 3-D joints are subject to high shear, out-of-plane tension, and out-of-plane bending loads, all of which are matrix property dominated properties in a traditional composite design. Since loading the matrix with large primary loads is a totally unacceptable design practice, metallic fittings are used to attach composite structure to metallic bulkheads with mechanical fasteners. With the introduction of high performance 3-D textiles, both woven and braided, this barrier to composite designs has the potential to be eliminated. While 3-D woven preforms show potential for being able to fabricate complex net-shaped preforms, the set-up time is extensive and the weaving process is slow. In addition, small tow sizes that generally increase cost must be used to achieve high fiber volume percents, and eliminate large resin pockets, that are susceptible to matrix microcracking during cure, or later when the part is placed in-service.

7.11.3 Multiaxial Warp Knits

Knitting can be effectively used to produce multiaxial warp knits (MWKs), also called stitch bonding, that combines the mechanical property advantages of unidirectional tape with the handling advantages and low cost fabrication advantages of fabrics. MWKs, as shown in Fig. 7.44, consist of unidirectional tows of strong, stiff fibers woven together with fine yarns of glass or polyester thread. The glass or polyester threads, which normally amounts to only 2% of the total weight, serve mainly to hold the unidirectional tows together during subsequent handling. An advantage of this process is that the x- and y-tows remain straight and do not suffer as much strength degradation as woven materials, in which the tows are crimped during the weaving process.

The multiaxial warp knit process is used to tie tows of unidirectional fibers together in layers with 0°, 90°, and $\pm\theta°$ orientations. During knitting, the polyester threads are passed around the primary yarns, and one another, in interpenetrating loops. Selecting the tow percentage in each of the orientations can be used to tailor the mechanical properties of the resulting stack. The MWK stacks form building blocks 2–9 layers thick that can be laminated to form the thickness desired the structure. Multiple layers of MWKs are often stitched together in a secondary operation to form stacks of any desired thickness and can be stacked, folded, and stitched into net shapes. The stitching operation also greatly improves the durability and damage tolerance of the cured composite. MWK has the advantages of being fairly low cost, having uniform thicknesses, can be capable of being ordered in prefabricated blanket-like preforms, and being very amendable to gentle or no contour parts, such as large skins.

7.11.4 Stitching

Stitching has been used for more than 20 years to provide through-the-thickness reinforcement in composite structures, primarily to improve damage tolerance.

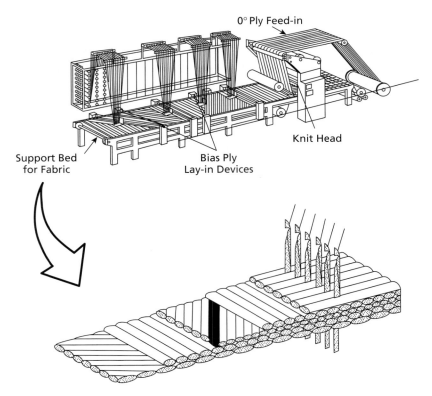

Fig. 7.44. Multi-axial Warp Knitting Machine and Typical Product Form Produced[24]

The major manufacturing advancement in recent years has been the introduction of liquid molding processes which allows stitching of dry preforms (Fig. 7.45), rather than prepreg material. This enhances speed, allows stitching through thicker material, and greatly reduces damage to the in-plane fibers. As well as enhancing damage tolerance, stitching also aids fabrication. Stitching provides a mechanical connection between the preform elements before the resin is introduced, allowing the completed preform to be handled without shifting or damage. In addition, stitching compacts (debulks) the fiber preform closer to the final desired thickness. Therefore, less mechanical compaction needs to be applied to the preform in the tool. Various stitching materials have been successfully used, including carbon, glass, and aramid, with Kevlar 29 (aramid) being the most popular. Yarn weights for Kevlar of between 800 and 2000 denier have been used. However, one disadvantage of aramid is that it absorbs moisture and can sometimes exhibit leaks through the skin at the stitch locations.

Polymer Matrix Composites

Fig. 7.45. Stitching Complex Dry Preform[24]

7.11.5 Braiding

Braiding is a commercial textile process dating from the early 1800s. In braiding, shown in Fig. 7.46, a mandrel is fed through the center of the machine at a uniform rate, and fiber yarns from moving carriers on the machine braid over the mandrel at a controlled rate. The carriers work in pairs to accomplish an

Fig. 7.46. Braiding Composite Preform[24]

333

over/under braiding sequence. Two or more systems of yarns are intertwined in the bias direction to form an integrated structure. Braided preforms are known for their high level of conformity, torsional stability, and damage resistance. Either dry yarns or prepregged tows can be braided, with typical fibers including glass, aramid, and carbon. Braiding normally produces parts with lower fiber volume fractions than filament winding but is much more amendable to intricate shapes. The mandrel can vary in cross-section with the braided fabric conforming to the mandrel shape. The total thickness of a braided part can be controlled by overbraiding, in which multiple passes of the mandrel are made through the braiding machine, laying down a series of nearly identical layers, similar to a lay-up. Possible fiber orientations are $\pm\theta°$ or $0°/\pm\theta°$ with no 90° layers unless the braider is fitted with filament winding capability. State-of-the-art braiding equipment provides full control over all of the braiding parameters, including translational and rotational control of the mandrel, vision systems for in-process inspection, laser projection systems to check braid accuracy, and even integrated circumferential filament winding.

Due to the material conformity inherent in a braided product form, braided "socks" can be removed from the braiding mandrel and formed over a mandrel of a different shape for curing. Cutting the cylindrical sheet from the mandrel and stretching it out flat can form a flat braided sheet. In other situations, the braided part is cured directly on the mandrel. Permanent, water soluble, or breakout mandrels can be used. Fixed, straight axial yarns (0°) can also be introduced at the center of orbit of the braider yarn carriers. The braider yarns lock the axial yarns into the fabric, forming a triaxial braid, i.e. a braid reinforced in three in-plane directions.

Three-dimensional braiding can produce thick, net section preforms, in which the yarns are so intertwined that there may be no distinct layers. 3-D braided socks can also be shaped into a preform suitable for use in joints and stiffeners. Due to the nature of braiding, a bias or 45° fiber orientation is inherent in the braided preform and, theoretically, allows the preform to carry high shear loads without the necessity of 45° hand layed-up overwrap plies. A 3-D braiding machine can be set-up to produce a near net shape to the cross-section of the final part. The disadvantages of 3-D braiding are similar to 3-D weaving, i.e. complicated set-ups and slow throughput. Again, resin microcracking can be a problem with maximum fiber volumes of 45–50% obtainable.

7.11.6 Preform Handling

Since the stiffness and strength of polymeric composites are dominated by the reinforcing fibers, maintaining accurate positioning of the fibers during all steps of manufacturing process is paramount. Poor handling and processing after preforming can destroy fiber uniformity. Uncontrolled material handling, draping the material over curved tools, debulking, and tool closure can spread or distort the fibers. Manufacturing prove-out parts should be examined to establish that the minimum fiber volume fractions have been obtained, with particular attention paid to geometric details such as joints. The problem of maintaining the desired fiber content is most challenging when fabrics are draped. The draping characteristics of a fabric over a singly curved surface are a direct function of the shear flexibility of the weave. Satin weaves have fewer crossover points than plain weaves, and have lower shear rigidity, and are therefore more easily draped. Draping over a complex compound contour also depends on the in-plane extensibility and compressibility. This is difficult for fabrics containing high volume fractions of more or less straight in-plane fibers, as required for most structural applications. For these products, only mild double curvature can be accommodated by draping without a significant loss of fiber regularity. However, compound contours can be achieved through net shape processes, such as braiding onto a mandrel, thus avoiding the problems of draping.

There are several reasons that preforming is conducted prior to the injection process. First, preforming does not tie-up the expensive matched die tool, i.e. the tool can be used to cure parts while the preforming operations are done ahead of time and off-line. Second, a well-constructed preform will be rather stiff and rigid as opposed to laying-up loose fabric directly into the mold. Therefore, preforming improves the fiber alignment of the resultant part and reduces part-to-part variability.

Planar fabric preforms can be stitched together or held together with a tackifier. A tackifier is usually an uncatalyzed thermoset resin that is applied as a thin veil, a solvent spray, or a powder. Veils can be placed between adjacent plies of fabric, followed by fusing the ply stacks with heat and pressure, to form the preform. Tackifiers can also be thinned with a solvent and then sprayed on the fabric plies. A third method is to apply powders to the surface followed by heating to melt the powder and allowing it to impregnate the fabric. Tackified fabric can be thought of as a low resin content prepreg (usually in the range of 4–6%) that can be made into ply kits using conventional automated broadgoods cutting equipment. It is important to keep the tackifier content as low as possible, because it reduces the permeability of the preform and makes resin filling more difficult.[27] It is also important that the tackifier and the resin to be injected are chemically compatible, preferably the same base resin system.[28] Once the tackifier has been applied to the fabric layers, they are formed to the desired shape on a low cost preforming tool, and then heat-set by heating to approximately 200°F for 30–60 s. The compaction behavior of a preform depends on the preform method used, the type of reinforcement, the tackifier used, the compaction pressure, and the compaction temperature. A tackifier can act as a lubricant and increase compaction, but this will also decrease the preform permeability and make injection more difficult. For any preform construction, it is important that the preform be dried prior to resin injection to remove all surface moisture that may have condensed on the surface from the atmosphere.

7.12 Resin Injection

Resin injection follows Darcy's law of flow through a porous media, that predicts that the flow rate per unit area (Q/A) is proportional to the preform permeability (k) and the pressure gradient (ΔP), and inversely proportional to the viscosity (η) of the resin and the flow length (L):

$$\frac{Q}{A} = \frac{k \Delta P}{\eta L}$$

Therefore, for a short injection time (high Q/A), one would want a preform with a high permeability (k), a high pressure (ΔP), a low resin viscosity (η), and a short flow length (L). Using this equation can provide useful guidelines for RTM: (1) use resins with low viscosity; (2) use higher pressures for faster injections; and (3) use multiple injection ports and vents for faster injections.

The ideal resin for RTM will have (1) a low viscosity to allow flow through the mold and complete impregnation of the fiber preform; (2) a sufficient pot life where the viscosity is low enough to allow complete injection at reasonable pressures; (3) a low volatile content to minimize the occurrence of voids and porosity; and (4) a reasonable cure time and temperature to produce a fully cured part.

Resin viscosity is a major consideration when selecting a resin system for RTM. Low viscosity resins are desirable with an ideal range being in the 100–300 cP range with about 500 cP being the upper limit. Although resins with higher viscosities have been successfully injected, high injection pressures or temperature are required, which results in more massive tools to prevent tool deflection. Normally, the resin is mixed and catalyzed before it is injected into the mold, or if the resin is a solid at room temperature with a latent curing agent, it must melted by heating. Vacuum degassing in the injection pot (Fig. 7.47) is a good practice to remove entrained air from mixing and low boiling point volatiles. Both epoxies and bismaleimides are amenable to RTM, with preformulated resins available from a large number of suppliers. Similar to prepreg resins, it is important to understand the resin viscosity and cure kinetics of any resin used for RTM.

Although resin injection pressures can range from vacuum only up to 400–500 psi applied pressure, applied pressures are normally 100 psi or lower. Although high pressures are often needed to fully impregnate the preform, the higher the injection pressure, the greater the chance of preform migration, i.e. the pressure front can actually cause the dry preform to migrate and move out of its desired location. Resin transfer molding dies are normally either designed so that they are stiff enough to react the injection pressures or they may be placed in a platen press under pressure to react the injection pressure. As a rule of thumb, the higher the injection pressure, the higher the tooling cost. Heating the resin or tool prior to, or during, injection can be used to reduce the viscosity but

Polymer Matrix Composites

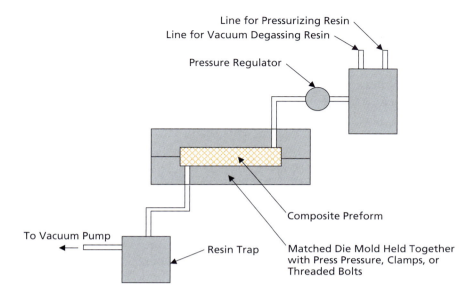

Fig. 7.47. Schematic of a Typical RTM Process[1]

will also reduce the working or pot life of the resin. A vacuum is also frequently used during the injection process to remove entrapped air from the preform and mold. The vacuum also helps to pull resin into the mold and preform, helps to remove moisture and volatiles, and aids in reducing voids and porosity. It has been reported that the use of a vacuum is a significant variable in improving product quality by reducing the occurrence of voids and porosity.[29]

The time it takes for the resin to fill the mold is a function of the resin viscosity, the permeability of the fiber preform, the injection pressure, the number and location of the injection ports, and the size of the part. The injection strategy usually consists of one of three main types: (1) point injection, (2) edge injection, or (3) peripheral injection. Point injection is usually done by injecting at the center of the part and allowing the resin to flow radially into the reinforcement, as air is vented along the part periphery. Edge injection consists of injecting the resin at one end of the part and allowing the resin to flow unidirectionally down the length, as air is vented at the opposite end. Finally, in peripheral injection, the resin is injected into a channel around the part and the flow is radially inward, as air is vented at the center of the part. Also, the locations of the injection and venting ports are important considerations in the ability to effectively achieve complete filling without entrapped air pockets or unimpregnated dry spots. Although there are several ways that the time to fill the mold can be reduced, such as using lower viscosity resins or higher injection pressures, the most effective method is to design an injection and porting system that minimizes the distance the resin has to flow. However, in designing an injection and porting system, the most important consideration is to have a system that will minimize any entrapped air pockets, as these will result in dry unimpregnated areas in the cured part. In peripheral injection, a phenomena known as "race tracking" can occur in which the resin runs around the peripheral injection channel, and then migrates inward, but traps air pockets resulting in dry spots. This can usually be avoided by the judicious selection of the location and number of the porting vents.

Vacuum assistance during injection will usually help to reduce the void content significantly. However, it is important that the mold be vacuum tight (sealed) if vacuum assistance is going to be used. If the mold leaks, air will actually be sucked into the mold, causing a potentially higher void content. During the injection process, when the mold is almost full, resin will start flowing out through the porting system. If there is evidence of bubbles in the exiting resin, the resin should be allowed to continue to bleed out until the bubbles disappear. To further reduce the possibility of voids and porosity, once the injection is complete, the ports can be sealed, while the pumping system is allowed to build-up hydrostatic resin pressure within the mold.

7.12.1 RTM Curing

Curing can be accomplished using several methods:

- matched die molds with integral heaters: electric, hot water, or hot oil;
- matched die molds placed in an oven;
- matched die molds placed between a heated platen press that provides the heat and reaction pressure on the mold; and
- for liquid molding processes that use vacuum injection only, such as VARTM and SCRIMP, a single sided tool with only a vacuum bag is used for pressure application. In this case, heat can be provided by integral heaters, ovens, or even heat lamps.

As opposed to autoclave curing, where the operator can control the variables time, temperature, and pressure (t, T, P), in RTM the P variable is often predetermined by the pressure applied to the resin during the injection process, or in VARTM, it is limited to the pressure that can be developed by a vacuum (≤ 14.7 psia). In some match mold applications, the vent ports can be sealed off and pressure can continue to be applied by the pump. To improve productivity, RTM parts are frequently cured in their molds, demolded, and then given free-standing post-cures in ovens.

7.12.2 RTM Tooling

Tooling is probably the single most important variable in the RTM process. A properly designed and built mold will normally yield a good part, while a poorly designed or fabricated mold will almost certainly produce a deficient part. Conventional RTM tooling consists of matched molds, usually machined from tool steel. Steel dies yield long lives for large production runs and are resistant to handling damage. The dies are usually blended and buffed to a fine surface finish that will yield good surface finishes on the RTM part. Many matched metal molds are built with sufficient rigidity that they do not need to be placed in a platen press during injection and cure to react the resin injection pressures. Since these molds necessarily become extremely heavy, attachment fittings are built into the mold to provide hoisting capability for cranes. They are held together with a system of heavy bolts and are often designed with internal ports for heating with hot water or oil. Hot water heaters are effective to about 280° F. Above that temperature, hot oil must be used. Electric heaters can also be placed within the mold, but are generally less reliable than hot oil because of the maintenance problems of replacing burned-out heaters. RTM molds can also be placed in convectively heated ovens, but for large tools the heat-up rates will be extremely slow.

Steel-matched metal molds have two disadvantages: (1) they are expensive, and (2) the heat-up and cool-down rates are slow. Matched metal molds have also been fabricated from Invar 42 to match the coefficient of thermal expansion of carbon composites, and from aluminum because it is easier to machine (less costly), has a high coefficient of thermal expansion which can be useful in some applications, but is much more prone to wear and damage than steel or Invar. For prototype and short production runs, matched molds can be made of high temperature resins that are frequently reinforced with glass or carbon fibers. Prototype dies can be NC machined directly from mass cast blocks laminated on a master model and finished by NC machining the surface.

Much lighter weight and less expensive tooling is a distinct advantage of processes that use only vacuum pressure for injection and cure, such as the VARTM process. In fact, most of these processes use single-sided hard tooling on one side and a vacuum bag on the other side. A porous media is almost always used on top of the fiber preform to aid in resin filling during injection.

7.13 Vacuum Assisted Resin Transfer Molding

Since VARTM processes use only vacuum pressure for both injection and cure, the single biggest advantage of VARTM is that the tooling cost is much less, and simpler to design, than for conventional RTM. In addition, since an autoclave is not required for curing, the potential exists to make very large structures using the VARTM process. Also, since much lower pressures are used in VARTM processes, lightweight foam cores can easily be incorporated into the lay-ups. VARTM type processes have been used for many years to build fiberglass boat hulls, but have only recently attracted the attention of the aerospace industry.

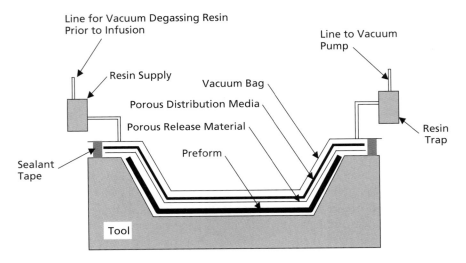

Fig. 7.48. Typical VARTM Process Setup[1]

A typical VARTM process, shown in Fig. 7.48, consists of single-sided tooling with a vacuum bag. VARTM processes normally use some type of porous media on top of the preform to facilitate resin distribution. The porous distribution media should be a highly permeable material that allows resin to flow through the material with ease. When a porous distribution media is used, the resin typically flows through the distribution media and then migrates down into the preform. Typical distribution media include nylon screens and knitted polypropylene. Since resin infiltration is in the through-the-thickness direction, race tracking and resin leakage around the preform are largely eliminated.[28]

Since the VARTM process uses only vacuum pressure for both injection and cure, autoclaves are not required and very large part sizes can be made. Ovens and integrally heated tools are normally used, and since the pressures are low (i.e., ≤ 14.7 psia), low-cost lightweight tools can be used. Some manufacturers use double vacuum bags to minimize variations in compaction pressure, and guard against potential vacuum leaks in the primary vacuum bag. A layer of breather between the two bags increases the ability to remove any air from leak locations. Reusable vacuum bags can also be used to reduce the cost of bagging complex shapes.

The resins used for VARTM processing should have even a lower viscosity than those used for traditional RTM. Resin viscosities less than 100 cP are desirable to give the flow needed to impregnate the preform at vacuum pressure. Vacuum degassing prior to infusion is normally used to help remove entrained air from the mixing operation. Some resins may be infused at room temperature, while others require heating. It is desirable to keep the resin source and vacuum trap away from the heated tool. This makes it easier to control the temperature of the resin at the source and minimizes the chance of an exotherm at the trap.

For large part sizes, multiple injection and venting ports are utilized. As a rule of thumb, resin feed lines and vacuum sources should be placed about 18 in. apart. As the resin moves away from its source, its velocity decreases in accordance with Darcy's law, and the final thickness and the resin content decreases. It is also more difficult to obtain high fiber volume contents in thick preforms. Since perfect fiber bundle nesting does not occur, there is an increase in free volume with every additional layer, which results in lower fiber volume contents in thick parts.

Since the pressure is much lower than that normally used in the conventional RTM or autoclave processes, it is more difficult to obtain as high a fiber volume percent as with the higher pressure processes; however, this process disadvantage is being overcome with near net preforms. In addition, the VARTM processes cannot hold as tight dimensional tolerances as conventional RTM, and the bag side surface finish will not be as good as a hard tooled surface. Thickness control is generally a function of the perform lay-up, the number of plies, the fiber volume percent, and the amount of vacuum applied during the process.

7.14 Pultrusion

Pultrusion is a rather mature process that has been used in commercial applications since the 1950s. In the pultrusion process, a continuous fibrous reinforcement is impregnated with a matrix that is continuously consolidated into a solid composite. While there are several different variations of the pultrusion process, the basic process for thermoset composites is shown in Fig. 7.49. The reinforcement, usually glass rovings, is pulled from packages on a creel stand and gradually brought together and pulled into an open resin bath, where the reinforcement is impregnated with liquid resin. After emerging from the resin bath, the reinforcement is first directed through a preform die that aligns the rovings to the part shape and then guides it into a heated constant cross-section die, where the part cures as it progresses through the die. Curing takes place from the outside of the part toward the interior. Although the die initially heats the resin, the exotherm resulting from the curing resin can also provide a significant amount of the heat required for cure. The temperature peak caused by the exotherm should occur within the confines of the die and allow the composite to shrink away from the die at the exit. The composite part emerges from the die as a fully cured part that cools as it is being pulled by the puller mechanism. Finally, the part is cut to the required length by a cut-off saw.

While pultrusion has the advantage of being an extremely cost-effective process for making long constant cross-section composite parts, it is definitely a high volume process, as the set-up time for a production run can be rather costly. In addition, there are limitations in that the part must be of constant

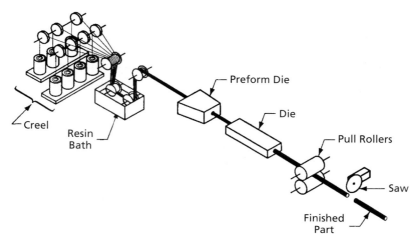

Fig. 7.49. Pultrusion Process[30]

cross-section, and the flexibility in defining reinforcement orientation is somewhat limited. While glass fiber/polyester materials dominate the market, a considerable amount of work has been done to develop the process for the aerospace industry with higher performance carbon/epoxy materials. Floor beams for commercial aircraft are a potential application. Pultrusion is capable of making a wide variety of structural shapes as shown in Fig. 7.50, including hollow sections when an internal mandrel is used.

The major advantages of the pultrusion process are low production costs due to the continuous nature of the process, low raw material costs and minimal scrap, uncomplicated machinery, and a high degree of automation. Disadvantages include: the process is limited to constant cross-section shapes; set-up times and initial process start-up is labor intensive; parts can have higher void contents than allowed for some structural applications; the majority of the reinforcement is oriented in the longitudinal direction; the resin used must have a low viscosity and a long pot life; and in the case of polyesters, styrene emissions can create worker health concerns.

Due to the nature of the pultrusion process, continuous reinforcement must be used, in the form of either rovings or rolls of fabric; however, discontinuous mats and veils can be incorporated. To facilitate set-up, creel stands are often placed on wheels so that a majority of the set-up can be done off-line, reducing the down time for the pultruder. A consideration for the preimpregnation guide mechanism is that the reinforcement is usually fragile, and in the case of glass and carbon abrasive. Dry rovings are often guided by ceramic eyelets to reduce wear on both the fibers and guidance mechanism. Fabrics, mats, and veils can be guided with plastic, or steel, sheets with machined slots or holes. Chopped

Polymer Matrix Composites

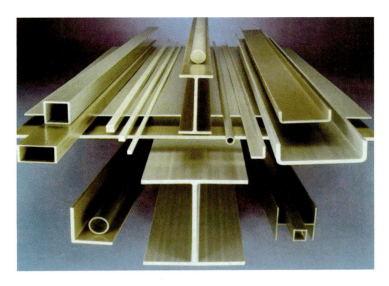

Fig. 7.50. Pultruded Parts

strand mat is often used at an areal weight of $1.5\,oz/yd^2$ in rolls up to 300 ft, with a minimum width of 4 in. Several sets of guidance mechanisms may be required to gradually shape the reinforcements prior to impregnation. In a process called pull-winding, moving winding units are used to overwrap the primarily unidirectional reinforcement, thereby providing additional torsional stiffness.

7.15 Thermoplastic Composites

During the 1980s and early 1990s, government agencies, aerospace contractors, and material suppliers invested hundreds of millions of dollars in developing thermoplastic composites to replace thermosets. In spite of all of this investment and effort, continuous fiber thermoplastic composites account for only a handful of production applications on commercial and military aircraft.

Before considering the potential advantages of thermoplastic composite materials, it is necessary to understand the difference between a thermoset and thermoplastic. As shown in Fig. 7.51, a thermoset crosslinks during cure to form a rigid intractable solid. Prior to cure, the resin is a relatively low molecular weight semi-solid that melts and flows during the initial part of the cure process. As the molecular weight builds during cure, the viscosity increases until the resin gels, and then strong covalent bond crosslinks form during cure. Due to the high crosslink densities obtained for high performance thermoset systems, they are inherently brittle unless steps are taken to enhance toughness. On the other hand, thermoplastics are high molecular weight resins that are fully reacted prior to processing. They melt and flow during processing but do not form

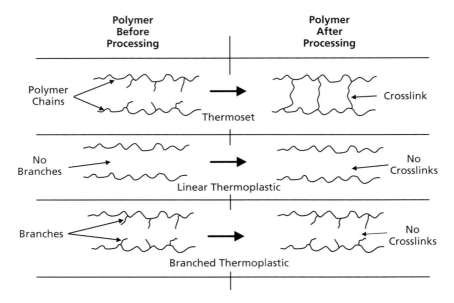

Fig. 7.51. Comparison of Thermoset and Thermoplastic Polymer Structures[1]

crosslinking reactions. Their main chains are held together by relatively weak secondary bonds. However, being high molecular weight resins, the viscosities of thermoplastics during processing are orders of magnitude higher than that of thermosets (e.g., 10^4–10^7 P for thermoplastics vs. 10 P for thermosets).[31] Since thermoplastics do not crosslink during processing, they can be reprocessed, for example they can be thermoformed into structural shapes by simply reheating to the processing temperature. On the other hand, thermosets, due to their highly crosslinked structures, cannot be reprocessed and will thermally degrade, and eventually char, if heated to high enough temperatures. However, there is a limit to the number of times a thermoplastic can be reprocessed. Since the processing temperatures are close to the polymer degradation temperatures, multiple reprocessing will eventually degrade the resin and in some cases it may crosslink.

The structural difference between thermosets and thermoplastics yields some insight into the potential advantages of thermoplastics. Since thermoplastics are not crosslinked, they are inherently much tougher than thermosets. Therefore, they are much more damage tolerant and resistant to low velocity impact damage than the untoughened thermoset resins used in the early to mid-1980s. However, as a result of improved toughening approaches for thermoset resins, primarily with thermoplastic additions to the resin, the thermosets available today exhibit toughness approaching thermoplastic systems.

Since thermoplastics are fully reacted high molecular weight resins that do not undergo chemical reactions during cure, the processing for these materials is theoretically simpler and faster. Thermoplastics can be consolidated and thermoformed in minutes (or even seconds), while thermosets require long cures (hours) to build molecular weight and crosslink through chemical reactions. However, since thermoplastics are fully reacted, they contain no tack, and the prepreg is stiff and boardy. In addition, competing thermoset epoxies are usually processed at 250–350° F, while high performance thermoplastics require temperatures in the range of 500–800° F. This greatly complicates the processing operations requiring high temperature autoclaves, or presses, and bagging materials that can withstand the higher processing temperatures. Another advantage of thermoplastic composites involves health and safety issues. Since these materials are fully reacted, there is no danger to the worker from low molecular weight unreacted resin components. In addition, thermoplastic composite prepregs do not require refrigeration, as do thermoset prepregs. They have essentially an infinite shelf life, but may require drying to remove surface moisture prior to processing.

Another potential advantage of thermoplastics is low moisture absorption. Cured thermoset composite parts absorb moisture from the atmosphere that lowers their elevated temperature (hot-wet) performance. Since many thermoplastics absorb only very little moisture, the design does not have to take as severe a structural "knock down" for lower hot-wet properties. However, since thermosets are highly crosslinked, they are resistant to most fluids and solvents encountered in service. Some amorphous thermoplastics are very susceptible to solvents and may even dissolve in methylene chloride, a common base for many paint strippers, while others, primarily semi-crystalline thermoplastics, are quite resistant to solvents and other fluids.

Since thermoplastics can be reprocessed by simply heating above their melting temperature, they offer potential advantages in forming and joining applications. For example, large flat sheets of thermoplastic composite can be autoclave or press consolidated, cut into smaller blanks, and then thermoformed into structural shapes. Unfortunately, this has proven to be much more difficult in practice than originally anticipated. Press-forming processes are limited to relatively simple geometric shapes, because of the extensible nature of the continuous fiber reinforcement. If a defect (e.g., an unbond) is discovered, the part can often be reprocessed to heal the defect, but in practice, such repairs are rarely practical without undesirable fiber distortion and the associated structural property degradation. The melt fusible nature of thermoplastics also offers a number of attractive joining options such as melt fusion, resistance welding, ultrasonic welding, and induction welding, in addition to conventional adhesive bonding and mechanical fastening.

7.15.1 Thermoplastic Consolidation

Consolidation of melt fusible thermoplastics consists of heating, consolidation, and cooling, as depicted schematically in Fig. 7.52. As with thermoset

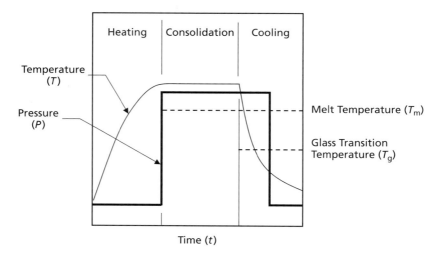

Fig. 7.52. *Typical Thermoplastic Composite Process Cycle*[32]

composites, the main processing variables are time (t), temperature (T), and pressure (P). Heating can be accomplished with infrared heaters, convection ovens, heated platen presses, or autoclaves. Since time for chemical reactions is not required, the time required to reach consolidation temperature is a function of the heating method and the mass of the tooling. The consolidation temperature depends on the specific thermoplastic resin, but should be above the T_g for amorphous resins, or above the melt temperature T_m for semi-crystalline materials.

As a general rule of thumb, the processing temperature for an amorphous thermoplastic composite should be 400° F above its T_g, and for a semi-crystalline material, it should be 200° F or less above its melt temperature T_m.[33] However, heating most thermoplastics above 800° F will result in degradation. The time at temperature for consolidation is primarily a function of the product form used. For example, well-consolidated hot melt impregnated tape can be successfully consolidated in very short times (minutes if not seconds), while woven powder coated, or comingled, prepregs require longer times for the resin to flow and impregnate the fibers. Occasionally, a process called film stacking is used, in which alternating layers of thermoplastic film and dry woven cloth are layed-up and consolidated. The time for successful consolidation for film stacked lay-ups becomes even longer, since the high viscosity resin has even longer distances to flow. A typical processing cycle to achieve fiber wet-out and full consolidation for a film stacked laminate would be 1 h at 150 psi applied pressure. Like heat-up, the cool-down rate from consolidation is a function of the processing method used and the mass of the tooling. The only caveat on cooling is that semi-crystalline thermoplastics should not be cooled so quickly (i.e., quenched)

Polymer Matrix Composites

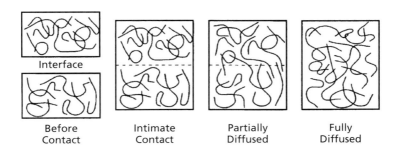

Fig. 7.53. Autohesion at Thermoplastic Interfaces[34]

that they fail to form the desired semi-crystalline structure, that provides optimal elevated temperature performance and solvent resistance. During cooling, the pressure should be maintained until the temperature falls well below the T_g of the resin. This restricts the nucleation of voids, suppresses the elastic recovery of the fiber bed, and helps to maintain the desired dimensions.[35] Finally, pressure during the process provides the driving force to put the layers in intimate contact, push them together, and further helps to impregnate the fiber bed. It should be noted that the properties of solvent impregnated prepreg, powder coated, comingled, and film stacked laminates are not as good as those made from hot melt impregnated prepreg, due to the superior fiber-to-matrix bond formed during the hot melt impregnation process.

Thermoplastic consolidation occurs by a process called autohesion, as depicted in Fig. 7.53. When two interfaces come together, they must obtain intimate contact before the polymer chains can diffuse across the interface and obtain full consolidation. Due to the low flow and tow height non-uniformity of thermoplastic prepregs, the surfaces must be physically deformed under heat and pressure to provide the intimate contact required for chain migration at the ply interfaces. To obtain intimate contact and autohesion, the material must be heated above the T_g if it is amorphous, and above the T_m if it is semi-crystalline. In general, higher pressures and higher temperatures lead to shorter consolidation times. Autohesion is a diffusion-controlled process, in which the polymer chains move across the interface and entangle with neighboring chains. As the contact time increases, the extent of polymer entanglement increases, and results in the formation of a strong bond at the ply interfaces.[35] Consolidation times are usually longer for amorphous thermoplastics since they do not melt and generally maintain higher viscosities at the processing temperature;[36] however, shorter times can be used if higher pressures are employed. The time required for autohesion is directly proportional to the polymer viscosity.[37] Therefore, a certain amount of bulk consolidation must occur at the interfaces prior to the initiation of autohesion. Consolidation is also aided by resin flow due to the applied pressure that aids in ply contact and eventually leads to 100% autohesion. The process is essentially complete when the fiber bed is compressed to the point it reacts the applied processing pressure.

There are several methods employed to consolidate thermoplastic composites. Flat sheet stock can be pre-consolidated for subsequent forming in a platen presses. Two press processes are shown in Fig. 7.54. In the platen press method, pre-collated ply packs are preheated in an oven and then rapidly shuttled into the pressure application zone for consolidation. If the material requires time for resin flow for full consolidation or crystallinity control, the press may require heating. If a well-consolidated prepreg is used, then rapid cooling in a cold platen press may suffice. It should be pointed out that this process still requires collation of the ply packs or layers, usually a hand lay-up operation. Since the material contains no tack, soldering irons, heated to 800–1200° F, are frequently used to tack the edges to prevent the material from slipping. Hand-held ultrasonic guns have also been used for ply tacking. A continuous consolidation process is the double belt press that contains both pressurized heating and cooling zones. This

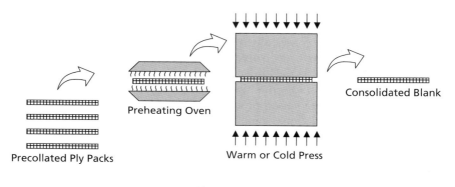

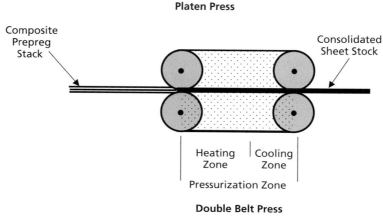

Fig. 7.54. Fabrication Methods for Sheet Stock[1]

process is widely used in making glass mat thermoplastic (GMT) prepreg for the automotive industry, with polypropylene as the resin and random glass mat as the reinforcement.

If the part configuration is complex, an autoclave is certainly an option for part consolidation. However, there are several disadvantages to autoclave consolidation. First, it may prove difficult even finding an autoclave that is capable of attaining the 650–750°F temperatures and 100–200 psi pressures required for some advanced thermoplastics. Second, at these temperatures, the tooling is going to be expensive and may be massive, dictating slow heat-up and cool-down rates. Third, since high processing temperatures are required, it is very important that the coefficient of thermal expansion of the tool match that of the part. For carbon fiber thermoplastics, monolithic graphite, cast ceramic, and Invar 42 are normally used. Fourth, the bagging materials must be capable of withstanding the high temperatures and pressures. In a typical bagging operation, the materials required include high temperature polyimide bagging material, glass bleeder cloth, and silicone bag sealant. The polyimide bagging materials (e.g., Kapton or Uplilex) are more brittle and harder to work with than the nylon materials used for 250–350°F curing thermosets. In addition, the high temperature silicone rubber sealants have minimal tack and tend not to seal very effectively at room temperature. Clamped bars are often placed around the periphery to help get the seal to take at room temperature. As the temperature is increased, the sealant develops tack under pressure and the seal becomes much more effective. A typical autoclave consolidation cycle for carbon/polyetheretherketone (PEEK) prepreg would be 680–750°F at 50–100 psi pressure for 5–30 min; however, the actual cycle time to heat and cool large tools is normally in the range of 5–15 h. In spite of all of these disadvantages, autoclaves nevertheless have a place in thermoplastic composite part fabrication for parts that are just too complex to make by other methods.

Autoconsolidation, or in-situ placement of melt fusible thermoplastics, is a series of processes that include hot tape laying, filament winding, and fiber placement. In the autoconsolidation process, only the area that is being immediately consolidated is heated above the melt temperature, the remainder of the part is held at temperatures well below the melt temperature. Two processes are shown in Fig. 7.55; a hot tape laying process that relies on conduction heating and cooling from hot shoes, and a fiber placement process that uses a focused laser beam at the nip point for heating. Other forms of heating include hot gas torches, quartz lamps, and infrared heaters. The mere fact that autoconsolidation is possible illustrates that the contact times for many thermoplastic polymers, at normal processing temperatures, can be quite short. Provided full contact pressure is made at the ply interfaces, autoconsolidation can occur in less than 0.5 of a second.[38]

A potential problem with autoconsolidation is lack of consolidation due to insufficient diffusion time. If a well-impregnated prepreg is used, then only the

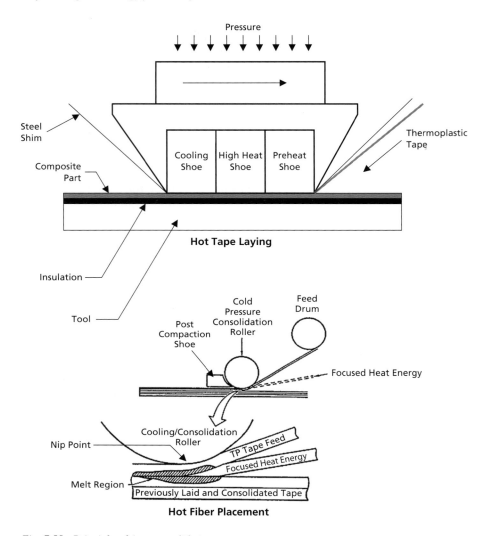

Fig. 7.55. Principle of Autoconsolidation

ply interfaces need be consolidated. However, if there are intraply voids, then the process time is so short that there is insufficient time to heal and consolidate these voids, and a post-consolidation cycle will be required to achieve full consolidation. Previous studies have shown that the interlaminar shear strength of a composite is reduced about 7% for each 1% of voids up to a maximum of 4%. A reasonable goal is 0.5% or less porosity.[39] It has been reported[38] that hot taping laying operations usually result in 80–90% consolidation, indicating the necessity of secondary processing to obtain full consolidation. However, productivity gains for processes such as hot tape laying of 200–300% have been cited compared to traditional hand lay-up methods.[40]

7.15.2 Thermoforming

One of the main advantages of thermoplastic composites is their ability to be rapidly processed into structural shapes by thermoforming. The term "thermoforming" encompasses quite a broad range of manufacturing methods. But, thermoforming is essentially a process that uses heat and pressure to form a flat sheet or ply stack into a structural shape. A typical thermoforming process for a melt fusible semi-crystalline PEEK thermoplastic part, shown schematically in Fig. 7.56, consists of (1) collating the plies, (2) press consolidating a

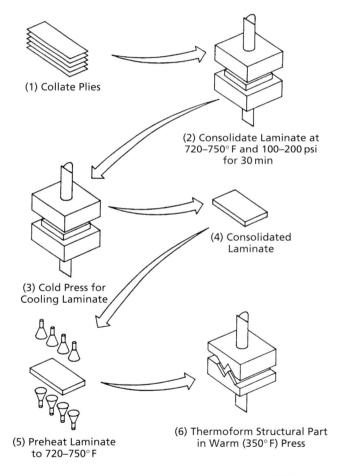

Fig. 7.56. *Typical Thermoforming Sequence for Carbon/PEEK Part*[1]

flat blank, (3) placing the blank in a second press for cooling, (4) trimming the blank to shape if required, (5) reheating the blank to above its melt temperature, and then (6) quickly transferring to a press containing dies of the desired shape. The part must be held under pressure until it cools below its T_g to avoid inducing residual stresses and part warpage.

The primary preheating methods used for press thermoforming are infrared (IR) heater banks, convection ovens, and heated platen presses. In IR heating, the heating time is typically short (i.e., 1–2 min) but temperature gradients can form within thick ply stacks. Since the surface heats considerably faster than the center, there is the danger of overheating unless the temperature is carefully controlled. In addition, it is difficult to obtain uniform heating of complex contours. Still, IR heating is a good choice for thin pre-consolidated blanks of moderate contour. On the other hand, convection heating takes longer (i.e., 5–10 min) but is generally more uniform through the thickness.[41] It is the preferred method for unconsolidated blanks and blanks containing high contour. Impingement heating is a variation of convection heating that uses a multitude of high velocity jets of heated gas that impinge on the surfaces, greatly enhancing the heat flow and reducing the time required to heat the part.[42]

Although matched metal dies can be used for thermoforming, they are expensive and unforgiving, i.e. if the dies are not precisely made there will be high and low pressure points that will result in defective parts. The dies can be made with internal heating and/or cooling capability. Facing one of the die halves with a heat resistant rubber, typically a silicone rubber, can help in equalizing the pressure. Similarly, one of the die halves can be made entirely from rubber, either as a flat block (Fig. 7.57) or a block that is cast to the shape of the part. Although the flat block is simpler and cheaper to fabricate, the shaped block provides a more uniform pressure distribution and better part definition.[2] Silicone rubber of 60–70 Shore A hardness is commonly used. For deep draws, it is usually better to make the male tool half metal and the female half rubber. If the female tool is metal, as is customary for moderate draws, it should incorporate draft angles of 2–3° to facilitate part removal.[38] Another method of applying pressure during the forming process is hydroforming, in which an elastomeric bladder is forced down around the part and lower die half using fluid pressure. Typical thermoforming pressures are 100–500 psi; however, some hydroforming presses are capable of pressures as high as 10 000 psi.[41]

In any thermoforming operation, the transfer time from the heating station to the press is critical. The part must be transferred or shuttled to the press and formed before it cools below its T_g for amorphous resins or below its T_m for semi-crystalline resins. This usually dictates a transfer time of 15 seconds or less. For optimum results, presses with fast closing speeds (e.g., 200–500 in./min) are preferred, that are capable of producing pressures of 200–500 psi. There is some disagreement as to whether it is better to use pre-consolidated blanks or loose unconsolidated ply packs for thermoforming. Pre-consolidated blanks offer the

Polymer Matrix Composites

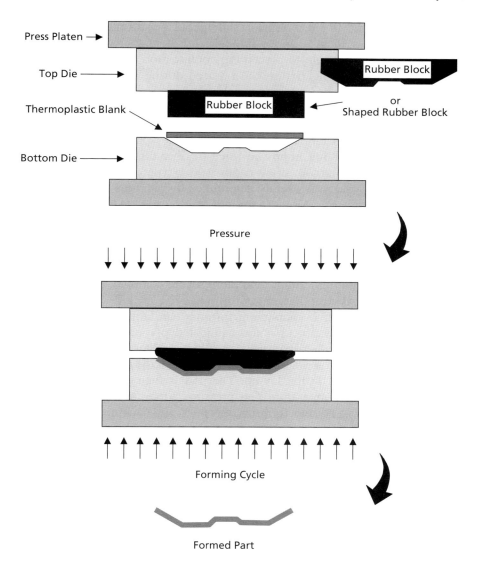

Fig. 7.57. Rubber Block Forming[1]

advantage of being well consolidated with no voids or porosity but do not slip as well during the forming operation as loose unconsolidated ply packs.

While these approaches seem fairly straightforward on the surface, they are actually quite complicated because of the inextensibility of the continuous fiber reinforcement. Carbon fibers in a viscous or near liquid thermoplastic resin are still extremely strong in tension but will buckle and wrinkle readily when placed in compression. Therefore, either the part shape or the die has to be designed to keep the fibers in tension throughout the forming process, but at the same time allow them to move through slip. If neither the part shape nor the die design is amendable to preventing compression buckling of the fibers, a special holding/clamping fixture can be used during the forming operation. These fixtures can be as simple as peripheral clamping fixtures for the blank that allow the material to slip as necessary during forming, or they can be rather sophisticated mechanisms, involving springs located at strategic positions to provide variable tension. Properly designed, the springs allow the part to rotate out-of-plane yielding improved force to fiber directional alignment and allowing greater variations in draw depth.[38] The type of holding fixture, and the location of its springs, is usually determined by previous experience and by considerable trial and error. Slower forming speeds also help to reduce wrinkling and buckling. It has been shown that fiber buckling and waviness can reduce part strength by up to 50%, and that tension pressures of 40–100 psi are often sufficient to suppress fiber buckling.

Diaphragm forming is a rather unique process that is capable of making a wider range of part configurations, and more severe contours, than can be made by press forming. A typical diaphragm-forming cycle for a PEEK thermoplastic part is shown in Fig. 7.58. Diaphragm forming can be done in either a press or an autoclave. In this process, unconsolidated ply packs, to more readily promote ply slippage, rather than pre-consolidated blanks, are placed between two flexible diaphragms. A vacuum is drawn between the diaphragms to remove air and provide tension on the lay-up. The part is then placed in a press and heated above the melt temperature. Gas pressure is used to form the pack down over the tool surface. During forming, the plies slide within the diaphragms, creating tensile stresses that reduce the tendency for wrinkling. The gas pressure both forms and consolidates the part to the tool contour. Pressures usually range from 50 to 150 psi with cycle times of 20–100 min; however, for more massive tools, cycle times of 4–6 h are not unusual. Slow pressurization rates are recommended to avoid out-of-plane buckling. Diaphragm materials include Supral superplastic aluminum and high temperature polyimide films (Upilex-R and -S). Supral aluminum sheet is more expensive than the polyimide films but is less susceptible to rupturing during the forming cycle. Polyimide films work well for thin parts with moderate draws, while Supral sheet is preferred for thicker parts with complex geometries. Typical diaphragm-forming temperatures are 750° F for Supral and 570–750° F for the polyimide films.[43] One disadvantage of this process is that the materials that can be formed must comply with the forming temperatures of the available diaphragm materials. In addition, the diaphragm materials are expensive and can be used only once.

Many other processes have been evaluated for fabricating thermoplastic composite structural shapes, including roll forming, pultrusion, and even resin transfer molding. Thermoplastic composites have also been successfully pultruded, but due to the high melt temperatures and viscosities, the process is much

Polymer Matrix Composites

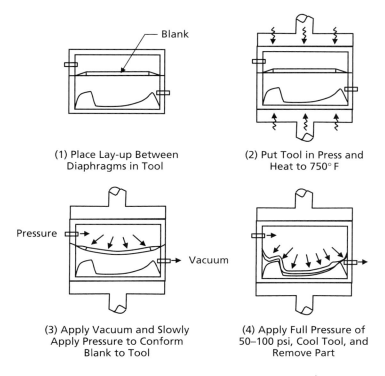

Fig. 7.58. Diaphragm Forming Method for Carbon/PEEK Parts[1]

more difficult and expensive than that for thermosets. Resin transfer molding of thermoplastics is not a feasible process with the materials discussed so far. The viscosity is just too high for the long flow paths required for RTM, and total wet-out of a dry reinforced fiber bed is rarely achieved. However, a relatively new class of materials called "cyclics" offers great potential for RTM. These materials initially melt and flow like thermosets, and then undergo a ring opening mechanism to form a linear thermoplastic on further heating.[43] The molecular weight increases during heating in the presence of an anionic catalyst.[38] At the present time, the technology is applicable only to low temperature thermoplastics, such as nylon and polybutylene terephathalate (PBT). Currently, these materials show great promise for commercial industries such as the automotive industry. In the future, if this technology can be extended to high temperature thermoplastics, it could drastically alter the approach to thermoplastic composite processing and usage.

7.15.3 Thermoplastic Joining

Another unique advantage of thermoplastic composites is the rather extensive joining options available. While thermosets are restricted to either cocuring, adhesive bonding, or mechanical fastening, thermoplastic composites can be joined by melt fusion, dual resin bonding, resistance welding, ultrasonic welding, or induction welding, as well as by conventional adhesive bonding and mechanical fastening.

Adhesive Bonding. In general, structural bonds using thermoset (e.g., epoxy) adhesives produce lower bond strengths with thermoplastic composites than with thermoset composites. This is believed to be due primarily to the differences in surface chemistry between thermosets and thermoplastics. Thermoplastics contain rather inert, non-polar surfaces that impede the ability of the adhesive to wet the surface. A number of different surface preparations have been evaluated including the following:[38] sodium hydroxide etch, grit blasting, acid etching, plasma treatments, silane coupling agents, corona discharge, and Kevlar (aramid) peel plies. While a number of these surface preparations, or combinations of them, give acceptable bond strengths, the long-term service durability of thermoplastic adhesively bonded joints has not been established.

Mechanical Fastening. Thermoplastic composites can be mechanically fastened in the same manner as thermoset composites. Initially, there was concern that thermoplastics would creep excessively, resulting in a loss of fastener preload and thus lower joint strengths. Extensive testing has shown that this was an unfounded fear and mechanically fastened thermoplastic composite joints behave very similar to thermoset composite joints.

Melt Fusion. Since thermoplastics can be processed multiple times by heating above their T_g for amorphous or T_m for semi-crystalline resins with minimal degradation, melt fusion essentially produces joints as strong as the parent resin. An extra layer of neat (unreinforced) resin film can be placed in the bondline for gap filling purposes and to insure that there is adequate resin to facilitate a good bond. However, if the joint is produced in a local area, adequate pressure must be provided over the heat affected zone (HAZ) to prevent the elasticity of the fiber bed from producing delaminations at the ply interfaces.

Dual Resin Bonding. In this method, a lower melting temperature thermoplastic film is placed at the interfaces of the joint to be bonded. As shown in Fig. 7.59, in a process called amorphous bonding or the Thermabond process, a layer of amorphous polyetherimide (PEI) is used to bond two PEEK composite laminates together. To provide the best bond strengths, a layer of PEI is fused to both PEEK laminate surfaces prior to bonding to enhance resin mixing. In addition, an extra layer of film may be used at the interfaces for gap filling purposes. Since the processing temperature for PEI is below the melt temperature of the PEEK laminates, the danger of ply delamination within the PEEK substrates is avoided. Like the melt fusion process, dual resin bonding would normally be used to join large sections together, such as bonding stringers to skins.

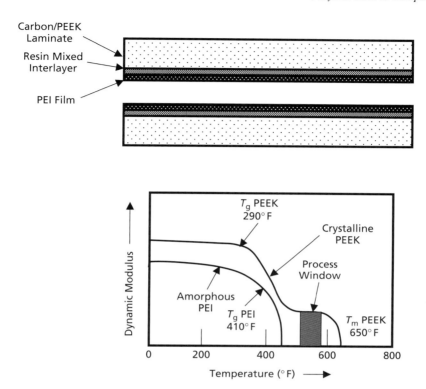

Fig. 7.59. Amorphous or Dual-Resin Bonding[1]

Resistance Welding. Two approaches have been used to join thermoplastic composite parts using resistance heating. As shown in Fig. 7.60, a carbon ply can be used as a resistance heater, or a separate metallic heater can be embedded in the bondline. The advantages of the carbon ply method are that there is no foreign object left in the bondline after bonding, and the thermoplastic resin may adhere better to the carbon fibers than to a metallic material. In general, the resin is removed from the ply ends where it is clamped to the electrical bus bars. However, the carbon ply method does not supply heat to the joint as effectively as an embedded metallic heater and is more prone to electrical shorts during the bonding process. Thin layers of polymer film are usually added to both sides of the resistance heater to provide electrical insulation from the carbon fibers, and additional matrix material may be used to fill any gaps. For all fusion welding operations with thermoplastic composites, it is necessary to maintain adequate pressure at all locations that are heated above the melt temperature. If pressure is not maintained at all locations that exceed the melt temperature, deconsolidation due to fiber bed relaxation will likely occur. The pressure should be maintained until the part is cooled below its T_g. Typical processing times for resistance welding are 30 s to 5 min at 100–200 psi pressure.[44]

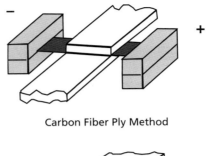

Carbon Fiber Ply Method

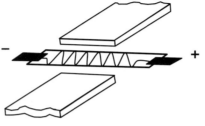

Embedded Heating Element

Fig. 7.60. Resistance Heating Methods for Joining Thermoplastic Composites[1]

Ultrasonic Welding. Ultrasonic welding is used extensively in commercial processes to join lower temperature unreinforced thermoplastics and can also be used for advanced thermoplastic composites. As shown in Fig. 7.61, an ultrasonic horn, also known as a sonotrode, is used to produce ultrasonic energy at the composite interfaces. Electrical energy is converted into mechanical energy. The sonotrode is placed in contact with one of the pieces to be joined. The second piece is held stationary while the vibrating piece creates frictional heating at the interface. Ultrasonic frequencies of 20–40 kHz are normally used. The process works best if one of the surfaces has small asperities that act as energy directors or intensifiers. The asperities have a high energy per unit volume and melt before the surrounding material. The quality of the bond is increased with increasing time, pressure, and amplitude of the signal.[38] Again, it is common practice to incorporate a thin layer of neat resin film to provide gap filling. Typical weld parameters are less than 10 s at 70–200 psi pressure.[44] This process is somewhat similar to spot welding of metals.

Induction Welding. Similar to resistance welding, induction welding techniques have been developed in which a metallic susceptor may, or may not, be placed in the bondline. It is generally accepted that the use of a metallic susceptor produces superior joint strengths. A typical induction set-up, shown in Fig. 7.62, uses an induction coil to generate an electromagnetic field that results

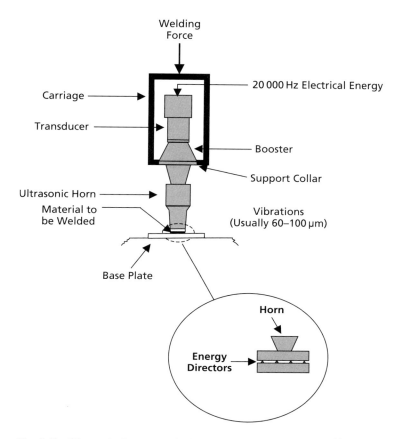

Fig. 7.61. Ultrasonic Joining Method for Thermoplastic Composites[44]

in eddy current heating in the conductive susceptor and/or by hysteresis losses in the susceptor. Susceptor materials evaluated include iron, nickel, carbon fibers, and copper meshes. As with resistance heating, it is normal practice to place a layer of polymer film on each side of the metallic susceptor. Typical welding parameters are 5–30 min at 50–200 psi pressure.[44]

A comparison of single lap shear strengths produced in thermoplastic composites, using the various techniques described above, is given in Fig. 7.63. Note that adhesive bonding yields lower joint strengths than the fusion bonding techniques and is very dependent on the surface preparation method used. Autoclave co-consolidated (melt fusion) joint strengths approach virgin autoclave molded strengths. Typically, resistance and induction welding strengths exhibit similar properties, both of which are superior to those of ultrasonic welding.

Manufacturing Technology for Aerospace Structural Materials

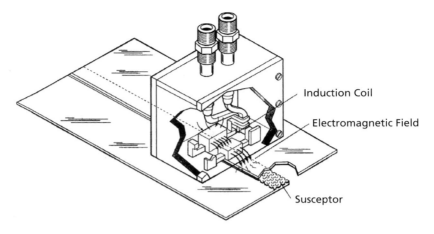

Fig. 7.62. *Induction Joining Method for Joining Thermoplastic Composites*[44]

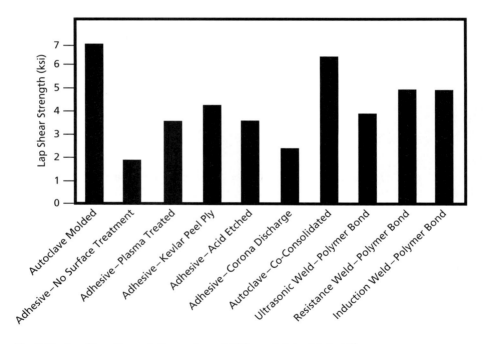

Fig. 7.63. *Lap Shear Strength Comparison of Different Joining Methods*[44]

Thermoplastic composites do offer some definite advantages compared to thermoset composites; however, in spite of large investments since the mid-1980s, very few continuous fiber thermoplastic composites have made it into production applications. Compared to thermoset composites, thermoplastic composites offer the potential for short processing times, but their inherent characteristics have prevented them from replacing thermoset composites in the aerospace industry, namely:

- High processing temperatures (500–800° F) increase the cost of both the prepreg and complicates the use of conventional processing equipment.
- The lack of tack and boardiness of the prepreg results in expensive manual handling operations.
- Thermoforming of continuous fiber reinforced thermoplastics has proven to be much more difficult than first anticipated, due to the tendency of the fibers to wrinkle and buckle if not maintained under tension during the forming operation.
- The early claims of superior toughness and damage tolerance have largely been negated by the development of much tougher thermoset resins.
- Solvent and fluid resistance properties remain major barriers to the use of amorphous thermoplastic composites.

In the author's opinion, two criteria must be satisfied to take advantage of continuous fiber thermoplastic composites: (1) the demand for a large quantity of parts, and (2) the process must be automated to remove almost all manual operations. Unfortunately, it is difficult to meet these criteria in the aerospace industry, where lot sizes are small and production rates cannot usually justify the investment in highly sophisticated automated equipment. It is both interesting and insightful that discontinuous GMT have made significant inroads in the automotive industry, where the demand for parts is large and the process has been almost totally automated.

7.16 Trimming and Machining Operations

Composites are more prone to damage during trimming and machining than conventional metals. Composites contain strong and very abrasive fibers held together by a relatively weak and brittle matrix. During machining, they are prone to delaminations, cracking, fiber pullout, fiber fuzzing (aramid fibers), matrix chipping, and heat damage. It is important to minimize forces and heat generation during machining. During metallic machining, the chips help to remove much of the heat generated during the cutting operation. Due to the much lower thermal conductivity of the fibers (especially glass and aramid), heat build up can occur rapidly and degrade the matrix, resulting in matrix cracking and even delaminations. When machining composites, generally high speeds, low feed rates, and small depths of cuts are used to minimize damage.[45]

Most composite parts require peripheral edge trimming after cure. Edge trimming is usually done either manually with high speed cut-off saws or automatically with NC abrasive water jet machines. Lasers have been often proposed for trimming of cured composites, but the surfaces become charred due to the intense heat and are unacceptable for most structural applications.

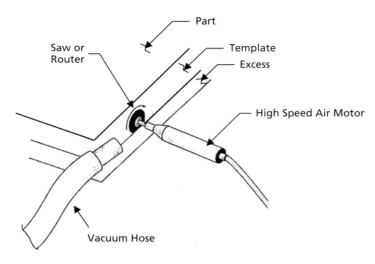

Fig. 7.64. Hand Trimming[46]

Carbon fibers are very abrasive and quickly wear out conventional steel cutting blades; therefore, trimming operations should be conducted using either diamond-coated circular saw blades, carbide router bits, or diamond-coated router bits. A typical manual edge trimming operation, shown in Fig. 7.64, can be conducted with a high speed air motor (e.g., 20 000 rpm) with either a diamond-impregnated cut-off wheel or more typically a carbide router bit. Fiberglass laminate trim templates are often clamped to the part to insure that the true trim path is followed and provide edge support to help prevent delaminations. Typical feed rates are 10–14 in./min. Hand trimming is a dirty job. The operator should wear a respirator, have eye and ear protection, and wear heavy duty gloves. Many facilities have installed ventilated trim booths to help control the noise and fine dust generated by this operation. Being a hand operation, the quality of the cut is very dependent on the skill of the operator. Too fast a feed rate can cause excessive heat leading to matrix overheating and ply delaminations.

Abrasive water jet trimming has emerged as probably the most accepted method for trimming cured composites; however, these are large and expensive NC machine tools (Fig. 7.65). The advantages of abrasive water jet cutting are that consistent delamination-free edges are produced, and since the cutting path is NC controlled, the requirement for tooling is much simpler. Abrasive water jet cutting is primarily an erosion process, rather than a true cutting process, so there is very little force exerted on the part during trimming; therefore, only simple holding fixtures are required to support the part during cutting. In addition, no heat is generated during cutting, negating the concern for possible matrix degradation. Water pumped at low volume (1–2 gal/min) enters the top of the head and is then mixed with garnet grit that is expelled through a 0.040 in.

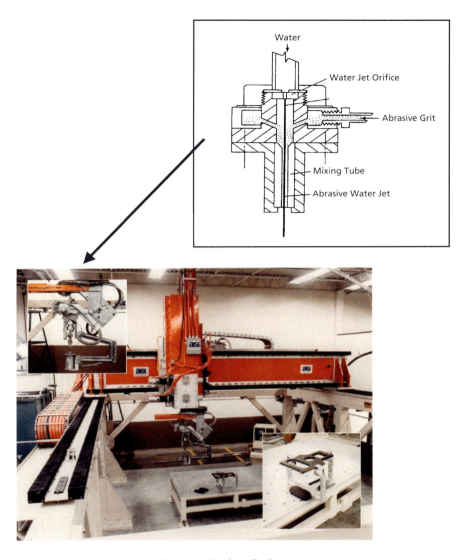

Fig. 7.65. Abrasive Water Jet Trimming Machine Tool
Source: The Boeing Company

diameter sapphire nozzle at 40 000–45 000 psi.[47] In general, higher grit size numbers (smaller grain diameters) produce better surface finishes, with a typical grit size being #80. Once the abrasive slurry has penetrated the composite laminate, there is a catcher filled with steel balls that spin to dissipate the flow. Other than the expense of these tools, the other main disadvantage is the noise level generated during the process. It is not unusual for trimming operations to exceed 100 dB; therefore, ear protection is required, and many units are isolated within their own sound-proof rooms.

If edge sanding is required, die grinders at speeds of 4000–20 000 rpm can be used along with 80 grit aluminum oxide paper for roughing and 240–320 grit silicon carbide paper for finishing.

Summary

Advanced polymer composites consist of either continuous or discontinuous fibers embedded in a resin matrix. Common fibers include glass, aramid, and carbon. Glass fibers are used extensively in commercial applications because of their good balance of properties and low cost. Aramid fibers, being organic, have low densities and outstanding toughness. Carbon fibers have the best combination of strength and stiffness but are also the most expensive of the three.

Manual lay-up is a labor-intensive process where each ply is layed-up on the tool one ply at a time. To make the job less costly, wide material product forms (broadgoods) were developed that can be cut on NC-programmed ply cutters, such as reciprocating knife or ultrasonically driven knife cutters. In addition, laser ply projection systems are used to show the exact location of each ply. Flat ply collation, followed by vacuum forming to final shape, can further reduce costs if the stacks can be formed to the contour of the tool.

Automated tape laying is a process that is amenable to large flat or mildly contoured parts. Two types of machines are currently in use: (1) flat tape laying machines and (2) contour tape laying machines capable of handling contours approaching 15°. Tape laying machines normally use either 3, 6, or 12 in. wide unidirectional tape. Filament winding is a process in which a fiber band is placed on a rotating mandrel. It is capable of placing as much as 100–400 lb/h It is an excellent lay-up process for bodies of revolution, such as cylinders, and can produce parts smaller than 1 in. in diameter up to about 20 ft in diameter. However, it cannot be used to lay material into concave cavities because of fiber tension. The three methods of filament winding are helical, polar, and hoop winding. Fiber placement is a hybrid process combining some of the features of automated tape laying and filament winding. Since there is only minimal tension on the fibers during fiber placement, material can be placed into concave cavities and true 0° plies can be placed.

Manual ply collation remains one of the highest cost areas of composite part fabrication, normally representing 40–60% of the part fabrication cost. Although several automated processes, such as tape laying, filament winding, and fiber placement have been developed to reduce the manual labor costs, manual or hand collation of plies still remains the primary method for high performance composite parts. Even though hand collation may at first seem the most expensive of the available methods, wide broadgoods combined with automated ply cutting, flat ply collation methods, and laser projection systems have helped to reduce the costs. Prior to investing or committing a part to an automated manufacturing method, a careful analysis should be conducted on a part-by-part basis to determine the most cost-effective process.

Autoclave processing remains the mainstay for processing continuous fiber-reinforced thermoset composite parts. Maintaining the resin hydrostatic pressure above the potential void pressure is a key to minimizing void formation. The resin pressure, however, is typically lower than the autoclave pressure due to resin flow, bagging, tooling, and support materials, such as honeycomb core. Void formation in addition curing composite laminates is primarily due to entrapped volatiles and air. High temperatures result in high volatile pressures. Void growth will occur if the potential void pressure (volatile vapor pressure) exceeds the hydrostatic resin pressure while the resin is still a liquid. The use of a partially impregnated prepreg greatly facilitates removing trapped air and volatiles from the lay-up and is another key in producing void free laminates. Condensation curing systems are much more difficult to process than addition curing systems, since they can give off water, alcohols, and solvents during cure. If condensation systems must be used, then volatile management during cure becomes the main challenge.

Resin transfer molding (RTM) is the most mature of the high performance composite liquid molding processes. In RTM, a preform is placed in a rigid matched die mold and injected with a liquid resin under pressure. The part is then cured in the mold. RTM is an excellent process for making highly dimensionally accurate parts that have complex geometries that would be difficult to hand lay-up on a consistent basis. However, the major disadvantage is the cost of the tooling. A well-designed and fabricated tool is a prerequisite for obtaining a good part. When combined with 3-D textile geometries, liquid molding processes offer the potential to introduce composites into new areas that were previously restricted to metals, such as fittings and bulkheads that have multi-directional complex load paths not amendable to 2-D reinforced designs.

Preforms can be fabricated from a number of processes including weaving, knitting, stitching, and braiding. Woven materials can be supplied as either conventional 2-D woven fabrics or more complex 3-D constructions. Knitting for structural composites is the MWK material, in which a number of unidirectional plies are knitted or stitch bonded together with minimal 3-D reinforcement to provide handling capabilities. MWK materials, along with 2-D woven or braided constructions, are often given true 3-D strength improvements by stitching operations. Braided 2-D and 3-D reinforced parts are ideal for bodies of revolution, or they can be cut off the mandrel and formed into substructure parts. The main advantages of preforms are the reduced labor required for manual ply-by-ply collation of complex shapes and the potential for 3-D reinforcement. These advantages must be weighed against the lower in-plane mechanical properties compared to unidirectional tapes, caused by the somewhat lower fiber volume contents for preforms and the damage done to the fibers by mechanical twisting and abrasion.

Vacuum assisted resin transfer molding (VARTM) uses a vacuum to draw the resin through the preform rather than injection pressure. Normally, single-sided tooling is used along with a vacuum bag and a porous media material to aid in resin infiltration. The major advantage of VARTM type processes is the potential for lower tooling costs and the ability to make large parts because an autoclave is not required for curing. While VARTM processes have been used for years in the commercial boat building industry, the technology for high performance composites is still evolving but shows great potential for cost savings.

Thermoplastic composites offer some definite advantages compared to thermoset composites; however, in spite of large investments since the mid-1980s, very few continuous fiber thermoplastic composites have made it into production applications. Compared to thermoset composites, thermoplastic composites offer the potential for short processing times, but their inherent characteristics have prevented them from replacing thermoset composites in the aerospace industry, namely: high processing temperatures increase the cost of both the prepreg and complicate the use of conventional processing equipment; the lack of tack and boardiness of the prepreg results in expensive manual handling operations, thermoforming of continuous fiber-reinforced thermoplastics has proven to be much more difficult than first anticipated due to the tendency of the fibers to wrinkle and buckle if not maintained under tension during the forming operation; the early claims of superior toughness and damage tolerance have largely been negated by the development of much tougher thermoset resins; and solvent and fluid resistance properties remain major barriers to the use of amorphous thermoplastic composites.

Recommended Reading

[1] Campbell, F.C., *Manufacturing Processes for Advanced Composites*, Elsevier Ltd, 2004.
[2] Miracle, D.P., Donaldson, S.L. (eds), *ASM Handbook Vol. 21 Composites*, ASM International, 2001.
[3] Peters, S.T., *Handbook of Composites*, 2nd edition, Chapman & Hall, 1998.

References

[1] Campbell, F.C., *Manufacturing Processes for Advanced Composites*, Elsevier Ltd, 2004.
[2] Taylor, A., "RTM Material Developments for Improved Processability and Performance", *SAMPE Journal*, Vol. 36, No. 4, July/August 2000, pp. 17–24.
[3] Walsh, P.J., "Carbon Fibers", in *ASM Handbook Vol. 21 Composites*, ASM International, 2001, p. 38.
[4] Prime, R.B., Chapter 5 in *Thermal Characterization of Polymeric Materials*, ed. E.A. Turi, Academic Press, 1981.
[5] SP Systems "Guide to Composites".
[6] Virtek LaserEdge product literature.

[7] Young, M., Paton, R., "Diaphragm Forming of Resin Pre-Impregnated Woven Carbon Fibre Materials", 33rd International SAMPE Technical Conference, November, 2001.
[8] Grimshaw, M.N., "Automated Tape Laying", in *ASM Handbook Vol. 21 Composites*, ASM International, 2001.
[9] Grimshaw, M.N., Grant, C.G., Diaz, J.M.L., "Advanced Technology Tape Laying for Affordable Manufacturing of Large Composite Structures", 46th International SAMPE Symposium, May 2001, pp. 2484–2494.
[10] Mantel, S.C., Cohen, D. "Filament Winding", in *Processing of Composites*, Hanser, 2000.
[11] Peters, S.T., Humphrey, W.D., Foral, R.F., *Filament Winding Composite Structure Fabrication*, SAMPE, 2nd edition, 1999.
[12] Grover, M.K., *Fundamentals of Modern Manufacturing: Materials, Processes, and Systems*, Prentice-Hall Inc., 1996.
[13] Evans, D.O., "Fiber Placement", in *ASM Handbook Vol. 21 Composites*, ASM International, 2001, pp. 477–479.
[14] Adrolino, J.B., Fegelman, T.M., "Fiber Placement Implementation for the F/A-18 E/F Aircraft", 39th International SAMPE Symposium, April 1994, pp. 1602–1616.
[15] Griffith, J.M., Campbell, F.C., and Mallow, A.R., "Effect of Tool Design on Autoclave Heat-Up Rates", Society of Manufacturing Engineers, Composites in Manufacturing 7 Conference and Exposition, 1987.
[16] Kardos, J.L., "Void Growth and Dissolution", in *Processing of Composites*, Hanser, 1999, pp. 182–207.
[17] Campbell, F.C., Mallow, A.R., Browning, C.E., "Porosity in Carbon Fiber Composites An Overview of Causes", *Journal of Advanced Materials*, Vol. 26, No. 4, July 1995, pp. 18–33.
[18] Brand, R.A., Brown, G.G. and McKague, E.L., "Processing Science of Epoxy Resin Composites", Air Force Contract No. F33615-80-C-5021, Final Report for August 1980 to December 1983.
[19] Mace, W.C., "Curing Polyimide Composites", in *ASM Vol. 1 Engineered Materials Handbook Composites*, ASM International, 1987, pp. 662–663.
[20] Thompkins, S.S., Shen, J.Y., Lavoie, *Proceedings of the 4th International Conference on Engineering, Construction, and Operations in Space*, 1994, p. 326.
[21] Swanson, S.R., *Introduction to Design and Analysis with Advanced Composite Materials*, Prentice-Hall, Inc., 1997.
[22] Beckwith, S.W., Hyland, C.R., "Resin Transfer Molding: A Decade of Technology Advances", *SAMPE Journal*, Vol. 34, No. 6, November/December 1998, pp. 7–19.
[23] Cox, B.N., Flanagan, G., "Handbook of Analytical Methods for Textile Composites", NASA Contractor Report 4750, March 1997.
[24] Dow, M.B., Dexter, H.B., "Development of Stitched, Braided and Woven Composite Structures in the ACT Program and at Langley Research Center (1985–1997)", Summary and Bibliography, NASA/TP-97-206234, November 1997.
[25] Karal, M., "AST Composite Wing Program- Executive Summary", NASA/CR-20001-210650, March 2001.
[26] Palmer, R., "Techno-Economic Requirements for Composite Aircraft Components", Fiber-Tex 1992 Conference, NASA Conference Publication 3211, 1992.
[27] Kittleson, J.L., Hackett, S.C., "Tackifier/Resin Compatibility is Essential for Aerospace Grade Resin Transfer Molding", 39th International SAMPE Symposium, April 1994, pp. 83–96.
[28] Loos, A.C., Sayre, J., McGrane, R., Grimsley, B., "VARTM Process Model Development", 46th International SAMPE Symposium, May 2001, pp. 1049–1060.
[29] Hayward, J.S., Harris, B., "Effect of Process Variables on the Quality of RTM Mouldings", *SAMPE Journal*, Vol. 26, No. 3, May/June 1990, pp. 39–46.

[30] Groover, M.P., *Fundamentals of Modern Manufacturing – Materials, Processes, and Systems*, Prentice-Hall, Inc., 1996.
[31] Strong, A.B., *Fundamentals of Composite Manufacturing: Materials, Methods, and Applications*, Society of Manufacturing Engineers, 1989.
[32] Muzzy, J., Norpoth, L., Varughese, B., "Characterization of Thermoplastic Composites for Processing", *SAMPE Journal*, Vol. 25, No. 1, January/February 1989, pp. 23–29.
[33] Leach, D.C., Cogswell, F.N., Nield, E., "High Temperature Performance of Thermoplastic Aromatic Polymer Composites", 31st National SAMPE Symposium, 1986, pp. 434–448.
[34] Muzzy, J.D., Colton, J.S., "The Processing Science of Thermoplastic Composites", in *Advanced Composites Manufacturing*, John Wiley & Sons, Inc., 1997.
[35] Astrom, B.T., *Manufacturing of Polymer Composites*, Chapman & Hall, 1997.
[36] Loos, A.C., Min-Chung, L., "Consolidation During Thermoplastic Composite Processing", in *Processing of Composites*, Hanser/Gardner Publications, Inc., 2000.
[37] Cogswell, F.N., *Thermoplastic Aromatic Polymer Composites*, Butterworth-Heinemann Ltd, 1992.
[38] Strong, A.B., *High Performance and Engineering Thermoplastic Composites*, Technomic Publishing Co., Inc., 1993.
[39] Strong, A.B., "Manufacturing", in *International Encyclopedia of Composites*, Stuart Lee ed, VCH Publishers Inc., 1990, pp. 102–126.
[40] Harper, R.C., "Thermoforming of Thermoplastic Matrix Composites – Part II", *SAMPE Journal*, Vol. 28, No. 3, May/June 1992, pp. 9–17.
[41] Okine, R.L., "Analysis of Forming Parts from Advanced Thermoplastic Sheet Materials", *SAMPE Journal*, Vol. 25, No. 3, May/June 1989, pp. 9–19.
[42] Harper, R.C., "Thermoforming of Thermoplastic Matrix Composites – Part I", *SAMPE Journal*, Vol. 28, No. 2, March/April 1992, pp. 9–18.
[43] Dave, R.S., Udipi, K., Kruse, R.L., "Chemistry, Kinetics, and Rheology of Thermoplastic Resins Made by Ring Opening Polymerization", in *Processing of Composites*, Hanser/Gardner Publications, Inc., 2000.
[44] McCarville, D.A., Schaefer, H.A., "Processing and Joining of Thermoplastic Composites", in *ASM Handbook Vol. 21 Composites*, ASM International, 2001, pp. 633–645.
[45] Astrom, B.T., *Manufacturing of Polymer Composites*, Chapman & Hall, 1997.
[46] Price, T.L., Dalley, G., McCullough, P.C., Choquette, L., "Handbook: Manufacturing Advanced Composite Components for Airframes", Report DOT/FAA/AR-96/75, Office of Aviation Research, April 1997.
[47] Kuberski, L.F., "Machining, Trimming, and Routing of Polymer-Matrix Composites", in *ASM Handbook Vol. 21 Composites*, ASM International, 2001, pp. 616–619.

Chapter 8

Adhesive Bonding and Integrally Cocured Structure

Adhesive bonding is a method of joining structure together that eliminates some, or all, of the cost and weight of mechanical fasteners. In adhesive bonding, cured composites or metals are adhesively bonded to other cured composites, honeycomb core, foam core, or metallic pieces. Cocuring is a process in which uncured composite plies are cured and bonded simultaneously during the same cure cycle to either core materials or other composite parts. The ability to make large bonded and cocured unitized structure can eliminate a significant portion of the assembly costs.

Adhesive bonding is a widely used industrial joining process in which a polymeric material (the adhesive) is used to join two separate pieces (the adherends or substrates). There are many types of adhesives; some are strong and rigid while others are weak and flexible. Adhesives used for structural bonding are always cured at either room or elevated temperatures and must possess adequate strength to transfer the loads through the joint. There are many types of structural adhesives; however, epoxies, nitrile phenolics, and bismaleimides are the most prevalent. In addition to fabricating large bonded components, adhesive bonding is frequently used for repairing damaged structural parts.

Bonded joints may be preferred if thin composite sections are to be joined where bearing stresses in bolted joints would be unacceptably high, or when the weight penalty for mechanical fasteners is too high. In general, thin structures with well-defined load paths are good candidates for adhesive bonding, while thicker structures with complex load paths are better candidates for mechanical fastening.[1]

8.1 Advantages of Adhesive Bonding

The advantages of adhesive bonding include:[2]

- Bonding provides a more uniform stress distribution than mechanical fasteners by eliminating the individual stress concentration peaks caused by mechanical fasteners. As shown in Fig. 8.1, the stress distribution across the joint is much more uniform for the adhesive bonded joint than for the mechanical joint, leading to better fatigue life than that for mechanically fastened joint. Bonded joints also provide superior vibration and damping capability.
- Due to the elimination of mechanical fasteners, bonded joints are usually lighter than mechanically fastened joints and are less expensive in some applications.
- Bonded joints enable the design of smooth external surfaces and integrally sealed joints with minimum sensitivity to fatigue crack propagation. Dissimilar materials can be assembled with adhesive bonding and the joints are electrically insulating, preventing galvanic corrosion of metal adherends.
- Bonded joints provide a stiffening effect compared to riveted or spot welded constructions. While rivets or spot welds provide local point stiffening,

Adhesive Bonding and Integrally Cocured Structure

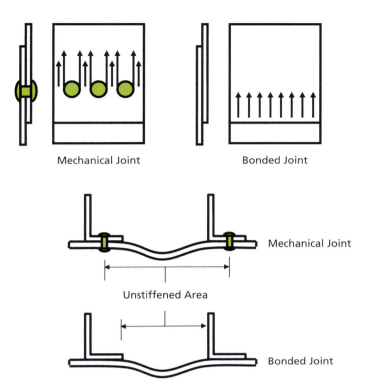

Fig. 8.1. *Load Distribution Comparison for Mechanically Fastened and Bonded Joints*[2]

bonded joints provide stiffening over the entire bonded area. The significance of this effect is shown in Fig. 8.1, where bonded joints may increase the buckling strength of the structure by as much as 30–100%.

8.2 Disadvantages of Adhesive Bonding

Adhesive bonding also has some disadvantages, including:

- Bonded joints should be considered permanent joints. Disassembly is not easy and often results in damage to the adherends and surrounding structure.
- Adhesive bonding is much more sensitive to surface preparation than mechanical fastening. Proper surface preparation is absolutely essential to producing a strong, durable bond. For field repair applications, it can be extremely difficult to execute proper surface preparation. For original manufacturing, adhesive bonding requires clean rooms with temperature and humidity control.
- Adhesively bonded joints can be non-destructively tested for the presence of voids and unbonds; however, at this time there is no reliable

non-destructive test method for determining the strength of a bonded joint. Therefore, traveler or process control test specimens must be fabricated and destructively tested using the same surface preparation, adhesive, and bond cycle as the actual structure.
- Adhesive materials are perishable. They must be stored according to the manufacturer's recommended procedures (often refrigerated). Once mixed or removed from the freezer, they must be assembled and cured within a specified time.
- Adhesives are susceptible to environmental degradation. Most will absorb moisture and exhibit reduced strength and durability at elevated temperature. Some are degraded by chemicals such as paint strippers or other solvents.

8.3 Theory of Adhesion

There are a number of theories on the nature of adhesion during adhesive bonding; however, there is some general agreement about what leads to a good adhesive bond. Surface roughness plays a key role. The rougher the surface, the more surface area available for the liquid adhesive to penetrate and lock onto. However, for this to be effective, the adhesive must wet the surface, a function of adherend cleanliness, adhesive viscosity, and surface tension. The importance of surface cleanliness cannot be overemphasized; surface cleanliness is one of the cornerstones of successful adhesive bonding.

In metals, coupling effects as a result of chemical etchants/anodizers, or other treatments, can also play a role in adhesion by providing chemical end groups that attach to the metal adherend surface and provide other end groups that are chemically compatible with the adhesive.

Therefore, for the best possible adhesive joint, the following areas must be addressed: the surface must be clean; the surface should have maximum surface area through mechanical roughness; the adhesive must flow and thoroughly wet the surface; and the surface chemistry must be such that there are attractive forces on the adherend surface to bond to the adhesive.

8.4 Joint Design[1]

In a structural adhesive joint, the load in one component is transferred through the adhesive layer to another component. The load transfer efficiency depends on the joint design, the adhesive characteristics, and the adhesive/substrate interface. To effectively transfer loads through the adhesive, the substrates (or adherends) are overlapped so that the adhesive is loaded in shear. Typical joint designs are shown in Fig. 8.2.

As shown in the shear stress distribution for a typical joint (Fig. 8.3), the loads peak at the joint ends while the center portion of the joint carries a much lower

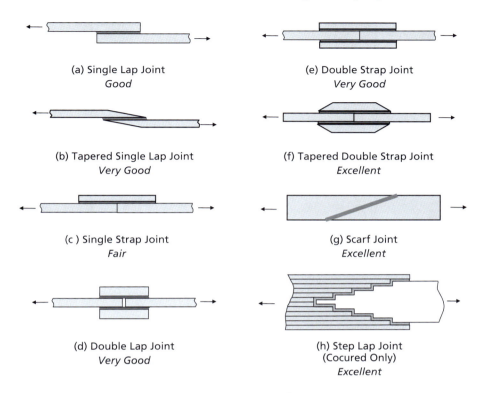

Fig. 8.2. Typical Adhesively Bonded Joint Configurations[3]

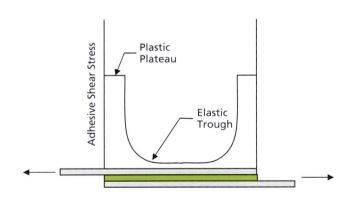

Fig. 8.3. Typical Bondline Shear Stress Distribution[1]

portion of the load. Therefore, adhesives designed to carry high loads need to be strong and tough, especially if there is any bending in the joint that would induce peel loads. In order to improve fracture toughness and fatigue life, adhesives are frequently modified with rubber or other elastomers that reduces the adhesive

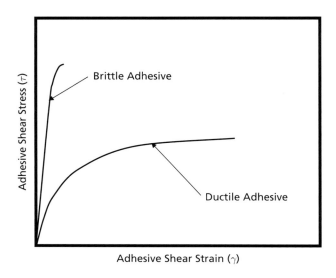

Fig. 8.4. Typical Stress–Strain Behavior for Brittle and Ductile Adhesives[4]

modulus. A comparison of a "brittle" high strength, high modulus adhesive with a "ductile" lower strength, lower modulus adhesive is shown in Fig. 8.4. While the brittle high strength adhesive has the highest strength, the tough ductile adhesive, which has a much larger area under the shear stress–strain curve, would be a much more forgiving adhesive, particularly in structural joints that often experience peel and bending loads. The joint design must insure that the adhesive is loaded in shear as much as possible. Tension, cleavage, and peel loading (Fig. 8.5) should be avoided when using adhesives. Actually, tension loading is acceptable as long as there is appreciable surface area, but certainly not in the butt joint shown. Some further considerations for joint design are summarized in Table 8.1.

Bonding to composites rather than metals introduces significant differences in criteria for adhesive selection for two reasons: (1) composites have a lower interlaminar shear stiffness compared to metals, and (2) composites have much lower shear strength than metals. This occurs because the interlaminar shear

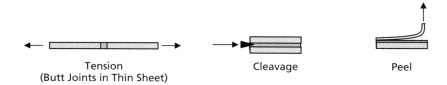

Fig. 8.5. Load Paths to Avoid in Bonded Structure[1]

Adhesive Bonding and Integrally Cocured Structure

Table 8.1 Considerations for Designing Adhesively Bonded Joints[1]

- The adhesive must be compatible with the adherends and be able to retain its required strength when exposed to in-service stresses and environmental factors.
- The joint should be designed to ensure a failure in one of the adherends rather than a failure within the adhesive bondline.
- Thermal expansion of dissimilar materials must be considered. Due to the large thermal expansion difference between carbon composite and aluminum, adhesively bonded joints between these two materials have been known to fail during cool down from elevated temperature cures as a result of the thermal stresses induced by their differential expansion coefficients.
- Proper joint design should be used, avoiding tension, peel, or cleavage loading whenever possible. If peel forces cannot be avoided, a lower modulus (nonbrittle) adhesive having a high peel strength should be used.
- Tapered ends should be used on lap joints to feather out the edge-of-joint stresses. The fillet at the end of the exposed joint should not be removed.
- Selection tests for structural adhesives should include durability testing for heat, humidity (and/or fluids), and stress, simultaneously.

---

stiffness and strength depends on the matrix properties and not on the higher properties of the fibers. The exaggerated deformations in a composite laminate bonded to a metal sheet under tension loading are shown in Fig. 8.6. The adhesive passes the load from the metal into the composite until, at some distance "L," the strain in each material is equal. In the composite, the matrix resin acts as an adhesive to pass load from one fiber ply to the next. Because the matrix shear stiffness is low, the composite plies deform unequally in tension as shown. Failure tends to initiate in the composite ply next to the adhesive near the beginning of the joint, or in the adhesive in the same neighborhood. The

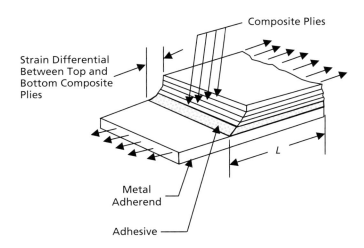

Fig. 8.6. Uneven Strain Distribution in Composite Plies[1]

highest failure loads are achieved by an adhesive with a low shear modulus and high strain to failure, as previously shown in Fig. 8.4.

It should also be noted that there is a limit to the thickness of the composite that can be loaded by a single bondline. However, multiple steps in the composite thickness, giving multiple bondlines, can be used for thick material, as in a step-lap joint. The effects of adherend thickness and joint configuration on failure mode are shown in Fig. 8.7. For thick adherends, it is necessary to use either a cocured scarf or a step-lap joint to carry the load. The other option for thick joints is to use mechanical fasteners. Note that the double scarf joint shown in Fig. 8.7, while extremely efficient, is rarely used because it is extremely difficult to fabricate. The step-lap joint configuration, while not easy to fabricate, contains discrete steps that can be used for accurate ply location during fabrication.

Basic design practice for adhesive bonded composite joints should include making certain that the surface fibers in a joint are parallel to the load direction to minimize interlaminar shear, or failure, of the bonded adherend or substrate layer. In designs in which joint areas have been machined to a step-lap configuration, for example, it is possible to have a joint interface composed of fibers

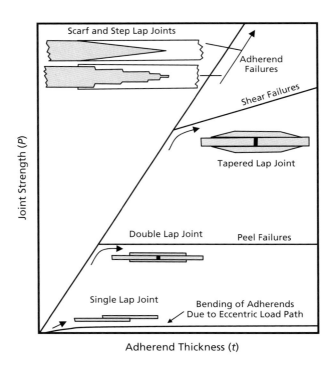

Fig. 8.7. Effect of Adherend Thickness on Failure Modes of Adhesively Bonded Joints[4]

at an orientation other than the optimal 0° orientation to the load direction. This tends to induce substrate failure more readily than would otherwise occur.

8.5 Adhesive Testing

Adhesive bond strength is usually measured by the simple single lap shear test shown in Fig. 8.8. The lap shear strength is reported as the failure stress in the adhesive, which is calculated by dividing the failing load by the bond area. Since the stress distribution in the adhesive is not uniform over the bond area (it peaks at the edges of the joint as previously shown in Fig. 8.3), the reported shear stress is lower than the true ultimate strength of the adhesive. While this test specimen is relatively easy to fabricate and test, it does not give a true measure of the shear strength, due to adherend bending and induced peel loads. In addition, there is no method of measuring the shear strain and thus calculating the adhesive shear modulus that is required for structural analysis. To measure the shear stress vs. shear strain properties of an adhesive as previously shown in Fig. 8.4, an instrumented thick adherend test can be run where the adherends are so thick that the bending forces are negligible. However, the single lap shear test is an effective screening and process control test for evaluating adhesives, surface preparations, and for in-process control. There are many other tests for characterizing adhesive systems.[5]

When testing or characterizing adhesive materials, there are several important points that should be considered: (1) all test conditions must be carefully controlled including the surface preparation, the adhesive, and the bonding cycle; (2) tests should be run on the actual joint(s) that will be used in production; and (3) a thorough evaluation of the in-service conditions must be tested, including temperature, moisture, and any solvents or fluids that the adhesive will be exposed to during its service life. The failure modes for all test specimens should be examined. Some acceptable and unacceptable failure modes are shown in Fig. 8.9. For example, if the specimen exhibits an adhesive failure at the adherend-to-adhesive interface, rather than a cohesive failure within the

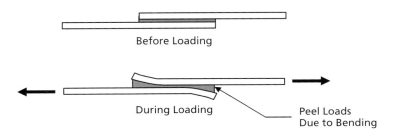

Fig. 8.8. Typical Single Lap Shear Test Specimen[3]

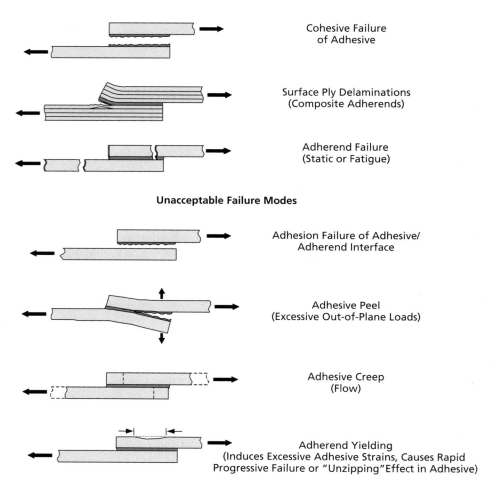

Fig. 8.9. Typical Failure Modes of Bonded Joints[3]

adhesive, it may be an indication of a surface preparation problem that will result in decreased joint durability.

8.6 Surface Preparation[1]

Surface preparation of a material prior to bonding is the keystone upon which the adhesive bond is formed. Extensive field service experience with structural adhesive bonds has repeatedly demonstrated that adhesive durability and longevity depends on the stability and bondability of the adherend surface.

Adhesive Bonding and Integrally Cocured Structure

In general, high performance structural adhesive bonding requires that great care be exercised throughout the bonding process to insure the quality of the bonded product. Chemical composition control of the adhesive, strict control of surface preparation materials and process parameters, and control of the adhesive lay-up, part fit-up, tooling, and the curing process are all required to produce durable structural assemblies.

The first consideration for preparing a composite part for secondary adhesive bonding is moisture absorption of the laminate itself. Absorbed laminate moisture can diffuse to the surface of the laminate during elevated temperature cure cycles, resulting in weak bonds or porosity or voids in the adhesive bondline, and in extreme cases, where fast heat-up rates are used, actual delaminations within the composite laminate plies. If honeycomb is used in the structure, moisture can turn to steam resulting in node bond failures or blown core. Relatively thin composite laminates (0.125 in. or less in thickness) can by effectively dried in an air-circulating oven at 250° F for 4 h minimum. Drying cycles for thicker laminates should be developed empirically using the actual adherend thicknesses. After drying, the surface should be prepared for bonding and then the actual bonding operation conducted as soon as possible. It should be noted that prebond thermal cycles, such as those using encapsulated film adhesive to check for part fit-up prior to actual bonding, can also serve as effective drying cycles. In addition, storage of dried details in a temperature and humidity controlled lay-up room can extend the time between drying and curing.

Numerous surface preparation techniques are currently used prior to the adhesive bonding of composites. The success of any technique depends on establishing comprehensive material, process, and quality control specifications and adhering to them strictly. One method that has gained wide acceptance is the use of a peel ply. In this technique, a closely woven nylon or polyester cloth is used as the outer layer of the composite during lay-up; this ply is torn or peeled away just before bonding or painting. The theory is that the tearing, or peeling process, fractures the resin matrix coating and exposes a clean, virgin, roughened surface for the bonding process. The surface roughness attained can, to some extent, be determined by the weave characteristics of the peel ply. Some manufacturers advocate that this is sufficient, while others maintain that an additional hand sanding or light grit blasting is required to adequately prepare the surface. The abrasion increases the surface area of the surfaces to be bonded and may remove residual contamination, as well as removing fractured resin left behind from the peel ply. The abrading operation should be conducted with care, however, to avoid exposing or rupturing the reinforcing fibers near the surface.

The use of peel plies on composite surfaces that will be structurally bonded certainly deserves careful consideration. Factors that need to be considered include: the chemical makeup of the peel ply (e.g., nylon vs. polyester), as well as its compatibility with the composite matrix resin; the surface treatment used on the peel ply (e.g., silicone coatings that make the peel ply easier to remove can also leave residues that inhibit structural bonding); and the final surface preparation (e.g., hand sanding vs. light grit blasting) employed. The reader is referred to References[6,7] for a more in-depth analysis of the potential pitfalls of using peel plies on surfaces to be bonded. The authors of References 6 and 7 maintain that the only truly effective method of surface preparation is a light grit blast after peel ply removal. Nevertheless, peel plies are very effective at preventing gross surface contamination that could occur between laminate fabrication and secondary bonding.

A typical cleaning sequence would be to remove the peel ply and then lightly abrade the surface with a dry grit blast at approximately 20 psi. After grit blasting, any remaining residue on the surface can be removed by dry vacuuming or wiping with a clean dry chessecloth. Although hand sanding with 120–240 grit silicon carbide paper can be substituted for grit blasting, hand sanding is not as effective as grit blasting in reaching all of the surface impressions left by the weave of the peel ply. In addition, the potential for removing too much resin and exposing the carbon fibers is actually higher for hand sanding than it is for grit blasting.

If it is not possible to use a peel ply on a surface requiring adhesive bonding, the surface can be precleaned (prior to surface abrasion) with a solvent such as methyl ethyl ketone to remove any gross organic contaminants. In cases where a peel ply is not used, some type of light abrasion followed by a dry wipe (or vacuum) is then required to break the glazed finish on the matrix resin surface. The use of solvents to remove residue after hand sanding or grit blasting is discouraged, due to the potential of recontaminating the surface.

Another method can be used to avoid abrasion damage to fibers. When the carbon composite is first laid-up, a ply of adhesive film is placed on the surface where the secondary bond is to take place. This adhesive is then cured together with the laminate. To prepare for the secondary bond, the surface of this adhesive ply is abraded with minimal chance of fiber damage; however, this sacrificial adhesive ply adds weight to the structure.

All composite surface treatments should have the following principles in common: (1) the surface should be clean prior to abrasion to avoid smearing contamination into the surface; (2) the glaze on the matrix surface should be roughened without damaging the reinforcing fibers, or forming subsurface cracks in the resin matrix; (3) all residue should be removed from the abraded surface in a dry process (i.e., no solvent); and (4) the prepared surface should be bonded as soon as possible after preparation.

Aluminum and titanium are often bonded in composite assemblies. Although aluminum should not be bonded directly to carbon/epoxy because the large differences in the coefficients of thermal expansion will result in significant residual stresses, and because carbon fiber in contact with aluminum will form a galvanic cell that will corrode the aluminum. Although seemingly adequate bond strength can often be obtained with rather simple surface treatments (e.g., surface

Adhesive Bonding and Integrally Cocured Structure

abrasion or sanding of aluminum adherends), long-term durable bonds under actual service environments can suffer significantly, if the metal adherend has not been processed using the proper chemical surface preparation.

Several different methods are used to prepare aluminum alloys for adhesive bonding. All have advantages and disadvantages that should be considered, including cost, cycle time, bond durability, performance, and environmental compliance. Aluminum alloys can be precleaned by vapor degreasing followed by alkaline cleaning. The main objective of aluminum etching, or anodizing procedures, is to create a clean surface that contains a porous oxide layer that the adhesive can flow into and become mechanically interlocked. The surface morphologies of the three most prevalent commercial processes are shown in Fig. 8.10. Forest Products Laboratory (FPL) etching is a chromic–sulfuric

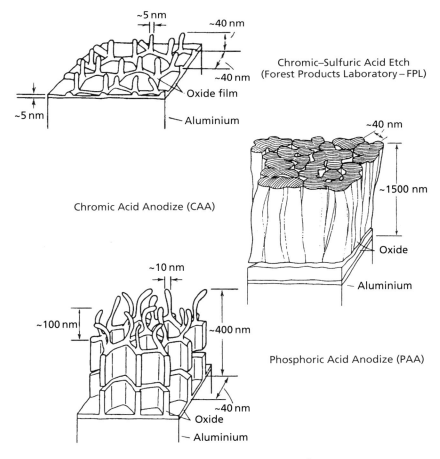

Fig. 8.10. Surface Morphologies of Etched Aluminum Surfaces[8]

381

acid etch and is one of the earliest modern methods developed for preparing aluminum for bonding. Chromic acid anodizing is a later method and is perhaps more widely used than the FPL etch. Chromic acid etching produces a thicker, more robust oxide film than the FPL process. Different manufacturers use minor variations of this method, usually in the sealing steps after anodizing. Phosphoric acid anodize (PAA) is the most recent of the well-established procedures and has an excellent service record for environmental durability. It also has the advantage of being very forgiving of minor variations in procedure. The PAA process produces a more open oxide film and thinner oxide film than that produced by the CAA process. It also results in a bound phosphate that improves the durability of the bond.

Several methods are also used with titanium. Any method developed for titanium should undergo a thorough test program prior to production implementation, and then must be monitored closely during production usage. A typical process used in the aerospace industry involves:

- Solvent wiping to remove all grease and oils,
- Liquid honing at 40–50 psi pressure,
- Alkaline cleaning in an air agitated solution maintained at 200–212° F for 20–30 min,
- Thoroughly rinsing in tap water for 3–4 min,
- Etching for 15–20 min in a nitric–hydrofluoric acid solution maintained at a temperature below 100° F,
- Thoroughly rinsing in tap water for 3–4 min followed by rinsing in deionized water for 2–4 min,
- Inspecting for a water break free surface,
- Oven drying at 100–170° F for 30 min minimum, and
- Adhesive bonding, or applying primer, within 8 h of cleaning.

The combination of liquid honing, alkaline cleaning, and acid etching results in a complex chemically activated surface topography, which contains a large amount of surface area that the adhesive can penetrate and adhere to. The adhesive bond strength is a result of both mechanical interlocking and chemical bonding. Other methods, such as dry chromic acid anodizing are also used.

Because metallic cleaning is such a critical step, dedicated processing lines are normally constructed, and chemical controls, as well as periodic lap shear cleaning control specimens, are employed to insure in-process control. Automated overhead conveyances are used to transport the parts from tank-to-tank under computer controlled cycles to insure the proper processing time in each tank.

Due to the rapid formation of surface oxides on both titanium and aluminum, the surfaces should be bonded within 8 h of cleaning, or primed with a thin protective coat (0.000 1–0.000 5 in.) of epoxy primer. Primer thickness is important. Actually, thinner coatings within this range give better long-term durability than thicker coatings. Color chips are often used in production to determine primer thickness. For parts that will undergo a severe service environment, priming is always recommended, because today's primers contain corrosion inhibiting compounds (strontium chromates) that enhance long-term durability. The two critical variables in corrosion of metal bonds are the metal surface preparation treatment and the chemistry of the primer. Some primers also contain phenolics, which have been found to produce outstanding bond durability.[9] Once the primer has been cured (e.g., 250°F), the parts may be stored in an environmentally controlled clean room for quite long periods of time (e.g., up to 50 days or longer would not be unusual).

All cleaned and primed parts should be carefully protected during handling or storage to prevent surface contamination. Normally, clean white cotton gloves are used during handling and wax free Kraft paper may be used for wrapping and longer storage. Gloves, which are used to handle cleaned and/or primed adherends, should be tested to insure that they are not contaminated with silicones or hydrocarbons, which can contaminate the bondline, or sulfur which can inhibit the cure of the adhesive.

8.7 Epoxy Adhesives[1]

Epoxy based adhesives are by far the most commonly used materials for bonding or repair of aircraft structures. The existence of a large variety of materials, to fit nearly any handling, curing, or performance requirement, results in an extensive list from which to choose. Epoxy adhesives impart high strength bonds and long-term durability over a wide range of temperatures and environments. The ease with which formulations can be modified makes it fairly easy for the epoxy adhesive fabricator to employ various materials to control specific performance properties, such as density, toughness, flow, mix ratio, pot life/shelf life, shop handling characteristics, cure time/temperature, and service temperature.

Advantages of epoxy adhesives include excellent adhesion, high strength, low or no volatiles during cure, low shrinkage, and good chemical resistance. Disadvantages include cost, brittleness unless modified, moisture absorption that adversely affects properties, and relatively long cure times. A wide range of one- and two-part epoxy systems are available. Some systems cure at room temperature, while others require elevated temperatures.

Epoxy resins used as adhesives are generally supplied as liquids or low melting temperature solids. The rate of the reaction can be adjusted by adding accelerators to the formulation, or by increasing the cure temperature. To improve structural properties, particularly at elevated temperatures, it is common to use cure temperatures close to (or preferably above) the maximum use temperature. Epoxy resin systems are usually modified by a wide range of additives that control particular properties, including accelerators, viscosity modifiers and other flow control additives, fillers and pigments, flexibilizers, and toughening agents. Epoxy based adhesives are available in two basic cure chemistries: room temperature and elevated temperature cure.

8.7.1 Two-part Room Temperature Curing Epoxy Liquid and Paste Adhesives

These systems are most commonly used when a room temperature cure is desired. They are available as clear liquids, or as filled pastes, with a consistency ranging from low viscosity liquids to heavy duty putties. Typical cure times are 5–7 days; however, in most cases 70–75% of the ultimate cure can be achieved within 24 h and, if needed, the pressure can usually be released at that point. Under normal bondline thickness conditions (0.005–0.010 in.), cure can be accelerated with heat without fear of exotherm. A typical cure would be 1 h at 180° F.

Two-part systems require mixing a Part A (the resin and filler portion) with a Part B (the curing agent portion) in a predetermined stoichiometric ratio. Two-part epoxy adhesives usually require mixing in precise proportions to avoid a significant loss of cured properties and environmental stability. The amount of material to be mixed should be limited to the amount needed to accomplish the task. In general, the larger the mass, the shorter the pot, or working, life of the material. Pot life is defined as the period between the time of mixing the resin and curing agent, and the time at which the viscosity has increased to the point when the adhesive can no longer be successfully applied as an adhesive.

Two-part resin systems are frequently used to repair damaged aircraft assemblies. Low viscosity versions can be used to impregnate dry carbon cloth for repair patches or injected into cracked bondlines or delaminations. Thicker pastes are used to bond repairs where more flow control is required. For example, if the material has too low a viscosity and is cured under high pressure, the potential for bondline starvation exists due to excessive flow and squeeze out. Viscosity control of two-part adhesives is usually done with metallic and/or non-metallic fillers. Fumed silica is frequently added to provide slump and flow control.

Many adhesives are of the same resin and curing chemistry family. However, different versions are manufactured (non-filled, metallic or non-metallic filled, thixotroped, low density and toughened) for specific performance requirements. For example, a non-metallic filled adhesive may be preferred over a metallic filled adhesive if there is concern for possible galvanic corrosion in the joint. In thin structures where bending or flexing is a concern, a toughened adhesive is usually warranted. In addition to bonding and repair applications, two-part epoxy paste adhesives are also used for liquid shim applications during mechanical assembly operations. The ability to tailor flow, cure time, and compressive strength has made these materials ideal for use in areas of poor fit-up.

8.7.2 Epoxy Film Adhesives

Structural adhesives for aerospace applications are generally supplied as thin films supported on a release paper and stored under refrigerated conditions (i.e., 0° F). Film adhesives are preferred to liquids and pastes because of their uniformity and reduced void content. In addition, since film adhesives contain latent curing agents that require elevated temperatures to cure, the adhesives are stable at room temperature for as long as 20–30 days. Film adhesives are available using high temperature aromatic amine or catalytic curing agents, with a wide range of flexibilizing and toughening agents. Rubber toughened epoxy film adhesives are widely used in the aircraft industry. The upper temperature limit of 250–350° F is usually dictated by the degree of toughening required and by the overall choice of resins and curing agents. In general, toughening an adhesive results in a lower usable service temperature.

Film materials are frequently supported by fibers (scrim cloth) that serve to improve handling of the films prior to cure, control adhesive flow during bonding, assist in bondline thickness control, and provide galvanic insulation. Fibers can be incorporated as short-fiber mats with random orientation or as a woven cloth. Commonly used fibers are polyesters, polyamides (nylon), and glass. Adhesives containing woven cloth may have slightly degraded environmental properties because of wicking of water by the fiber. Random mat scrim cloth is not as efficient for controlling film thickness as woven cloth, because the unrestricted fibers move during bonding, although spun bonded non-woven scrims do not move and are therefore widely used.

8.8 Bonding Procedures[1]

Some general guidelines for adhesive bonding are summarized in Table 8.2. The basic steps in the adhesive bonding process are:

- collection of all the parts in the bonded assembly, which are then stored as a kit;
- verification of the fit to bondline tolerances;
- cleaning the parts to promote good adhesion;
- application of the adhesive;
- mating of the parts and adhesive to form the assembly;
- application of force concurrent with application of heat to the adhesive to promote cure, if required; and
- inspection of the bonded assembly.

8.8.1 Prekitting of Adherends

Many adhesives have a limited working life at room temperature, and adherends, especially metals, can become contaminated by exposure to the environment.

Table 8.2 General Considerations for Adhesive Bonding[1]

- When received, the adhesive should be tested for compliance with the material specification. This may include both physical and chemical tests.
- The adhesive should be stored at the recommended temperature.
- Cold adhesive should always be warmed to room temperature in a sealed container.
- Liquid mixes should be degassed, if possible, to remove entrained air.
- Adhesives which evolve volatiles during cure should be avoided.
- The humidity in the lay-up area should be below 40% relative humidity for most formulations. Lay-up room humidity can be absorbed by the adhesive and is released later during heat cure as steam, yielding porous bondlines and possibly interfering with the cure chemistry.
- Surface preparation is absolutely critical and should be conducted carefully.
- The recommended pressure and the proper alignment fixtures should be used. The bonding pressure should be great enough to ensure that the adherends are in intimate contact with each other during cure.
- The use of a vacuum as the method of applying pressure should be avoided whenever possible, since an active vacuum on the adhesive during cure can lead to porosity or voids in the cured bondline.
- Heat curing systems are almost always preferred, because they yield bonds that have a better combination of strength and resistance to heat and humidity.
- When curing for a second time, such as during repairs, the temperature should be at least 50° F below the earlier cure temperature. If this is not possible, then a proper and accurate bond form must be used to maintain all parts in proper alignment and under pressure during the second cure cycle.
- Traveler coupons should always be made for testing. These are test coupons that duplicate the adherends to be bonded in material and joint design. The coupon surfaces are prepared by the same method and at the same time as the basic bond. Coupons are also bonded together at the same time with the same adhesive lot used in the basic joint and subjected to the same curing process simultaneously with the basic bond. Ideally, traveler coupons are cut from the basic part, on which extensions have been provided.
- The exposed edges of the bond joint should be protected with an appropriate sealer, such as an elastomeric sealant or paint.

Thus, it is normal practice to kit the adherends so that application of the adhesive and buildup of the bonded assemblies can proceed without interruption. The kitting sequence is determined by the product and production rate. Prefitting of the details is also useful in determining locations of potential mismatch such as high and low spots. A prefit check fixture is often used for complex assemblies containing multiple parts. This fixture simulates the bond by locating the various parts in the their exact relationship to one another, as they will appear in the actual bonded assembly. Prefitting is usually conducted prior to cleaning so that the details can be reworked if necessary.

8.8.2 Prefit Evaluation

For complex assemblies, a prefit evaluation ("verifilm") is frequently conducted, as depicted in Fig. 8.11. The bondline thickness is simulated by placing a vinyl plastic film, or the actual adhesive encased in plastic film, in the bond lines. The assembly is then subjected to the heat and pressure normally used for curing. The

Adhesive Bonding and Integrally Cocured Structure

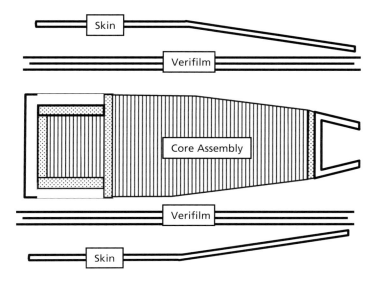

Fig. 8.11. *Prefit of Details Using Verifilm*[3]

parts are disassembled, and the vinyl film or cured adhesive is then visually or dimensionally evaluated to see what corrections are required. These corrections can include sanding the parts to provide more clearance, reforming metal parts to close the gaps, or applying additional adhesive (within permissible limits) to particular locations in the bondline. Verification of bondline thickness may not be required for all applications. However, the technique can be used to validate the fit of the mating parts prior to the start of production, or to determine why large voids are produced in repetitive parts. Once the fit of mating parts has been evaluated, any necessary corrections can be made. For cases in which the component parts can be dimensionally corrected, it is much more efficient to make the correction than risk having to scrap the bonded assembly, or, worse yet, having it fail in service.

8.8.3 Adhesive Application

The most commonly used adhesives are supplied as liquids, pastes, or prefabricated films. The liquid and paste systems may be supplied as one-part or two-part systems. The two-part systems must be mixed before use and thus require scales and a mixer. The amount of material to be mixed should be limited to the amount needed to accomplish the task. The larger the mass, the shorter the pot or working life of the mixed adhesive. To prevent potential exotherm conditions, excess mixed material should be removed from the container and spread out in a thin film. This will prevent the risk of mass-related heat build-up and the possibility of a fire or the release of toxic fumes.

One factor that must be considered in adhesive application is the time interval between adhesive preparation and final assembly of the adherend. This factor, which is referred to as pot, open, out-time, or working life, must be matched to the production rate. Obviously, materials that are ready to bond quickly are needed for high rate applications, such as those found in the automotive and appliance industries. It should be noted that many two-part systems that cure by chemical reaction often have a limited working life before they become too viscous to apply. Application of liquid adhesives can be accomplished using brushes, rollers, manual sprays, or robotically controlled sprays. Application of paste adhesives can be accomplished by brush, by spreading with a grooved tool, or by extrusion from cartridges or sealed containers using compressed air.

Film adhesives are high quality but costly and thus are used mainly in aircraft applications. They consist of an epoxy, a bismaleimide or polyimide resin film, and a fabric carrier. The fabric guarantees a minimum bondline thickness, because it prevents adherends from contacting each other directly. These adhesives are manually cut to size, usually with knives, and placed in the bondlines. When applying film adhesives, it is important to prevent or eliminate entrapped air pockets between the adherend and adhesive film by pricking bubbles or "porcupine" rolling over the adhesive prior to application.

8.8.4 Bond Line Thickness Control

Controlling the thickness of the adhesive bondline is a critical factor in bond strength. This control can be obtained by matching the quantity of available adhesive to the size of the gap between the mating surfaces under actual bonding conditions (heat and pressure). For liquid and paste adhesives, it is a common practice to embed nylon or polyester fibers in the adhesive to prevent adhesive starved bondlines. Applied loads during bonding tend to reduce bondline thickness. A slight overfill is usually desired to insure that the gap is totally filled. Conversely, if all of the adhesive is squeezed out of a local area due to a high spot in one of the adherends, a disbond can result.

For highly loaded bondlines and large structures, film adhesives are used that contain a calendared film with a thin fabric layer. The fabric maintains the bondline thickness by preventing contact between the adherends. In addition, the carrier acts as a corrosion barrier between carbon skins and aluminum honeycomb core. In the most common case, the bondline thickness can vary from 0.002 to 0.010 in. Extra adhesive can be used to handle up to 0.020 in. gaps. Larger gaps must be accommodated by reworking the parts, or by producing hard shims to bring the parts within tolerance.

8.8.5 Bonding

Theoretically, only contact pressure is required so that the adhesive will flow and wet the surface during cure. In reality, somewhat higher pressures are usually required to (1) squeeze out excess adhesive to provide the desired bondline thickness, and/or (2) provide sufficient force to insure all of the interfaces obtain intimate contact during cure.

The position of the adherends must be maintained during cure. Slippage of one of the adherends before the adhesive gels will result in the need for costly reworking, or the entire assembly might be scrapped. When a paste or liquid adhesive is used, it is usually helpful to have a load applied to the joint to deform the adhesive to fill the bondline. C-clamps, spring loaded clamps, shot bags, and jack screws are frequently used for simple configurations. But, if elevated temperature curing is required, some care is required that these pressure devices do not become heat sinks.

Liquid and paste adhesives that are cured at room temperature will normally develop enough strength after 24 h. so that the pressure can be removed. For these adhesives that require moderate cure temperatures (e.g., 180°F), heat lamps or ovens are frequently used. When using heat lamps, some degree of caution is necessary to insure that the part does not get locally overheated. If the contour is complex, it may be necessary to bag the part and employ the isostatic pressure of an autoclave. Instead of using the positive pressure of a vented bag in an autoclave, a vacuum bag (<15 psia) in an oven is quite commonly used. The disadvantage of this process is that the vacuum tends to cause many adhesives to release volatiles and form porous and weak bondlines.[10]

When elevated temperature (e.g., 250–350°F) curing film adhesives are used, autoclave pressures of 15–50 psi are normally used to force the adherends together. The majority of these adhesive systems cure in 1–2 h at elevated temperature. Autoclave bonded parts are made on bond tools very similar to the ones used for cure tooling. The bagging procedures for autoclave bonding are also very similar to those used for composite curing except that bleeder is not required since we are not trying to remove any excess resin during cure. Both straight heat-up and ramped (intermediate hold) cure cycles are used. A typical autoclave cure cycle for a 350°F curing epoxy film adhesive would be:

- Pull a 20–29 in. of Hg vacuum on the assembly and check for leaks. If the assembly contains honeycomb core, do not pull more than 8–10 in. of Hg vacuum.
- Apply autoclave pressure, usually in the range of 15–50 psi. Vent the bag to atmosphere when the pressure reaches 15 psi.
- Heat to 350°F at a rate of 1–5°F/min (Option: an intermediate hold at 240°F for 30 min is sometimes used to allow the liquid resin to thoroughly wet the adherend surfaces).[10]
- Cure at $350 \pm 10°$F for 1–2 h under 15–50 psi.
- Cool to 150°F before releasing autoclave pressure.

During cure, the adhesive flows and forms a fillet at the edge of the bond. It is important not to remove this fillet during clean-up after bonding. Testing has shown that the presence of the fillet significantly improves the joint strength.[11]

8.9 Sandwich Structures

Because it is an extremely lightweight structural approach that exhibits high stiffness and strength-to-weight ratios, sandwich construction is used extensively in both aerospace and commercial industries. The basic concept of a sandwich panel[12] is that the facesheets carry the bending loads (tension and compression), while the core carries the shear loads, much like the I-beam comparison shown in Fig. 8.12. As shown in Fig. 8.13, sandwich construction, especially honeycomb core construction, is extremely structurally efficient,[12] particularly in stiffness critical applications. Doubling the thickness increases the stiffness over 7X with only a 3% weight gain, while quadrupling thickness increases stiffness over 37X with only a 6% weight gain. Little wonder that structural designers like to use sandwich construction whenever possible. Sandwich panels are typically used for their structural, electrical, insulation, and/or energy absorption characteristics.

Facesheet materials that are normally used include aluminum, glass, carbon, or aramid. Typical sandwich structure has relatively thin facesheets (0.010–0.125 in.) with core densities in the range of 1–30 pcf (pounds per cubic foot). Core materials include metallic and non-metallic honeycomb core, balsa wood, open and closed cell foams, and syntactics. A cost versus performance comparison[13] is given in Fig. 8.14. Note that, in general, the honeycomb cores are more expensive than the foam cores but offer superior performance. This

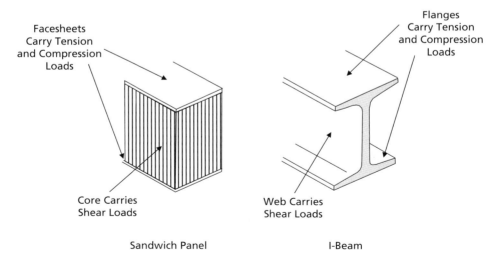

Fig. 8.12. Why Sandwich Structures Are So Efficient[12]

	Solid Material	Sandwich Construction	Thicker Sandwich
Stiffness	1.0	7.0	37.0
Flexural Strength	1.0	3.5	9.2
Weight	1.0	1.03	1.06

Fig. 8.13. Efficiency of Sandwich Structure[12]

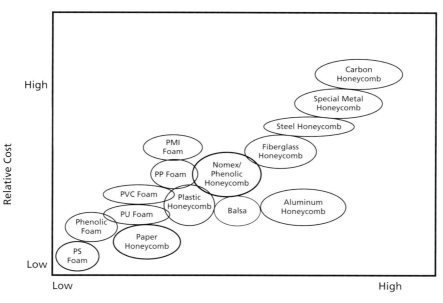

PS – Polystyrene
PU – Polyurethane
PP – Polypropylene
PMI – Polymethacrylimide

Fig. 8.14. Cost vs. Performance for Core Materials[13]

explains why many commercial applications use foam cores, while aerospace applications use the higher performance but more expensive honeycombs. It should also be noted that the foam materials are normally much easier to work with than the honeycombs. A relative strength and stiffness comparison of different core materials is given in Fig. 8.15.

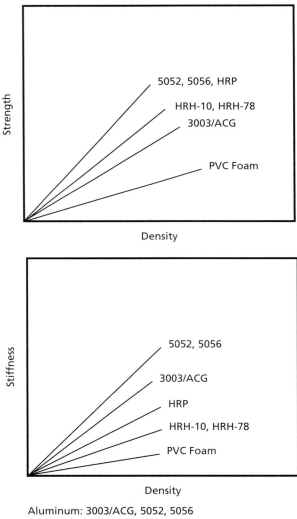

Aluminum: 3003/ACG, 5052, 5056
Nomex: HRH-10, HRH-78
Fiberglass: HRP

Fig. 8.15. Strength and Stiffness of Various Core Materials[12]

Foam core sandwich assemblies can be bonded together with supported film adhesives, but the more common case is either to use liquid/paste adhesives or to do wet lay-up of the skin plies directly on the foam surface. More recently, foam cores with dry composite skins are impregnated and bonded with liquid molding techniques, such as RTM or low pressure VARTM. Supported film adhesives are normally used to bond composite structural honeycomb assemblies.

8.9.1 Honeycomb Core

The details of a typical honeycomb core panel are shown in Fig. 8.16. Typical facesheets include aluminum, glass, aramid, and carbon. Structural film adhesives are normally used to bond the facesheets to the core. It is important that the adhesive provide a good fillet at the core-to-skin interface. Typical honeycomb core terminology is given in Fig. 8.17. The honeycomb itself can be manufactured from aluminum, glass fabric, aramid paper, aramid fabric, or carbon fabric. Honeycomb manufactured for use with organic matrix composites is bonded together with an adhesive, called the node bond adhesive. The "L" direction is the core ribbon direction and is stronger than the width (node

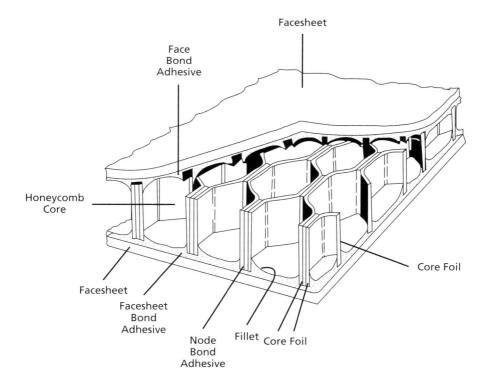

Fig. 8.16. Honeycomb Panel Construction[1]

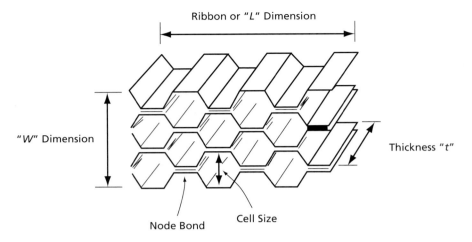

Fig. 8.17. Honeycomb Core Terminology[12]

bond) or "W" direction. The thickness is denoted by "t" and the cell size is the dimension across the cell as shown in the figure.

Although there are a variety of cell configurations available,[14] the three most prevalent (Fig. 8.18) are hexagonal, flexible-core, and overexpanded core. Hexagonal core is by far the most commonly used core configuration. It is available in aluminum and all non-metallic materials. Hexagonal core is structurally very efficient, and can even be made stronger by adding longitudinal reinforcement (reinforced hexagonal) in the "L" direction along the nodes in the ribbon direction. The main disadvantage of the hexagonal configuration is limited formability; aluminum hexagonal core is typically rolled formed to shape, while non-metallic hexagonal core must be heated formed. Flexible-core was developed to provide much better formability. This configuration provides for exceptional formability on compound contours without cell wall buckling. It can be formed around tight radii in both the "L" and the "W" directions. Another configuration with improved formability is overexpanded core. This configuration

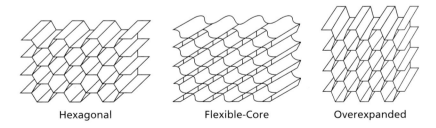

Fig. 8.18. Types of Honeycomb Core Cell Configurations[12]

is hexagonal core that has been overexpanded in the "W" direction, providing a rectangular configuration that facilitates forming in the "L" direction. The "W" direction is about twice the "L" direction. This configuration, as compared to regular hexagonal core, increases the "W" shear properties but decreases the "L" shear properties. An excellent source of more detailed information on honeycomb core is available.[15]

Honeycomb core is normally made by either the expansion or the corrugation process shown in Fig. 8.19. The expansion process is the one that is the most prevalent for lower density (≤ 10 pcf) honeycomb core used for bonded assemblies. The foil is cleaned, corrosion protected if it is aluminum, printed with layers of adhesive, cut to length, stacked, and then placed in a press under heat

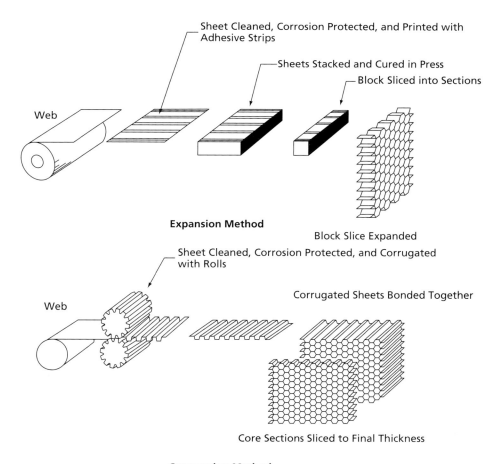

Fig. 8.19. Fabrication Methods for Honeycomb Core[14]

and pressure to cure the node bond adhesive. After curing, the block, or HOBE (honeycomb before expansion), is sliced to the correct thickness and expanded by clamping and then pulling on the edges. Expanded aluminum honeycomb retains its shape at this point due to yielding of the aluminum foil during the expansion process. Non-metallic cores, such as glass or aramid, must be held in the expanded position and dipped in a liquid epoxy, polyester, phenolic, or polyimide resin, which then must be cured before the expansion force can be released. Several dip and cure sequences can be required to produce the desired density.[16] Since phenolics and polyimides are high temperature condensation curing resins, it is important that they are thoroughly cured to drive off all volatiles. If the volatiles are not totally removed during core manufacturing, they can evolve during sandwich curing, creating enough pressure to potentially split the node bonds. Therefore, after the initial cure, it is common practice to post-cure the phenolic or polyimide core at higher temperatures to insure that the reactions are complete. Corrugation is a more expensive process reserved for materials that cannot be made by expansion, or for higher density cores such as ≥ 10 pcf. For example, high temperature metallic core (e.g., titanium) is made by corrugation and then welded together at the nodes to make the completed core sections.

The comparative properties of some of the commercial honeycomb cores are given in Table 8.3. Aluminum honeycomb has the best combination of strength and stiffness. The higher performance aerospace grades are 5052-H39 and 5056-H39, while the commercial grade is 3003 aluminum. Cell sizes range from 1/16 to 3/8 in. but 1/8 and 3/16 in. are the ones most frequently used for aerospace applications. Glass fabric honeycomb can be made from either a normal bi-directional glass cloth or a bias weave ($\pm 45°$) cloth. It is usually impregnated with phenolic resin, but for high temperature applications a polyimide resin is used. The advantage of the bias weave is that it enhances the shear modulus and improves the damage tolerance of the core. There are currently three types of aramid core. The original Nomex core is made by impregnating aramid paper with either a phenolic or a polyimide resin. However, an issue with Nomex is that the resin cannot fully impregnate the paper. Therefore, Dupont developed Korex paper that is thinner and more easily saturated leading to better impregnation, resulting in a core material that has improved mechanical properties and less moisture absorption.[17] Kevlar honeycomb is made by impregnating Kevlar 49 fabric. Finally, bias weave carbon fabric core is a high performance but expensive material that was developed for special applications requiring high specific stiffness and thermal stability when bonded with carbon reinforced facesheets.

The good news about honeycomb core is that it does offer superior performance compared to other sandwich cores. A comparison of strength and stiffness for several core types was previously shown in Fig. 8.15. Note that aluminum core has the best combination of strength and stiffness, followed by the non-metallic honeycombs, and then polyvinyl chloride (PVC) foam. The

Table 8.3 Characteristics of Typical Honeycomb Core Materials[18]

Name and Type of Core	Strength/Stiffness	Maximum Temperature (°F)	Typical Product Forms	Density (pcf)
5052-H39 and 5056-H39 Al Core	High/High	350	Hexagonal Flex-Core	1–12 2–8
3003 Al Commercial Grade Hexagonal Core	High/High	350	Hexagonal	1.8–7
Glass Fabric Reinforced Phenolic	High/High	350	Hexagonal Flex-Core OX	2–12 2.5–5.5 3–7
Bias Weave Glass Fabric Reinforced Phenolic	High/Very High	350	Hexagonal OX	2–8 4.3
Bias Weave Glass Fabric Reinforced Polyimide	High/High	500	Hexagonal	3–8
Aramid Paper Reinforced Phenolic (Nomex)	High/Moderate	350	Hexagonal Flex-Core OX	1.5–9 2.5–5.5 1.8–4
Aramid Paper Reinforced Polyimide (Nomex)	High/Moderate	500	Hexagonal OX	1.5–9 1.8–4
High Performance Aramid Paper Reinforced Phenolic (Korex)	High/High	350	Hexagonal Flex-Core	2–9 4.5
Aramid Fabric Reinforced Epoxy	High/Moderate	350	Hexagonal	2.5
Bias Weave Carbon Fabric Reinforced Phenolic	High/High	350	Hexagonal	4

bad news about honeycomb core is that it is expensive and difficult to fabricate complex assemblies, and the in-service experience, particular with aluminum honeycomb, has not always been good. It can also be very difficult to make major repairs to honeycomb assemblies.

Aluminum honeycomb assemblies have experienced serious in-service durability problems, the most severe being moisture migrating into the assemblies and causing corrosion of the aluminum core cells. Honeycomb suppliers have responded by producing corrosion inhibiting coatings that have improved durability. The newest corrosion protection system, called PAA core, is shown in Fig. 8.20. The core foil is first cleaned and then phosphoric acid anodized. It is then coated with a corrosion inhibiting primer before printing with node bond adhesive. PAA core has demonstrated an approximate three-fold (3X) increase in corrosion protection when compared to typical (non-PAA) corrosion resistant aluminum honeycomb. However, even the most rigorous corrosion protection methods will not stop core corrosion but only delay its onset.

If liquid water is present in the honeycomb cells, freeze-thaw cycles encountered during a typical aircraft flight can cause node bond failures.[19] At high

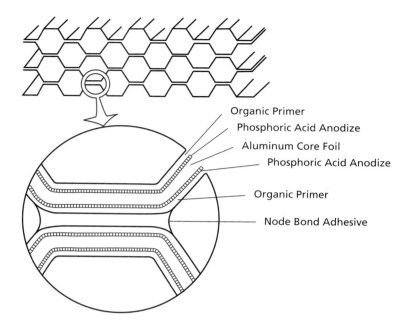

Fig. 8.20. Construction of Phosphoric Acid Anodize Honeycomb Core[1]

altitudes, the standing water in the core freezes, expands, and stresses the cell walls. After landing, the water thaws and the cell walls relax. After a number of these freeze-thaw cycles, the node bonds fail and the damage propagates. It should be noted that this freeze-thaw cyclic damage is not confined to aluminum honeycomb but can occur in the non-metallic cores. In addition, water in the honeycomb can also cause disbonds and delaminate the facesheets, particularly if the temperatures exceed the boiling point of water (212° F), as can happen during operation or repair.

Liquid water normally enters the core through exposed edges, such as panel edges, closeouts, door and window sills, attachment fittings, or almost any location that the skin and core bond terminates. The majority of the damage is often found at the edges of panels.[20] Adhesive bond degradation will lower the skin-to-core bond strength, the fillet bond strength, and the node bond strength. Node bond degradation can reduce the core shear strength so that the assembly fails prematurely by core failure. In addition, water will enter the assembly through any puncture in the facesheets. Since some honeycomb assemblies contain extremely thin skins, water has been known to pass through the skins and then condense on the cell walls. Interconnected microcracks in thin skin honeycomb panels can also allow water ingression.[21] Although absorbed moisture affects the properties of any composite assembly, it is the presence of liquid water in the cells that does the majority of the damage. Many field reports blame water ingression on "poor" sealing techniques. While there is a great deal of truth to the statement that good sealing practices are important, it is the author's opinion that it is just a matter of time before water will find its way into the core of most honeycomb designs and initiates the damage process.

8.9.2 Honeycomb Processing

Honeycomb processing before adhesive bonding includes: perimeter trimming, mechanical or heat forming, core splicing, core potting, contouring, and cleaning.

Trimming. The four primary tools used to cut honeycomb to dimensions are serrated knife, razor blade knife, band saw, and a die. The serrated and razor edge knives and die cutter are used on light density cores, white heavy density cores and complex shaped cores are usually cut with a band saw.

Forming. Metallic hexagonal honeycomb can be roll or brake formed into curved parts. The brake forming method will crush the cell walls and densify the inner radius. Overexpanded honeycomb can be formed to a cylindrical shape on assembly. Flexible core usually can be shaped to compound curvatures on assembly. Non-metallic honeycomb can be heat formed to obtain curved parts. Usually the core is placed in an oven at high temperature for a short period of time (e.g., 550°F for 1–2 min). The heat softens the resin and allows the cell walls to deform more easily. Upon removal from the oven, the core is quickly placed on a shaped tool and held in place until it cools.

Splicing. When large pieces of core are required, or when strength requirements dictate different densities, smaller pieces or different densities of core can be spliced together to form the finished part. This is usually accomplished with a foaming adhesive, as shown in Fig. 8.21. Core splice adhesives normally contain blowing agents that produce gases (e.g., nitrogen) during heat-up to provide the expansion necessary to fill the gaps between the core sections. Foams are one-part epoxy pastes that expand during heating. They are used for core joining, insert potting, and edge filling applications. Foaming film adhesives are thick unsupported films (0.04–0.06 in.) that expand 1.5–3 times their original thickness when cured. Although some of these products can damage the core by overexpanding if too much material is used in the joint, most just expand to fill the gap and then stop when they meet sufficient resistance. A normal practice is to allow up to three layers of foaming adhesive to fill gaps between core sections. Larger gaps call for rework or replacement of the core sections. It is also important to not process some foaming adhesives under a vacuum, or excessive frothing of the foam bondline may occur.

Potting. Potting compounds are frequently required for fitting attachments where fasteners must be put through the honeycomb assembly. As shown in Fig. 8.22, the cells are potted with a high viscosity paste that is cured either during core splicing operations or during final bonding. These compounds usually contain fillers, such as milled glass or aramid fibers, silica, or glass or phenolic

Manufacturing Technology for Aerospace Structural Materials

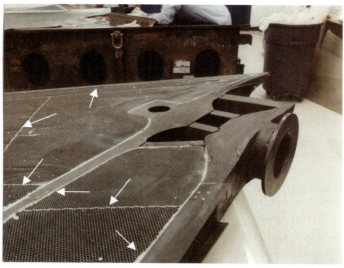

↙ – Locations of Core Splice Bonds

Fig. 8.21. *Complex Structure with Different Core Densities*
Source: The Boeing Company

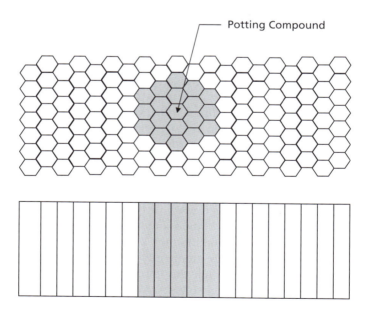

Fig. 8.22. *Core Potting in Honeycomb Core*[3]

microballoons. They can be formulated to cure at room temperature, 250, or 350° F depending upon the intended use temperature for the structure.

Machining. In most applications, honeycomb must have its thickness machined to some contour. This is normally accomplished using valve stem type cutters on expanded core. Occasionally, before expansion, the solid honeycomb block is machined using milling cutters. Typical machines used for contour machining (carving) are gantry, apex, 3-D tracer, or NC five-axis. With five-axis NC machining, the cutting head is controlled by computer programs, and almost any surface that can be described by x-, y-, and z-coordinates can be produced. These machines can carve honeycomb at speeds of up to 3000 in./min with extreme accuracy. A standard contour tolerance of an NC machine is ± 0.005 in. Many core suppliers will supply core machined to contour ready for final bonding.

Cleaning and Drying. It is preferable to keep honeycomb core clean during all manufacturing operations prior to adhesive bonding; however, aluminum honeycomb core can be cleaned effectively by solvent vapor degreasing. Some manufacturers require vapor or aqueous degreasing of all aluminum core prior to bonding; however, most part manufacturers accept "Form B" core from the honeycomb suppliers, and bond without further cleaning. Non-metallic core, such as Nomex or Korex (aramid), fiberglass, and graphite core, readily absorbs moisture from the atmosphere. Similar to composite skins, non-metallic core sections should be thoroughly dried prior to adhesive bonding. A further complication is that since the cell walls are relatively thin and contain a lot of surface area, they can reabsorb moisture rather rapidly after drying, and therefore should be bonded into assemblies as soon as possible after drying.

Honeycomb Bonding. Honeycomb bonding procedures are similar to regular adhesive bonding with a few special considerations. Unlike many composite assemblies, honeycomb assemblies require special closeouts, several of which are shown in Fig. 8.23. During bonding, these require filler blocks in cavities and ramp areas to prevent edge crushing during the cure cycle. Again, closeouts are areas for potential water ingression so special care is required during both the design and the manufacturing process.

Pressure selection is an important consideration during honeycomb bonding. The pressure should be high enough to push the parts together, but not be so high that there is danger of crushing or condensing the core. The allowable pressure depends on both the core density and the part geometry. Common bonding pressures can range anywhere from 15 to 50 psi for honeycomb assemblies. As previously discussed for adhesive bonding, the positive pressure of an autoclave, with a vented bag, gives superior quality to that of a bond produced in an oven under vacuum bag pressure. The amount of pressure, as well as the adhesive selected, are important in forming fillets at the core-to-skin bondlines. The degree of filleting to a large extent determines the strength of the assembly.

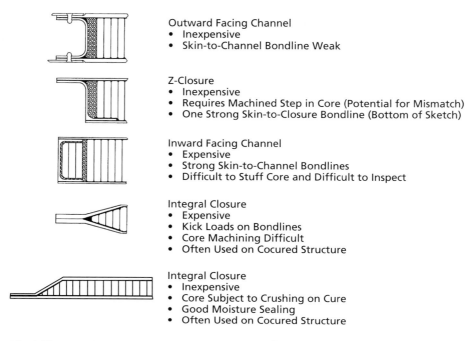

Fig. 8.23. Examples of Honeycomb Structure Close-outs[3]

Pressures that are applied on the sides of the core can easily condense the cells. Since honeycomb is stronger in the longitudinal ("L") direction than the width ("W") direction, the core is more prone to crushing in the "W" direction. Even when the initial vacuum is pulled, vacuum pressure alone has been known to cause core migration and cell crushing. Some manufacturers limit the vacuum level to 8–10 in. of Hg to help in preventing differential pressures within the cells. Autoclave processing of honeycomb assemblies is more sensitive to bag leaks than regular adhesive bonding. If pressure enters the bag through a leak, it can literally blow the honeycomb apart due to the large differential pressure.

Honeycomb assemblies can also be made by cocuring the composite plies onto the core. In this process, the composite skin plies are consolidated and cured at the same time they are bonded to the core. Although a film adhesive is normally used at the skin-to-core interface, self-adhesive prepreg systems are available that do not require a film adhesive. To prevent core crushing and migration, this process is normally conducted at approximately 40–50 psi, as opposed to the normal 100 psi used for regular laminate processing. This can produce skins that are somewhat more porous than those processed under higher pressures, but the biggest drawback is the pillowing or dimpling that occurs in the skins (Fig. 8.24), due to the skin being only supported at the cell walls. Although the schematic is somewhat exaggerated in the amount of pillowing

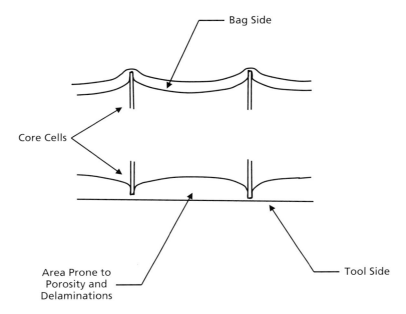

Fig. 8.24. *Pillowing Effect in Composite Cocured Honeycomb Panels*[3]

usually experienced, pillowing does create a serious knockdown in mechanical properties, as much as 30% in some cases.[22] The amount of pillowing can be reduced by using a smaller cell size (e.g., 1/8 vs. 3/16 in.).

Although core migration and crushing can be a problem when bonding precured composite skins, it is an even bigger problem with cocured skins. A considerable amount of work has been done to solve this problem.[23–25] Potential solutions include: (1) reducing the ramp angle (20° or less is recommended); (2) increasing the core density; (3) using grip, or hold down, strips to restrain the plies; (4) potting the cells in the ramp area to increase the core rigidity; (5) encapsulating the core with a layer of adhesive prior to cocuring; (6) bonding fiberglass plies into the center of the core (septums) to increase core rigidity; (7) adjusting the temperature and pressure during heat-up; and (8) even using "high friction" prepregs[26,27] that minimize ply movement during curing.

8.9.3 Balsa Wood

Balsa wood is one of the oldest forms of core materials and was used in early aircraft but now primarily in the boat industry. The core is manufactured by first cutting sections transverse to the grain direction, which are then cut into rectangles, and adhesively bonded together. This results in the grains running perpendicular to the facesheets and is known as end-grained balsa. Typical cell sizes are 0.002 in. in diameter with densities in the range of 6–19 pcf.

End-grain balsa has good mechanical properties, is fairly inexpensive, and is easy to machine and bond to facesheets. The main disadvantages include severe moisture sensitivity, lack of formability, and variable mechanical properties since the block is made by bonding together smaller sections of variable density. Since balsa can absorb large amounts of resin during cure, it is a common practice to seal the surface prior to the lay-up operations.

8.9.4 Foam Cores

A third type of core material frequently used in adhesively bonded structure is foam core. While the properties of foam cores are not as good as honeycomb core, they are used extensively in commercial applications such as boat building and light aircraft construction. The term "polymer foam" or "cellular polymer" refers to a class of materials that are two-phase gas–solid systems in which the polymer is continuous and the gaseous cells are dispersed through the solid. These polymeric foams can be produced by several methods including extrusion, compression molding, injection molding, reaction injection molding, and other solid-state methods.[28] Foam cores are made by using a blowing or foaming agent that expands during manufacture to give a porous, cellular structure. The cells may be open and interconnected or closed and discrete. Usually, the higher the density, the greater the percentage of closed cells. Almost all foams used for structural applications are classified as closed cell, meaning almost all of their cells are discrete. Open cell foams, while good for sound absorption, are weaker than the higher density closed cell foams and also absorb more water, although water absorption in both open and closed cell foams can be problematic. Both uncrosslinked thermoplastic and crosslinked thermoset polymers can be foamed with the thermoplastic foams exhibiting better formability, and the thermoset foams better mechanical properties and higher temperature resistance. Almost any polymer can be made into a foam material by adding an appropriate blowing or foaming agent.

The blowing agents used to manufacture foams are usually classified as either physical or chemical blowing agents. Physical blowing agents are usually gases, mixed into the resin, that expand as the temperature is increased, while chemical blowing agents are often powders that decompose on heating to give off gases, usually nitrogen or carbon dioxide. Although there are foams that can be purchased as two-part liquids that expand after mixing for foam-in-place applications, the majority of structural foams are purchased as pre-expanded blocks that can be bonded together to form larger sections. Sections may be bonded together using either paste or adhesive films. Sections can also be heat formed to contour using procedures similar to those for non-metallic honeycomb core. Although the uncrosslinked thermoplastic foams are easier to thermoform, many of the thermoset foams are only lightly crosslinked and exhibit some formability. Core densities normally range from about 2 to 40 pcf. The most

Table 8.4 Characteristics of Some Foam Sandwich Materials[3]

Name and Type of Core	Density (pcf)	Maximum Temperature (°F)	Characteristics
Polystyrene (Styrofoam)	1.7–3.5	165	Low density, low cost, closed cell foam capable of being thermoformed. Used for wet or low temperature lay-ups. Susceptible to attack by solvents.
Polyurethane Foam	3–29	250–350	Low to high density close cell foam capable of thermoforming at 425–450°F. Both thermoplastic and thermoset foams are available. Used for cocured and secondarily bonded sandwich panels with both flat and complex curved geometries.
Polyvinyl Chloride Foam (Klegecell and Dinvinycell)	1.8–26	150–275	Low to high density foam. Low density can contain some open cells. High density is closed cell. Can be either thermoplastic (better formability) or thermoset (better properties and heat resistance). Used for secondarily bonded or cocured sandwich panels with both flat and complex curved geometries.
Polymethacrylimide Foam (Rohacell)	2–18.7	250–400	Expensive high performance closed cell foam that can be thermoformed. High temperature grades (WF) can be autoclaved at 350°F/100 psi. Used for secondarily bonded or cocured high performance aerospace structures.

widely used structural foams are summarized in Table 8.4. It is important to thoroughly understand the chemical, physical, and mechanical properties of any foam considered for a structural application, particularly with respect to solvent and moisture resistance, and long-term durability. Depending on their chemistry, foam core materials can be used in the temperature range 150–400°F.

Polystyrene cores are lightweight, of low cost, and easy to sand but are rarely used in structural applications due to their low mechanical properties. They cannot be used with polyester resins, because the styrene in the resin will dissolve the core; therefore, epoxies are normally employed.

Polyurethane foams are available as either thermoplastics or thermosets, with varying degrees of closed cells. There are polyurethane foams that are available as finished blocks, and formulations that can be mixed and foamed in place. Polyurethane foams exhibit only moderate mechanical properties and the resin-to-core interface bond tends to deteriorate with age, which can lead to skin delaminations. Polyurethane foams can be readily cut and machined to contours but hot wires should be avoided for cutting since harmful fumes can be released.

Polyvinyl chloride (PVC) foams are one of the most widely used core materials for sandwich structures. PVC foam can be either uncrosslinked (thermoplastic) or crosslinked (thermoset). The uncrosslinked versions are tougher, more damage resistant, and are easier to thermoform, while the crosslinked materials have higher mechanical properties, are more resistant to solvents, and have better temperature resistance. However, the crosslinked foams are more brittle and more difficult to thermoform than the uncrosslinked materials. Because they are not highly crosslinked like the normal thermoset adhesives and matrix systems, they can be thermoformed to contours. The crosslinked systems can be toughened with plasticizers, in which some of the mechanical properties of the normal crosslinked systems are traded for some of the toughness of the uncrosslinked materials. PVC foams are often given a heat stabilization treatment to improve their dimensional stability and reduce the amount of off-gassing during elevated temperature cures. Styrene acrylonitrile (SAN) foams are also available that have mechanical properties similar to crosslinked PVC but have the toughness and elongation of the uncrosslinked PVCs. Patterns of grooves can be scribed in the surfaces of foams to act as infusion aids for resin transfer molding processing.

Polymethylmethacrylimides (PMIs) are lightly crosslinked close-cell foams that have excellent mechanical properties and good solvent and heat resistance. They can be thermoformed to contours and are capable of withstanding autoclave curing with prepregs. These foams are expensive and are usually reserved for high performance aerospace applications.

Due to the inherently lower mechanical properties of foam compared to honeycomb core, manufacturer's are looking at methods of reinforcing foam cores to improve the mechanical properties. Two of these methods include stitching through both the skins and foam core,[29] and the insertion of high strength pultruded pins (carbon, glass, or ceramic) into the foam core to form a truss configuration.[30]

8.9.5 Syntactic Core

Syntactic core consists of a matrix (e.g., epoxy) that is filled with hollow spheres (e.g., glass or ceramic microballoons), as shown in Fig. 8.25. Syntactics can be supplied as pastes for filling honeycomb core or as B-staged formable sheets for core applications. Syntactic cores are generally of much higher density than honeycomb, with densities in the range of 30–80 pcf. The higher the percentage of the microballoon filler, the lighter but weaker the core becomes. Syntactic core sandwiches are used primarily for thin secondary composite structures where it would be impractical or too costly to machine honeycomb to thin gages. When cured against precured composite details, syntactics do not require an adhesive. However, if the syntactic core is already cured and requires adhesive bonding, it should be scuff sanded and then cured with a layer of adhesive.

Glass microballoons are the most prevalent filler used in syntactic core, ranging in diameter from 1 to 350 μm, but typically in the range of 50–100 μm.

Adhesive Bonding and Integrally Cocured Structure

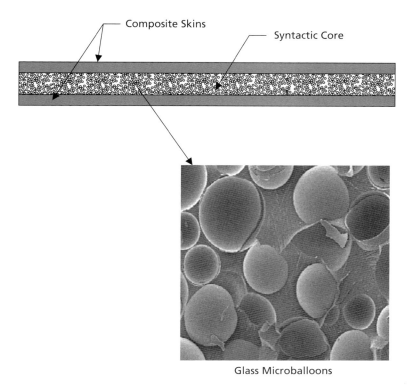

Fig. 8.25. Syntactic Core Construction[1]

Glass microballoons have specific gravities 18 times lower than fillers like $CaCO_3$; however, there have been issues in the past with moisture absorption into the glass microballoons. Ceramic microballoons have properties similar to glass but better elevated temperature properties, while polymeric microballoons (e.g., phenolic) are lower density than either glass or ceramic but have lower mechanical properties. The properties of the microballoons can be improved by increasing the wall thickness at the expense of higher densities. In most commercial applications, the microballoons have a size distribution to improve the packing density. Packing densities as high as 60–80% have been achieved.

8.9.6 Inspection

Adhesively bonded joints and assemblies are normally non-destructively inspected after all bonding operations are completed. Radiographic and ultrasonic inspection methods are typically used to look for defects in both the bondlines and the honeycomb core portions of the assemblies. In addition, it is quite common practice to leak check honeycomb bonded assemblies by immersing the assembly for a short time in a tank of hot water (e.g., 180°F). The hot water heats the residual air inside the honeycomb core and any leaks can be detected by air bubbles escaping from the assembly.

8.10 Integrally Cocured Structure

Integrally cocured or unitized structure is another manufacturing approach that can greatly reduce the part count and final assembly costs for composite structures. The process flow for an integral cocured control surface[31] is shown in Fig. 8.26. In this particular structure, the spars are cocured to the lower skin. The upper skin is cured at the same time as the spars are cocured to the lower but is separated from the lower skin and spar assembly by a layer of release film. This is necessary to allow removal of the upper skin for mechanical installation of the ribs and center control box components. After substructure installation, the upper skin is attached to the spar caps with mechanical fasteners.

The plies for the spars are collated and then hot pressure debulked on their individual tooling details. Hot pressure debulking is required to remove excess bulk from the plies so that all of the tooling details will fit together. While the spars are being prepared, both the upper and the lower skins are collated and hot

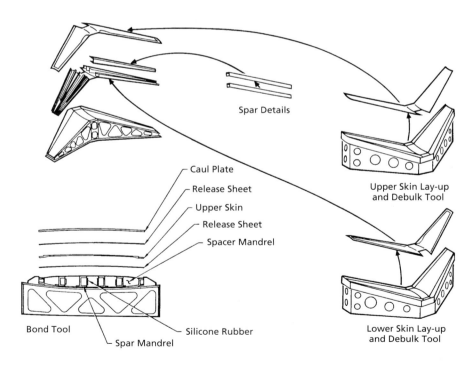

Fig. 8.26. Process Flow for Integrally Cocured Control Surface[31]

pressure debulked on separate plastic lay-up mandrels (PLMs). The lower skin is placed on the tool first, followed by the spar details and the tooling filler blocks that go in the bays between the spars. As previously mentioned, the upper skin is separated from the lower skin and spar assembly by a layer of release material. The completed assembly is bagged, leak checked, and then autoclave cured. Several of the key lay-up and tool assembly sequences are shown in Fig. 8.27. Note that since this type of structure does not contain honeycomb core, drain holes can be drilled in strategic locations to allow any water to drain out of the assembly while in-service.

Pressure is provided by both the autoclave and the expansion of the aluminum substructure blocks, as shown in Fig. 8.28. The autoclave applies pressure to the skins and spar caps, while the expansion of the aluminum substructure blocks applies pressure to the spar webs. If required, the expansion of the aluminum substructure blocks can be supplemented by the presence of silicone rubber intensifiers. The unit after cure and removal from the tool is shown in Fig. 8.29.

Composite Skin Lay-up

Center Box Tooling on Skin

Locating Spar on Skin

Spars and Filler Blocks Located on Skin

Fig. 8.27. Key Process Steps for Integrally Cocured Control Surface
Source: The Boeing Company

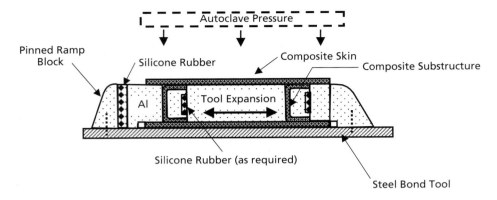

Fig. 8.28. *Pressure Application for Unitized Cocured Structure*

As shown in Fig. 8.30, the upper skin is mechanically fastened to the spar caps after the substructure is completed.

The advantages of this type of structure are obvious: fewer detail parts, fewer fasteners, and fewer problems with part fit-up on final assembly. The main disadvantages are the cost and accuracy of the tooling required, and the complexity of the lay-up that requires a highly skilled workforce. To help control tool accuracy, the substructure tooling is usually machined as a single block (Fig. 8.31) and then sectioned into the individual tooling details.

One potential problem with this type of structure is spring-in. As shown in Fig. 8.32, both the spar cap and the web will spring-in during cool down after cure. The web spring-in can largely be eliminated by placing a couple of plies on the backside to support the web. However, the spar cap cannot be compensated by increasing the angle on the tooling block, because it would make an indentation in the concurrently cured upper skin. It is therefore necessary to shim this joint during final assembly.

Another type of structure that is frequently cocured is skins with cocured hats,[32] such as the one shown in lay-up in Fig. 8.33. Although matched die tooling can be used to make this type of structure, the more common practice is to use localized tooling only at the hat stiffener locations. A typical bagging arrangement, shown in Fig. 8.34, contains an elastomeric mandrel to support the hat during cure; an elastomeric pressure intensifier to insure that the radii obtain sufficient pressure; and some plastic shims to minimize mark-off on the skin from the pressure intensifier. In this figure, the mandrel contains a hole in the center to equalize the pressure. However, some mandrels are solid elastomer or frequently elastomer reinforced with carbon or glass cloth. If the stiffeners require exact location, it may be necessary to use cavity tooling similar to that shown in Fig. 8.35.

Adhesive Bonding and Integrally Cocured Structure

Lower Skin and Spars

Matching Upper Skin

Fig. 8.29. Cocured Unitized Control Surface
Source: The Boeing Company

One of the key design areas for cocured stiffeners are the terminations. Since the bond holding the stiffener to the skin is essentially a resin or an adhesive bond, any peel loads induced at the stiffener ends could cause the bondline to "unzip" and fail. The most prevalent method for preventing this is by installing mechanical fasteners near the stiffener terminations, as shown in the hat design of Fig. 8.36. Note that the hat is also thicker at the ends and scarfed to further help reduce the tendency for bondline peeling. Other methods include stitching and pinning in the transverse (Z) direction. However, stitching of pregreg lay-ups is expensive and can damage the fibers. Z-pining is a technology in which small

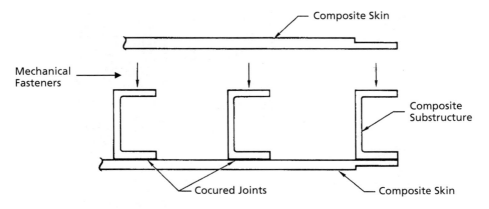

Fig. 8.30. Final Assembly for Integrally Cocured Control Surface

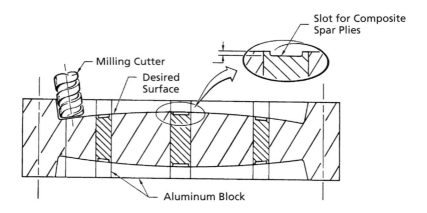

Fig. 8.31. Machining Substructure Spar Mandrels and Filler Blocks

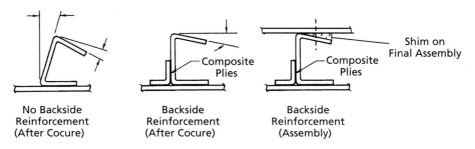

Fig. 8.32. Spring-in for Cocured Joints

Adhesive Bonding and Integrally Cocured Structure

*Fig. 8.33. Lay-up of Fuselage Panel with Cocured Hats
Source: The Boeing Company*

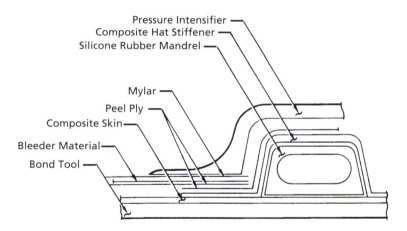

Fig. 8.34. Bagging Procedure for Cocured Hat Section[32]

diameter precured carbon pins are driven through the prepreg lay-up before cure with an ultrasonic gun.

Cobonding is a hybrid process combining cocuring and adhesive bonding. As shown in the joint in Fig. 8.37, a series of precured stiffeners are adhesive bonded to a skin at the same time the skin is cured. The surfaces of the precured composite parts must, of course, be prepared for adhesive bonding just like any other adhesive bonding process. The advantage of this process is that, in

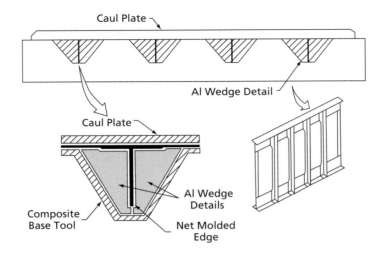

Fig. 8.35. *Cavity Tool for Precise Location of Substructure*

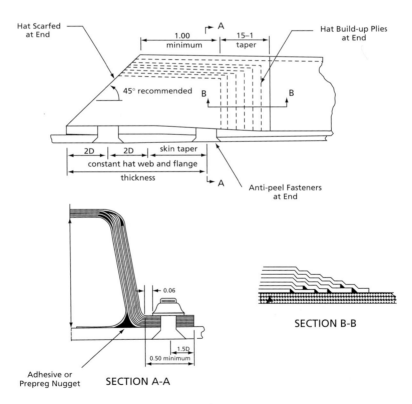

Fig. 8.36. *Methods of Reducing Hat Stiffener Peel*

Adhesive Bonding and Integrally Cocured Structure

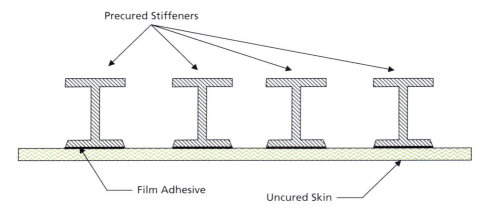

Fig. 8.37. Principle of Cobonding

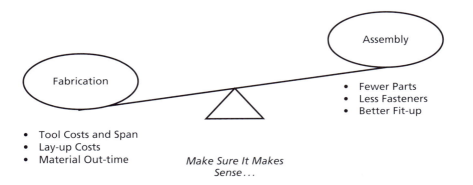

Fig. 8.38. Advantages and Disadvantages of Unitized Cocured Structure

certain situations, the amount of tooling required for the curing operation can be reduced.

Cocured unitized structure is a trade-off. As shown in Fig. 8.38, the advantages include fewer parts, less fasteners, and better fit-up at assembly. These advantages must be weighed against the additional tooling costs and span time to produce the tooling, the lay-up costs, and personnel requirements for large complex assemblies, and material out-time issue if the lay-up requires a long time.

Summary

Adhesive bonding is often preferred if thin sections are to be joined where bearing stresses in bolted joints would be unacceptably high, or when the weight penalty for mechanical fasteners is too high. In general, thin structures with stiff and well-defined load paths are good candidates for adhesive bonding, while thicker structures with complex load paths are better candidates for mechanical fastening. It is important to load the adhesive in shear. Tension, cleavage, and peel loading should be avoided to prevent premature failures.

Adhesively bonded structure offers the potential to reduce the number of parts and fasteners required for final assembly, and therefore lowers the cost of structures. The most critical fabrication aspect of bonding is surface preparation. Without proper surface preparation, the long-term durability of the joint will not be realized. Fortunately, the surface preparation for composites is quite a bit simpler than that for aluminum or titanium. When adhesive bonding composite parts, it is important that all cured parts be thoroughly dried prior to bonding. In addition it is important that prefit operations are conducted to make sure that the parts to be bonded will fit together without excessive high or low areas; otherwise, thick glue lines and unbonds may result.

Sandwich construction is very structurally efficient, particularly in stiffness critical applications. Unfortunately, honeycomb assemblies are difficult to build, have frequently experienced in-service durability problems, and can be difficult to repair. Aluminum honeycomb has been prone to severe corrosion problems. If the honeycomb contains liquid water, all honeycomb assemblies can be damaged from repeated freeze-thaw cycles. In addition, if liquid water is present and the assembly is heated above 212° F, there is the danger of steam pressure delamination.

Foam core sandwich structures are somewhat easier to build than honeycomb structures, but the current foams do not have as good mechanical properties as the honeycombs. In addition, foam cores are subject to quite high moisture saturation levels. Both types of core materials can be fabricated into composites structures be either precuring the composite details and then bonding them to the core, or by cocuring the assembly with prepreg plies at one time.

Cocured unitized structure is yet another option for reducing assembly costs. Although it eliminates some of the potential durability costs associated with honeycomb, it is more difficult to design and more costly to tool than honeycomb assemblies. A variation of cocuring, called cobonding, can reduce some of the tooling costs for certain types of assemblies.

Recommended Reading

[1] Campbell, F.C., "Secondary Adhesive Bonding of Polymer-Matrix Composites", in *ASM Handbook Vol. 21 Composites*, ASM International, 2001.
[2] Campbell, F.C., *Manufacturing Processes for Advanced Composites*, Elsevier Ltd, 2004.
[3] Scardino, W.M., "Adhesive Specifications", in *ASM Engineered Materials Handbook Vol. 1 Composites*, ASM International, 1987.
[4] Bitzer, T., *Honeycomb Technology – Materials, Design, Manufacturing, Applications and Testing*, Chapman & Hall, 1997.

References

[1] Campbell, F.C., "Secondary Adhesive Bonding of Polymer-Matrix Composites", in *ASM Handbook Vol. 21 Composites*, ASM International, 2001.
[2] "Redux Bonding Technology", Hexcel Composites, December 2001.
[3] Campbell, F.C., "Adhesive Bonding and Integrally Cocured Structure", in *Manufacturing Processes for Advanced Composites*, Elsevier Ltd, 2004, pp. 242–299.
[4] Heslehurst, R.B., Hart-Smith, L.J., "The Science and Art of Structural Adhesive Bonding", *SAMPE Journal*, Vol. 38, No. 2, March/April 2002, pp. 60–71.
[5] Scardino, W.M., "Adhesive Specifications", in *ASM Engineered Materials Handbook Vol. 1 Composites*, ASM International, 1987, pp. 689–701.
[6] Hart-Smith, L.J., Brown, D., Wong, S., "Surface Preparations for Ensuring that the Glue Will Stick in Bonded Composite Structures", 10th DOD/NASA/FAA Conference on Fibrous Composites in Structural Design, 1–4 November 1993, Hilton Head Island, SC.
[7] Hart-Smith, L.J., Redmond, G., Davis, M.J., "The Curse of the Nylon Peel Ply", 41st SAMPE International Symposium and Exhibition, 25–28 March 1996, Anaheim, CA.
[8] Venables, J.D., McNamara, D.K., Chen, J.M., Sun, T.S., Hopping, J.L., *Applied Surface Science*, Vol. 3, 1979, p. 88.
[9] Krieger, R.B., "A Chronology of 45 Years of Corrosion in Airframe Structural Bonds", 42nd International SAMPE Symposium, 4–8 May 1997, pp. 1236–1242.
[10] Hinrichs, R.J., "Vacuum and Thermal Cycle Modifications to Improve Adhesive Bonding Quality Consistency", 34th International SAMPE Symposium, 8–11 May 1989, pp. 2520–2529.
[11] Gleich, D.M., Tooren, M.J., Beukers, A., "Structural Adhesive Bonded Joint Review", 45th International SAMPE Symposium, 21–25 May 2000, pp. 818–832.
[12] "HexWeb Honeycomb Sandwich Design Technology", Hexcel Composites, 2000.
[13] Kindinger, J., "Lightweight Structural Cores", in *ASM Handbook Vol. 21 Composites*, ASM International, 2001.
[14] Corden, J., "Honeycomb Structures", in *ASM Engineered Materials Handbook Vol. 1 Composites*, ASM International, 1987.
[15] Bitzer, T., *Honeycomb Technology – Materials, Design, Manufacturing, Applications and Testing*, Chapman & Hall, 1997.
[16] Danver, D., "Advancements in the Manufacture of Honeycomb Cores", 42nd International SAMPE Symposium, 4–8 May 1997, pp. 1531–1542.
[17] Black, S., "Improved Core Materials Lighten Helicopter Airframes", *High-Performance Composites*, May 2002, pp. 56–60.
[18] "HexWeb Honeycomb Selector Guide", Hexcel Composites, 1999.
[19] Radtke, T.C., Charon, A., Vodicka, R., "Hot/Wet Environmental Degradation of Honeycomb Sandwich Structure Representative of F/A-18: Flatwise Tension Strength", Australian Defence Science & Technology Organization (DSTO), Report DSTO-TR-0908.
[20] Whitehead, S., McDonald, M., Bartholomeusz, R.A., "Loading, Degradation and Repair of F-111 Bonded Honeycomb Sandwich Panels – A Preliminary Study", Australian Defence Science & Technology Organization (DSTO), Report DSTO-TR-1041.
[21] Loken, H.Y., Nollen, D.A., Wardle, M.W., Zahr, G.E., "Water Ingression Resistant Thin Faced Honeycomb Cored Composite Systems with Facesheets Reinforced with Kevlar Aramid Fiber and Kevlar with Carbon Fibers", E.I. DuPont de Nemours & Company.
[22] Stankunas, T.P., Mazenko, D.M., Jensen, G.A., "Cocure Investigation of a Honeycomb Reinforced Spacecraft Structure", 21st International SAMPE Technical Conference, 25–28 September 1989, pp. 176–188.

[23] Brayden, T.H., Darrow, D.C., "Effect of Cure Cycle Parameters on 350° F Cocured Epoxy Honeycomb Panels", 34th International SAMPE Symposium, 8–11 May 1989, pp. 861–874.
[24] Zeng, S., Seferis, J.C., Ahn, K.J., Pederson, C.L., "Model Test Panel for Processing and Characterization Studies of Honeycomb Composite Structures", *Journal of Advanced Materials*, January 1994, pp. 9–21.
[25] Renn, D.J., Tulleau, T., Seferis, J.C., Curran, R.N., Ahn, K.J., "Composite Honeycomb Core Crush in Relation to Internal Pressure Measurement", *Journal of Advanced Materials*, October 1995, pp. 31–40.
[26] Hsiao, H.M., Lee, S.M., Buyny, R.A., Martin, C.J., "Development of Core Crush Resistant Prepreg for Composite Sandwich Structures", 33rd International SAMPE Technical Conference, 5–8 November 2001.
[27] Harmon, B., Boyd, J., Thai, B., "Advanced Products Designed to Simplify Co-Cure Over Honeycomb Core", 33rd International SAMPE Technical Conference, 5–8 November 2001.
[28] Weiser, E., Baillif, F., Grimsley, B.W., Marchello, J.M., "High Temperature Structural Foam", 43rd International SAMPE Symposium, 31 May–4 June 1998, pp. 730–740.
[29] Herbeck, I.L., Kleinberg, M., Schoppinger, C., "Foam Cores in RTM Structures: Manufacturing Aid or High-Performance Sandwich?", 23rd International Europe Conference of SAMPE, 9–11 April 2002, pp. 515–525.
[30] Carstensen, T., Cournoyer, D., Kunkel, E., Magee, C., "X-Cor™ Advanced Sandwich Core Material", 33rd International SAMPE Technical Conference, 5–8 November 2001.
[31] Moors, G.F., Arseneau, A.A., Ashford, L.W., Holly, M.K., "AV-8B Composite Horizontal Stabilator Development", 5th Conference on Fibrous Composites in Structural Design, 27–29 January 1981.
[32] Watson, J.C., Ostrodka, D.L., "AV-8B Forward Fuselage Development", 5th Conference on Fibrous Composites in Structural Design, 27–29 January 1981.

Chapter 9

Metal Matrix Composites

Metal matrix composites offer a number of advantages compared to their base metals, such as higher specific strengths and moduli, higher elevated temperature resistance, lower coefficients of thermal expansion, and, in some cases, better wear resistance. On the down side, they are more expensive than their base metals and have lower toughness. Metal matrix composites also have some advantages compared to polymer matrix composites, including higher matrix dependent strength and moduli, higher elevated temperature resistance, no moisture absorption, higher electrical and thermal conductivities, and nonflammability. However, metal matrix composites (MMCs) are normally more expensive than even polymer matrix composites, and the fabrication processes are much more limited, especially for complex structural shapes. Due to their high cost, commercial applications for metal matrix composites are sparse. There are some limited uses for discontinuously reinforced MMCs but almost no current applications for continuously reinforced MMCs.

Metal matrix composites can be subdivided according to the type of reinforcement shown in Fig. 9.1. The reinforcement can be particulates (particles which are approximately equiaxed); high strength single crystal whiskers; short

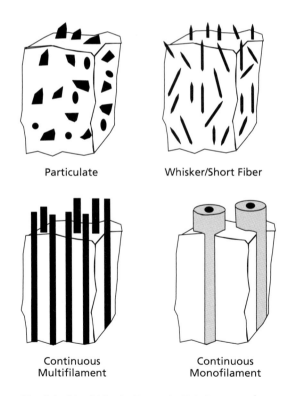

Fig. 9.1. Metal Matrix Composite Reinforcements[1]

Fig. 9.2. Silicon Carbide Particulate

fibers which are usually random but can contain some degree of alignment; or long aligned multifilament or monofilament fibers. Particulate reinforced composites (Fig. 9.2), primarily silicon carbide (SiC) or alumina (Al_2O_3) ceramic particles in an aluminum matrix, are known as discontinuously reinforced aluminum (DRA). They exhibit high stiffness, low density, high hardness, adequate toughness at volume percentages less than 25%, and relatively low cost. Normal volume percentages are 15–25% with SiC_p particle diameters of 3–30 μm. DRA is usually manufactured by melt incorporation during casting, or by powder blending and consolidation. MMCs can also be reinforced with single crystal SiC_w whiskers (Fig. 9.3), made by either vapor deposition processes or from rice hulls. They have better mechanical properties than particulate reinforced MMCs, but whiskers are more expensive than particles; it is more difficult to obtain a uniform dispersion in the matrix; and there are health hazard concerns for whiskers. Short fibers are also used to reinforce MMCs; for example, Saffil short alumina fibers are used in aluminum matrices. Short fiber reinforced MMCs can also be produced by melt infiltration, squeeze casting, or by powder blending/consolidation. Typical fiber diameters are a few micrometers to several hundred micrometers in length but are broken-up during processing so that their aspect ratios range from 3 to about 100.

Large stiff boron and silicon carbide monofilament fibers (Fig. 9.4) have been evaluated for high value aerospace applications. Early work was conducted in the 1970s with boron, or Borsic (a silicon carbide-coated boron fiber), reinforced aluminum for aircraft/spacecraft applications. Later work in the 1990s concentrated on SiC monofilaments in titanium (Fig. 9.5) for the National

Fig. 9.3. Silicon Carbide Whiskers

Fig. 9.4. Silicon Carbide Monofilament

Aerospace Plane. Potential applications are extremely high temperature airframes and engine components. Multifilament fibers, such as carbon and the ceramic fibers Nextel (alumina based) and Nicalon (silicon carbide), have also been used in aluminum and magnesium matrices; however, the smaller and more numerous multifilament tows are difficult to impregnate using solid state processing techniques, such as diffusion bonding, because of their small size and tightness of their tow construction. In addition, carbon fiber readily reacts

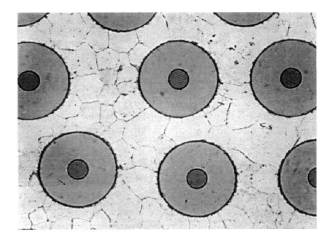

Fig. 9.5. Silicon Carbide Monofilament/Titanium Matrix Composite

with aluminum and magnesium during processing and can cause these matrices to galvanically corrode during service.

Most of the commercial work on MMCs has focused on aluminum as the matrix metal. Aluminum's combination of lightweight, environmental resistance, and useful mechanical properties are attractive. The melting point is high enough to meet many applications, yet low enough to allow reasonable processing temperatures. Also, aluminum can accommodate a variety of reinforcing agents. Although much of the early work on aluminum MMCs concentrated on continuous fibers, most of the present work is focused on discontinuously reinforced (particulate) aluminum MMCs, because of their greater ease of manufacture, lower production costs, and relatively isotropic properties. The most common reinforcement materials in discontinuously reinforced aluminum composites are SiC and Al_2O_3, although silicon nitride (Si_3N_4), TiB_2, graphite, and others have also been used in some specialized applications. For example, graphite/aluminum MMCs have been developed for tribological applications due to their excellent anti-friction properties, wear resistance, and anti-seizure characteristics. Typical fiber volumes of discontinuous DRAs are usually limited to 15–25%, because higher volumes result in low ductility and fracture toughness.

The primary MMC fabrication processes are often classified as either liquid phase or solid state processes. Liquid phase processing is generally considerably less expensive than solid state processing. Characteristics of liquid phase processed discontinuous MMCs include low cost reinforcements, such as silicon carbide particles, low temperature melting matrices such as aluminum and magnesium, and near net shaped parts. Liquid phase processing results in intimate interfacial contact and strong reinforcement-to-matrix bonds, but also can result in the formation of brittle interfacial layers as a result of interactions with the high temperature liquid matrix. Liquid phase processes include various casting processes, liquid metal infiltration, and spray deposition. However, since continuous aligned fiber reinforcement is normally not used in liquid state processes, the strengths and stiffness are lower.

Solid state processes, in which no liquid phase is present, are usually associated with some type of diffusion bonding to produce final consolidation, whether the matrix is in a thin sheet or powder form. Although the processing temperatures are lower for solid state diffusion bonding, they are often still high enough to cause significant reinforcement degradation. In addition, the pressures are almost always higher for the solid state processes. The choice of a fabrication process for any MMC is dictated by many factors, the most important being: preservation of reinforcement strength; preservation of reinforcement spacing and orientation; promotion of wetting and bonding between the matrix and reinforcement; and minimization of reinforcement damage, primarily due to chemical reactions between the matrix and reinforcement.

9.1 Discontinuously Reinforced Metal Matrix Composites

Processing methods for discontinuous reinforced aluminum MMCs include various casting processes, liquid metal infiltration, spray deposition, and PM. A wide range of conventional, as well as specialized, aluminum casting processes have been explored for DRAs. Casting is currently the most inexpensive method of producing MMCs and lends itself to the production of large ingots, which can be further worked by extrusion, hot rolling, or forging. A relative comparison of composite performance with materials and process technologies for a number of MMCs is shown in Fig. 9.6.

9.2 Stir Casting

In casting of metal matrix composites, the reinforcement is incorporated as loose particles or whiskers into the molten metal matrix. Because most metal reinforced systems exhibit poor wetting, the mechanical force produced by stirring is required to combine the phases. In stir casting, shown schematically in Fig. 9.7, the particulate/whisker/short fiber reinforcement is mechanically mixed into a molten metal bath. A heated crucible is used to maintain the molten matrix at the desired temperature, while a motor drives a mixing impeller that is submerged in the molten matrix. The reinforcement is slowly poured into molten matrix at a controlled rate to insure a smooth and continuous feed. As the impeller rotates, it generates a vortex that draws the reinforcement particles into the melt from the surface. The impeller is designed to create high shear to strip adsorbed gases from the reinforcement surfaces. Shearing also helps to cover the reinforcement with molten matrix, promoting reinforcement

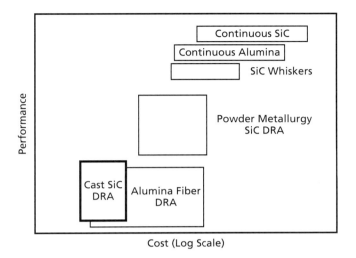

Fig. 9.6. Performance/Cost Trade-offs for MMCs[2]

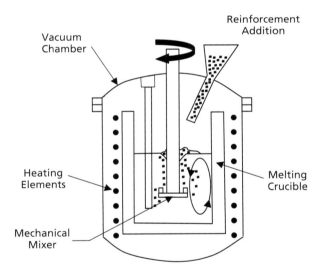

Fig. 9.7. Stir Casting[3]

wetting. Proper mixing techniques and impeller design are required to produce adequate melt circulation and a homogeneous reinforcement distribution. The use of an inert atmosphere or a vacuum is essential to avoid the entrapment of gases.[4] A vacuum prevents the uptake of gases; eliminates the gas boundary layer at the melt surface; and eases the conditions for particle entry into the melt. Difficulties can include segregation/settling of the secondary phases in the matrix, particle agglomeration, particle fracture during stirring, and excessive interfacial reactions.

Good particle wetting is important. Adding particles, and especially fibers, to the melt increases the viscosity. In general, ceramic particles are not wetted by the metallic matrix.[5] Wetting can be enhanced by coating the reinforcement with a wettable metal which serves three purposes: (1) protects the reinforcement during handling, (2) improves wetting, and (3) reduces particle agglomeration. For example, the addition of magnesium to molten aluminum has been used to improve the wettability of alumina, and silicon carbide particles have been coated with graphite to improve their weattability in A356 aluminum.[5] Effective wetting also becomes more difficult as the particle size decreases due to increases in surface energy, greater difficulty in dispersing due to increased surface area, and a greater tendency to agglomerate.

Microstructural inhomogeneities can occur due to particle agglomeration and sedimentation in the melt. Redistribution, as a result of the reinforcements being pushed by an advancing solidification front, can also cause segregation. After mixing, reinforcement segregation can occur due to both gravity effects and solidification effects. The reinforcement, when it encounters a moving liquid/solid interface, may either be engulfed in the metal or it may be pushed by the interface into areas that solidify last, such as interdendritic regions. Since stir casting usually involves prolonged liquid-reinforcement contact, substantial interfacial reactions can result that degrades the composite properties and also increases the viscosity of the melt, making casting more difficult. In SiC_p/Al, the formation of Al_4C_3 and silicon can be extensive. The rate of reaction can be reduced, and even become zero, if the melt is silicon-rich, either by prior alloying or as a result of the reaction.[6] Therefore, stir casting of SiC_p/Al is well suited to high silicon content aluminum casting alloys but not to most wrought alloys.

Unreinforced liquid metals generally have viscosities in the range of 0.1–1.0 P.[7] Adding particles to a liquid metal increases the apparent viscosity because the particles interact with the liquid metal, and each other, resulting in more resistance to shear. Typical values are in the range of 10–20 P for aluminum reinforced with 15 volume percent SiC_p. Since viscosity is a function of reinforcement percentage, shape and size, an increase in volume fraction, or a decrease in size, will increase the viscosity of the slurry, often limiting the reinforcement level to about 30 volume percent.

Porosity in cast parts usually results from gas entrapment during mixing, hydrogen evolution, and/or shrinkage during solidification. Preheating the reinforcement before mixing can help in removing moisture and trapped air between the particles. During casting, porosity can be reduced by: (1) casting in a vacuum, (2) bubbling inert gas through the melt, (3) casting under pressure, and (4) deformation processing after casting to close the porosity. It has been observed that porosity in cast composites increases almost linearly with particle content.

9.3 Slurry Casting – Compocasting

When a liquid metal is vigorously stirred during solidification, it forms a slurry of fine spherical solids floating in the liquid. Stirring at high speeds creates a high shear rate, which tends to reduce the viscosity of the slurry, even at solid fractions as high as 50–60% volume. The process of casting a slurry, where the liquid metal is held between the liquidus and solidus temperatures, is called rheocasting. The slurry can also be mixed with particulates, whiskers, or short fibers before casting. This modified form of rheocasting to produce near net shaped MMC parts is called compocasting. The particulate reinforcement becomes trapped in the slurry, helping to minimize segregation. Continued stirring then reduces the viscosity, resulting in a mutual interaction between the matrix melt and the reinforcement, which enhances wetting and bonding between the two.[4]

Since reinforcements have a tendency to either float to the top or segregate near the bottom of the melt because their densities differ from that of the melt, the advantages of this process are the increase in apparent viscosity of the slurry and the prevention of settling by the buoyant action of the liquid metal.[8] The melt reinforced slurry can be cast by gravity casting, die casting, centrifugal casting or squeeze casting. A careful choice of casting technique, as well as the mold configuration, is important in obtaining a uniform distribution of reinforcements in compocast MMCs.

Continuous stirring of the slurry helps create intimate contact between the reinforcement and the matrix. Good bonding is achieved by reducing the slurry viscosity, as well as by increasing the mixing time. The slurry viscosity can be reduced by increasing the shear rate and by increasing the slurry temperature. Compocasting can be performed at temperatures lower than those conventionally employed in foundry practice during pouring, resulting in reduced thermochemical degradation of the reinforcement surface. The slurry can be transferred directly to a shaped mold prior to complete solidification, or it can be allowed to solidify in a billet or rod shape so that it can be reheated into a slurry for further processing by techniques such as die casting.

9.4 Liquid Metal Infiltration (Squeeze Casting)

As shown in Fig. 9.8, squeeze casting is a metal forming process in which solidification is accomplished under high pressure to help eliminate shrinkage porosity and reduce porosity by keeping gases dissolved in solution. Squeeze cast parts are usually fine grained with excellent surface finishes and almost no porosity.[9] To produce metal matrix composites, porous preforms of reinforcement material are infiltrated by molten metal under pressure. Reinforcement

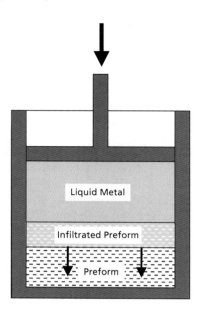

Fig. 9.8. Squeeze Casting[8]

forms can be continuous fiber, discontinuous fiber, or particulate with aluminum or magnesium alloys. The volume fraction of reinforcement in the MMC can vary from 10 to 70 volume percent, depending on the particular application for the material. Pressure is applied to the solidifying system by a hydraulically actuated ram. This process is similar to conventional die casting, except that the ram continues to apply pressure during solidification and the pressures are higher (usually in the range of 1.5–15 ksi) and applied slowly. High pressure helps in increasing processing speed, producing finer matrix microstructures, and producing sounder castings by minimizing solidification shrinkage.[8] To help minimize solidification shrinkage, the pressure is maintained until solidification is complete. If cold dies and reinforcements are used along with high pressures, chemical reactions between the liquid metal and reinforcement can be minimized due to the shorter processing cycles. Squeeze casting is one of the most economical processes for fabricating MMCs and allows relatively large size parts to be made.[5]

Preforms for infiltration are usually prepared by sedimentation of short fibers or particles from a suspension.[1] A Saffil alumina preform is shown in Fig. 9.9. Binders are frequently required to maintain preform integrity for handling. Binders (5–10 weight percent) can either be fugitive that are burned off during casting or high temperature silica or alumina compounds requiring firing before casting.[1] The binding agent is normally introduced through a suspension liquid, so that it deposits or precipitates out on the fibers, often forming preferentially at fiber contact points, where it serves to lock the fiber array into a strong network.

Metal Matrix Composites

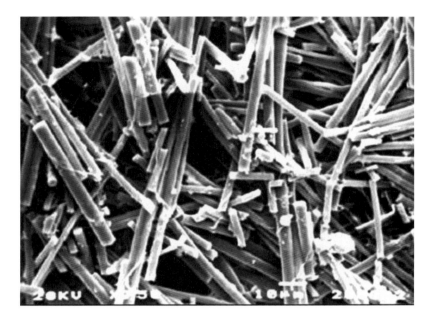

Fig. 9.9. Saffil Alumina Fiber Preform

Composites produced by squeeze casting are usually porosity free; however, the high pressures used may damage the preform, especially during the early stages of infiltration. A major advantage of infiltration processes is that they allow for near net shaped parts with either total reinforcement or areas with selective reinforcement. A limitation is the need for the reinforcement preform to be self-supporting, either as a preform or a dense pack of reinforcement. Indeed, preforming is one of the highest cost areas for this process. In addition, preform deformation during infiltration can result in lower-than-desired mechanical properties. Tooling may also be expensive for complex part shapes.

A pressure-less metal infiltration process, called the Primex process, allows an aluminum alloy to infiltrate a reinforcement preform without the application of pressure or vacuum. The reinforcement level can be controlled by the starting density of the preform being infiltrated. As long as interconnected porosity and appropriate infiltration conditions exist, the liquid aluminum will spontaneously infiltrate the preform. Key process characteristics include an aluminum alloy, the presence of magnesium, and a nitrogen atmosphere. During heating to the infiltration temperature (1380° F), magnesium reacts with the nitrogen atmosphere to form magnesium nitride (Mg_3N_2). Magnesium nitride is the infiltration enhancer that allows the aluminum alloy to infiltrate the reinforcing phase without the necessity of applied pressure or vacuum. During infiltration, Mg_3N_2 is reduced by the aluminum to form a small amount of aluminum nitride

(AlN).[2] The AlN forms small precipitates and a thin film on the surface of the reinforcing phase. Reinforcement loading can be as high as 75 volume percent, given the right combination of particle shape and size. The most widely used cast composite produced by liquid metal infiltration is an Al-10Si-1Mg alloy reinforced with 30 volume percent SiC_p. The 1% magnesium in this alloy is obtained during infiltration by the reduction of the Mg_3N_2. The only restriction in the selection of an aluminum alloy is the presence of magnesium to allow the formation of the Mg_3N_2. For SiC_p containing systems, silicon must also be present in sufficient quantity to suppress the formation of Al_4C_3.

9.5 Pressure Infiltration Casting

Pressure infiltration casting (PIC) is similar to squeeze infiltration except that gas, rather than mechanical pressure, is used to promote consolidation. In PIC, an evacuated particulate or fiber preform with molten metal is subjected to isostatically applied gas pressure, usually an inert gas such as argon in the range of 150–1500 psi. The most common composite fabricated using the PIC process is DRA. Although there are several variations to the PIC process, all involve the infiltration of molten aluminum into a freestanding evacuated preform by an externally applied isostatic inert gas. Both the mold and the preform are evacuated, generally by placing the entire mold assembly in a vacuum/pressure vessel and evacuating both the mold and vacuum/pressure chamber. After the mold is preheated and the aluminum melt has reached the desired superheat temperature, inert gas pressure in the range of 150–1500 psi is applied. As liquid aluminum infiltrates the preform, the pressure acting on the mold quickly approaches the isostatic state; therefore, the mold only supports the pressure difference for a very short period of time, so that large expensive and cumbersome molds are not needed. To minimize porosity, the mold is cooled directionally, and pressure is maintained until the entire casting is solidified.

Discontinuously reinforced aluminum fabricated using a preform provides a uniform reinforcement distribution, with no segregation of the reinforcement during solidification. In PIC, the preform acts as a nucleation site for solidification and inhibits grain growth during solidification and cool down, resulting in a very fine cast microstructure. In addition, because the preform is evacuated and the mold is directionally cooled, properly designed and processed components can be produced without porosity. Preforms have been fabricated and pressure infiltration cast in a range of reinforcement levels varying from 30% to greater than 70%. Current technology does not allow for lower reinforcement volume fractions, because the preform must have sufficient reinforcement content to produce a stable geometry. PIC is used to make DRA electronic packages for integrated circuits and multi-chip modules. The benefits of DRA for this application include a controlled CTE, which closely matches that of the integrated circuits, and a high thermal conductivity, which aids in the removal of the heat generated by the electronics.

9.6 Spray Deposition

Spray atomization and deposition is a process in which a stream of molten metal is atomized into fine droplets (300 μm or less) using a high velocity inert gas, usually argon or nitrogen, which is then deposited on a mold or substrate. When the particles impact the substrate, they flatten and weld together to form a high density preform, which can be subsequently forged to form a fully dense product.[10] The production of MMCs by spray deposition can be accomplished by introducing particulates into the metal spray, leading to codeposition with the atomized metal onto the substrate. The process is a rapid solidification process because the metal experiences a rapid transition through the liquidus to the solidus, followed by slower cooling from the solidus to room temperature. Inherent in spray processes are composites with minimal reinforcement degradation, little segregation, and fine grained matrices. The critical process parameters are the initial temperature, size distribution, and velocity of the metal droplets; the velocity, temperature, and feed rate of the reinforcement; and the temperature of the substrate.[8] In general, spray deposition methods are characterized by significant porosity levels. Careful control of the atomizing and particulate feeding conditions are required to insure that a uniform distribution of particulates is produced within a typically 90–98% dense aluminum matrix. A number of aluminum alloys containing SiC_p have been produced by spray deposition, including aluminum–silicon casting alloys and the 2XXX, 6XXX, and 7XXX wrought alloys.

Spray deposition was developed commercially by Osprey, Ltd as a method of building-up bulk material by atomizing a molten stream of metal with jets of cold gas. It has been adapted to particulate MMC production by injection of ceramic powder into the spray. The four stages in the Osprey method, shown in Fig. 9.10, are (1) melting and dispensing, (2) gas atomization, (3) deposition, and (4) collector manipulation.[10] Induction heating is used to produce the melt which flows into a gas atomizer. Melting and dispensing is carried out in a vacuum chamber. The atomized stream of metal is collected on a substrate placed in the line of flight. Overspray is separated by a cyclone and collected. Among the notable microstructural features of Osprey MMC material are strong interfacial bonds, little or no interfacial reaction layers, and very low oxide contents. A major attraction of the process is its high rate of metal deposition.

The advantages of spray deposition processes include fine grain sizes and minimal reinforcement degradation. Disadvantages include high porosity levels and the resulting need to further process these materials to achieve full consolidation. In general, spray processes are more expensive than casting due to the longer processing times, the high costs of the gases used, and the large amount of waste powder that results from spraying.

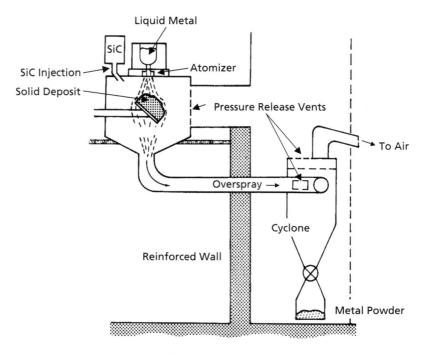

Fig. 9.10. Osprey Spray Process[1]

9.7 Powder Metallurgy Methods

When higher strength discontinuous MMCs are required, PM processes are often used because segregation, brittle reaction products, and high residual stresses from solidification shrinkage can be minimized. In addition, with the advent of rapid solidification and mechanical alloying technology, the matrix alloy can be produced as a prealloyed powder, rather than starting with elemental blends. PM processing can be used to make aluminum MMCs with both SiC particulates and whiskers, although Al_2O_3 particles and Si_3N_4 whiskers have also been employed. Processing, as shown in Fig. 9.11, involves (1) blending of the gas atomized matrix alloy and reinforcement in powder form; (2) compacting (cold pressing) the homogeneous blend to roughly 75–80% density; (3) degassing the preform (which has an open interconnected pore structure) to remove volatile contaminants (lubricants and mixing and blending additives), water vapor, and gases; and (4) consolidation by vacuum hot pressing or hot isostatic pressing. Then, the consolidated billets are normally extruded, rolled, or forged.

The matrix and reinforcement powders are blended to produce a homogeneous distribution. Prealloyed atomized matrix alloy powder is mixed with the particulate or whisker reinforcement and thoroughly blended. Since the metal powder is usually 25–30 μm and the ceramic particulates are often much smaller

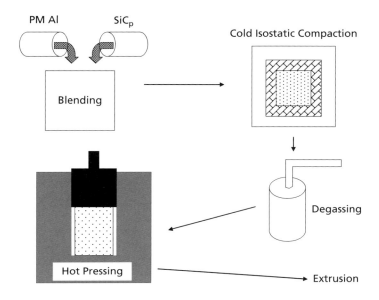

Fig. 9.11. Powder Metallurgy Processing

(1–5 μm), the large size difference creates agglomeration of the particulates in the blend. Ultrasonic agitation can be used to break up the agglomerations. Dry blending is normally conducted in an inert environment. The blending stage is followed by cold pressing to produce what is called a green body, which is about 75–80% dense and can be easily handled.

The next step is hot pressing, uniaxial or isostatic, to produce a fully dense billet. The hot pressing temperature can be either below or above that of the matrix alloy solidus. The powder blend is loaded into a metal can and evacuated at 750–930° F to remove all air and volatiles. Evacuation times are long; times as long as 10–30 h are not unusual. Out-gassing removes adsorbed water from the reinforcement and matrix powder, chemically combined water, and other volatile species.[10] After out-gassing, the can is sealed. If a vacuum hot press is used, the canning and sealing processes are not necessary and the powder can be loaded directly into the press. Hot pressing or hot isostatic pressing is normally conducted at as high a temperature as possible so that the matrix will be in its softest condition. Although liquid-phase sintering is normally not used due to grain boundary segregation and reinforcement degradation, some small amount of liquid phase allows the use of lower pressures. Another method is to consolidate the blend in a can to about 95% density by hot pressing, and then remove the can, and HIP to produce full density. Consolidation occurs by creep of the matrix material into the interstices between the reinforcement particles. When all the interconnected porosity is closed at about 90% density, diffusion closes the remaining porosity at triple points and grain boundaries.

PM processed billets are then normally extruded, forged, and/or hot rolled to produce useful shapes.

The PM technique generally produces properties superior to those obtained by casting and liquid metal infiltration (squeeze casting) techniques. Because no melting and casting are involved, the powder process offers several advantages. Conventional wrought alloys such as 2XXX, 6XXX, and 7XXX can be processed. A lower temperature can be used compared to casting. This results in less interaction between the matrix and the reinforcement, minimizing undesirable interfacial reactions, which leads to improved mechanical properties. The distribution of particulate or whiskers is usually better when the PM method is used than in casting processes. It is popular because it is reliable compared with alternative methods, but it also has some disadvantages. The blending step is a time-consuming, expensive, and a potentially dangerous operation. In addition, it is difficult to achieve an even distribution of particulate throughout the product, and the use of powders requires a high level of cleanliness; otherwise, inclusions will be incorporated into the product, with a deleterious effect on fracture toughness and fatigue life.

9.8 Secondary Processing of Discontinuous MMCs

After full consolidation, many discontinuous reinforced MMCs are subjected to additional deformation processes to improve their mechanical properties and produce part shapes. Secondary deformation processing of discontinuously reinforced MMCs helps to break up reinforcement agglomerates, reduce or eliminate porosity, and improve the reinforcement-to-matrix bond.

Extrusion can generate fiber alignment parallel to the extrusion axis, but often at the expense of progressive fiber fragmentation. Due to the presence of 15–25% hard non-deformable particulates or whiskers, fracture during extrusion can occur with conventional extrusion dies. Therefore, streamlined dies which gradually neck down are used to help minimize fracture. The degree of fiber fracture decreases with increasing temperature and decreasing local strain rate. Particulate segregation can also be a problem, in which particulate clusters occur next to regions containing little or no reinforcement. For whisker reinforced composites, whisker rotation and alignment with the metal flow direction can occur. In addition, the aspect ratio (i.e., length) of whiskers is generally reduced due to the shearing action. To prevent extrusion related defects, it is important that the material remain in compression during the extrusion process. Another microstructural feature of extruded MMCs is the formation of ceramic-enriched bands parallel to the extrusion axis. The mechanism of band formation is still unclear, but it appears to involve the concentration of shear strain in regions where ceramic particles or fibers accumulate.[1] However, extrusion of consolidated MMCs, such as castings, can reduce the level of clustering and inhomogeneities in the material.

Rolling is used after extrusion to produce sheet and plate products. Because the compressive stresses are lower in rolling than extrusion, edge cracking during rolling can occur. To minimize edge cracking, discontinuous composites are rolled at temperatures around 0.5T, where T is the absolute melting temperature, using relatively slow speeds. Isothermal rolling using light reductions and large diameter rolls can also reduce cracking. Isothermal rolling is often conducted as a pack rolling operation in which the MMC is encapsulated by a stronger metal. During rolling, the outside of the pack cools during rolling, while the interior containing the MMC remains at a fairly constant hot temperature. When the rolling temperature is low, light reductions and intermediate anneals may be required. As for extrusion, rolling further breaks up the particulate agglomerates. In heavily rolled sheet that has undergone about a 90% reduction in thickness, the particulate clusters are completely broken up and the matrix has flowed between individual particles that were previously in contact.

Processes such as rolling and forging, which involve high strains being applied quickly, can cause damage such as cavitation, particle fracture, and macroscopic cracking, particularly at low temperatures. In addition, very high temperatures increase the chance of matrix liquation, resulting in hot tearing or hot shortness. In contrast, HIP generates uniform stresses and so is unlikely to give rise to either microstructural or macroscopic defects.[1] It is an attractive method for removing residual porosity, which can include surface connected porosity as long as some form of encapsulation is provided. However, it can be very difficult to remove residual porosity in regions of very high ceramic content, such as within particle clusters, and the absence of any macroscopic shear stresses means that such clusters are not readily dispersed during HIP.

Machining discontinuous MMCs can be accomplished using circular saws and router bits. For straight cuts, diamond grit impregnated saws with flood coolant produces good edges. For contour cuts, solid carbide router bits with a diamond-shaped chisel cut geometry, or diamond-coated router bits, can be used with good results. Due to the hard reinforcement particles, speeds and feeds for all machining operations must be adjusted. In general, the speed is reduced to minimize tool wear, while the feed is increased to obtain productivity before the tools wear. The higher the reinforcement content, the greater the tool wear, with wear generally being a bigger problem than excessive heat generation. Polycrystalline diamond cutting tools exhibit longer lives than either solid or diamond-coated carbide tools but are also more expensive. Other methods, such as abrasive water jet cutting and wire electrical discharge machining (EDM) can also be used to produce acceptable cuts.

9.9 Continuous Fiber Aluminum Metal Matrix Composites

Aluminum MMCs reinforced with continuous fibers provides high strength and stiffness. However, because of their high cost, most applications have been limited to aerospace. Boron/aluminum (B/Al) was one of the first systems evaluated.

Applications include the tubular truss members in the mid-fuselage structure of the Space Shuttle orbiter and cold plates in electronic microchip carrier boards. Boron was developed in the early 1960s and was initially used successfully in an epoxy matrix on the F-14 and F-15 aircraft. Boron is made as a single monofilament using chemical vapor deposition, as shown schematically in Fig. 9.12. A 0.5 mil tungsten wire is drawn through a long glass reactor where boron trichloride (BCl_3) gas is chemically reduced to deposit boron on the tungsten core. Since this process produces a single 4.0 mil monofilament per reactor, many reactors (Fig. 9.13) are needed to produce production quantities. In addition, the process is inherently more expensive than those used for multifilament fiber forms, such as carbon fibers, in which many fibers (e.g., 12 000) are made with a single pass through a reactor. Due to the high cost of continuous fiber aluminum matrix composites and their limited temperature capabilities, the emphasis has shifted to continuous fiber titanium matrix composites that have potential payoffs in hypersonic airframes and jet engine components. However, since many of the processing procedures originally developed for aluminum MMCs have been transitioned to titanium, a brief review of these methods will be given.

Early work with boron monofilament/aluminum matrix composites in the 1970s was primarily done using diffusion bonding. Several methods, shown in Fig. 9.14, were developed for producing single layer B/Al sheets called

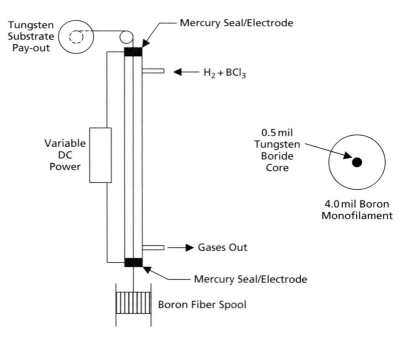

Fig. 9.12. Boron Monofilament Manufacture

Metal Matrix Composites

Fig. 9.13. Boron Monofilament Reactor Banks
Source: Speciality Materials Inc.

monotapes that were provided by material suppliers. The original single ply monotapes were generally made by winding the boron fiber on a drum, applying aluminum foil over the wound fibers, and then securing the fibers with an organic binder. The material supplier then diffusion bonded each monotape layer in a vacuum hot press. These diffusion bonded monotapes were supplied to the user who layed up multilayer laminates and diffusion bonded them together. The lay-up was placed in a welded stainless steel bag using an arrangement similar to that shown in Fig. 9.15. A typical diffusion bonding cycle for B/Al is 950–1000° F at 1000–3000 psi for 60 min. It was soon recognized that the boron fiber could react with aluminum during processing, forming brittle intermetallic compounds at the fiber interface that degraded the fiber strength properties. Strong interfacial bonding led to reduced longitudinal tensile strengths but higher transverse strengths, while weaker interfacial bonding resulted in higher longitudinal but lower transverse strengths.

Other methods of making monotape included placing a layer of aluminum foil on a mandrel, followed by filament winding the boron fiber over the foil in a collimated manner. An organic fugitive binder, such as an acrylic adhesive, was used to maintain the fiber spacing and alignment once the preform was cut from the mandrel. In this method often called the "green tape" method, the fibers were normally wound onto a foil-covered rotating drum, over sprayed with resin, followed by cutting the layer from the drum to provide a flat sheet

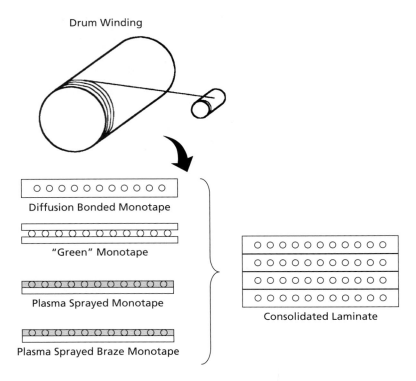

Fig. 9.14. MMC Monotape Product Forms

of monotape. Since this product form contains a fugitive organic binder, it must be removed by outgassing before diffusion bonding. A typical process would be to place the vacuum pack between the platens of a hot press, apply a small load to hold the fibers in place, and heat under vacuum to an intermediate level and hold at that temperature until the binder is outgassed.

Plasma spraying was another method developed for B/Al montoapes. In this process, a mandrel was again covered with a thin layer of aluminum and fibers were wound onto the drum. The fibers were then plasma sprayed with the aluminum matrix material to produce a somewhat porous montoape. The spraying operation was carried out in a controlled atmosphere to minimize matrix oxidation. The advantages of this process were that no organic binders were required which could potentially lead to contamination problems, and no diffusion bond cycle was required to produce the monotape, which lessened the potential of fiber degradation.

With the advent of boron fiber coated with a thin layer of silicon carbide (Borsic) to minimize fiber degradation during processing, a variant of the plasma sprayed material was offered in which the thin aluminum foil next to the drum surface was replaced with a thin layer of 713 aluminum alloy braze foil. Although the

Metal Matrix Composites

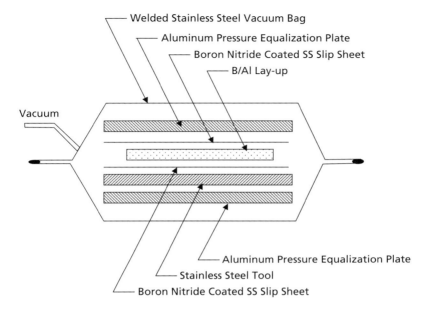

Fig. 9.15. Stainless Steel Bagging B/Al for Diffusion Bonding[11]

processing temperatures required for braze bonding were somewhat higher than for diffusion bonding, the pressures were a lot less.[11] A typical consolidation cycle was 1080° F for 15 min at <300 psi. A disadvantage of both plasma sprayed product forms was that the monotape was extremely stiff and had a tendency to curl due to the residual stresses introduced during plasma spraying, and therefore required vacuum annealing prior to forming structural shapes.

In the 1980s, continuous SiC monofilaments largely replaced boron because they have similar properties and are not degraded by hot aluminum during processing. The SCS-2 SiC monofilament, which was tailored specifically for aluminum matrices, has a 1 μm thick carbon-rich coating that increases in silicon content toward its outer surface. Hot molding[12] was a low pressure, hot pressing process designed to fabricate SiC/Al parts at significantly lower cost than possible with solid state diffusion bonding. Because the SCS-2 fibers can withstand molten aluminum for long periods, the molding temperature could be raised into the liquid-plus-solid region of the alloy to insure aluminum flow and consolidation at low pressure, thereby eliminating the need for high pressure die molding equipment.[13] A plasma sprayed aluminum preform was laid into the mold, heated to near molten aluminum temperature, and pressure consolidated in an autoclave by a metallic vacuum bag. SiC/Al MMCs exhibit increased strength and stiffness as compared with unreinforced aluminum, with no weight penalty. In contrast to the base metal, the composite retains its tensile strength at temperatures up to 500° F.

Graphite/aluminum (Gr/Al) MMCs were developed primarily for space applications, where high modulus carbon or graphite multifilaments could be used to produce structures with high stiffness, low weight, and little or no thermal expansion over large temperature swings. Unidirectional high modulus graphite P100 Gr/6061 aluminum tubes exhibit an elastic modulus in the fiber direction significantly greater than that of steel, with a density approximately one-third that of steel.[14] In addition, carbon and graphite fibers quickly became less expensive than either boron or silicon carbide monofilaments. However, Gr/Al composites are difficult to process: (1) the carbon fiber reacts with the aluminum matrix during processing forming Al_4C_3 which acts as crack nucleation sites leading to premature fiber failure;[15] (2) molten aluminum does not effectively wet the fiber; and (3) the carbon fiber oxidizes during processing.[14] In addition, Gr/Al parts can be subject to severe galvanic corrosion if used in a moist environment. Two processes have been used for making Gr/Al MMCs: liquid metal infiltration of the matrix on spread tows and hot press bonding of spread tows sandwiched between thin sheets of aluminum.

Alumina (Al_2O_3)/aluminum MMCs have been fabricated by a number of methods, but liquid or semi-solid state processing techniques are commonly used. The 3M Company produces a material by infiltrating Nextel 610 alumina fibers with an aluminum matrix at a fiber volume fraction of 60%. A fiber reinforced aluminum MMC is used in pushrods for high performance racing engines. Hollow pushrods of several diameters are produced, where the fibers are axially aligned along the pushrod length. Hardened steel end caps are then bonded to the ends of the MMC tubes.

9.10 Continuous Fiber Reinforced Titanium Matrix Composites

Continuous monofilament titanium matrix composites (TMC) offer the potential for strong, stiff, lightweight materials at usage temperatures as high as about 1500° F. The principal applications for this class of materials would be for hot structure, such as hypersonic airframe structures, and for replacing superalloys in some portions of jet engines. The use of TMCs has been restricted by the high cost of the materials, fabrication, and assembly procedures.

Specialty Materials SCS-6 silicon carbide fiber is the most prevalent fiber used in continuous reinforced titanium matrix composites. SCS-6 is made in a manner very similar to boron fiber. A small diameter carbon substrate (1.3 mil in diameter) is resistively heated as it passes through a long glass reactor, and silicon carbide is chemically vapor deposited. A gradated carbon-rich SiC protective coating is then applied to help slow the interaction between the fiber and the titanium matrix, both during processing and later during elevated temperature service. If the metal matrix and the fiber surface interact extensively during processing or elevated temperature usage, the fiber surfaces can develop brittle intermetallic compounds, and even surface notches, which drastically

Metal Matrix Composites

lowers the fiber tensile strength. Typical properties of 5.6 mil diameter SCS-6 fiber include a tensile strength of 550 ksi, a modulus of 58 msi and a density of 0.11 lb/in.^{3}

Two other smaller diameter SiC fibers, SCS-9 and Sigma, have also been evaluated as reinforcements for titanium matrix composites. The SCS-9 fiber, also made by Specialty Materials, is basically the same as SCS-6 except for its smaller diameter (3.2 mil). It is also deposited on a 1.3 mil carbon core. By way of comparison, the core comprises about 16% of the cross-sectional area of SCS-9 but only about 5% of the SCS-6 cross-section, so the SCS-9 has lower mechanical properties (tensile strength of 500 ksi and modulus of 47 msi) but, being smaller in diameter, has better formability and its density is also lower (0.09 lb/in.^{3}). Sigma is a 4 mil diameter fiber currently produced Defence Evaluation and Research Agency (DERA) in the UK. Unlike the SCS fibers, Sigma is deposited on a tungsten core rather than carbon. Sigma has a tensile strength of 500 ksi, a modulus of 60 msi, and a density of 0.1 lb/in.^{3} Sigma SM1240, which contains a 1 μm coating of TiB_2 over a 1 μm inner coating of carbon, was developed for titanium aluminide matrices, while SM1140+, which contains a thicker 4.5 μm coating of only carbon, was developed originally for beta titanium alloys.[16] Cross sections of an SCS-6 fiber and a Sigma fiber are shown in Fig. 9.16.

Ti-6Al-4V, the most prevalent titanium alloy used in the aerospace industry, was one of the first alloys to be evaluated as a matrix for TMC composites. However, Ti-6Al-4V, being a lean alpha-beta alloy has at least two serious shortcomings: (1) it has only moderate strength at elevated temperature; and (2) it is not very amenable to cold rolling into thin foil for foil-fiber-foil lay-ups, or for cold forming into structural shapes. To overcome the forming problem, a considerable amount of work was conducted with the beta alloy Ti-15V-3Cr-3Sn-3Al. Testing showed that the Ti-15-3-3-3 alloy performs well in all respects but one: its oxidation resistance at temperatures approaching 1300–1500° F is poor. The need for a matrix alloy with the good forming characteristics

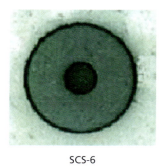
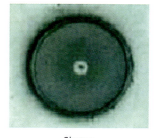

SCS-6   Sigma

Fig. 9.16. Silicon Carbide Monofilaments

of Ti-15-3-3-3 but with improved oxidation resistance led Titanium Metals Corporation (Timet) to initiate a program to develop an alloy with improved oxidation resistance. Ti-15Mo-2.8Nb-3Al-0.2Si alloy was selected as the most promising. This alloy, subsequently designated Beta 21S, was found to be far superior not only to Ti-15-3-3-3 but also to alloys such as Commercially Pure and Ti-6Al-4V, from the standpoint of oxidation resistance. It should be noted that the addition of alloying elements such as vanadium, molybdenum, and aluminum has been found to reduce the tendency of the titanium matrix to degrade the fiber.[17] Titanium aluminide matrices offer the potential of even higher temperature usage. However, the aluminides are both difficult to process and very expensive, so they may be prohibitive from a cost standpoint except where they are required to meet the most stringent temperature requirements.

The most prevalent method used to fabricate continuous silicon carbide fiber titanium matrix composites is the "foil-fiber-foil" method. In this method, a silicon carbide fiber mat is held together with a cross-weave of either molybdenum, titanium, or titanium–niobium wire or ribbon. The fabric is a uniweave system in which the relatively large diameter SiC monofilaments are straight and parallel, and held together by a cross-weave of metallic ribbon. The titanium foil is normally cold rolled down to a thickness of 0.0045 in. An example of a SiC uniweave and two pieces of titanium foil are shown in Fig. 9.17. The foil surfaces must be cleaned prior to lay-up to remove all volatile contaminates. The thin foils should also be uniform in thickness to avoid uneven matrix flow during

Fig. 9.17. SiC Uniweave with Titanium Foils

Metal Matrix Composites

diffusion bonding. A fine grain size in the foil will enhance diffusion bonding by facilitating creep and possibly superplastic deformation. Since extremely thin foils with good surface finishes are required, the final rolling conditions are cold rolling. For TMC this dictates the use of beta titanium alloys that can be cold rolled to thin gauges. The plies are cut, layed-up on a consolidation tool as shown in Fig. 9.18, and then consolidated by either vacuum hot pressing or HIP. A disadvantage the foil-fiber-foil process is the rather poor fiber distribution with some fibers touching, which has a detrimental effect on mechanical properties, especially fatigue crack nucleation.[15]

Plasma spraying has also been evaluated for SiC/Ti composites. In vacuum plasma spraying, metallic powders of 20–100 μm are fed continuously into a plasma where they are melted and propelled at high velocity onto a single layer of fibers, which have been wound onto a drum. One potential disadvantage of plasma spraying is that titanium, being an extremely reactive metal, can pickup oxygen from the atmosphere potentially leading to embrittlement problems. This method has been primarily evaluated for titanium aluminide matrix composites, due to the extreme difficulty of rolling these materials into thin foil.

The two primary consolidation procedures for continuous fiber TMCs are VHP and HIP. Diffusion bonding is attractive for titanium because titanium dissolves its own oxide at temperatures above about 1300° F[18] and exhibits extensive plastic flow at diffusion bonding temperatures. High temperature/short time roll bonding was evaluated some years ago but only to a very limited extent. Typical fiber contents for continuous fiber TMC laminates range from 30 to 40 volume percent.

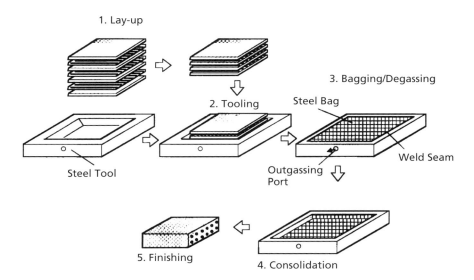

Fig. 9.18. TMC Lay-up Process

In the VHP technique, the lay-up is sealed in a stainless steel envelope and placed in a vacuum hot press. After evacuation, a small positive pressure is applied by the press platens. This pressure acts to hold the filaments in place during the initial 800–1000° F soak used to decompose any volatile organics and remove them under the action of a dynamic vacuum. The temperature is then gradually increased to a level where the titanium flows around the fibers under full pressure and the foil interfaces are diffusion bonded together. A typical VHP cycle is 1650–1750° F at 6–10 ksi pressure for 60–90 min.

Hot isostatic pressing has largely replaced vacuum hot pressing as the consolidation technique of choice. The primary advantages of HIP consolidation are: (1) the gas pressure is applied isostatically, alleviating the concern about uneven platen pressure; and (2) the HIP process is much more amenable to making complex structural shapes. Typically, the part to be HIPed is canned (or a steel bag is welded to a tool), evacuated, and then placed in the HIP chamber. A typical HIP facility is shown in Fig. 9.19. For TMC, typical HIP parameters are 1600–1700° F at 15 ksi gas pressure for 2–4 h. Since HIP processing is a fairly expensive batch processing procedure, it is normal practice to load a number

Fig. 9.19. Loading Large HIP Furnace

of parts into the HIP chamber for a single run. The vertical HIP units can have large thermal gradients from top to bottom. The temperature is slowly ramped up and held until all points on the tool are at a uniform temperature. The gas pressure is then increased after the tool has reached this uniform temperature. The pressure then deforms the steel bag and plastically consolidates the laminate when everything is soft. The hold time must be sufficient to insure complete diffusion bonding and consolidation.

Diffusion bonding consists of creep flow of the matrix between the fibers to make metal-to-metal contact, and then diffusion across the interfaces to complete the consolidation process.[19] As shown in Fig. 9.20, obtaining complete flow in the interstices between the fiber mid-plane and the foil segments on either side is difficult. Fine grained foil materials and high temperatures where the matrix is either very soft, or superplastic, can help. However, high processing temperatures can cause fiber-to-matrix reactions which cause degradation of the

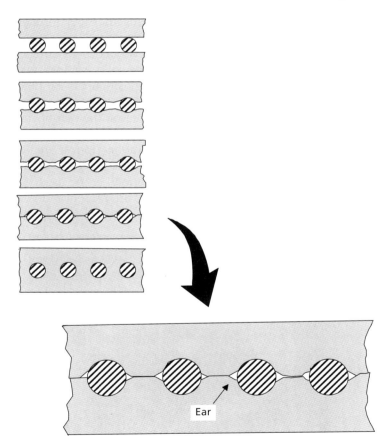

Fig. 9.20. Diffusion Bonding Progression[19]

fiber strength. Thermal expansion mismatches between the fiber and matrix can also cause high residual stresses, resulting in matrix cracking during cool down.[19]

The selection of the diffusion bonding parameters can have an effect on the occurrence of structural defects such as fiber breakage, matrix cracking, and interfacial reactions.[20] For example, reducing the levels of fiber breakage and matrix cracking would dictate high processing temperatures and low pressures, while minimizing interfacial reactions would dictate exactly the opposite: low processing temperatures and high pressures. Higher consolidation temperatures promote creep and diffusion processes that contributes to void closure but, at the same time, can result in excessive interfacial reactions and grain growth in the matrix. A low consolidation temperature leads to long processing times and requires higher pressures, which can result in fiber cracking.

Another method for fabricating TMCs is to apply the matrix directly to the SiC fibers using physical vapor deposition (PVD). A single PVD matrix coated SiC fiber is shown in Fig. 9.21. An evaporation process used for fabrication of monofilament reinforced titanium involves passing the fiber through a region having a high vapor pressure of the metal to be deposited, where condensation takes place to produce a coating. The vapor is produced by directing a high power ($\sim 10\,\text{kW}$) electron beam onto the end of a solid bar feedstock. Typical deposition rates are $\sim 5\text{--}10\,\mu\text{m}/\text{min}$. Alloy composition can be tailored, since differences in evaporation rates between different solutes are compensated by changes in composition of the molten pool formed on the end of the bar, until a steady state is reached, in which the alloy content of the deposit is the same as that of the feedstock.[1] Electron beam evaporation from a single source is possible if the vapor pressures of the elements in the alloy are relatively close to each other, otherwise multiple source evaporation can be used if low vapor pressure elements, such as niobium or zirconium, are present in the alloy.[21]

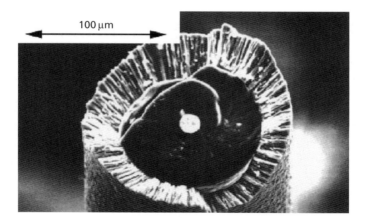

Fig. 9.21. PVD Coated SiC Monofilament[22]

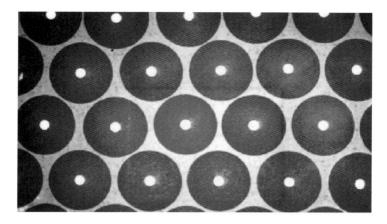

Fig. 9.22. PVD Coated SiC/Ti Composite[22]

Composite fabrication is completed by assembling the coated fibers into a bundle and consolidating by hot pressing or HIP. Very uniform fiber distributions can be produced in this way, with fiber contents of up to about 80%, as shown in the Fig. 9.22 photomicrograph. The fiber volume fraction can be accurately controlled by the thickness of the deposited coatings, and the fiber distribution is always very homogeneous. The main advantages of this process are: (1) the fiber distribution and volume percentage are readily controllable, (2) the time required for diffusion bonding is shorter, and (3) the coated fiber is relatively flexible and can be wound into complex part shapes.

9.11 Secondary Fabrication of Titanium Matrix Composites[23]

Successful joining of TMC components by diffusion bonding can be accomplished at pressures and temperatures lower than normal HIP runs. To process preconsolidated C-channels into spars, as shown in Fig. 9.23, they can be joined together in a back-to-back fashion. For secondary diffusion bonding in a HIP chamber, the parts are assembled and encapsulated in a leak free steel envelope or bag. Since the bag is subjected to the same high pressure and temperature as the parts, the ability of the steel bag to withstand the HIP forces is critical for success. If the bag develops a leak, the isostatic pressure is lost, and so is the bonding pressure. To minimize the risk associated with the extreme pressures and temperatures required for initial consolidation, lower temperatures and pressures can be used for secondary diffusion bonding. Lower temperatures and pressures reduce the risk of bag failures during the secondary HIP bonding cycle. However, since TMC is inherently stiffer than conventional sheet metal

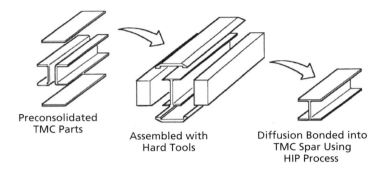

Fig. 9.23. Secondary Diffusion Bonding of TMC Spars

components, it generally requires higher pressures to achieve secondary diffusion bonding. A machined filler is placed in the corner to fill the void created by the C-channels. By not pinning the TMC components together, and allowing them to slide a little in relation to one another, complete bonding is achieved. Incomplete bonding or damage can occur if movement is not permitted. Successful bonding can be achieved with HIP pressures as low as 5 ksi at 1600° F for two hours.

Superplastic forming and diffusion bonding (SPF/DB) can be used to take advantage of elevated temperature characteristics inherent in certain titanium alloys. Structural shapes, which combine superplastic forming and bonding of the parent metal, can be fabricated from thin titanium alloy sheets to achieve high levels of structural efficiency. Components fabricated from TMC do not lend themselves to superplastic forming. However, since they retain the diffusion bonding capability of the matrix alloy, TMC components can be readily bonded with superplastically formed sheet metal substructures. A TMC reinforced SPF/DB structural panel retains the structural efficiency of TMC, while possessing the fabrication simplicity of a superplastically formed part. Two potential methods of fabricating TMC stiffened SPF/DB panels are shown in Fig. 9.24. The core pack, which forms the substructure, can be fabricated from a higher temperature titanium alloy, such as Ti-6Al-2Sn-4Zr-2Mo, rather than Ti-6Al-4V which is normally used in SPF/DB parts. Ti-6Al-2Sn-4Zr-2Mo has good SPF characteristics at a moderate processing temperature (1650° F). The final shape and size of the substructure is determined by the resistance seam welding pattern of the core pack prior to the SPF/DB cycle, which is inflated during the SPF/DB cycle with argon gas to form the substructure. More detailed information on SPF/DB of titanium can be found in Chapter 4 on Titanium.

Conventional non-destructive testing (NDT) techniques, including both through-transmission and pulse echo ultrasonics, can be used for the detection of manufacturing defects. Conventional ultrasonic and X-ray have the ability to find many TMC processing defects, including lack of consolidation,

Metal Matrix Composites

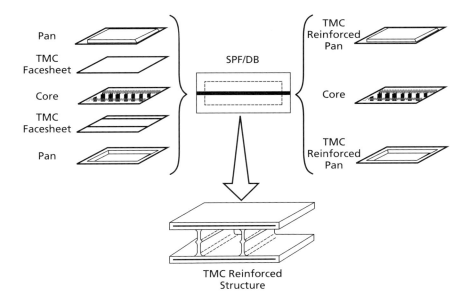

Fig. 9.24. Methods for Making SPF/DB TMC Reinforced Parts

delaminations, fiber swimming, and fiber breakage. Ultrasonics can find defects as small as 3/64 in. in diameter, with through-transmission giving the best resolution. Normal radiographic X-ray inspection can also be used to examine TMC components.

Titanium matrix composite is extremely difficult to machine. The material is highly abrasive, and tool costs can be high. Improper cutting not only damages tools but can also damage the part as well. Abrasive waterjet cutting, illustrated in Fig. 9.25, has been found to work well on TMC. Typical machining parameters are 45 000 psi water pressure, dynamically mixed with #80 garnet grit, at a feed rate of 0.5–1.2 ipm. The waterjet cutter has multi-axis capability and can make uninterrupted straight and curved cuts. A diamond cut-off wheel also produces excellent cuts. However, this method is limited to straight cuts and is rather slow. The diamond cut-off wheel is mounted on a horizontal mill, and cutting is done with controlled speeds and feeds. Because this method is basically a grinding operation, cutting rates are typically slow; however, the quality of the cut edge is excellent.

Wire EDM is also a flexible method of cutting TMC. EDM is a non-contact cutting method which removes material through melting or vaporization by high frequency electric sparks. Brass wire (0.010 in. in diameter) can be used at a feed rate of 0.020–0.050 ipm. The EDM unit is self-contained with its own coolant and power system. EDM has the advantage of being able to make small diameter cuts, such as small scallops and tight radii. Like waterjet cutting, the EDM method is programmable and can make uninterrupted straight or curved cuts.

Fig. 9.25. Abrasive Waterjet Cutting of TMC
Source: The Boeing Company

Several methods can be used to generate holes in TMC. For thin components (e.g., 3 plies), with only a few holes, the cobalt grades of high speed steel (e.g., M42) twist drills can be used with power feed equipment; however, tool wear is so rapid that several drills may be required to drill a single hole. Punching has also been used successfully on thin TMC laminates. Using conventional dies, punching is fast and clean, with no coolant required. Punching results in appreciable fiber damage and metal smearing. However, most of the disturbed metal can be removed and the hole cleaned up by reaming. However, some holes may require several passes, and several reamers, to sufficiently clean up the hole. In some instances, the diameter of the reamer is actually reduced due to wear of the reamer cutting edges, rather than increasing the diameter of the hole. Punching is not used extensively because of the large number of fastener holes required in the internal portion of structures. In addition, load-bearing sections are normally too thick for punching.

Neither conventional twist drills nor punching will consistently produce high quality holes in TMC, especially in thicker material. The use of diamond core drills has greatly improved hole drilling quality in thick TMC components. High quality holes can consistently be produced with diamond core drills. The core drills are tubular with a diamond matrix built-up on one end. This construction is similar to some grinding wheels. A typical core drill and coolant chuck is shown in Fig. 9.26. The important parameters for successful diamond

Metal Matrix Composites

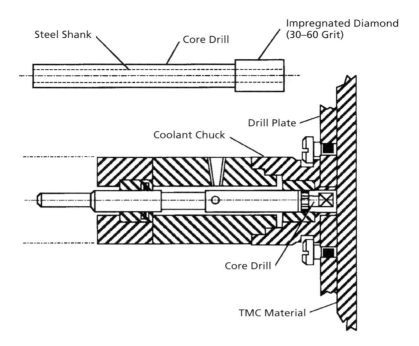

Fig. 9.26. Diamond Impregnated Core Drill and Coolant Chuck

core drilling are drill design, coolant delivery system, drill plate design, type of power feed drilling equipment used, and the speeds and feeds used during drilling. During drilling, the core drill abrasively grinds a cylindrical core plug from the material. Some fabricators even mix an abrasive grit with the water coolant to improve the material removal rate. Multiple drill set-ups on a TMC structure are shown in Fig. 9.27. The drill plate can hold several drill motors, which allows the operator to operate more than one at a time. The drill plates must be stiff enough to produce a rigid setup. A properly drilled hole can hold quite good tolerances, depending on the thickness of the TMC: ± 0.0021 in. in 4 plies, ± 0.0030 in. in 15 plies, and ± 0.0055 in. in 32 plies.

Another method for joining thin TMC components into structures is resistance spot welding. Conventional 50 kW resistance welding equipment (Fig. 9.28), with water cooled copper electrodes, has been successfully used to spot weld thin TMC. Fabricators often use conventional titanium (e.g., Ti-6Al-4V) to set the initial welding parameters. As with any spot welding operation, it is important to thoroughly clean the surfaces before welding. Initial welding parameters should be verified by metallography and lap shear testing.

Fig. 9.27. Diamond Core Drilling of TMC Component
Source: The Boeing Company

Fig. 9.28. Spot Welding TMC Hat Stiffeners
Source: The Boeing Company

9.12 Fiber Metal Laminates[24]

Fiber metal laminates are laminated materials consisting of thin layers of metal sheet and unidirectional fiber layers embedded in an adhesive system. (GLARE) Glass Laminate Aluminum Reinforced is a type of aluminum fiber metal laminate,

Metal Matrix Composites

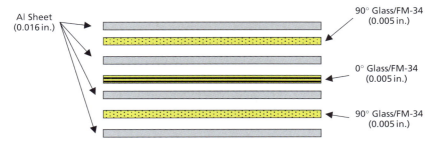

Fig. 9.29. Typical GLARE Construction[24]

in which unidirectional S-2 glass fibers are embedded in FM-34 epoxy structural film adhesive. A typical construction is shown in Fig. 9.29. The S-2 glass is bonded to the aluminum sheets with the film adhesive. The aluminum metal layers are chemically cleaned (chromic acid anodized or phosphoric acid anodized) and primed with BR-127 corrosion inhibiting primer. The adhesion between the FM 94 adhesive and treated metal surface, and between FM 94 and S-2 glass fibers, is so high that these bond lines often remain intact until cohesive adhesive failure occurs. The primary advantages of GLARE are: better fatigue crack propagation resistance than aluminum, superior damage tolerance compared to aluminum, higher bearing strengths than carbon/epoxy, 10% lighter weight than aluminum, and lower cost than composite but higher cost than aluminum.

The GLARE is normally available in six different standard grades, as outlined in Table 9.1. They are all based on unidirectional S-2 glass fibers embedded

Table 9.1 Commercial GLARE Grades[24]

Glare Grade	Sub Grade	Material Sheet Thickness (in.) and Alloy	Prepreg Orien. in Each Layer	Principal Benefits
Glare 1	–	0.012–0.016, 7475-T761	0/0	Fatigue, Strength, Yield Strength
Glare 2	Glare 2A	0.0008–0.020, 2024-T3	0/0	Fatigue, Strength
	Glare 2B	0.0008–0.020, 2024-T3	90/90	Fatigue, Strength
Glare 3	–	0.0008–0.020, 2024-T3	0/90	Fatigue, Impact
Glare 4	Glare 4A	0.0008–0.020, 2024-T3	0/90/0	Fatigue, Strength in 0° Direction
	Glare 4B	0.0008–0.020, 2024-T3	90/0/90	Fatigue, Strength in 90° Direction
Glare 5	–	0.0008–0.020, 2024-T3	0/90/90/0	Impact
Glare 6	Glare 6A	0.0008–0.020, 2024-T3	+45/−45	Shear, Off-Axis Properties
	Glare 6B	0.0008–0.020, 2024-T3	−45/+45	Shear, Off-Axis Properties

with FM 94 adhesive, resulting in a 0.005 in. thick prepreg with a nominal fiber volume fraction of 59%. The prepreg is layed-up in different orientations between the aluminum alloy sheets. From 1990 to 1995, GLARE laminates were produced only as flat sheets. It was believed that the aircraft manufacturer would use these flat sheets to manufacture fuselage panels by applying the curvature, thickness steps, and joints, using conventional methods developed for metal structures (forming, bonding, riveting, etc.). Several studies showed the benefits in performance and weight of these GLARE shells, but they also indicated the high cost of these parts in comparison with conventional aluminum structures.

To reduce manufacturing costs, the self-forming technique (SFT), shown in Fig. 9.30, was developed. Autoclave pressure forms the laminate over external and internal doublers, because the stiffness of the package of thin aluminum layers and still uncured adhesive and fiber layers is low. Additional adhesive, of the same type as that used to impregnate the glass fibers in the prepreg, is also added at certain locations to adhere interrupted metal sheets to each other; adhere thin aluminum internal or external doublers to the aluminum layers of the laminate; and to fill gaps in the laminate which would otherwise remain unfilled. Two doubler splice concepts are shown in Fig. 9.31. The additional adhesive in the laminate has a much higher shear strength than the prepreg layers. This means that the splices manufactured using the SFT, where adhesive is added at those locations where load transfer from one metal layer to another occurs, are not critical regarding delamination during static or fatigue loading. In other words, the splice is not the weakest link in the panel strength.

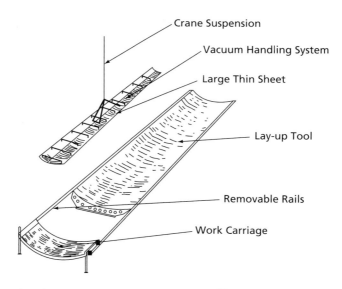

Fig. 9.30. GLARE Self-Forming Technique[24]

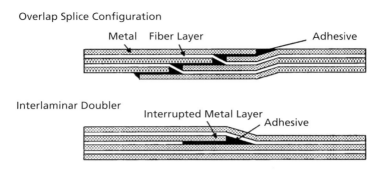

Fig. 9.31. GLARE Splice Concepts[24]

Summary

Metal matrix composites offer a number of advantages compared to their base metals, such as higher specific strengths and moduli, higher elevated temperature resistance, lower coefficients of thermal expansion, and, in some cases, better wear resistance. On the down side, they are more expensive than their base metals and have lower toughness. MMCs are even more expensive than polymer matrix composites and the fabrication processes are much more limited, especially for complex structural shapes. Despite 30 years of R&D, there are very few industrial applications for MMCs, primarily as a result of their high costs. There are some limited uses for discontinuously reinforced MMCs but almost no current applications for continuously reinforced MMCs.

The MMCs can be subdivided according to the type of reinforcement. The reinforcement can be particulates, high strength single crystal whiskers, short fibers, long aligned multifilaments, or monofilament fibers.

The primary MMC fabrication processes can be classified as either liquid phase or solid state processes. Liquid phase processing is generally considerably less expensive than solid state processing. Characteristics of liquid phase processed discontinuous MMCs include low cost reinforcements, such as silicon carbide particles, low temperature melting matrices such as aluminum and magnesium, and near net shaped parts. Liquid phase processes include various casting processes, liquid metal infiltration, and spray deposition. However, since continuous aligned fiber reinforcement is normally not used in liquid state processes, the strengths and stiffness are lower.

Solid state processes, in which no liquid phase is present, are usually associated with some type of diffusion bonding to produce final consolidation, whether the matrix is in a thin sheet or powder form. Although the processing temperatures are lower for solid state diffusion bonding, they are often still high enough to cause significant reinforcement degradation. In addition, the pressures are almost always higher for the solid state processes.

The majority of applications are for discontinuous MMCs with aluminum reinforced with SiC_p. Discontinuous MMCs are normally made either by liquid phase techniques or by PM blending followed by consolidation. Casting processes for MMCs include stir casting, slurry casting or compocasting, squeeze casting and pressure infiltration casting. Powder metallurgy techniques are also used to make discontinuous MMCs. Spray deposition is a hybrid process in which the matrix is both liquid and solid during the spray operation. The Osprey process is the most prevalent spray deposition process.

Continuous monofilament SiC reinforced titanium offers performance advantages for high temperature structures up to 1500° F for moderate times, with possibly higher temperatures for titanium aluminide matrix composites. Although liquid casting is often used with aluminum matrix alloys, continuous fiber MMCs are usually consolidated using solid state diffusion bonding. Diffusion bonding can be done using either hot pressing or hot isostatic pressing. Continuous fiber preforms can be prepared by filament winding, followed by either applying a fugitive organic binder (green tape), weaving with metallic fibers, or plasma spraying the matrix material. For woven preforms, the foil-fiber-foil lay-up method is normally used prior to consolidation. Titanium matrix composites have also been prepared by coating SiC monofilaments with titanium using PVD, followed by diffusion bonding.

Fiber metal laminates, in particular GLARE, are another form of composite material. The primary advantages of GLARE are: better fatigue crack propagation resistance than aluminum, superior damage tolerance compared to aluminum, higher bearing strengths than carbon/epoxy, 10% lighter weight than aluminum, and lower cost than composite but higher cost than aluminum.

Recommended Reading

[1] Clyne, T.W., Withers, P.J., *An Introduction to Metal Matrix Composites*, Cambridge University Press, 1993.
[2] Suresh, S., Martensen, A., Needleman, A., eds, *Fundamentals of Metal-Matrix Composites*, Butterworth-Heinemann, 1993.
[3] Vlot, A., Gunnick, J.W., *Fibre Metal Laminates An Introduction*, Kluwer Academic Publishers, 2001.

References

[1] Clyne, T.W., Withers, P.J., "Fabrication Processes", in *An Introduction to Metal Matrix Composites*, Cambridge University Press, 1993, pp. 318–360.
[2] "Processing of Metal-Matrix Composites", in *ASM Handbook Vol. 21 Composites*, ASM International, 2001, pp. 579–588.
[3] Herling, D.R., Grant, G.J., Hunt, W., "Low-Cost Aluminum Metal Matrix Composites", *Advanced Materials & Processes*, July 2001, pp. 37–40.
[4] Ejiofor, J.U., Reddy, R.G., "Developments in the Processing and Properties of Particulate Al-Si Composites", *Journal of Metals*, Vol. 49 (11), 1997, pp. 31–37.

[5] Hashim, J., Looney, L., Hashmi, M.S.J., "Metal Matrix Composites: Production by the Stir Casting Method", *Journal of Materials Processing Technology*, Vol. 92–93, 1999, pp. 1–7.
[6] Kaczmar, J.W., Pietrzak, K., Wlosinski, W., "The Production and Application of Metal Matrix Composites", *Journal of Materials Processing Technology*, Vol. 106, 2000, pp. 58–67.
[7] Hashim, J., Looney, L., Hashmi, M.S.J., "The Wettability of SiC Particles by Molten Aluminum Alloy", *Journal of Materials Processing Technology*, Vol. 119, 2001, pp. 324–328.
[8] Michaud, V.J., "Liquid State Processing", in *Fundamentals of Metal-Matrix Composites*, Butterworth-Heinemann, 1993, pp. 3–22.
[9] Ghomashchi, M.R., Vikhrov, A., "Squeeze Casting: An Overview", *Journal of Materials Processing Technology*, Vol. 101, 2000, pp. 1–9.
[10] Srivatsan, T.S., Sudarshan, T.S., Lavernia, E.J., "Processing of Discontinuously-Reinforced Metal Matrix Composites by Rapid Solidification", *Progress in Materials Science*, Vol. 39, 1995, pp. 317–409.
[11] Bilow, G.B., Campbell, F.C., "Low Pressure Fabrication of Borsic/Aluminum Composites", 6th Symposium of Composite Materials in Engineering Design, May 1972.
[12] Mittnick, M.A., "Continuous SiC Fiber Reinforced Materials", 21st International SAMPE Technical Conference, September 1989, pp. 647–658.
[13] McElman, J.A., "Continuous Silicon Carbide Fiber MMCs", in *Engineered Materials Handbook Vol. 1 Composites*, ASM International, 1987, pp. 858–866.
[14] "Metal-Matrix Composites", in *Metals Handbook-Desk Edition*, ASM International, 2nd edition, 1998, pp. 674–680.
[15] Vassel, A., "Continuous Fibre Reinforced Titanium and Aluminum Composites: A Comparison", *Materials Science and Engineering*, A263, 1999, pp. 305–313.
[16] Shatwell, R.A., "Fibre-Matrix Interfaces in Titanium Matrix Composites Made With Sigma Monofilament", *Materials Science and Engineering*, A259, 1999, pp. 162–170.
[17] Lindroos, V.K., Talvitie, M.J., "Recent Advances in Metal Matrix Composites", *Journal of Materials Processing Technology*, Vol. 53, 1995, pp. 273–284.
[18] Hull, D., Clyne, T.W., "Fabrication", in *An Introduction to Composite Materials*, 2nd edition, Cambridge University Press, 1992, pp. 280–286.
[19] Ghosh, A.K., "Solid State Processing", in *Fundamentals of Metal-Matrix Composites*, Butterworth-Heinemann, 1993, pp. 232–241.
[20] Guo, Z.X., Derby, B., "Solid-State Fabrication and Interfaces of Fibre Reinforced Metal Martrix Composites", *Progress in Materials Science*, Vol. 39, 1995, pp. 411–495.
[21] Ward-Close, C.M., Chandraesekaran, L., Robertson, J.G., Godfrey, S.P., Murgatroyde, D.P., "Advances in the Fabrication of Titanium Matrix Composite", *Materials Science and Engineering*, A263, 1999, pp. 314–318.
[22] Ward-Close, C.M., Partridge, P.O., "A Fibre Coating Process for Advanced Metal Matrix Composites", *Journal of Materials Science*, Vol. 25, 1990, pp. 4315–4325.
[23] Sullivan, S., "Machining, Trimming and Drilling Metal Matrix Composites for Structural Applications", *ASM International Materials Week 92*, Chicago, Illinois, 2–5 November 1992.
[24] Vlot, A., Gunnick, J.W., *Fibre Metal Laminates An Introduction*, Kluwer Academic Publishers, 2001.

Chapter 10
Ceramic Matrix Composites

Monolithic ceramic materials contain many desirable properties, such as high moduli, high compression strengths, high temperature capability, high hardness and wear resistance, low thermal conductivity, and chemical inertness. As shown in Fig. 10.1, the high temperature capability of ceramics makes them very attractive materials for extremely high temperature environments. However, due to their very low fracture toughness, ceramics are limited in structural applications. While metals plastically deform due to the high mobility of dislocations (i.e., slip), ceramics do not exhibit plastic deformation at room temperature and are prone to catastrophic failure under mechanical or thermal loading. They have a very low tolerance to crack-like defects, which can result either during fabrication or in-service. Even a very small crack can quickly grow to critical size leading to sudden failure.

While reinforcements such as fibers, whiskers, or particles are used to strengthen polymer and metal matrix composites, reinforcements in ceramic matrix composites are used primarily to increase toughness. Some differences in polymer matrix and ceramic matrix composites are illustrated in Fig. 10.2. The toughness increases afforded by ceramic matrix composites are due to energy dissipating mechanisms, such as fiber-to-matrix debonding, crack deflection, fiber bridging, and fiber pull-out. A notional stress–strain curve for a monolithic ceramic and a ceramic matrix composite is shown in Fig. 10.3. Since the area under the stress–strain curve is often considered as an indication of toughness, the large increase in toughness for the ceramic matrix composite is evident. The

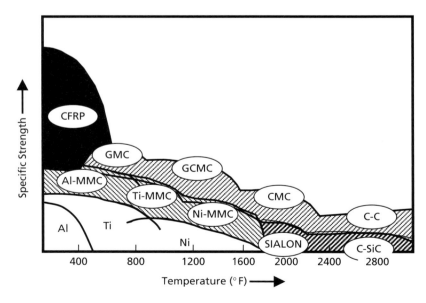

Fig. 10.1. Relative Material Temperature Limits

Ceramic Matrix Composites

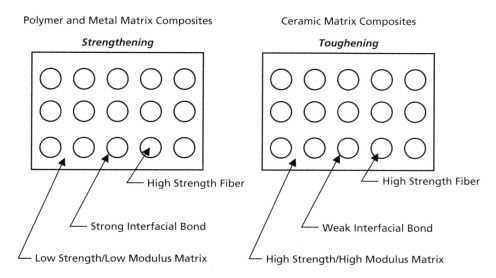

Fig. 10.2. *Comparison of Polymer and Metal with Ceramic Matrix Composites*

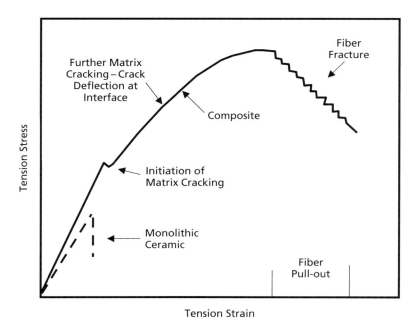

Fig. 10.3. *Stress–Strain for Monolithic and Ceramic Matrix Composites*

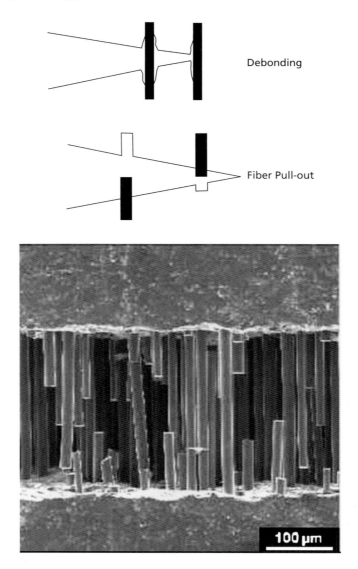

Fig. 10.4. Crack Dissipation Mechanisms

mechanisms of debonding and fiber pull-out are shown in Fig. 10.4. For these mechanisms to be effective, there must be a relatively weak bond at the fiber-to-matrix interface. If there is a strong bond, the crack will propagate straight through the fibers, resulting in little or no energy absorption. Therefore, proper control of the interface is critical. Coatings are often applied to protect the fibers during processing and to provide a weak fiber-to-matrix bond.

Carbon–carbon (C–C) composites[1] are the oldest and most mature of the ceramic matrix composites. They were developed in the 1950s by the aerospace industry for use as rocket motor casings, heat shields, leading edges, and thermal protection. It should be noted that C–C composites are often treated as a separate material class from other ceramic matrix composites, but their usage and fabrication procedures are similar and overlap other ceramic matrix composites. A relative comparison of C–C with other ceramic matrix composites is given in Table 10.1. For high temperature applications carbon–carbon composites offer exceptional thermal stability (>4000° F) in non-oxidizing atmospheres, along with low densities (0.054–0.072 lb/in.^{3}). Their low thermal expansion and range of thermal conductivities provides high thermal shock resistance. In vacuum and inert gas atmospheres, carbon is an extremely stable material, capable of use to temperatures exceeding 4000° F. However, in oxidizing atmospheres, it starts oxidizing at temperatures as low as 950° F. Therefore, C–C composites for elevated temperature applications must be protected with oxidation resistant coating systems, such as silicon carbide that is over-coated with glasses. The silicon carbide coating provides the basic protection, while the glass over-coat melts and flows into coating cracks at elevated temperature. In addition, oxidation inhibitors, such as boron, are often added to the matrix to provide additional protection.

Ceramic matrix materials include the element carbon, glasses, glass-ceramics, oxides (e.g., alumina – Al_2O_3) and non-oxides (e.g., silicon carbide – SiC). The majority of ceramic materials are crystalline with predominately ionic bonding,

Table 10.1 Carbon–Carbon and Ceramic Composite Comparison

Carbon–Carbon	Continuous CMCs	Discontinuous CMCs
Exceptional High Temp Mech Properties	Excellent High Temp Mech Properties	Excellent High Temp Mech Properties
High Specific Strength and Stiffness Low to Moderate Toughness	High Specific Strength and Stiffness Low to Moderate Toughness	Lower Specific Strength and Stiffness
Dimensional Stability Low Thermal Expansion	Dimensional Stability Low Thermal Expansion	Fracture Toughness Good but Lower than Continuous CMCSs
High Thermal Shock Resistance Graceful Failure Modes	Good Thermal Shock Resistance Graceful Failure Modes	Thermal Shock Resistance Lower than Continuous CMCs
Tailorable Properties Machinability Poor Oxidation Resistance	Oxidation Resistance Machining More Difficult Processing More Complicated and Expensive	Amendable to Lower Cost Conventional Processes Machining Expensive

along with some covalent bonding. These bonds, in particular the strongly directional covalent bond, provide a high resistance to dislocation motion and go a long way in explaining the brittle nature of ceramics. Since ceramics and carbon–carbon composites require extremely high processing temperatures compared to polymer or metal matrix composites, ceramic matrix composites are difficult and expensive to fabricate.

Reinforcements for ceramic matrix composites are usually carbon, oxide or non-oxide ceramic fibers, whiskers, or particulates. Carbon fiber is used primarily in carbon–carbon composites, while oxide fibers (such as alumina) or non-oxide fibers (such as silicon carbide) are used in glass, glass-ceramic and crystalline ceramic matrices. Most high performance oxide and non-oxide continuous fibers are expensive, further leading to the high cost of ceramic matrix composites. The cost and great difficulty of consistently fabricating high quality ceramic matrix composites has greatly limited their applications to date.

10.1 Reinforcements

Fibers used in ceramic matrix composites, classified according to their diameters and aspect ratios, fall into three general categories: whiskers, monofilaments, and textile multifilament fibers. Reinforcements in the form of particulates and platelets are also used. A summary of a number of oxide and non-oxide continuous ceramic fibers is given in Table 10.2.

Whiskers are nearly perfect single crystals with strengths approaching the theoretical strength of the material. They are usually 1 μm in diameter, or less, and up to 200 μm long. As reinforcements, it is their size and aspect ratio (length/diameter) that determines their strengthening effect.[2] SiC, Si_3N_4, and Al_2O_3 are the most commonly used whiskers for ceramic matrix composites.

Monofilament SiC fibers are produced by chemical vapor deposition of silicon carbide onto a 1.3 mil diameter amorphous carbon substrate, resulting in a large 5.5 mil diameter fiber. A carbon substrate is preferred to a tungsten substrate, because, above 1500°F, silicon carbide reacts with tungsten, resulting in fiber strength degradation. During manufacture, a 1 μm thick layer of pyrolytic graphite is deposited on a resistance heated carbon substrate to provide a smooth surface and control its electrical conductivity. The coated substrate is then chemically vapor deposited (CVD) using a mixture of silane and hydrogen gases. On exiting the reactor, a thin layer of carbon and silicon carbide is applied to provide improved handleability; act as a diffusion barrier for reducing the reaction between the fiber and the matrix; and heal surface flaws for improved fiber strength.[3] Since these monofilaments are large, they can tolerate some surface reaction with the matrix without a significant strength loss. However, their large diameter also inhibits their use in complex structures due to their large diameter and high stiffness, which limits their ability to be formed over tight radii.

Ceramic textile multifilament fibers, in tow sizes ranging from 500 to 1000 fibers, are available that combine high temperature properties with small

Table 10.2 Properties of Selected Continuous Ceramic Fibers

Fiber	Composition	Tensile Strength (ksi)	Tensile Modulus (msi)	Density (g/cm^3)	Diameter (mil)	Critical Bend Radii (mm)
SCS-6	SiC on C Monofilament	620	62	3.00	5.5	7.0
Nextel 312	62Al$_2$O$_3$-14B$_2$O$_3$-15SiO$_2$	250	22	2.7	0.4	0.48
Nextel 440	70Al$_2$O$_3$-2B$_2$O$_3$-28SiO$_2$	300	27	3.05	0.4–0.5	–
Nextel 480	70Al$_2$O$_3$-2B$_2$O$_3$-28SiO$_2$	330	32	3.05	0.4–0.5	–
Nextel 550	73Al$_2$O$_3$-27SiO$_2$	290	28	3.03	0.4–0.5	0.48
Nextel 610	99 α-Al$_2$O$_3$	425	54	3.88	0.6	–
Nextel 720	85Al$_2$O$_3$-15SiO$_2$	300	38	3.4	0.4–0.5	–
Almax	99 α-Al$_2$O$_3$	260	30	3.60	0.4	–
Altex	85Al$_2$O$_3$-15SiO$_2$	290	28	3.20	0.6	0.53
Nicalon NL200	57Si-31C-12O	435	32	2.55	0.6	0.36
Hi-Nicalon	62Si-32C-0.5O	400	39	2.74	0.6	–
Hi-Nicalon-S	68.9Si-30.9C-0.2O	375	61	3.10	0.5	–
Tyranno LOX M	55.4Si-32.4C-10.2O-2Ti	480	27	2.48	0.4	0.27
Tyranno ZM	55.3Si-33.9C-9.8O-1Zr	480	28	2.48	0.4	–
Sylramic	66.6Si-28.5C-2.3B-2.1Ti-0.8O-0.4N	465	55	3.00	0.4	–
Tonen Si$_3$N$_4$	58Si-37N-4O	360	36	2.50	0.4	0.80

diameters (0.4–0.8 mil), allowing them to be used for a wide range of manufacturing options, such as filament winding, weaving, and braiding. A useful measure of the ability of a fiber to be formed into complex part shapes is the critical bend radius ρ_{cr}, which is the smallest radius that the fibers can be bent before they fracture. The critical bend radius ρ_{cr} can be calculated by multiplying the fiber failure strain by the fiber radius. High strength, low modulus, and small diameters all contribute to fibers that can be processed using conventional textile technology. For example, while SiC monofilaments have a critical bend radii of only 7 mm, many ceramic textile multifilament fibers are less than 1 mm.

Both oxide and non-oxide fibers are used for ceramic matrix composites. Oxide based fibers, such as alumina, exhibit good resistance to oxidizing atmospheres, but, due to grain growth, their strength retention and creep resistance at high temperatures is poor. Oxide fibers can have creep rates of up to two orders of magnitude greater than non-oxide fibers. Non-oxide fibers, such as C and SiC, have lower densities and much better high temperature strength and creep retention than oxide fibers but have oxidation problems at high temperatures.

Ceramic oxide fibers are composed of oxide compounds, such as alumina (Al_2O_3) and mullite ($3Al_2O_3$-$2SiO_2$). Unless specifically identified as single crystal fibers, oxide fibers are polycrystalline. 3M's Nextel family of fibers are by far the most prevalent. Nextel is produced by a sol-gel process, in which a sol-gel solution is dry spun into fibers, dried, and then fired at 1800–2550° F. Nextel 312, 440, and 550 were designed primarily as thermal insulation fibers. Both Nextel 312 and 440 are aluminosilicate fibers containing 14% boria (B_2O_3) and 2% boria, respectively, which means that both of these fibers contain both crystalline and glassy phases. Although boria helps to retain high temperature short time strength, the glassy phase also limits its creep strength at high temperatures. Since Nextel 550 does not contain boria, it does not contain a glassy phase and exhibits better high temperature creep resistance, but lower short time high temperature strength. For composite applications, Nextel 610 and 720 do not contain a glassy phase and have more refined α-Al_2O_3 structures, which allows them to retain a greater percentage of their strength at elevated temperatures. Nextel 610 has the highest room temperature strength due to its fine grained single phase composition of α-Al_2O_3, while Nextel 720 has better creep resistance due to the addition of SiO_2 that forms α-Al_2O_3/mullite, which reduces grain boundary sliding.[4] As a class, oxide fibers are poor thermal and electrical conductors, have higher CTE, and are denser than non-oxide fibers. Due to the presence of glass phases between the grain boundaries, and as a result of grain growth, oxide fibers rapidly lose strength in the 2200–2400° F range.

Ceramic non-oxide fibers are dominated by silicon carbide based compositions. All of the fibers in this category contain oxygen. Nippon's Nicalon series of SiC fibers are the most prevalent. Nicalon fibers are produced by a polymer pyrolysis process that results in a structure of ultra fine β-SiC particles (~1–2 nm) dispersed in a matrix of amorphous SiO_2 and free carbon.

Fiber manufacture consists of synthesizing a spinnable polymer; spinning the polymer into a precursor fiber; curing the fiber to crosslink it so that it will not melt during pyrolysis; and then pyrolyzing the cured precursor fiber into a ceramic fiber.[5] Nicalon's high oxygen content (12%) causes an instability problem above 2200° F, by producing gaseous carbon monoxide. Therefore, a low oxygen content (0.5%) variety, called Hi-Nicalon, was developed that has improved thermal stability and creep resistance. The oxygen content is reduced by radiation curing using an electron beam in a helium atmosphere. Their latest fiber, Hi-Nicalon-S, has an even lower oxygen content (0.2%) and a larger grain size (20–200 nm) for enhanced creep resistance.[6]

Another SiC type fiber with TiC in its structure is Tyranno, produced by Ube Industries. It contains 2 weight percent titanium to help inhibit grain growth at elevated temperatures. In the Tyranno ZM fiber, zirconium is used instead of titanium to enhance creep strength and improve the resistance to salt corrosion. A new silicon carbide fiber, Sylramic-iBN, contains excess boron in the fiber, which diffuses to the surface where it reacts with nitrogen to form an in situ boron nitride coating on the fiber surface. The removal of boron from the fiber bulk allows the fiber to retain its high tensile strength while significantly improving its creep resistance and electrical conductivity.[7]

Although the creep strengths of the stoichiometric fibers, such as Hi-Nicalon-S, Tyranno SA, and Sylramic, are better than that of the earlier non-stoichiometric silicon carbide fibers, their moduli are 50% higher and their strain-to-failures are 1/3 lower, which adversely impacts their ability to toughen ceramic matrices.[8] However, of the commercial fibers currently available, the advanced Nicalon and Tyranno fibers are the best in terms of as-produced strength, diameter, and cost for ceramic matrix composites for service temperatures up to $\sim 2000°$ F.[3]

The oxide based fibers are typically more strength limited at high temperatures than the non-oxide fibers; however, oxide fibers have a distinct advantage in having a greater compositional stability in high temperature oxidizing environments. While fiber creep can be a problem with both oxide and non-oxide fibers, it is generally a bigger problem with the oxide fibers. Fiber grain size is a compromise, with small grains contributing to higher strength, while large grains contribute to better creep resistance.

10.2 Matrix Materials

The selection of a ceramic matrix material is usually governed by thermal stability and processing considerations. The melting point is a good first indication of high temperature stability. However, the higher the melting point, the more difficult it is to process. Mechanical and chemical compatibility of the matrix with the reinforcement determines whether or not a useful composite can be fabricated. For some whisker reinforced ceramics, even moderate reactions with

Table 10.3 Select Ceramic Matrix Materials

Matrix	Modulus of Rupture (ksi)	Modulus of Elasticity (msi)	Fracture Toughness (ksi $\sqrt{in.}$)	Density (g/cm^3)	Thermal Expansion ($10^{-6}/°C$)	Melting Point (°F)
Pyrex Glass	8	7	0.07	2.23	3.24	2285
LAS Glass-Ceramic	20	17	2.20	2.61	5.76	–
Al$_2$O$_3$	70	50	3.21	3.97	8.64	3720
Mullite	27	21	2.00	3.30	5.76	3360
SiC	56–70	48–67	4.50	3.21	4.32	3600
Si$_3$Ni$_4$	72–120	45	5.10	3.19	3.06	3400
Zr$_2$O$_3$	36–94	30	2.50–7.70	5.56–5.75	7.92–13.5	5000

Note: Values depend on exact composition and processing.

the matrix during processing can consume the entire reinforcement. Likewise, large differences in thermal expansion between the fibers and matrix can result in large residual stresses and matrix cracking. Several important matrix materials are listed in Table 10.3.

Carbon[1] is an exceptionally stable material in the absence of oxygen, capable of surviving temperatures greater than 4000°F in vacuum and inert atmospheres. In addition, carbon is lightweight with a density of approximately 0.072 lb/in.^{3}. However, monolithic graphite is brittle, of low strength, and cannot be easily formed into large complex shapes. To overcome these limitations, carbon–carbon composites were developed, in which high strength carbon fibers are incorporated into a carbon matrix. For high temperature applications, carbon–carbon composites offer exceptional thermal stability (>4000°F) in non-oxidizing atmospheres along with low densities (0.054–0.072 lb/in.^{3}). Carbon–carbon composites are used in rocket nozzles, nosecones for reentry vehicles, leading edges, cowlings, heat shields, aircraft brakes, brakes for racing vehicles, and high temperature furnace setters and insulation. These applications utilize the following nominal properties of carbon–carbon composites (which depends on fiber type, fiber architecture, and matrix density):

Ultimate Tensile Strength > 40 ksi
Modulus of Elasticity > 10 msi
Thermal Conductivity = 0.9–19 Btu-in./(s-ft^2-°F)
Thermal Expansion = 1.1 ppm/K
Density < 0.072 lb/in.^{3}

The low thermal expansion and range of thermal conductivities give carbon–carbon composites high thermal shock resistance. As previously mentioned, the one major shortcoming of carbon–carbon composites is their oxidation susceptibility. At temperatures above 950°F, both the matrix and the fiber are vulnerable to oxidation if they are not protected from oxygen exposure. The two primary oxidation protection methods are external coatings and internal oxidation inhibitors. Surface coatings, such as silicon carbide, provide an external barrier to oxygen penetration. The addition of internal oxidation inhibitors acts either as internal barriers to oxygen ingress or as oxygen sinks (forming a protective barrier). In high temperature oxidizing environments, the time–temperature-cycle capabilities of these oxidation barriers are the primary limit to the temperature capabilities of current carbon–carbon composites.

Because they can be consolidated at lower temperatures and pressures than polycrystalline ceramic materials, glass-ceramics are potentially attractive matrix materials. Since all glass-ceramics contain some residual glass after ceraming, the upper use temperature is controlled by the softening point of the residual glass; however, silica based glass-ceramics can be used for moderate temperature applications. The most common glass-ceramic systems are LAS, MAS, MLAS, CAS, and BMAS where $L = Li_2O$, $A = Al_2O_3$, $S = SiO_2$, $M = MgO$, $C = CaO$, and $B = BaO$.[9] Glass-ceramics have the advantage that, being glasses, they can be melted to relatively low viscosities during hot pressing to impregnate the fiber bundle, and the processing temperatures are lower than for traditional crystalline ceramics, thereby reducing fiber degradation. After hot pressing, they are converted to a glass-ceramic by a heat treatment (i.e., ceraming). The amount of crystalline phase can be as high as 95–98%. However, the presence of the residual glassy phase limits their elevated temperature creep resistance.

Non-oxide matrices include carbon (C), silicon carbide (SiC), and silicon nitride (Si_3N_4). Carbon is extremely refractory and has a low density; however, for elevated temperature applications it must be protected from oxidation. Silicon carbide has a high melting point and excellent mechanical properties at elevated temperature. Silicon carbide is a little less refractory than carbon and has a slightly higher density, with oxidation resulting in the formation of silica (SiO_2); however, it can be used up to 2700°F in air. Silicon nitride matrices have properties similar to silicon carbide, except that they are less thermally stable and exhibit lower conductivities.

Oxide matrices include alumina (Al_2O_3), mullite ($3Al_2O_3$-$2SiO_2$), cordierite ($2MgO$-$2Al_2O_3$-$5SiO_2$), and zirconia (ZrO_2). Oxide matrices are of relatively low cost, exhibit rapid sintering at moderate temperatures, and exhibit high temperature oxidation resistance. Limitations include poor thermal expansion matches with many fibers, intermediate strength, and low high-temperature properties. Alumina is the most prevalent oxide matrix and contains the best balance of properties. Mullite has a lower thermal expansion than alumina, processes well with sol-gel methods, and exhibits good toughness. Cordierite has very low thermal expansion and is often used in conjunction with other oxides, such as mullite for matrices. Ziconia, which has excellent toughness when partially stabilized, loses much of its toughness at elevated temperatures.[10]

In selecting a fiber and matrix combination for a ceramic matrix composite, several factors need to be considered. First, the constituents need to be compatible from the standpoint of CTE. If the CTE of the matrix is greater than the radial CTE of the fiber, the matrix, on cooling from the processing temperature, will clamp the fibers resulting in a strong fiber-to-matrix bond and will exhibit brittle failures in service. On the other hand, if the CTE of the matrix is less than the radial CTE of the fibers, the fibers may debond from the matrix on cooling.[9] Chemical compatibility of the constituents is also an important factor. Due to the high processing temperatures for ceramic matrix composites, reactions between the fiber and matrix resulting in reduced fiber strengths are an ever present concern. For example, SiC fibers react with silica based glass-ceramics, so the fibers must be coated with a protective interfacial coating.

For temperatures exceeding 1800° F, candidate matrices are carbon, silicon carbide, silicon nitride, and alumina. Although these compositions are possible, the performance requirement that the matrix material have a CTE very close to that of the commercially available carbon, silicon carbide, and alumina based fibers effectively eliminates silicon nitride and alumina as matrix choices for the silicon carbide fibers, and silicon carbide and silicon nitride as matrix choices for the alumina based fibers. The lack of high thermal conductivity and the availability of oxide based fibers that are creep resistant for long times above 1800° F are two factors currently limiting the commercial viability of oxide/oxide ceramic matrix composites.

10.3 Interfacial Coatings

Interfacial, or interphase, coatings are often required to: (1) protect the fibers from degradation during high temperature processing, (2) aid in slowing oxidation during service, and (3) provide the weak fiber-to-matrix bond required for toughness. The coatings, ranging in thickness from 0.1 to 1.0 μm, are applied directly to the fibers prior to processing, usually by CVD. CVD produces coatings of relatively uniform thickness, composition and structure, even with preforms of complex fiber architecture. Carbon and boron nitride (BN) are typical coatings, used either alone or in combination with each other. Frequently, in addition to the interfacial coatings, an over-coating is also applied, such as a thin layer (~0.5 μm) of SiC that becomes part of the matrix during processing. The SiC over-coating helps to protect the interfacial coating from reaction with the matrix during processing. Since C and BN interfacial coatings will degrade in an ambient environment, the over-coating is usually applied immediately after the interfacial coating. During service, the over-coating also acts to protect the fibers and interfacial coatings from aggressive environments, such as oxygen and water vapor. The interfacial and over-coating are sometimes repeated as multilayer coatings to provide environmental protection layers in the presence of in-service generated matrix cracks.

10.4 Fiber Architectures

Fiber architecture is similar to that used for polymer matrix composites. Either unidirectional or woven cloth can be prepregged, or textile techniques, such as weaving or braiding, can be used to form a near net preform.[11]

To form unidirectional prepreg, tows are precoated with the interfacial protection system and filament wound on a drum, which can be prepregged with the ceramic matrix precursor material or held together with a fugitive binder. The sheets are then cut from the drum and layed-up in the desired orientation. For fabric constructions, the tows are usually woven prior to applying the interfacial coatings to minimize damage to the coatings. These prepregging approaches are illustrated in Fig. 10.5.

Near net shaped preforms can also be constructed using textile technologies, such as weaving, stitching, and braiding. Again, to avoid coating damage, the interfacial coating system is normally applied after the textile forming operations are complete, as shown in Fig. 10.6.

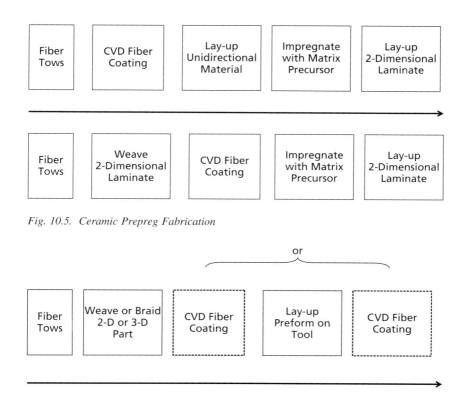

Fig. 10.5. Ceramic Prepreg Fabrication

Fig. 10.6. Ceramic Preform Fabrication

10.5 Fabrication Methods

Fabrication of ceramic matrix composites can be conducted using solid, liquid, or gas phase processing to infiltrate the matrix onto the reinforcement. In any process, the objective is to achieve minimum porosity, obtain a uniform dispersion of reinforcement, and control the fiber-to-matrix bonding. Although many fabrication routes have been explored, the most prevalent fabrication approaches for ceramic matrix composites include:

- Powder processing for discontinuous matrix composites,
- Slurry infiltration and consolidation for glass and glass-ceramic composites,
- Polymer infiltration and pyrolysis (PIP),
- Chemical vapor infiltration (CVI),
- Directed metal oxidation (DMO), and
- Liquid silicon infiltration (LSI).

10.6 Powder Processing[2]

In a manner similar to that used for monolithic ceramic materials, the powder processing route can be used to fabricate discontinuously reinforced ceramic matrix composites. While this process works well for very small reinforcements, such as whiskers and particulates, longer discontinuous fibers will be broken into shorter fibers during the mixing and consolidation processes. The basic processing steps are:

- Powder mixing (matrix and reinforcement),
- Green body fabrication,
- Machining if required,
- Binder removal,
- Consolidation and densification, and
- Inspection.

The process flow for the powder route is shown in Fig. 10.7. A uniform fine dispersion of the reinforcement and matrix powder helps in minimizing voids in the consolidated composite. When the constituents are not effectively packed, densification becomes more difficult, requiring higher pressures and longer times. Optimum packing occurs when the particle size distribution contains about 30% by volume of small particles and 70% by volume of large particles.[12] Constituent mixing is often accomplished by ball milling.

Short fibers or whiskers are often mixed with a ceramic powder slurry, dried, and hot pressed. Short fibers and whiskers often undergo some orienting during hot pressing. Whisker agglomeration in the green body is a major problem. Mechanical stirring and adjustment of the pH level of the suspension (whiskers and matrix powder in water) can help. Addition of whiskers to a slurry can result in unacceptable increases in viscosity. Also, whiskers with a large aspect

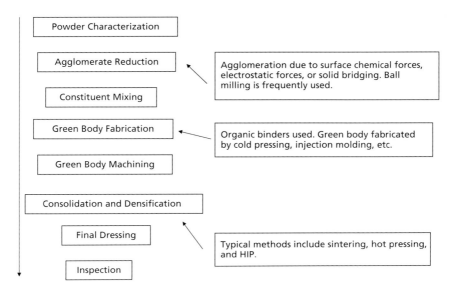

Fig. 10.7. Fabrication Sequence for Powder Processing

ratio (>50) tend to form bundles and clumps. Obtaining well-separated and deagglomerated whiskers is important for a uniform composite. Organic dispersants and techniques, such as agitation mixing assisted by ultrasonics, can be used along with deflocculation by proper pH control.[13]

Organic binders are usually mixed with the reinforcement and matrix so that near net shaped parts can be produced by cold forming processes, such as uniaxial pressing, cold isostatic pressing, tape casting, extrusion, and injection molding. After cold consolidation, the green body can be handled and even machined without damage. All organic binders must be burned out either before or during the consolidation process.

While sintering without pressure is often used to fabricate monolithic ceramics, the presence of reinforcements significantly hinders the sintering process. Although SiC whisker reinforced alumina, with whisker contents less than 10% by volume, can be pressure-less sintered to greater than 95% theoretical densities, higher volume fractions will result in unacceptable porosity levels.[14] To improve consolidation and provide acceptable levels of porosity, very fine ceramic particle sizes are used as well as pressure applied by hot pressing or hot isostatic pressing (HIP).[15] Both hot pressing and HIP have limitations. Hot pressing is restricted by press size and tonnage and is not practical for complex shapes, while HIP can be expensive and difficult if the part must be "canned" with a metal bag for parts containing porosity open to the surface. Hot pressing is used to make cutting tools, such as those shown in Fig. 10.8, used to machine

Fig. 10.8. SiC_w/Al_2O_3 Composite Cutting Tools
Source: Greenleaf Corporation

difficult-to-machine metals, such as nickel based superalloys. The SiC_w/Al_2O_3 composites can be hot pressed at 2700–3450° F at pressures of 3–6 ksi.

10.7 Slurry Infiltration and Consolidation

Slurry infiltration and consolidation is the most prevalent process used to make glass and glass-ceramic composites, mainly because the processing temperatures used for glass and glass-ceramics are lower than those used for crystalline ceramics. The melting point of crystalline ceramics is so high, that even fibers with interfacial coatings would be either dissolved or severely degraded. Another problem is the large temperature difference between extremely high processing temperatures and room temperature, which can result in shrinkage and matrix cracking. In addition, crystalline ceramics heated past their melting points have such high viscosities that the infiltration of preforms is very difficult, if not impossible.

Glass-ceramics start as amorphous glasses that can be formed into a shape and then converted to crystalline ceramics by a high temperature heat treatment. During heat treatment, small crystallites (~ 1 nm) nucleate and grow until they impinge on adjacent crystallites. On further heating, very fine ($< 1 \mu m$) angular crystallites are formed. The resulting glass-ceramic is a fine polycrystalline material in a glassy matrix with a crystalline content as high as 95–98%.

It should be noted that only certain compositions of glasses are capable of forming glass-ceramics. For example, lithium aluminosilicate (LAS) is a glass-ceramic with the composition Li_2O-Al_2O_3-SiO_2. Titanium dioxide (TiO_2) is added as a nucleating agent. When this glass is heated to 1400° F for 1.5 h, TiO_2 precipitates nucleate in the glass matrix. When the temperature is further raised to 1750° F, crystallization of the glass matrix initiates at the TiO_2 precipitate particles.

In slurry infiltration, fiber tows or a preform are impregnated in a tank containing the liquid slurry matrix, as shown in Fig. 10.9. The slurry consists of the matrix powder, an organic binder, and a liquid carrier such as water or alcohol. The slurry composition is very important. Variables such as the powder content, particle size distribution, type and amount of binder, as well as the carrier medium, will have a significant impact on part quality. For example, using a matrix powder that is smaller than the fiber diameter will help in thorough impregnation, thereby reducing porosity. Wetting agents can also be used to help infiltration into the fiber tows or preform.

After infiltration, the liquid carrier is allowed to evaporate. The resulting prepreg can then be layed-up on a tool for consolidation. Prior to consolidation, the organic binder must be burned out. Consolidation is normally accomplished in a hot press; however, HIP is an option if complex shapes are required. Consolidation parameters (time, temperature, and pressure) will also affect part quality. While high temperatures, long times, and high pressures may help to reduce porosity, fiber damage can result from either high pressures (mechanical damage) or high temperatures and long times (interfacial reactions).[13]

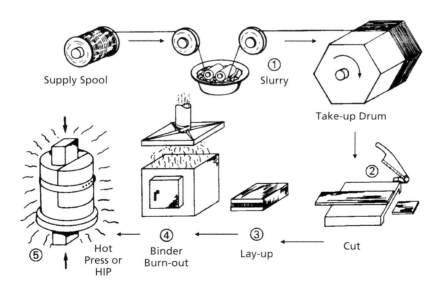

Fig. 10.9. Slurry Infiltration and Consolidation

The slurry infiltration process generally yields a composite with a fairly uniform fiber distribution and low porosity. The main disadvantage is that it is restricted to relatively low melting or low softening point matrix materials.

10.8 Polymer Infiltration and Pyrolysis (PIP)[1,16,17]

The PIP process is very similar to the processes used to make polymer matrix composites. Either a fiber reinforced prepreg is made with a matrix material that can be converted to a ceramic on heat treatment (pyrolysis) or a dry preform is infiltrated multiple times with a liquid organic precursor that can be converted by pyrolysis to a ceramic. In the case of prepreg, after the initial conversion to a ceramic, subsequent infiltrations are conducted with a liquid precursor material. Instead of a conventional thermoset resin (e.g., epoxy), an organometallic polymer is used. The polymer infiltration and pyrolysis process is shown schematically in Fig. 10.10. The process consists of:

- Infiltration of the preform with the polymer,
- Consolidate the impregnated preform,
- Cure of the polymer matrix to prevent melting during subsequent processing,
- Pyrolysis of the cured polymer to convert it to a ceramic matrix, and
- Repeating the infiltration and pyrolysis process "N" times to produce the desired density.

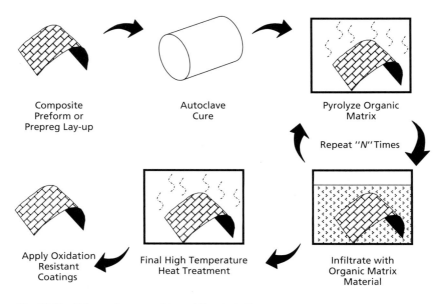

Fig. 10.10. Polymer Impregnation and Pyrolysis Process

Ceramic Matrix Composites

This section covers three types of PIP processes, the Space Shuttle carbon–carbon process, the conventional PIP process, and the sol-gel infiltration and pyrolysis process.

Space Shuttle C–C. The fabrication process[1] for the carbon–carbon Space Shuttle nose cap and wing leading edge components (Fig. 10.11) is a multi-step process, typical of the infiltration and pyrolysis technology used to produce C–C composites. As shown in Fig. 10.12, initial material lay up is similar to that used for thermoset composite parts. Plain weave carbon fabric, impregnated

Fig. 10.11. Space Shuttle Carbon–Carbon Applications

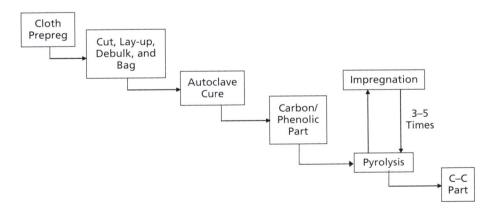

Fig. 10.12. Fabrication Sequence for Space Shuttle Carbon–Carbon Parts

with phenolic resin, is layed up on a fiberglass/epoxy tool. Laminate thickness varies from 19 plies in the external skin and web areas up to 38 plies at the attachment locations. The part is vacuum bagged and autoclave cured at 300° F for 8 h. The cured part is rough trimmed, X-rayed, and ultrasonically inspected. The part is then post-cured by placing the part in a graphite restraining fixture, loading it into a furnace, and heating it to 500° F very slowly to avoid distortion and delamination. The post-cure cycle alone can take up to 7 days.

The next step is initial pyrolysis. The part is loaded in a graphite restraining fixture and placed in a steel retort, which is packed with calcined coke. The part is slowly heated to 1500° F and held for 70 h to facilitate conversion of the phenolic resin to a carbon state. During pyrolysis, the resin forms a network of interconnected porosity that allows the escape of volatiles. This stage is extremely critical. Adequate volatile escape paths must be provided, and sufficient times must be employed, to allow the volatiles to escape. If the volatiles become entrapped and build up internal pressure, massive delaminations can occur in the relatively weak matrix. After this initial pyrolysis cycle, the carbon is designated reinforced carbon–carbon-0 (RCC-0), a state in which the material is extremely light and porous with a flexural strength of only 3000–3500 psi.

Densification is accomplished in three infiltration and pyrolysis cycles. In a typical cycle, the part is loaded in a vacuum chamber and impregnated with furfural alcohol. It is then autoclave cured at 300° F for 2 h and post-cured at 400° F for 32 h. Another pyrolysis cycle is then conducted at 1500° F for 70 h. After three infiltration/pyrolysis cycles, the material is designated RCC-3, with a flexural strength of 18 000 psi.

To allow usage at temperatures above 3600° F in an oxidizing atmosphere, it is necessary to apply an oxidation resistant coating system. The coating system consists of two steps: (1) applying a SiC diffusion coating to the C–C part, and (2) applying a glassy sealer to the SiC diffusion coating. The coating process (Fig. 10.13) starts with blending of the constituent powders, consisting of 60% silicon carbide, 30% silicon, and 10% alumina. In a pack cementation process, the mix is packed around the part in a graphite retort. The retort is loaded into a vacuum furnace, where it undergoes a 16 h cycle that includes drying at 600° F and then a coating reaction up to 3000° F in an argon atmosphere. During processing, the outer layers of the C–C are converted to SiC. The silicon carbide coated C–C part is removed from the retort, cleaned, and inspected. During cool down from 3000° F, the silicon carbide coating contracts slightly more than the C–C substrate, resulting in surface crazing (coating fissures). This crazing, together with the inherent material porosity, provides paths for oxygen to reach the C–C substrate.

To obtain increased life, it is necessary to add a surface sealer. The surface sealing process involves impregnating the part with TEOS (tetraethylorthosilicate). The part is covered with a mesh, placed in a vacuum bag, and the bag is filled with liquid TEOS. A five cycle TEOS impregnation process is

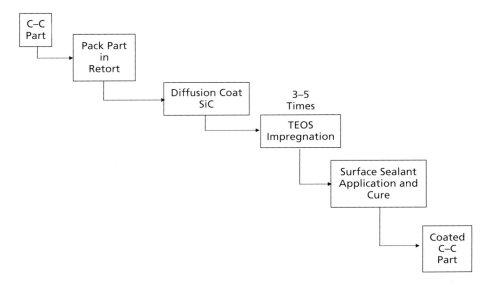

Fig. 10.13. Coating Sequence for Space Shuttle Carbon–Carbon Parts

then performed on the bagged part. After the fifth impregnation cycle, the part is removed from the bag and oven cured at 600° F to liberate hydrocarbons. This process leaves silica (SiO_2) in all of the microcracks and fissures, greatly enhancing the oxidation protection.

Conventional PIP Processes. For ceramic matrices other than carbon, silicon based organometallic polymers are the most common precursors, including Si-C, Si-C-O, Si-N, Si-O-N, and Si-N-C-O precursors. Al-O and BN have also been studied.[16] The polymeric precursor should produce a high char yield to obtain the desired density in as few a polymer infiltration and pyrolysis cycles as possible. Polymers containing highly branched and crosslinked structures, and those with high percentages of ring structures, are good candidates. Those that contain long chains tend to break up into low weight volatiles and are poor choices. Polymer branching and crosslinking increases the ceramic yield, by inhibiting chain scission and the formation of volatile silicon oligomers during pyrolysis. Initial pyrolysis produces an amorphous ceramic matrix, while high temperature treatments lead to crystallization and shrinkage as the amorphous matrix develops small domains of crystalline structure.[16] Fillers, such as silica or ceramic whiskers, can improve matrix properties by reducing and disrupting the matrix cracks that form during shrinkage. Typical filler loadings are 15–25 volume percent of the matrix.

If prepreg is going to be used, the first step is to impregnate the reinforcement with the precursor matrix material. Another alternative is to use a dry preform and to infiltrate the precursor resin directly into the preform before pyrolysis.

Resin transfer molding, vacuum impregnation, fiber placement, and filament winding have all been used to impregnate and form preceramic preforms. The composite part is initially autoclave cured at 300–500° F and 50–100 psi. The part is then put through the first pyrolysis cycle to convert the precursor matrix to ceramic. Pyrolysis in either argon or nitrogen atmospheres of at least 1300° F are required, with typical processing temperatures in the range of 1700–2200° F. During pyrolysis, large amounts of organic volatiles such as H_2 and CO are released. Therefore, pyrolysis cycles must be done slowly to allow the volatiles to escape without causing part delamination. As a result of pore formation and growth, pyrolysis gases can produce both micro- and macro-cracking. In addition, there is an extremely large reduction in volume during the matrix conversion from polymers to ceramics.[18] Cycles as long as 1–2 days are not uncommon.

Subsequent infiltrations are conducted with a low viscosity prepolymer. Infiltration is best done by vacuum impregnation in a vacuum bag.[16] After the first pyrolysis cycle, the matrix is amorphous and highly porous, with multiple matrix cracks and a void content of 20–30%. The infiltration and pyrolysis process is repeated, often between 5 and 10 times to reduce the porosity, fill the cracks, and obtain the desired density. After the last pyrolization cycle, the part can be heat treated at higher temperatures to convert the amorphous matrix to a crystalline phase, relieve residual stresses, and provide final consolidation.

The biggest advantage of the polymer infiltration and pyrolysis process is the use of the familiar methods employed in organic matrix composite fabrication. However, the multiple infiltration and pyrolysis cycles required to obtain high density parts are expensive and the lead times are long. In addition, it is almost impossible to fill all of the fine matrix cracks, which degrades mechanical properties.

Sol-Gel Infiltration. Sol-gel infiltration is a lower temperature infiltration and pyrolysis process that can be used for oxide based ceramic matrix composites. Densification still requires pressure and high temperatures, but the densification temperatures are usually less than those required in the slurry infiltrated matrices. Sol-gel techniques require minimal exposures to temperatures above 1800° F, helping to reduce thermal damage to the fibers. Infiltration can be performed at temperatures less than 600° F, using either vacuum infiltration or autoclave molding.[9] In addition, sintering aids, such as boria, are not required due to the lower processing temperatures.

In sol-gel infiltration, a chemical precursor is hydrolyzed, polymerized into a gel, and then dried and fired to produce a glass or ceramic composition. The precursors range from mixtures of water, alcohols, and metal oxides to commercially available stabilized colloids containing discrete ceramic particles.[17] Hydrolysis reactions form an organometallic solution, or sol, composed of polymer-like chains containing metallic ions and oxygen. Amorphous oxide particles form from the solution, producing a rigid gel. The gel is then dried and fired to provide sintering and densification of the final ceramic part.

As an example, in the sol-gel process for alumina, an organometallic precursor is hydrolyzed, converted to colloidal solution (i.e., peptized) with hydrochloric acid, and then fired at increasing temperatures to produce first γ-alumina and eventually α-alumina as follows:[17]

$$Al(OC_4H_9)_3 + H_2O \rightarrow Al(OC_4H_9)_2 + C_4H_9OH \text{ (Hydrolysis)}$$

$$2\ Al(OC_4H_9)_2(OH) + H_2O \xrightarrow{HCl} C_4H_9O-AlOH-O-AlOH-OC_4H_9 +$$

$$2C_4H_9OH \text{ (Peptization)}$$

$$2AlO(OH) \rightarrow \gamma\text{-}Al_2O_3 + H_2O \rightarrow \alpha\text{-}Al_2O_3 \text{ (Firing)}$$

The ideal sol should have as high a ceramic content as possible (>30 weight percent if possible), a low viscosity (15–20 cP), and a small particle size (<30 nm if possible). A neutral pH helps minimize fiber degradation. The sol should be capable of being processed at room temperature for several hours and should be stable enough to allow shipping and storage.[17]

Again, fillers can be used to help reduce shrinkage and subsequent matrix cracking. Ceramic filler powders can (1) reduce shrinkage and matrix cracking as it loses water and alcohols, (2) provide sites for the nucleation of grains, and (3) maximize matrix densification during the infiltration. Typically, silica particles are added to silica sols and alumina particles to alumina sols. However, loading the liquid precursor with powder increases the viscosity considerably and hinders the reimpregnation step.[19]

Impregnation of fiber tows can be accomplished by several methods: (1) the tows can be passed through a bath containing the sol and then hydrolyzed in humid air; (2) the tows can be passed through a bath containing a partially hydrolyzed sol and then wet wound; or (3) the tows can be dry wound and then pressure infiltrated with the sol. For woven 2-D and 3-D fiber architectures, it is normal practice to weave the fabric first and then infiltrate the woven preform with the sol. For 2-D woven fabrics that will require lay-up, polymeric binders are often used to provide tack and drape but require burn-off prior to densification.

A typical fabrication sequence would be to weave a preform and then impregnate it with the sol. Impregnation can be conducted by immersing the preform in the sol and pulling a vacuum to facilitate impregnation. An autoclave can also be used to provide positive pressure, resulting in better impregnation. The sol is then gelled by heating. Temperatures in the range of 200–400° F will remove water and alcohols, while higher temperatures (550–750° F) can be used to drive off any organics.

The infiltration and gelling cycle is repeated until the desired density is obtained. Processing under vacuum pressure usually yields a composite with about 20% porosity. Lower porosity levels can be achieved by using 50–100 psi autoclave pressure. After the desired density is achieved through multiple infiltration cycles, the part is fired at high temperatures to obtain the final ceramic microstructure.

While lower processing temperatures can be used in the sol-gel process, disadvantages are high shrinkage which results in matrix cracking, low yield which requires multiple infiltrations, and high precursor costs. High shrinkage of the gelled sols results from the large volume of water and alcohols that must be removed, often resulting in significant levels of porosity and extensive matrix micro- and macro-cracking. Excess water is typically used to insure complete hydrolysis, but large amounts of water also reduce the yield of the dried matrix obtained per infiltration cycle. In some cases, the polymeric precursor chemicals are very expensive, and most metal-organic compounds are very sensitive to moisture, heat, and light.[18]

10.9 Chemical Vapor Infiltration (CVI)[20,21]

Chemical vapor deposition is a well-established industrial process for applying thin coatings to materials. When it is used to infiltrate and form a ceramic matrix, it is called chemical vapor infiltration (CVI). In CVI, a solid is deposited within the open volume of a porous structure by the reaction or decomposition of gases or vapors. A porous preform of fibers is prepared and placed in a high temperature furnace, as shown in Fig. 10.14. Reactant gases or vapors are

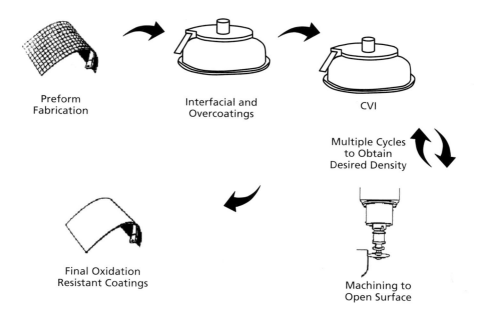

Fig. 10.14. Chemical Vapor Infiltration Fabrication Sequence

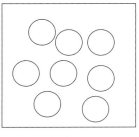  

Initial Fiber Array   Initial Fiber Coating   Continued Fiber Coating

Fig. 10.15. Chemical Vapor Infiltration Growth[21]

pumped into the chamber and flow around and diffuse into the preform. The gases decompose or react to deposit a solid onto and around the fibers. As the reaction progresses, the apparent diameter of the fibers increases and eventually fills the available porosity, as depicted in Fig. 10.15. Since this is essentially a deposition process, there is very little mechanical stress on the fibers. The primary processes involved are mass and heat transfer, with the objective to maximize the rate of matrix deposition while minimizing density gradients.[20]

One of the problems with CVI is that the reactions occur preferentially at, or near, the first surfaces contacted by the reactant gases, resulting in sealing-off the interior pores in the preform. To minimize this effect, it is necessary to run the process at low temperatures, reduced pressures, and to use dilute reactant concentrations, all of which translate into long processing times.

The most important uses of CVI are to produce carbon and silicon carbide matrix composites. A carbon matrix can be formed by the decomposition of methane (e.g., CH_4 + nitrogen + hydrogen) on a hot fibrous preform. For a silicon carbide matrix, methyltrichlorosilane (CH_3SiCl_3) is used, along with hydrogen at 1800° F, according to the following reaction:

$$CH_3SiCl_3(g) \rightarrow SiC(s) + 3HCl(g)$$

To obtain the desired microstructure and delay the closing of porosity at the preform surface, it is important to control parameters such as the temperature, pressure, and flow rate of the gases, and the preform temperature. It should be noted that the microstructure produced by CVI is normally not as fine as that produced by hot pressing. In addition, carbon fiber and silicon carbide are incompatible from a coefficient of thermal expansion stand point, and the silicon carbide matrix will develop microcracks allowing oxygen penetration into the structure during service. Therefore, SiC fibers are often used with silicon carbide matrices.

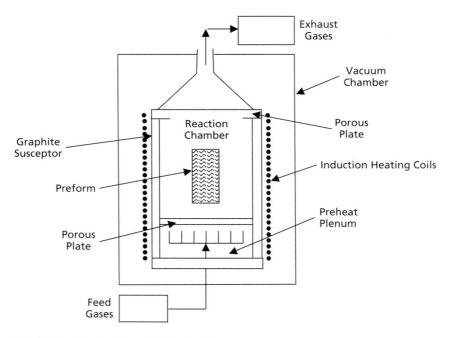

Fig. 10.16. Chemical Vapor Infiltration Reactor

The equipment used for CVI includes gas handling and distribution equipment, a reactor or furnace to heat the substrates, a pressure control system, and scrubbers or traps to remove hazardous materials from the effluent gases. The details of a reactor are shown in Fig. 10.16, and a schematic of the equipment layout is given in Fig. 10.17. Reactants may be gases, liquids, or solids at room temperature. Direct evaporation of liquids and solids can be accomplished; however, liquids are usually swept through the reactor as vapors, by bubbling carrier gases such as hydrogen, argon, nitrogen, or helium through the reactant liquid. The reactors are normally heated by either resistance or induction. Graphite fixtures are used to support the preforms during the initial stages of infiltration. Vacuum pumps are used to control the process pressure.[20]

There are a number of variations to the CVI process, the most important being the isothermal, forced gradient and pulse flow processes. The objectives of the forced gradient and pulse flow processes are to utilize temperature (or temperature and pressure) gradients, or by pulsing the reactant gases, to reduce the long cycle times inherent in the isothermal process.

The isothermal process is by far the most common process and the only widely used commercially process. The gases flow outside the preform by convection and inside the preform by diffusion. To delay sealing-off the interior of the preform by surface crusting, deposition is performed at relatively low temperatures

Ceramic Matrix Composites

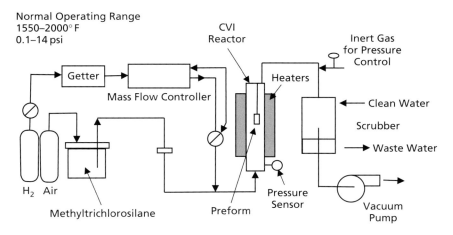

Fig. 10.17. *Chemical Vapor Infiltration Equipment*

and reduced pressures. The relatively low temperatures result in lower reaction rates, with the reduced pressures favoring diffusion. Since pore sealing on the surface occurs, sometimes called crusting or canning, it is necessary to periodically remove the part and machine the surface to allow further densification. An advantage of the isothermal process is that a furnace can be loaded with a number of parts with different configurations for processing at the same time.

In forced gradient CVI, a graphite tool that holds the preform is placed in contact with a water-cooled metallic gas distributor. The preform is heated on one side and the reactant gases are injected through the cooler side where almost no deposition occurs under pressure. The reactant gases pass unreacted through the preform because of the lower temperature. When the gases reach the hot zone, they decompose and deposit on the fibers to form the matrix. As the matrix material gets deposited in the hot portion of the preform, the preform density and thermal conductivity increases and the hot zone moves progressively from the hot side toward the cooler side. This process reduces the number of machining cycles required to obtain a dense part.

In the pressure pulse CVI process, the chamber is evacuated and reactant gases are then injected for a very short time followed by another evacuation. This cycle is repeated until the part is fully dense. The pulsing action speeds the deposition process by continually removing spent gases and supplying fresh reactant gases, but still does not eliminate all of the machining operations.

The CVI method offers several advantages. First, it is conducted at relatively low temperatures, so damage to the fibers is minimal. Since most interfacial coatings are applied using CVD, matrix infiltration can be conducted immediately after the interfacial coatings are applied. It can also be used to fabricate fairly large and complex near net shapes. The mechanical and thermal properties are good because high purity matrices with controlled microstructures can be obtained. In addition, a large number of ceramic matrices can be formed using CVI, a few of these are listed in Table 10.4. The major disadvantage of the CVI process is that it is not possible to obtain a fully dense part since the amount of residual porosity (Fig. 10.18) is around 10–15%, which adversely affects the

Table 10.4 CVI Reactions for Ceramic Matrices

Ceramic Matrix	Reactant Gases	Reactant Temperature (°F)
C	CH_4-H_2	1800–2200
TiC	TiC_4-CH_4-H_2	1650–1800
SiC	CH_3SiCl_3-H_2	1800–2550
B_4C	BCl_3-CH_4-H_2	2200–2550
TiN	$TiCl_4$-N_2-H_2	1650–1800
Si_3N_4	$SiCl$ – NH_3-H_2	1800–2550
BN	BCl_3-NH_3-H_2	1800–2550
AlN	$AlCl_3$-NH_3-H_2	1475–2200
Al_2O_3	$AlCl_3$-CO_2-H_2	1650–2010
SiO_2	SiH-CO_2-H_2	400–1110
TiO_2	$TiCl_3$-H_2O	1470–1830
ZrO_2	$ZrCl_4$-CO_2-H_2	1650–2200
TiB_2	$TiCl_4$-BCl_3-H_2	1470–1830
WB	WCl_6-BBr_3-H_2	2550–2900

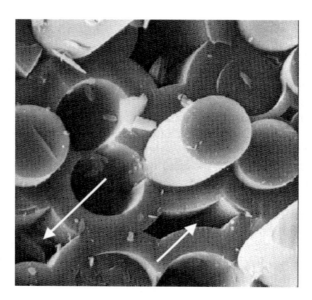

Fig. 10.18. CVI Residual Porosity

mechanical and thermal properties, particularly after matrix cracking occurs in an oxidizing atmosphere at elevated temperatures. The other big disadvantage is that long processing times, often greater than 100 h, and multiple machining cycles result in high costs.

10.10 Directed Metal Oxidation (DMO)[22]

Directed metal oxidation, or reactive melt infiltration, uses liquid aluminum that is allowed to react with air (oxygen) to form alumina (Al_2O_3), or with nitrogen to form aluminum nitride (AlN). A fiber preform, which has been precoated with a BN interfacial coating and a SiC protective over-coating, is brought in contact with molten aluminum alloy held in a suitable container, along with air or nitrogen, for the growth of an alumina or aluminum nitride matrix according to the following reactions:

$$2Al(l) + 3O_2(g) \rightarrow Al_2O_3(s)$$
$$Al(l) + N_2(g) \rightarrow AlN(s)$$

When the liquid aluminum is heated to 1650–2200° F in air, the alumina matrix grows by a complex process involving the dissolution of oxygen and the precipitation of alumina. Wicking of the molten aluminum along interconnected microscopic channels in the reaction product sustains the reaction, promoting outward growth from the original metal surface. To form near net shaped parts, the preform surface is coated on all surfaces, except the surface in contact with the molten metal. The coating is a gas permeable barrier layer that is applied by either spraying or dipping. Since the barrier layer is not wet by the molten aluminum, an impervious barrier is formed when the growth front comes in contact with the barrier. The process flow for fabricating a part by directed metal oxidation is shown in Fig. 10.19.

Important process parameters include alloy composition, growth temperature, oxygen partial pressure, and the presence of fillers, which can be used to control grain size and act as nucleation sites for the reaction. Magnesium, added as a minor alloying element, forms a thin layer of magnesia at the molten interface that prevents the formation of a dense protective scale of alumina that would otherwise stop the in-depth diffusion of oxygen. The matrix slowly builds up (~ 1 in./day) within the fiber preform and fills in the space between the fibers. Since the reaction product is not continuous and contains microscopic channels, the liquid metal wicks to the surface and reacts with the gaseous atmosphere.

For discontinuous composites, preforming can be conducted using conventional ceramic processes such as slip casting, pressing, and injection molding. For continuous fiber composites, fabric lay-up, weaving, braiding, or filament winding can all be used. Since thin layers of both carbon and boron nitride oxidize rapidly during matrix growth of alumina based composites, a duplex

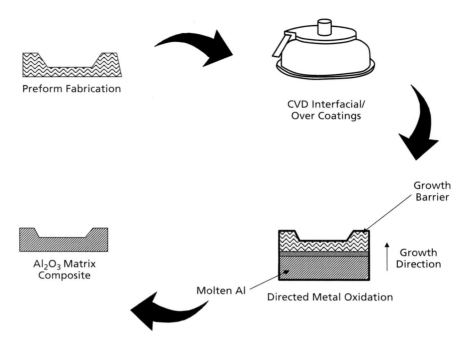

Fig. 10.19. Directed Metal Oxidation Process

coating of BN/SiC is used to protect the fiber and provide the weak interface necessary. The over-coating of SiC (3–4 μm) protects the thin inner BN layer (0.2–0.5 μm) from attack by the molten aluminum during processing.[22]

This process is relatively low cost with near net shape capabilities, and complex shaped parts can be fabricated. Only small dimensional changes occur during processing since the matrix fills the pores within the preform without disturbing the reinforcement. A disadvantage is the presence of residual aluminum phase ($\sim$5–10%) in the matrix that must be removed if the part is to be used above the melting point of aluminum (1220° F).[11] The residual metal can be leached away by an acid treatment that leaves behind open porosity with a residual metal content of $\sim$1%.

Matrix cracking also occurs due to the thermal expansion mismatch between the alumina matrix and the protective SiC over-coating. This cracking and the residual porosity from the removed aluminum phase reduces the strength and thermal conductivity. This process is conducted commercially as the proprietary Dimox process, which stands for *di*rected *m*etal *ox*idation.

10.11 Liquid Silicon Infiltration (LSI)[11]

Liquid silicon, or one of its lower melting point alloys, is used to infiltrate a fiberous preform to form a silicon carbide matrix. The fibers must contain

Ceramic Matrix Composites

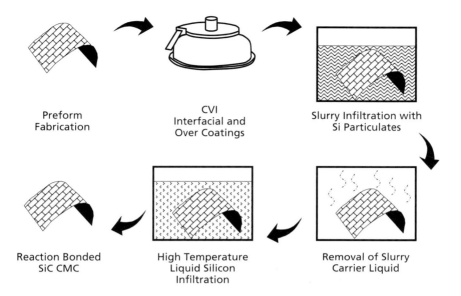

Fig. 10.20. Liquid Silicon Infiltration Process

an interfacial coating, along with a SiC over-coating, to protect them from the liquid silicon. Before infiltration, a fine grained silicon carbide particulate is slurry cast into the fiber preform. After removal of the slurry carrier liquid, melt infiltration is usually done at 2550° F, or higher, and is usually complete within a few hours, as illustrated in Fig. 10.20. The liquid silicon bonds the silicon carbide particulates together and forms a matrix that is somewhat stronger and denser than that obtained by CVI. Since the resulting matrix can contain up to 50% unreacted silicon, the long-term use temperature is limited to about 2200° F.[11] The amount of unreacted silicon can be reduced by infiltrating the preform with carbon slurries prior to the silicon infiltration process. However, some unreacted or free silicon will always be present in the matrix.

In another variant of the liquid metal infiltration process, liquid silicon is reacted with unprotected carbon fibers to form a silicon carbide matrix. After the initial step of using prepreg to form a highly porous carbon matrix part, liquid silicon is infiltrated into the structure, where it reacts with the carbon to form silicon carbide along with unreacted silicon and carbon. Since the carbon fibers are intended to react with the liquid silicon, no coatings are used on the fibers. However, the poor oxidation resistance of the carbon fibers means that the entire part will require an oxidation resistant coating.

The liquid metal infiltration processes have several advantages: (1) they produce a fairly dense SiC based matrix with a minimum of porosity; (2) the processing time is shorter than for most ceramic matrix composite fabrication processes; and (3) the dense and closed porosity on the surface can often eliminate the need for a final oxidation resistant coating. The major disadvantage is the high temperatures (>2550° F) required for liquid silicon infiltration that exposes the fibers to possible degradation, due to the high temperatures and corrosive nature of liquid silicon. In addition, the temperatures can be even higher due to the possibility of an exothermic reaction between silicon and carbon.

Summary

Monolithic high performance ceramics combine desirable characteristics, such as high strength and hardness, high temperature capability, chemical inertness, wear resistance, and low density. The greatest drawback of ceramics is their extremely low fracture toughness, which in practice means that these materials have a very low tolerance of crack-like defects. Incorporation of fibers, whiskers, or particles in a ceramic matrix can result in a tougher ceramic material, because the reinforcements introduce energy dissipating mechanisms, such as debonding at the fiber-to-matrix interface, crack deflection, fiber bridging, and fiber pullout. For these energy dissipating mechanisms to be effective, there must be a poor bond at the fiber-to-matrix interface.

Similar to MMCs, there are very few commercial applications for CMCs, due to their high costs and concerns for reliability; however, carbon–carbon has found applications in aerospace for thermal protection systems. Carbon–carbon is capable of withstanding temperatures in excess of 3000° F provided it is protected against oxidation. Both internal and external oxidation protection systems are used, with SiC and glass forming sealer compounds being the most prevalent. C–C can be made by either pyrolysis of organic compounds or CVI. Both procedures are complex and lengthy, resulting in long lead times and high costs.

The slurry infiltration process is the most important technique used to produce glass and glass-ceramic composites. The slurry infiltration process followed by hot pressing is well suited for glass or glass-ceramic matrix composites, mainly because the processing temperatures for these materials are lower than those used for crystalline ceramics. For discontinuous CMCs, short fibers, whiskers, or particles are mixed with a ceramic powder slurry, dried, and hot pressed. For continuous fiber crystalline CMCs, either prepregs or preforms are used to maintain the fiber architecture. To provide protection to the fibers during processing and service and to insure weak interface bonding, interfacial coatings of C and BN are applied by CVD.

The most prevalent fabrication approaches for continuous fiber crystalline ceramic matrix composites include: PIP, CVI, DMO, and LSI.

The PIP process consists of (1) infiltration of the preform with the polymer; (2) consolidation of the impregnated preform; (3) cure of the polymer matrix to prevent melting during subsequent processing; and (4) pyrolysis of the cured polymer to convert it to a ceramic matrix. A number of re-infiltration and pyrolysis cycles, often 5–10, are required to produce a high density part. The biggest advantage of the polymer infiltration and pyrolysis process is the use of the familiar methods employed in organic matrix composite fabrication. However, the multiple infiltration and pyrolysis cycles are expensive and the lead times are long. In addition, it is almost impossible to fill all of the fine matrix cracks which degrade the mechanical properties and thermal conductivity.

In CVI, a solid is deposited within the open volume of a porous structure by the reaction or decomposition of gases or vapors. A porous preform of fibers is prepared and placed in a high temperature furnace. Reactant gases or vapors are then pumped into the chamber and flow around and diffuse into the preform. The gases decompose, or react, to deposit a solid onto and around the fibers. As the reaction progresses, the apparent diameter of the fibers increases and eventually fills the available porosity. The CVI method offers several advantages: (1) it is conducted at relatively low temperatures so damage to the fibers is minimal; (2) since most interfacial coatings are applied using CVD, the matrix infiltration can be conducted immediately after the interfacial coatings are applied; (3) it can also be used to fabricate fairly large and complex near net shapes; and (4) the mechanical and thermal properties are good because high purity matrices with controlled microstructures can be obtained. The major disadvantage of the CVI process is that it is not possible to obtain a fully dense part since the amount of residual porosity is around 10–15%, which adversely affects the mechanical and thermal properties. The other big disadvantage is that long processing times, often greater than 100 h, and multiple machining cycles result in high costs.

Directed metal oxidation, or reactive melt infiltration, uses liquid aluminum that reacts with air (oxygen) to form alumina (Al_2O_3), or with nitrogen to form aluminum nitride (AlN). This process is relatively low cost with near net shape capabilities, and complex shaped parts can be fabricated. Only small dimensional changes occur during processing since the matrix fills the pores within the preform without disturbing the reinforcement. A disadvantage is the presence of residual aluminum phase ($\sim$5–10%) in the matrix that must be removed if the part is to be used above the melting point of aluminum.

In LSI, liquid silicon, or one of its lower melting point alloys, is used to infiltrate a fiberous preform to form a silicon carbide matrix. Before infiltration, a fine grained silicon carbide particulate is slurry cast into the fiber preform. After removal of the slurry carrier liquid, melt infiltration is usually done at 2550° F, or higher, and is usually complete within a few hours. The liquid silicon bonds the silicon carbide particulates together and forms a matrix that is somewhat stronger and denser than that obtained by CVI. Since the resulting matrix can contain up to 50% unreacted silicon, the long-term use temperature is limited to about 2200° F. The liquid metal infiltration processes have several advantages: (1) they produce a fairly dense SiC based matrix with a minimum of porosity; (2) the processing time is shorter than for most ceramic matrix composite fabrication processes; and (3) the dense and closed porosity on the surface can often eliminate the need for a final oxidation resistant coating. The major disadvantage is the high temperatures required for liquid silicon infiltration that exposes the fibers to possible degradation, due to the high temperatures and the corrosive nature of liquid silicon.

Recommended Reading

[1] *Handbook on Continuous Fiber-Reinforced Ceramic Matrix Composites*, Ceramics Information Analysis Center, 1995.
[2] Chawla, K.K., *Ceramic Matrix Composites*, Chapman & Hall, 1993.

References

[1] Buckley, J.D., "Carbon-Carbon Composites", in *Handbook of Composites*, Chapman & Hall, 1998, pp. 333–351.
[2] Amateau, M.F., "Ceramic Composites", in *Handbook of Composites*, Chapman & Hall, 1998, pp. 307–332.
[3] DiCarlo, J.A., Dutta, S., "Continuous Ceramic Fibers for Ceramic Matrix Composites", in *Handbook on Continuous Fiber-Reinforced Ceramic Matrix Composites*, Ceramics Information Analysis Center, 1995, pp. 137–183.
[4] 3M Nextel Ceramic Textiles Technical Notebook, 3M Ceramic Textiles and Composites, 2003.
[5] Marzullo, A., "Boron, High Silica, Quartz and Ceramic Fibers", in *Handbook of Composites*, Chapman & Hall, 1998, pp. 156–168.
[6] "State of the Art in Ceramic Fiber Performance", in *Ceramic Fibers and Coatings: Advanced Materials for the Twenty-First Century*, The National Academy of Sciences, 1998, pp. 20–36.
[7] DiCarlo, J.A., Yun, H.M., "New High-Performance SiC Fiber Developed for Ceramic Composites", NASA-Glenn Research & Technology Reports, 2002.
[8] Luthra, K.L., Corman, G.S., "Melt Infiltrated (MI) SiC/SiC Composites for Gas Turbine Applications", GE Research & Development Center, Technical Information Series, 2001.
[9] Naslain, R.R., "Ceramic Matrix Composites: Matrices and Processing", in *Encyclopedia of Materials: Science and Technology*, Elsevier Science Ltd, 2000.
[10] Zolandz, R., Lehmann, R.L., "Crystalline Matrix Materials for Use in Continuous Filament Fiber Composites", in *Handbook on Continuous Fiber-Reinforced Ceramic Matrix Composites*, Ceramics Information Analysis Center, 1995, pp. 111–136.
[11] DiCarlo, J.A., Bansal, N.P. "Fabrication Routes for Continuous Fiber-Reinforced Ceramic Composites (CFCC)", NASA/TM-1998-208819, 1998.
[12] Lange, F.F., Lam, D.C.C., Sudre, O., Flinn, B.D., Folsom, C., Velamakanni, B.V., Zok, F.W., Evans, A.G., "Powder Processing of Ceramic Matrix Composites", *Material Science and Engineering*, A144, 1991, pp. 143–152.
[13] Chawla, K.K., "Processing of Ceramic Matrix Composites", in *Ceramic Matrix Composites*, Chapman & Hall, 1993, pp. 126–161.

[14] Becker, P.F., Tiegs, T.N., Angelini, P., "Whisker Reinforced Ceramic Composites", in *Fiber Reinforced Ceramic Composites: Materials, Processing and Technology*, Noyes Publications, 1990, pp. 311–327.

[15] Lewis III, D., "Continuous Fiber-Reinforced Ceramic Matrix Composites: A Historical Overview", in *Handbook on Continuous Fiber-Reinforced Ceramic Matrix Composites*, Ceramics Information Analysis Center, 1995, pp. 1–31.

[16] French, J.E., "Ceramic Matrix Composite Fabrication and Processing: Polymer Pyrolysis", in *Handbook on Continuous Fiber-Reinforced Ceramic Matrix Composites*, Ceramics Information Analysis Center, 1995, pp. 269–299.

[17] Cullum, G.H., "Ceramic Matrix Composite Fabrication and Processing: Sol-Gel Infiltration", in *Handbook on Continuous Fiber-Reinforced Ceramic Matrix Composites*, Ceramics Information Analysis Center, 1995, pp. 185–204.

[18] Mah, T., Yu, Y.F., Hermes, E.E., Mazdiyasni, K.S., "Ceramic Fiber Reinforced Metal-Organic Precursor Matrix Composites", in *Fiber Reinforced Ceramic Composites: Materials, Processing and Technology*, Noyes Publications, 1990, pp. 278–310.

[19] Naslain, R., "Design, Preparation and Properties of Non-Oxide CMCs for Application in Engines and Nuclear Reactors: An Overview", *Composite Science and Technology*, Vol. 64, 2004, pp. 155–170.

[20] Lowden, R.A., Stinton, D.P., Besmann, T.M., "Ceramic Matrix Composite Fabrication and Processing: Chemical Vapor Infiltration", in *Handbook on Continuous Fiber-Reinforced Ceramic Matrix Composites*, Ceramics Information Analysis Center, 1995, pp. 205–268.

[21] Lackey, W.J., Starr, T.L., "Fabrication of Fiber-Reinforced Ceramic Composites by Chemical Vapor Infiltration: Processing, Structure and Properties", in *Fiber Reinforced Ceramic Composites: Materials, Processing and Technology*, Noyes Publications, 1990, pp. 397–450.

[22] Fareed, A.S., "Ceramic Matrix Composite Fabrication and Processing: Directed Metal Oxidation", in *Handbook on Continuous Fiber-Reinforced Ceramic Matrix Composites*, Ceramics Information Analysis Center, 1995, pp. 301–324.

Chapter 11

Structural Assembly

Assembly represents a significant portion of the total manufacturing cost. Assembly operations are labor intensive and involve many steps and, as shown in Fig. 11.1, can represent as much as 50% of the total delivered part cost.[1] For example, the wing shown in Fig. 11.2 requires:

(1) A framing operation in which all of the spars and ribs must be located in their proper location and connected together with shear ties.
(2) Each skin must then be located on the substructure, shimmed, holes drilled, and fasteners installed. During and after skin installation there are various sealing operations that must be performed.
(3) The final wing torque box must have the leading edges, wing tips, and control surfaces assembled.

This brief description is a gross over-simplification of the complexity involved in assembling a large structural component. The number of mechanical fasteners in a typical fighter aircraft might be in the range of 200 000–300 000, while a commercial airliner or transport aircraft can have as many as 1 500 000–3 000 000 fasteners, depending on aircraft size. A hole has to be drilled for each of these fasteners and then the fastener has to be installed in each hole. In this chapter, the basic assembly operations will be covered with an emphasis on hole preparation and the types of mechanical fasteners used in aircraft structures.

11.1 Framing

Framing operations, in which the substructure is located and fastened in its proper location, has made significant progress since the mid-1960s. In the 1960s and 1970s, substructure was primarily manually located using some hard tool located positions, usually supplemented with large pieces of clear plastic film (Mylar's) scribed with hole pattern locations. The design, tooling, and fabrication databases were not necessarily coordinated with each other, which created a lot

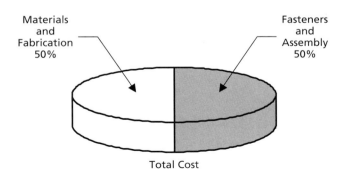

Fig. 11.1. The High Cost of Assembly Operations[1]

Structural Assembly

 Framing

 Skinning

 Completed Wing Torque Box

Fig. 11.2. Assembly Complexity
Source: The Boeing company

Manufacturing Technology for Aerospace Structural Materials

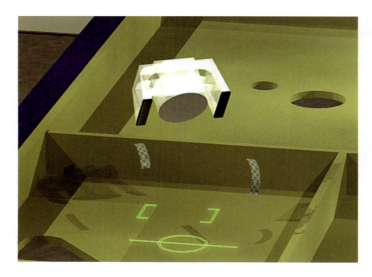

Fig. 11.3. Laser Projection Location
Source: Laser Projection Technologies, Inc.

of variability and poorly fitting parts. During the 1980s and 1990s, there was less use of Mylar's and a greater reliance on hard tooling to position parts. However, the extensive use of hard tooling to locate parts increased the non-recurring investment required at the start of a new program. With the advent of solid modeling and electronic master models in the 1990s, a process called determinant assembly emerged. In determinant assembly, coordinated undersize fastener holes are drilled in the parts during fabrication. These holes are then used to position the parts during assembly, eliminating the need for hard tooling locators. Another recent development is the use of laser projection units for establishing part and hole location. A typical application for a laser projection system is shown in Fig. 11.3.

11.2 Shimming

Prior to starting hole drilling and fastener installation, it is important to check all joints for the presence of gaps. The presence of gaps can unnecessarily preload metallic members when fasteners are installed, a condition that can initiate premature fatigue cracking and even stress corrosion cracking of aluminum. However, gaps in structure containing composites can cause even more serious problems than in metallic structure. Since composites do not yield, and are more brittle and less forgiving than metals, excessive gaps can result in delaminations when they are pulled out during fastener installation. The composite is put in bending due to the force exerted by the fastener drawing the parts together and can develop matrix cracks and/or delaminations around the holes. Cracks and delaminations usually occur on multiple layers through the thickness and can adversely affect the joint strength.[2] Gaps can also trap metal chips and contribute to backside hole splintering. If the skin is composite and the substructure is metal, and if an appreciable gap is present during fastener installation, the composite skin will often crack and delaminate. If both the skin and substructure are composite, then cracks can develop in either the skin or substructure, or both. Substructure cracking often occurs at the radius between the top of the stiffener and the web.

To prevent unnecessary preloading of metallic structure, and the possibility of cracking and delaminations in composite structure, it is important to measure all gaps and then shim any gaps greater than 0.005 in. Liquid shim, which is a filled thixotropic adhesive, can be used to shim gaps between 0.005 and 0.030 in. If the gap exceeds 0.030 in., then a solid shim is normally used, but engineering approval is often required for a gap this large. Solid shims can be made from solid metal, laminated metal that can be peeled to the correct thickness, or composite. When selecting a solid shim material, it is important to make sure there is no potential for galvanic corrosion within the joint.[3]

Liquid shimming can be accomplished by first drilling a series of undersize holes in the two mating surfaces, for installing temporary fasteners, to provide a light clamp-up during the shimming process. The liquid shim is usually bonded to one of the two mating surfaces. The surface that will be bonded should be clean and dry, to provide adequate adhesion. Composite surfaces should be scuff sanded. The other surface is covered with release tape or film. After the liquid shim is mixed, it is buttered onto one surface and the other surface is located, and then clamped up with mold-released temporary fasteners. The excess or squeeze-out is removed prior to gellation, which usually occurs within an hour of mixing. After the shim material is cured, typically for about 16 h, the part is disassembled and any voids or holes in the shim are repaired. After the repair, the parts are assembled.

11.3 Hole Drilling

It should be noted that there are some differences between fighter aircraft and commercial passenger aircraft. Fighter aircraft designs are highly tailored to performance and loads, leading to a lot of thickness variations in the skins and substructure to reduce weight. This results in a wide variety, but limited numbers, of fastener types, grip lengths, and diameters. Due to the smaller size of a fighter airframe, there are more areas of limited access during assembly. On the other hand, larger commercial aircraft have much more fastener commonality with regards to type, grip length, and diameter, but also, due to their size, many more fasteners. Skins and substructure tend to be more uniform in thickness. Limited access is not as much of a problem, but the shear size of the parts makes them difficult to handle. There are many types of drill motors and units that can be used to drill structures, but they can be broadly classified as either hand, power feed, automated drilling units, or automated riveting equipment.

11.3.1 Manual Drilling

Manual, or free hand, drilling using hand-held drill motors, such as the one shown in Fig. 11.4, has the least chance of making a close tolerance hole ($+0.003/-0.000$ in.). The only real control is the drill speed (rpm). It is up to the operator to make sure the drill: (1) is located in the proper location; (2) is perpendicular to the surface; and (3) is fed with enough pressure to generate the hole, but not too much pressure to damage the hole. Although free hand drilling is obviously not the best method, it is frequently used, because it requires no investment in tooling (i.e., drill templates), and, in many applications, where access is limited, it may be the only viable method. A typical tight access situation is shown in Fig. 11.5, where a mechanic is installing collars on Hi-Lok fasteners. For tight access areas, right-angle drill motors are available. If free hand drilling is used, it is recommended that the operators use a drill bushing, or tri-pod support, to insure normality, and that they be provided with detailed written instructions for hole generation and inspection.

Manual hole drilling during assembly is often done by drilling undersize holes (pilot holes), installing temporary fasteners to the hold the parts together, and then bringing the holes up to full size after all of the pilot holes are drilled. Pilot holes are usually drilled with small diameter drill bits (e.g., 0.90–0.125 in.). Hole diameters for aerospace structures nominally range from 0.164 to 0.375 in., with the predominate hole sizes being 0.188 and 0.250 in. diameter.[4] During drilling, it is important to use a sharp drill bit: (1) a sharp drill bit will not wander as

Fig. 11.4. Typical Free-hand Drill Motor
Source: Cooper Power Tools

Structural Assembly

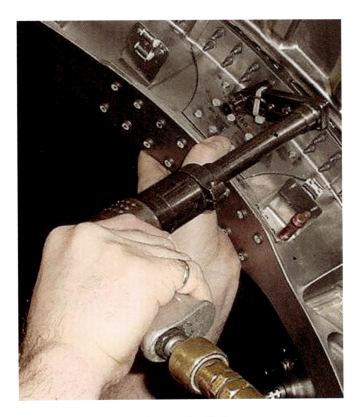

Fig. 11.5. Limited Access Fastener Installation
Source: The Boeing Company

easily as a dull one, (2) drilling is faster with a sharp bit, and (3) lower forces are required, minimizing the possibility of injury or part damage. When drilling multi-material stack-ups, it is important to make sure they are securely clamped together. A typical assembly with temporary fasteners installed to hold the parts in position and provide clamp-up is shown in Fig. 11.6.

The speeds used in manual drilling depend on both the materials and their thickness. Typical maximum speeds for aluminum range from 1000 rpm for 0.50 in. holes up to 5000 rpm for 0.16 in. holes. For titanium, they range from 150 rpm for 0.50 in. holes up to 700 rpm for 0.16 in. holes. As the material gets thicker, the maximum speed goes down, and as the material gets harder, the maximum speed allowed goes down. When drilling through stack-ups of dissimilar metals (e.g., aluminum and titanium), it is necessary to use a drill motor with the speed adjusted for the harder material.

After all the holes have been drilled, the assembly should be taken apart, and the holes deburred on both surfaces. Holes should only be deburred with

Manufacturing Technology for Aerospace Structural Materials

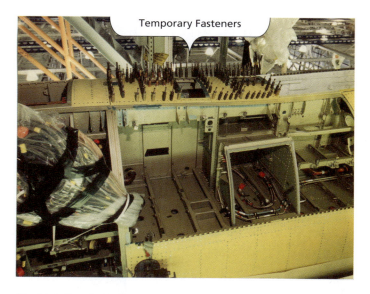

Fig. 11.6. Panel with Temporary Fasteners
Source: The Boeing Company

a deburring tool, or a drill bit of a larger size. Holes should be deburred by hand due to the danger of hole enlargement if a power drill is used. It is also important not to go too deep as shown in Fig. 11.7, as a knife edge can result. During all hole drilling operations, it is important that proper edge distances are maintained, to insure that the skins do not fail due to inadequate shear strength. The engineering drawing should specify edge distances but they are normally around $2-3D$, where D is the hole diameter.

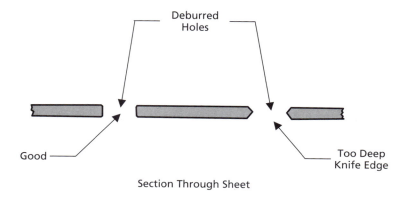

Fig. 11.7. Deburring of Drilled Holes

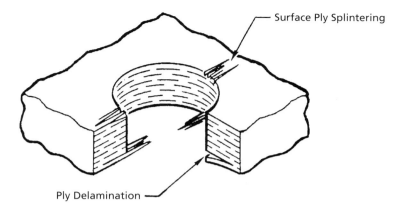

Fig. 11.8. Composite Hole Splintering[5]

Hole drilling of composites is more difficult than in metals, again due to their relatively low sensitivity to heat damage, and their weakness in the through-the-thickness direction. Composites are very susceptible to surface splintering (Fig. 11.8), particularly if unidirectional material is present on the surface. Note that splintering can occur at both the drill entrance and exit side of the hole.[6] As shown in Fig. 11.9, when the drill enters the top surface, it creates peeling forces on the matrix as it grabs the top plies. When it exits the hole, it induces punching forces that again creates peel forces on the bottom surface plies. If top surface splintering is encountered, it is usually a sign that the feed

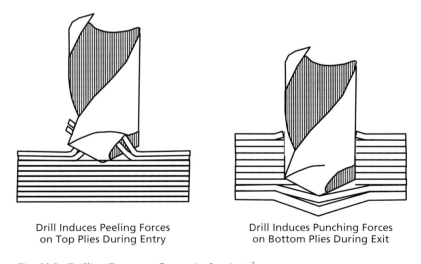

Fig. 11.9. Drilling Forces on Composite Laminate[2]

rate is too fast, while exit surface splintering indicates that the feed force is too high.[3] It is common practice to cure a layer of fabric on both surfaces of composite parts, which will largely eliminate the hole splintering problem, i.e. woven cloth is much less susceptible to splintering than unidirectional material. When drilling composites, a back-up material such as aluminum or composite, clamped to the backside, will frequently help to prevent backside hole splintering. Coolant is normally not used for carbon/epoxy laminates that are 0.250 in. thick or thinner. When drilling composites dry, operators should be provided with vacuum capability to suck up the dust and should always wear eye protection and a respirator.

Since epoxy matrix composites will start to degrade if heated above 400° F, it is important that heat generation be minimized during drilling. Typical drilling parameters are 2000–3000 rpm at feed rates of 0.002–0.004 in. per revolution (ipr), although this will vary depending on the drill geometry and the type of equipment used. Thermocouples and heat sensitive paints are often used during drilling parameter development tests to monitor the heat generated. Drilling parameters for composite-to-metal stack-ups are often controlled more by the metal than the composite. For example, when drilling C/E-to-aluminum, a speed of 2000–3000 rpm with a feed rate of 0.001–0.002 ipr might be used, while a stack-up of C/E-to-titanium would require a slower speed (e.g., 300–400 rpm) and a higher feed rate (0.004–0.005 ipr). Titanium (Ti-6Al-4V) is also very sensitive to heat build-up (hence the lower speed) and tends to rapidly work harden if light cuts are used (hence the higher feed rate).

To help reduce the variability in manual hole drilling, some manufacturers have produced detailed written instructions covering specific hole drilling operations, and provide kits for the mechanics, which have all of the correct tools needed for a specific operation.

11.3.2 Power Feed Drilling

Power feed drilling is much preferred to hand drilling. In power feed drilling, the drill unit is locked into a drill template that establishes both hole location and maintains drill normality. In addition, once the drilling operation starts, the unit is programmed to drill at a given speed and feed. Some units, such as the one shown in Fig. 11.10, can be programmed for different peck cycles. All of these controls lead to much better and more consistent hole quality, particularly when drilling composite-to-metal stack-ups. A typical peck drilling cycle for a 3/16 in. diameter hole through C/E-to-titanium would be a speed of 550 rpm with a feed rate of 0.002–0.004 ipr and 30–60 pecks per inch of thickness.[7]

When drilling into composite-to-metal stack-ups, a phenomenon called back counterboring can occur. As shown in Fig. 11.11, as the metal chips (i.e., aluminum or titanium) travel up the flutes, they tend to erode the softer liquid shim and composite matrix material, causing eroded and oversize holes. Back

Structural Assembly

Fig. 11.10. Power Feed Peck Drill
Source: Cooper Power Tools

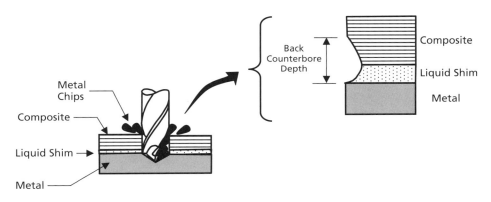

Fig. 11.11. Back Counterboring[5]

counterboring can be minimized by (1) eliminating all gaps, (2) using a drill geometry that produces small chips, (3) changing speeds and feeds, (4) providing better clamp-up, (5) reaming the hole to final diameter after drilling, or (6) by peck drilling.[3]

Peck drilling (Fig. 11.12) is a process in which the drill bit is periodically withdrawn to clear the chips from the flutes. Peck drilling is used almost exclusively when drilling composite-to-titanium stack-ups, due to the back counterboring potential of the hard titanium chips. The process also greatly reduces the heat build-up that can rapidly occur when drilling titanium.

11.3.3 Automated Drilling

For high volume hole generation, automated drilling equipment can be designed and built for specific applications.[8] Being large and sophisticated machine tools,

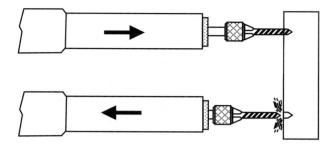

Fig. 11.12. Peck Drilling[3]

Fig. 11.13. Automated Wing Drilling System
Source: The Boeing Company

these units are expensive, so the number of holes drilled and the number of units produced needs to be large enough to justify the equipment investment. Examples of two of these large units are shown in Figures 11-13 and 11-14. These machines are extremely rigid and allow for accurate hole location and normality. They are NC so there is no need for drill templates. The one shown has a vision system that can scan the substructure, and software that will then adjust the hole location to match where the substructure is actually located, versus design nominal. All drilling parameters are automatically controlled with the capability to change speeds and feeds, when drilling through different materials. Due to the thick stack-ups that must be drilled in a wing, a water soluble flood or mist coolant is usually used during the hole drilling operations. All drilling data is automatically recorded and stored for quality control purposes. The drill holders contain bar codes that must match the

Structural Assembly

Four Independent Drill Columns

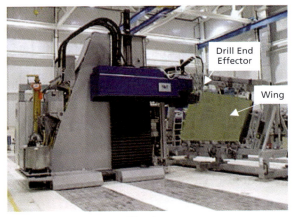

Drill Column

Fig. 11.14. Automated Wing Drilling System
Source: PaR Systems, Inc.

drilling program to make sure the correct drills are used for the correct holes. These machines can also install temporary fasteners to clamp the skins to the substructure during drilling, and frequently use integral drill-countersink cutters that drill the hole and then continue to countersink it, during the same operation.

The current trend in industry is to replace these large installations with smaller more flexible units.[9,10] An off-the-shelf commercial robot, with some modifications and a special drilling end effector, is used to drilling holes in the control surface shown in Fig. 11.15. Another approach is to integrate drilling

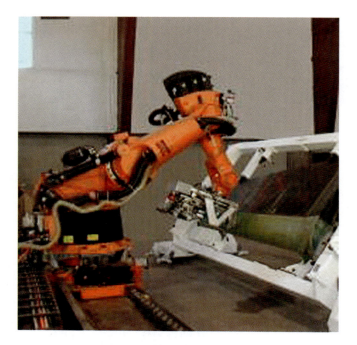

Fig. 11.15. Robotic Drilling of Control Surface
Source: The Boeing Company

units into the assembly fixture, an approach called Numerically Controlled Drill Jigs (NCDJ). Examples of these types of relatively low cost units are shown in Fig. 11.16 for a forward fuselage and Fig. 11.17 for a fighter aircraft outer wing.

11.3.4 Automated Riveting Equipment

A typical piece of automated riveting equipment is shown in Fig. 11.18. These machines will drill the hole, inspect the hole, select the correct grip length of rivet, install sealant on the rivet (if required), and then install the rivet by squeezing. This equipment is available in a wide variety of sizes, ranging from small units to very large computer numerically controlled (CNC) units, capable of installing stringers on full size commercial aircraft wing skins. Being an automated process, the quality of drilling and fastener installation is better and more consistent than with hand methods. There are also units that will install pin and collar fasteners, such as Lockbolts, where the collar is automatically swaged onto the collar portion of the pin. An example of one of these larger automated drill and fastening systems is shown in Fig. 11.19.

Structural Assembly

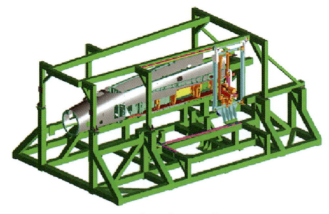

Forward Fuselage Tooling

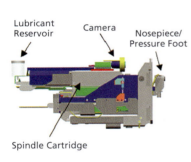

Drill Head

Drilling Operation

Fig. 11.16. Forward Fuselage NCDJ Drilling System[9]

11.3.5 Drill Bit Geometries

Many variations of twist drills (Fig. 11.20) are used in drilling metallic structure. Since specific drill bit geometries can influence both hole quality and the quantity of holes drilled, many geometries are proprietary to the various

509

Manufacturing Technology for Aerospace Structural Materials

Drill Head

NCDJ

Fig. 11.17. Outer Wing NCDJ Drilling System[10]

aerospace manufacturers. Examples of typical variations in the standard twist drill include step drills that drill an undersize hole and then a final hole size in one pass, and drill-reamers in which a hole is drilled and then reamed in the same pass. When drilling aluminum, standard high speed steels, such as M2 or M7, give satisfactory drill life. For harder materials, such as titanium, the cobalt grades of high speed

Structural Assembly

Fig. 11.18. Medium Size Automatic Riveter
Source: The Boeing Company

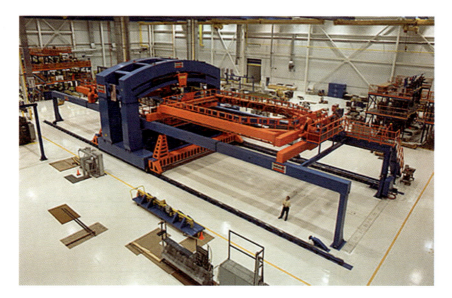

Fig. 11.19. Large Automated Fastening System
Source: The Boeing Company

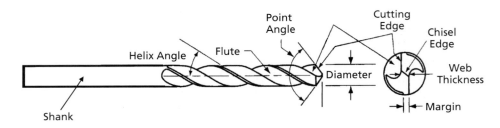

Fig. 11.20. Twist Drill Geometry

steel, such as M33 or M42, yield longer lives. Carbide drills, such as grade C-2, give even longer life in titanium, but are more prone to chipping on the cutting edges.

While standard twist drills are used for drilling metallic structure, a number of unique drill geometries have been developed for composites, several of which are shown in Fig. 11.21. The design of the drill and the drilling procedures are very dependent on the materials being drilled. For example, carbon and aramid fibers exhibit different machining behavior, and therefore require different drill geometries and procedures. In addition, composite-to-metal stack-ups will also require different cutters and procedures. The flat two-flute and four-flute dagger drills were developed specifically for drilling stack-ups of carbon/epoxy. The two-flute variety is normally run at 2000–3000 rpm, while the four-flute is run at 18 000–20 000 rpm.[3] When drilling through composite-to-metal stack-ups, the drill geometry is usually controlled more by the metal and standard twist drill geometries are often used. Due to their low compressive strengths, aramid fibers have a tendency to recede back into the matrix rather than being cleanly cut, resulting in fuzzing and fraying during drilling. Therefore, the aramid drill contains a "C" type cutting edge that grabs the fibers on the outside of the hole and

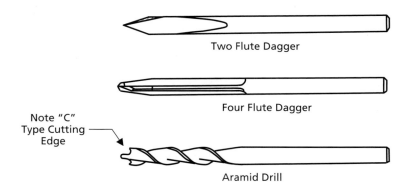

Fig. 11.21. Composite Drill Configurations[5]

keeps the fibers in tension during the cutting process. Typical drilling parameters for aramid fiber composites would be 5000 rpm and a feed rate of 0.001 ipr.[7]

While standard high speed steel (HSS) drills work well in glass and aramid composites, the extremely abrasive nature of carbon fibers requires carbide drills to obtain an adequate drill life. For example, a HSS drill may only be capable of drilling one or two acceptable holes in carbon/epoxy, while a carbide drill of the same geometry can easily generate 50 or more acceptable holes. For drilling carbon/epoxy in rigid automated drilling equipment, polycrystalline diamond (PCD) drills (Fig. 11.22) have exhibited outstanding productivity improvements. While

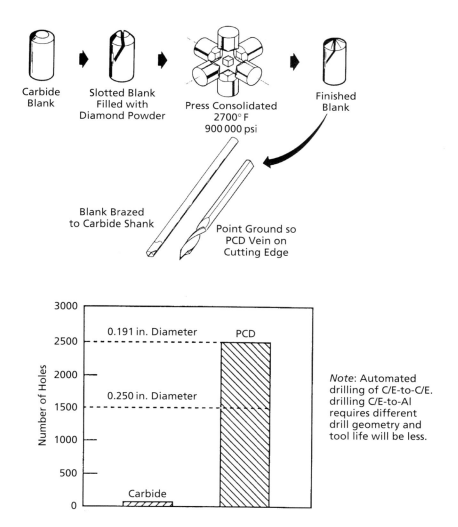

Fig. 11.22. Polycrystalline Diamond Drills[5]

PCD drills are very expensive, the number of holes obtained per drill and the fewer changes required make them cost effective. It should be noted that PCD drills cannot be used with free hand or non-rigid setups; the point will immediately chip and break if any vibration or chatter is present during drilling.

11.3.6 Reaming

Although it is desirable to drill the final hole size in one pass, it is often necessary to ream the hole to final diameter. Reaming is done with HSS or carbide reamers at about one half the drilling speed (e.g., 500–1000 rpm). In some composite-to-metal structures, fasteners are installed clearance fit in the composite and interference fit in the metallic structure for metallic fatigue life enhancement. In this situation, the final hole diameter would be drilled in the composite-to-metal stack-up, the composite skin would then be taken down, and the holes reamed to provide a clearance fit for the fastener. When the stack-up is reassembled, the fasteners would be installed clearance fit in the composite and interference fit in the metal.

11.3.7 Countersinking

Countersinking should only be used where protruding head fasteners will not satisfy design requirements. In general, countersinking reduces joint strength and fatigue life. During countersinking for flush head fasteners, it is important not to countersink too deep and create a knife-edge condition in the countersunk member. A knife-edge creates a significant stress riser, as well as allowing the fastener to tilt and rise up on the countersink surface, resulting in low joint yield strength and reduced fatigue life. As a general rule, at least 0.025 in. or no more than 80% of the sheet thickness, whichever is less, should not be countersunk, as shown in Fig. 11.23. Piloted countersinks are helpful in centering the countersink tool in the hole and depth control can be obtained with microstop cages.

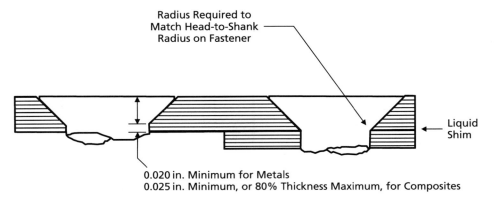

Fig. 11.23. Countersinking[5]

Countersinking of composite structures is similar to that done in metals with one additional caution – the area where the countersink transitions into the hole must have the same radius as the fastener head-to-shank radius. Again, due to the low interlaminar shear strength of composites, this condition can result in cracks and delaminations under the force of fastener installation. Countersinking cutters for composites are normally made of solid carbide, steel bodies with carbide inserts, or steel bodies with PCD inserts.

11.4 Fastener Selection and Installation

There are many types of fasteners used in aerospace structural assembly, the most prevalent being solid rivets, pins with collars, bolts with nuts, and blind fasteners, with examples shown in Fig. 11.24. There are also many other miscellaneous fasteners such as quick release multiple piece fasteners, latches, straight pins, headed pins, lock pins, cotter pins, threaded inserts, retaining rings and washers. A typical structural fastener usage for a fighter aircraft is shown in Fig. 11.25.

Since the number and types of fasteners used in aircraft construction is large and complex, aerospace companies have developed fastener usage policies for their various programs, which establish the policies and criteria for the selection and application of fasteners. Typical contents include usage limitations, selection criteria, hole size/callout information, strength allowables, material compatibility and protection, and lists of approved fasteners. Minimum edge distances and fastener spacing requirements are specified in the fastener usage policy or on the individual engineering drawings. Typical edge distances are $2–3D$ with typical fastener spacings of $4–6D$, where D is the hole diameter.

The selection of a specific fastener depends on its ability to satisfactorily transmit the expected design loads, be environmentally compatible with the materials it joins, and be amenable to installation in the intended joint. Environmental or corrosion compatibility depends on both the fastener material and the materials in the joint. Some examples of the compatibility of fastener materials with structural materials are shown in Table 11.1. It should also be noted that fasteners, especially steel fasteners, are often coated for corrosion protection (e.g., cadmium) and the compatibility of the coating with materials being joined needs to be considered. For example, cadmium-plated fasteners should not be used with titanium due to the potential of stress corrosion cracking.

Mechanical fastener material selection for composites is important in preventing potential corrosion problems. Aluminum and cadmium coated steel fasteners will galvanically corrode when in contact with carbon fibers. Titanium (Ti-6Al-4V) is usually the best fastener material for carbon fiber composites, based on its high strength-to-weight ratio and corrosion resistance. When higher strength is required, cold worked A286 iron–nickel or the iron–nickel based

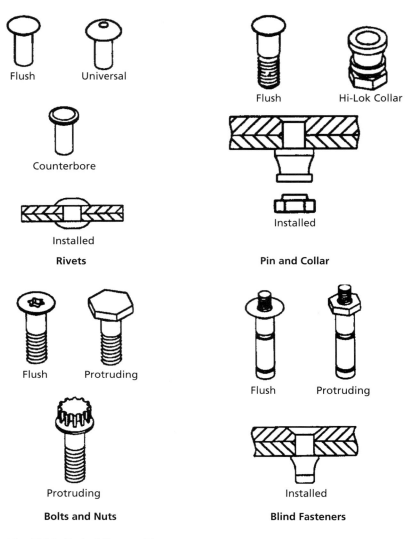

Fig. 11.24. Typical Fastener Types

alloy Inconel 718 can be used. If extremely high strengths are required for very highly loaded joints, the nickel–cobalt–chromium multi-phase alloys MP35N and MP159 are available. It should be noted that glass and aramid fibers, being non-conductive, do not cause galvanic corrosion with metallic fasteners.

Prior to installing fasteners in structure, it is important to measure the grip length of the fastener to be installed. There are commercial gages available, which can be placed through the hole to measure the correct grip length. In determining fastener grip length for threaded fasteners, it is important that the

Structural Assembly

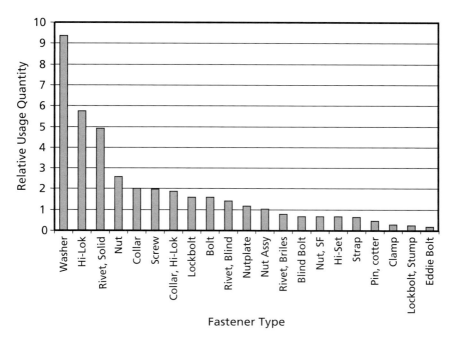

Fig. 11.25. Typical Fighter Aircraft Fastener Usage

Table 11.1 Fastener Material Compatibility

Structural Materials Being Joined	Fastener Material		
	Preferred	*Acceptable*	*Prohibited*
Aluminum to Aluminum	Anodized Aluminum	Titanium A286	Cadmium Plated Steel
Titanium to Titanium Austenitic Stainless Steel Nickel Base Alloys	Titanium	A286 Inconel 718	Alloy Steel Aluminum Aluminum Coated Fasteners
Titanium to Aluminum	Titanium	A286 Inconel 718	Aluminum Aluminum Coated Fasteners
Carbon/Epoxy	Titanium	Inconel 718 A286	Aluminum Aluminum Coated Fasteners

fastener is long enough, so that the threads are never loaded in bearing or shear, i.e. no threads should be allowed in the hole. In addition, no more than three, and no less than one thread, should be showing when the nut is installed and tightened to the proper torque. Fasteners exposed to the outer moldline of the aircraft are normally installed "wet" by dipping the end of the fastener in polysulfide sealant prior to installation. This helps in preventing moisture that can cause corrosion from intruding into the structure.

11.4.1 Special Considerations for Composite Joints

When a hole is placed in a composite laminate, it creates a stress concentration and the overall load-bearing capability of the laminate is severely reduced. Even a properly designed mechanically fastened joint exhibits only 20–50% of the basic laminate tensile strength.[11] The various failure modes for composite joints are shown in Fig. 11.26. The only acceptable failure mode is when the joint fails

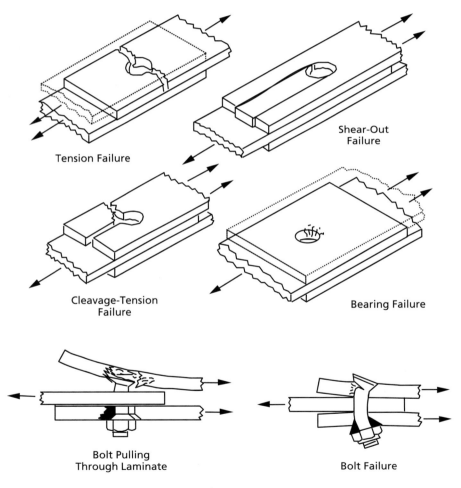

Fig. 11.26. Composite Joint Failure Modes[5]

Structural Assembly

in bearing, since the joined members do not separate catastrophically. Bearing failures are characterized by localized damage, such as delaminations and matrix crazing around the hole. Potential causes for the other failure modes shown include:

1. *Shearout*: Insufficient edge distance, or too many plies oriented in the load direction.
2. *Tension*: Insufficient width, or too few plies oriented in the load direction.
3. *Cleavage-Tension*: Insufficient edge distance and width, or not enough cross-plies (e.g., $+45°$ and $-45°$).
4. *Fastener Pull-through*: Countersink too deep, or use of a shear head fastener.
5. *Fastener Failure*: Fastener too small for laminate thickness, unshimmed gaps or excessive shimmed gaps in joint, or insufficient fastener clamp-up.

Like hole drilling, fastener installation in composites is more difficult and damage prone than for metallic structure. Some of the potential problems with fastener installation are shown in Fig. 11.27. As previously discussed, unshimmed gaps can cause cracking of either the composite skin or the composite

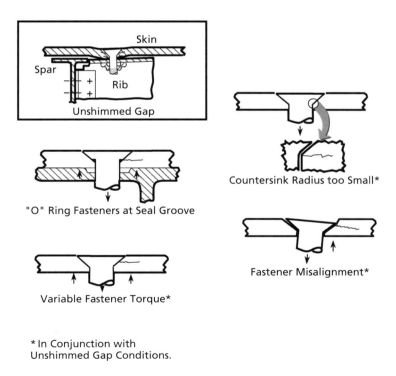

Fig. 11.27. Fastener Installation Defects[5]

substructure (or both) as the fastener is being installed and pulls the two pieces together. In fuel tanks, channel seal grooves are often used to help prevent fuel leakage. In addition, fasteners with O-ring seals can be used to further prevent leakage. It has been found through experience that these are potential areas for interlaminar cracking. While good clamp-up of the fastener is certainly desirable, over-torquing fasteners can also result in cracking. If the countersink radius is too small and does not match that of the fastener head-to-shank radius, the fastener can apply a concentrated point load and cause matrix cracking. Likewise, fastener misalignment, where the hole and the countersink are not properly aligned, can result in point loading and cracking. In addition, fastener cocking during loading (Fig. 11.28) can result in point loading and lead to progressive damage during fatigue cycling.

In any mechanically fastened joint, high clamp-up forces are beneficial to both static and fatigue strength. High clamp-up produces friction in the joint, delays fastener cocking and reduces joint movement, or ratcheting, during fatigue loading. Most holes eventually fail in bearing caused by fastener cocking and localized high bearing stresses.[12] To allow the maximum clamp-up in composites without locally crushing the surface, special fasteners have been designed for composites that have large footprints (large heads and nut areas that bear against the composite) to help spread the fastener clamp-up loads over as large an area as possible. Washers are also frequently used under the nut, or collar, to help spread the clamp-up loads. In general, the larger the bearing area, the greater the clamp-up that can be applied to the composite, resulting in improved joint strength.[13] In addition, tension head, rather than shear head, fasteners are normally used in composites, because they are not as susceptible to bolt bending during fatigue or fastener pull-through during installation in thin structure.

11.4.2 Solid Rivets

Traditionally, riveting has been the most prevalent method of building aerospace structure. However, with greater amounts of composite and titanium structure replacing aluminum structure on newer aircraft, the amount of riveted structure will decrease in the future, since rivets are not normally used in titanium or composite structure. However, rivets are, and will remain, an important fastening method for aerospace structure. Rivets should be restricted to joints that are primarily loaded in shear with some secondary tension loading allowed, but should not be used where the primary loads are in tension.

Solid rivets are one piece fasteners made from malleable metal installed by squeezing or vibration driving. They are used in permanent applications where the lowest cost, reliable fasteners are desired. Aluminum rivets are by far the most widely used, with 2117-T4 aluminum ("AD" rivet) the most prevalent. Where higher strengths are required, 7050-T73 or 2024-T4 aluminum rivets can

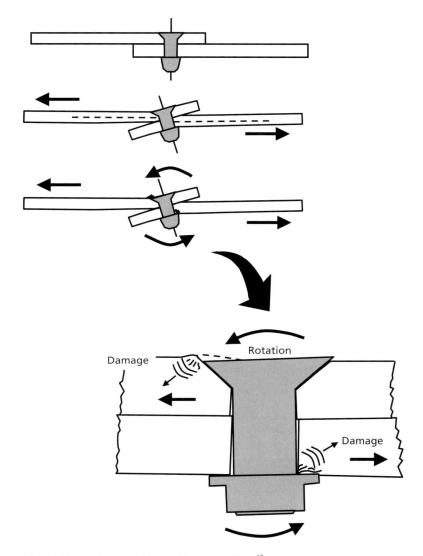

Fig. 11.28. Fastener Cocking in Single Lap Shear[12]

be used instead of the AD rivet. However, 2024-T4 rivets require refrigeration before driving, and they have somewhat lower stress corrosion resistance.

Universal and countersunk rivet heads are most common rivets used in aircraft construction. Universal, or protruding, head rivets are used internally in structures not exposed to the air stream. Since a universal head does not require countersinking the skin, the joint is capable of higher bearing loads than a countersunk joint. The standard countersunk rivet has a 100° included angle. Since the countersunk joint is inherently weaker, a larger number of rivets are usually required to compensate for the reduced bearing and shear strength.

During rivet installation, several physical changes take place: (1) the rivet diameter expands to fill the hole, (2) the hardness of the rivet increases due to work hardening, and (3) the manufactured head is formed through plastic deformation. One advantage of rivets is that they do expand during installation to give a tight fit in the hole.

Pneumatic rivet guns are driven by compressed air and are classified as light, medium, and heavy hitting. Light hitting guns are used to install 0.09–0.125 in. diameter rivets, while medium hitting guns are used for 0.15–0.19 in. diameter rivets. The heavy hitting guns are used for larger diameters. There are two types of gun sets, one for the universal head rivets and one for countersunk rivets. During manual riveting, a bucking, or backing, bar is used to form the manufactured head during the riveting process. Bucking bars come in different shapes, sizes, and weights. The weight of the bucking bar should be proportional to the size of the rivet being installed. A number of rivet defects are shown in Fig. 11.29. Perhaps the most serious is the clinched rivet due to improper bucking during installation. The rivet forms to one side which can later trap moisture and potentially cause a corrosion problem.

Rivets are rarely used in composites for two reasons: (1) aluminum rivets will galvanically corrode when in contact with carbon fibers, and (2) the vibration and expansion of the rivet during the driving process can cause delaminations. If rivets are used, they are usually a bimetallic rivet consisting of a Ti-6Al-4V pin with a softer titanium–niobium tail, which are installed by squeezing rather than

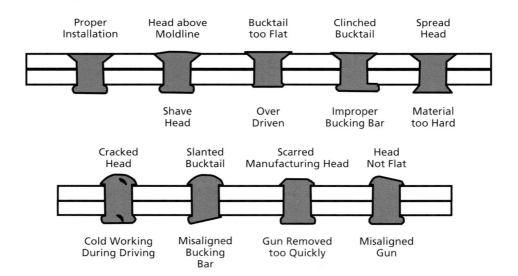

Fig. 11.29. Installed Rivet Defects

vibration driving. In addition, the head that is upset must be against metallic structure and not composite. There are also hollow end solid rivets designed to allow flaring of the ends without damaging expansion, when used in double countersunk holes in composites.[13]

11.4.3 Pin and Collar Fasteners

Pin and collar fasteners are the most commonly used fasteners for permanent installations, where there is no requirement to remove the fastener. The pin, similar to a bolt, is used with a self-locking or swaged-on-collar, which cannot be removed with typical tools without destroying the collar or pin. Pin and collar fasteners are high strength fasteners made from either Ti-6Al-4V, A286 iron–nickel, or Inconel 718. Typical head designs include protruding tension head, protruding shear head, 100° full countersink, and 100° reduced countersink.

A typical pin and collar fastener is the Hi-Lok fastener shown in Fig. 11.30. A Hi-Lok fastener consists of a threaded pin and a collar. The threaded pin is essentially a modified bolt, while the collar is basically a nut with a breakaway groove, which controls the amount of torque and preload on the pin. Hi-Lok fasteners are available in both flush and protruding heads, and as shear and tension-shear fasteners. Hi-Lok fasteners can be installed either clearance-fit or interference-fit. The fastener pin is usually made from Ti-6Al-4V with an A286 nut. Titanium nuts are occasionally used, but the threads tend to gall if they are not coated with an anti-galling lubricant, which then adversely affects long-term clamp-up. A hex key is inserted into the fastener stem to react the torque applied to the nut. The nut is tightened down until a predetermined torque level is achieved, and the top portion of the nut fractures. Washers can be used under the head to help spread the bearing load on the surface.

Lockbolts are another common pin and collar fastener that can be installed by either pulling or swaging the collar from the backside. A typical pull type Lockbolt installation sequence is shown in Fig. 11.31. Lockbolts differ from Hi-Loks in that Hi-Loks have true threads that the nut is threaded onto, while Lockbolts have a series of annular grooves that the collar is swaged into. Once swaged in place, they cannot back off (loosen) and have superior vibration resistance.[13] Lockbolts are available with flush or protruding heads, as shear or tension pull types, and as shear stump types. Lockbolts are normally lighter and cost less to install than bolt–nut combinations of the same diameter. There are two precautions that need to be followed when installing Lockbolts in composite structure: (1) pull type Lockbolts exert quite a bit of force on the composite, due to the pulling action necessary to swage the collar onto the pin. If the composite is thin, fastener pull-through is a real possibility, and if there are any unshimmed gaps, cracking and delaminations can occur when the fastener pin fractures; and (2) if backside Lockbolts (called stump Lockbolts) are installed in

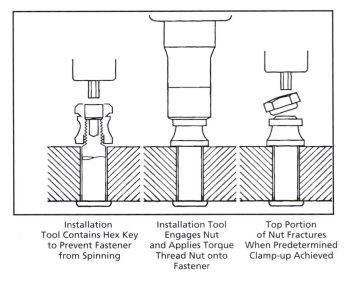

Fig. 11.30. Installation of Hi-Lok Fastener[12]

composites, they should be installed by a piece of automated equipment, where careful control of the swaging operation can be exercised.

A third type of pin and collar fastener is the Eddie bolt shown in Fig. 11.32. As shown in the installation sequence, the collar initially threads onto the pin, but then is swaged into flutes on the pin to provide a positive lock. The advantage of Eddie bolts is that they do provide a positive lock and will retain clamp-up loads better than Hi-Loks that rely on torque only. However, the fasteners and installation tools are expensive, the sockets are subject to wear, and the installation procedure is more difficult. They are often specified in inlet duct areas, where there is the potential for a fastener pin coming loose and flying into the engine blades and damaging the engine.

Structural Assembly

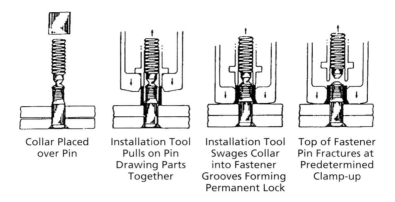

Fig. 11.31. *Installation of Pull Type Lockbolt*[12]

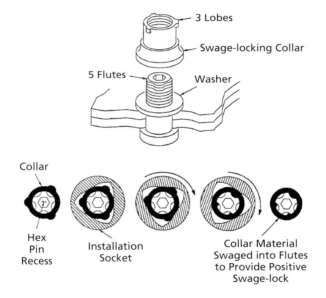

Fig. 11.32. *Eddie Bolt Positive Lock Fastener*[12,13]

11.4.4 Bolts and Nuts

Bolts, along with nuts and washers, are used to join highly loaded structural members which must be removable for service access. They are also used as permanent attachments for structure. Structural bolts are used in fatigue, shear, and tension critical joints. Nuts, which are tightened with wrenches, may be used when there is access to both sides. Nut plates and gang channels are used when one-sided access is applicable. The shanks of all structural bolts must be long enough to insure that there are no threads bearing in the joint. Extra washers may be used to adjust the grip length. However, the use of lock washers is often prohibited because they can damage the protective finish on the structure being joined. A washer should be used under both the bolt head of protruding bolts and the nut, to help distribute the load and prevent damaging the surface finish.

There are many head styles including protruding tension flange head, protruding shear head, 100° full countersink, and 100° reduced countersink. For structural bolts, the threads are rolled and the heads are forged for additional strength. Since both of these operations induced residual compression stresses in the fastener, they also improve their resistance to stress corrosion cracking. Bolts smaller than 0.190 in. in diameter are considered to be non-structural; they are used to attach brackets and other miscellaneous hardware. The threads on non-structural bolts can be either machined or rolled. Some nuts are self-locking while others are not. Cotter pins and safety wire are used to secure those without self-locking features. Self-aligning nuts (up to 8°) are also available for situations in which the structural members are not parallel. There are also nuts designed specifically for tension or shear loading applications.

When installing structural bolts, a high preload (i.e., high torque) is desirable from the standpoint of fatigue and vibration. However, since a high preload places the bolt under tension stress, too high a preload can increase the susceptibility of certain fastener materials to stress corrosion cracking. Therefore, bolt preload is usually restricted to around 50–60% of the bolt yield strength. It is also important that all of the fasteners in the joint are preloaded to close to the same torque, so that the fasteners share the load equally. If there is variable torque on the fasteners, those that are preloaded to higher values will carry a substantially higher portion of the load than those with lower torque. Low bolt preloads can result in joint rotation, misalignment, loosening, and the formation of gaps between mating parts. Low bolt preloads also reduce the fatigue life of the installed fastener. For highly loaded bolted joints, torque values are often specified on the engineering drawing.

High strength bolts, along with nut plates or gang channels, are used when there is a requirement for skin removal. A typical nut plate and gang channel are shown in Fig. 11.33. Note that three holes are required for each nut plate: two small holes for rivets to attach the nut plate to the structure and then the main fastener hole, which is threaded to accept the threads in the screw. There are a wide variety of nut plate configurations available for different installations, including self-aligning nut plates. Because they do not require two rivet holes per fastener, gang channels are frequently used where there are long rows of fasteners to be installed. They are attached at periodic points along the channel, thus saving installation labor. Bolts with nut plates or gang channels do not perform as well as blind fasteners, either in static or fatigue loading, due to increased joint deflection and compliance of the fastening system.[12]

Structural Assembly

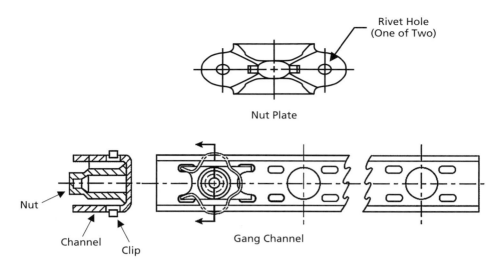

Fig. 11.33. Nut Plate and Gang Channel[14]

11.4.5 Blind Fasteners

Blind fasteners are used in areas where there is limited, or no access, to the backside of the structure. However, the solid pin and collar fasteners previously discussed are usually preferred, because they are stronger, provide better clamp-up, and have better fatigue resistance. Two types of blind fasteners are the threaded core bolt type and the pull type shown in Fig. 11.34. The threaded core bolt (Fig. 11.35) relies on an internal screw mechanism to deform the head and pull it up tight against the structure, while the pull type blind fastener uses a pure pulling action to form the backside head. Higher clamp-up forces and larger

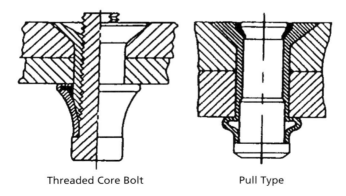

Fig. 11.34. Blind Fasteners[12]

527

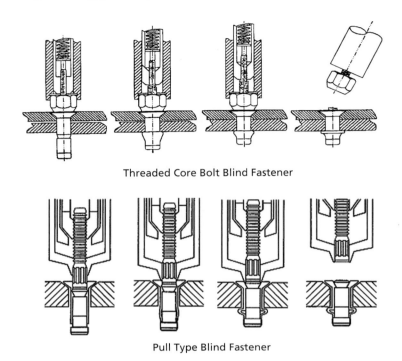

Fig. 11.35. Installation of Blind Fasteners[12]

footprints are obtainable with the threaded core bolt, leading to longer fatigue life. However, the pull types install quicker, are lighter, and are less expensive. The use of blind fasteners in composite materials is not recommended unless the upset portion of the fastener bears against metal structure. This being said, there are some special blind fasteners with large foot prints that can be placed directly against composite surfaces.

11.4.6 Fatigue Improvement and Interference Fit Fasteners

Since fatigue cracks often initiate at fastener holes in metallic structure, methods such as cold working fastener holes and interference fit fasteners have been developed to improve fatigue life. Both cold working and interference fit fasteners set up residual compressive stress fields in the metal immediately adjacent to the hole, as shown in Fig. 11.36. The applied tension stress during fatigue loading must then overcome the residual compressive stress field, before the hole becomes loaded in tension. The fatigue improvement due to cold working in 2024-T851 aluminum is shown in Fig. 11.37. Somewhat surprisingly, the fatigue strength of the material with a tight fastener in a coldworked hole is actually somewhat better than the base material.

Structural Assembly

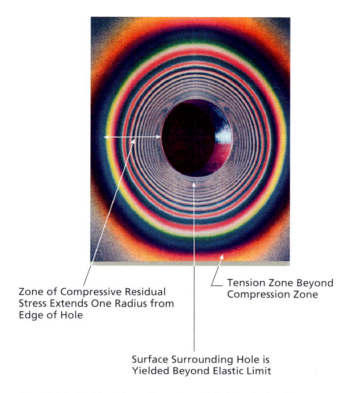

Fig. 11.36. Residual Stress State Around Coldworked Hole
Source: Fatigue Technology, Inc.

Cold working of holes is usually conducted using either the split-sleeve or split-mandrel method. Both methods involve pulling a mandrel through the hole that expands the hole diameter, creating plastic deformation of material around the hole and a resulting residual compressive stress field. The residual stress field, depending upon the material and the amount of expansion, will extend approximately one radius from the edge of the hole.[15] In the split-sleeve process (Fig. 11.38), a stainless steel split sleeve is placed over a tapered mandrel and inserted into the hole. The hole is coldworked when the largest part of the mandrel is drawn back through the sleeve. After cold working, the sleeve is removed and discarded. In the split-mandrel process, a collapsible mandrel is placed in the hole, and as the mandrel is withdrawn, it expands to coldwork the hole.

Interference fit fasteners are also frequently used in metallic structure to improve the fatigue life. When the interference fit fastener is installed in metal, it also plastically deforms a small zone around the hole, setting up a compressive stress field, which again is beneficial when fatigue loading is primarily in tension. The amount of interference can vary, depending on structural requirements, but it is usually in the range of 0.003–0.004 in. In some highly loaded holes, both

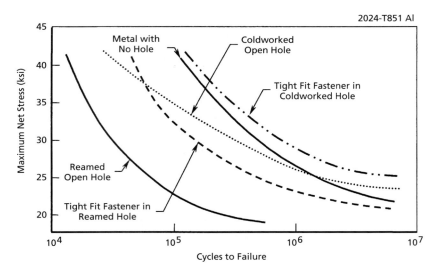

Fig. 11.37. Fatigue Life Improvement with Cold working

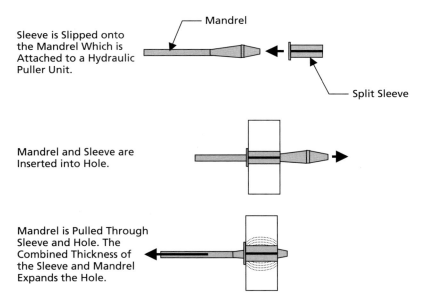

Fig. 11.38. Split-Sleeve Cold working[15]

cold working and interference fit fasteners are used. While both cold working and the use of interference fit fasteners are proven methods of improving fatigue resistance, both increase assembly costs and should only be specified when they are really needed.

Structural Assembly

In composite-to-metal assemblies, it is possible to have interference fit in the metallic structure and clearance fit in the composite. This is normally done by drilling the hole through the stack-up. The composite skin is then removed and opened up to a larger diameter with a reamer. The composite skin is then re-assembled to the metallic structure, and the fastener is installed clearance fit through the composite skin but interference fit in the metallic substructure. Even then, it is important to be careful when installing the interference fit fastener, i.e. excessive vibration from a rivet gun has been known to produce delaminations around the composite hole even though it is a clearance fit.

Since composites do not plastically deform, there is no fatigue life improvement with using interference fit fasteners. However, a potential benefit to having some interference fit fasteners in composite structure is that they will help "lock up" the structure and prevent any movement at the joint (called ratcheting) during fatigue loading. Previous work has shown that installing standard interference fit fasteners in composites, with as little as 0.0007 in. of interference, can lead to cracking and interlaminar delaminations.[12] To eliminate this problem, special sleeve type interference fit fasteners (Fig. 11.39) have been designed so that the sleeve spreads the load evenly during installation to

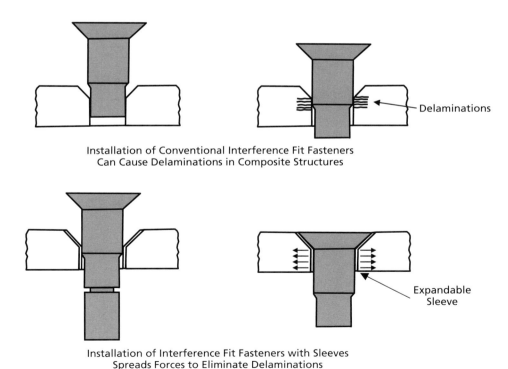

Fig. 11.39. Interference Fit Fasteners in Composites[12]

531

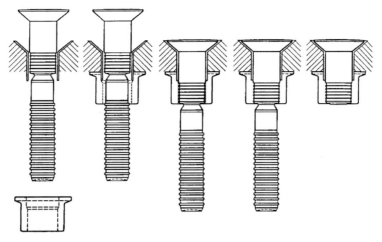

Interference Fit Sleeve Type Lockbolt

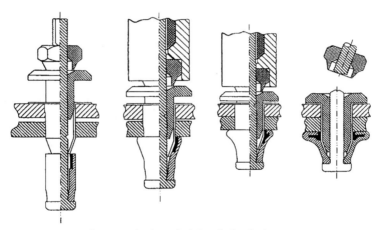

Interference Fit Threaded Corebolt Blind Fastener

Fig. 11.40. Installation of Sleeve Type Interference Fit Fasteners[12]

prevent the delamination problem. Interferences as high as 0.006 in. have been obtained without damaging the composite. Both pin and collar (Lockbolts), and threaded core blind bolt fasteners (Fig. 11.40), are available with sleeves. There are several potential advantages to using interference fit fasteners in composite structure:[12] lower joint deflection; reduction in fastener cocking that leads to high localized bearing stresses; locks up structure to prevent ratcheting during fatigue; and reduced assembly costs when interference fits are required in metallic structure (no disassembly and ream operation for the composite).

11.5 Sealing

Many structures require sealing for (1) corrosion protection, (2) to keep water out of the structure, or (3) to keep fuel in the structure. The typical wing fuel tank configuration shown in Fig. 11.41 will be used to explain the different sealing and corrosion protection methods. For joints between carbon/epoxy and aluminum parts, it is common practice to cocure, or bond, a thin layer of glass cloth to the surface of the carbon/epoxy part, which acts as an electrical isolation barrier to prevent galvanic corrosion of the aluminum.

A good sealant must have good adhesion properties, high elongation, and be resistant to both temperature and chemicals. Sealing is usually accomplished using polysulfide sealants, which are available in a variety of product forms, with a range of viscosities and cure times. Polysulfide sealants are usable in the temperature range of −65 to 250° F, with short-term capabilities to 350° F. They also contain leachable corrosion resistant compounds, which aid in preventing corrosion of aluminum structure. If higher temperatures are required, then silicone sealants can be used that have temperature capabilities as high as 500° F.[14] Moldline fasteners are usually installed "wet" by applying sealant to the fastener before installation. The nuts are often over coated after installation. During assembly, the faying surfaces are sealed, and then fillet seals are placed around

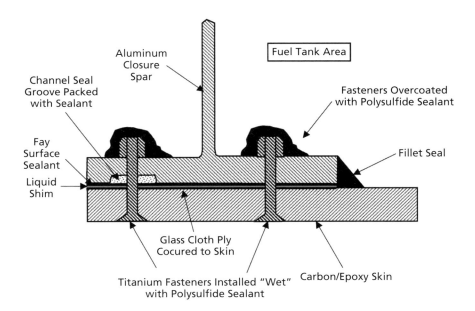

Fig. 11.41. Typical Wing Fuel Tank Sealing[5]

the periphery. Because the fay seal is extremely thin and may be separated due to structural deflections, it cannot be considered to be the primary seal. The fillet seal, applied after assembly, must be the primary seal. All potential leak paths must be fillet sealed. Fillet beads must be dressed with a filleting tool, to work out air bubbles and voids and provide the finished fillet shape. Adequate fillet size is important in preventing leakage. Selecting the right sealant working life is important, or the sealant may set up prior to finishing the job, or it may take too long to cure, affecting the production schedule.

Fuel tanks often have a channel groove that is packed with a fluorosilicone sealant, which contains around 10% of small microspheres that are graduated from 0.002 to 0.030 in. in diameter, for more effective gap filling. They assist in keeping the compound in the groove and are effective in helping to seal gaps, up to 0.010 in. wide. When fuel comes in contact with the sealant, it swells to help further seal the structure. Normally the channel is prepacked with sealant and then injected under pressure (up to 4000 psi) after assembly. Injection points, usually spaced at 4–6 in. apart, can be at fastener holes, or through specially designed fasteners that contain internal injection ports.

11.6 Painting

Metallic parts are usually chemically treated prior to painting. Aluminum, in particular, requires careful surface preparation prior to painting to prevent corrosion. Anodizing, as covered in Chapter 2 on Aluminum, and conversion coatings are the two most common surface treatments. After treatment, the detail part is usually primed with one or two layers of corrosion resistant epoxy primer. Paint adhesion to composite structure is actually not as difficult as it is with metallic structure. The surface should be cleaned of all dirt and grease. If the part contains a peel ply, it should be removed. Surface preparation can be accomplished by either scuff sanding with 150–180 grit sandpaper, or by lightly grit blasting.

For aerospace applications, the standard finishing system is epoxy primer, followed by one or two layers of polyurethane topcoat. Epoxy primers are addition curing polyamides that contain: (1) strontium chromate that is an exceptional corrosion inhibitor for aluminum; (2) titanium dioxide for enhanced durability and chemical resistance; and (3) fillers, such as silica, to control viscosity and reduce cost. After sanding, the part should be primed within 36 h. The primer is applied to a dry film thickness of 0.000 8–0.001 4 in. and then room temperature cured for a minimum of 6 h. Polyurethane topcoats are aliphatic ester based polyurethanes that exhibit good weathering, chemical resistance, durability, and flexibility. They are applied to a dry film thickness of around 0.002 in., with an initial cure within 2–8 h, and a full cure within 7–14 days.[16] Environmentally more friendly paint systems are being developed that are free of solvents, or low in solvents, called low volatile organic compounds (VOC) coatings. Also, toxic heavy metals (e.g., chromium) are being replaced with self-priming topcoats that are non-chromated high solids polyurethane coatings, which replace both the epoxy primer and traditional polyurethane topcoat.[16]

Summary

Due to the labor intensity associated with assembly operations and the resultant high costs, structures should be designed to eliminate as much assembly as possible. However, the requirement to mechanically fasten structures will not disappear in the foreseeable future.

Prior to hole drilling and fastener installation, it is critical to locate any gaps between structural members and shim them appropriately, before starting the actual assembly operations. Failure to properly address gaps will lead to excessive preloads and possibly cracks and delaminations in composites during fastener installation and clamp-up. In addition, preloaded metallic parts could be subject to fatigue or stress corrosion cracking.

There are many types of drill motors and units that can be used to drill structures, but they can be broadly classified as either hand, power feed, automated drilling units or automated riveting equipment. Free hand drilling of structures should be avoided if possible. Power feed or automated drilling equipment gives much better and more consistent hole quality. Although free hand drilling is obviously not the best method, it is frequently used because it requires no investment in tooling (i.e., drill templates) and in many applications, where access is limited, it may be the only viable method. Power feed drilling is much preferred to hand drilling. In power feed drilling, the drill unit is locked into a drill template that establishes both hole location and maintains drill normality. In addition, once the drilling operation starts, the unit is programmed to drill at a given speed and feed.

For high volume hole generation, automated drilling equipment can be designed and built for specific applications. Being large and sophisticated machine tools, these units are expensive, so the number of holes drilled and the number of units produced needs to be large enough to justify the equipment investment. Automated riveting equipment is also available that will drill the hole, inspect the hole, install sealant on the rivet (if required), and then install the rivet by squeezing. This equipment is available in a wide variety of sizes and price ranges, ranging from small units to very large units, capable of installing stringers on full size commercial aircraft wing skins.

Composite materials require more care when machining and drilling than comparable metals like aluminum. Their relative brittleness, low interlaminar shear and peel strengths, and low heat tolerance calls for special care in all machining and drilling operations. Special drill geometries are available for composites that reduce the occurrence of defects, such as hole splintering and fiber pull-out.

There are many types of fasteners used in aerospace structural assembly, the most prevalent being solid rivets, pins with collars, bolts with nuts, and blind fasteners. The selection of a specific fastener depends on its ability to satisfactorily transmit the expected design loads, be environmentally compatible with the materials it joins, and be amenable to installation in the intended joint. Environmental or corrosion compatibility depends on both the fastener material and the materials in the joint. For composite structures, fasteners with large footprints should be selected to spread the clamp-up loads across the composite surfaces. Fastener installation processes that induce vibration or sudden impact loads on the parts should never be used in fastening composites.

Solid rivets are one piece permanent fasteners made from malleable metal installed by vibration driving or squeezing. They are used primarily in aluminum structure as permanent fasteners, where low cost reliable fasteners are desired. Pin and collar fasteners are the most commonly used fasteners for permanent installations, where there is no requirement to remove the fastener. Typical pin and collar fasteners include Hi-Loks, Lockbolts, and Eddie Bolts. Pin and collar fasteners require two sided access.

Bolts, along with nuts and washers, are used to join highly loaded structural members which must be removable for service access. They are also used as permanent attachments for structure. Structural bolts are used in fatigue, shear, and tension critical joints. Nuts, which are tightened by wrenches, may be used when there is access to both sides. Nut plates and channel nuts are used when one sided access is applicable.

Blind fasteners are used in areas where there is limited or no access to the backside of the structure. However, the solid core pin and collar fasteners are usually preferred, because they are stronger and have better fatigue resistance. Two types of blind fasteners are the threaded core bolt type and the pull type.

Since fatigue cracks often initiate at fastener holes in metallic structure, methods such as cold working of fastener holes and interference fit fasteners have been developed to improve fatigue life. Both cold working and interference fit fasteners set up a residual compressive stress field in the metal immediately adjacent to the hole. The applied tension stress during fatigue loading must then overcome the residual compressive stress field before the hole becomes loaded in tension.

Aerospace structure is usually sealed to keep water out using polysulfide sealants. For corrosion protection, metallic structure will often be treated, such as anodizing or conversion coating treatments for aluminum. Paint primers with corrosion resistant compounds are used to provide additional protection. Outer moldline structure is then coated with polyurethane topcoats. Sealing and painting of composites is very similar to the processes used for metallic structure. In fact, since composites themselves are not prone to corrosion, in some cases, the job is far less complex, since special corrosion inhibiting compounds are not required. Likewise, paint adhesion to composites is as good, or better, than to metals, provided that proper surface preparation techniques are followed.

Recommended Reading

[1] Paleen, M.J., Kilwin, J.J., "Hole Drilling in Polymer-Matrix Composites", in *ASM Handbook Vol. 21 Composites*, ASM International, 2001, pp. 646–650.
[2] Parker, R.T., "Mechanical Fastener Selection", in *ASM Handbook Vol. 21 Composites*, ASM International, 2001, pp. 651–658.

References

[1] Taylor, A., "RTM Material Developments for Improved Processability and Performance", *SAMPE Journal*, Vol. 36, No. 4, July/August 2000, pp. 1–24.
[2] Fraccihia, C.A., Bohlmann, R.E., "The Effects of Assembly Induced Delaminations at Fastener Holes on the Mechanical Behavior of Advanced Composite Materials", 39th International SAMPE Symposium, 11–14 April 1994, pp. 2665–2678.
[3] Paleen, M.J., Kilwin, J.J., "Hole Drilling in Polymer-Matrix Composites", in *ASM Handbook Vol. 21 Composites*, ASM International, 2001, pp. 646–650.
[4] Born, G.C., "Single-Pass Drilling of Composite/Metallic Stacks", 2001 Aerospace Congress, SAE Aerospace Manufacturing Technology Conference, 10–14 September 2001.
[5] Campbell, F.C., "Assembly", in *Manufacturing Processes for Advanced Composites*, Elsevier Ltd, 2004, pp. 440–469.
[6] Astrom, B.T., *Manufacturing of Polymer Composites*, Chapman & Hall, 1997.
[7] Bolt, J.A., Chanani, J.P., "Solid Tool Machining and Drilling", in *Engineered Materials Handbook Vol. 1 Composites*, ASM International, 1987, pp. 667–672.
[8] Bohanan, E.L., "F/A-18 Composite Wing Automated Drilling System", 30th National SAMPE Symposium, 19–21 March 1985, pp. 579–585.
[9] McGahey, J.D., Schaut, A.J., Chalupa, E., Thompson, P., Williams, G., "An Investigation into the Use of Small, Flexible, Machine Tools to Support the Lean Manufacturing Environment", 2001 Aerospace Congress, SAE Aerospace Manufacturing Technology Conference, 10–14 September 2001.
[10] Jones, J., Buhr, M., "F/A-18 E/F Outer Wing Lean Production System", 2001 Aerospace Congress, SAE Aerospace Manufacturing Technology Conference, 10–14 September 2001.
[11] Niu, M.C.Y., *Composite Airframe Structures: Practical Design Information and Data*, Conmilit Press, Hong Kong, 1992.
[12] Parker, R.T., "Mechanical Fastener Selection", in *ASM Handbook Vol. 21 Composites*, ASM International, 2001, pp. 651–658.
[13] Armstrong, K.B., Barrett, R.T., *Care and Repair of Advanced Composites*, SAE International, 1998.
[14] Hoeckelman, L.A., "Environmental Protection and Sealing", in *ASM Handbook Vol. 21 Composites*, ASM International, 2001, pp. 659–665.
[15] Leon, A., "Developments in Advanced Cold working", SAE 982145, 1998.
[16] Spadafora, S.J., Eng, A.T., Kovalseki, K.J., Rice, C.E., Pulley, D.F., Dumsha, D.A., "Aerospace Finishing Systems for Naval Aviation", 42nd International SAMPE Symposium, 4–8 May 1997, pp. 662–676.

Appendix A

Metric Conversions

To Convert From	To	Multiply By
Area		
in.^{2}	mm^2	6.451 600E+02
in.^{2}	cm^2	6.451 600E+00
in.^{2}	m^2	6.451 600E−04
ft^2	m^2	9.290 304E−02
Force		
lbf	N	4.448 222E+00
kip (1000 lbf)	N	4.448 222E+03
Fracture Toughness		
ksi in.	Mpa (m)$^{1/2}$	1.098
Length		
mil	μm	2.540 000E+01
in.	mm	2.540 000E+01
in.	cm	2.540 000E+00
ft	m	3.048 000E−01
yd	m	9.144 000E−01
Mass		
oz	kg	2.834 952E−02
lb	kg	4.535 924E−01
Mass Per Unit Area (Areal Weight)		
oz/in.^{2}	kg/m^2	4.395 000E+01
oz/ft^2	kg/m^2	3.051 517E−01
oz/yd^2	kg/m^2	3.390 575E−02
lb/ft^2	kg/m^2	4.882 428E+00
Mass Per Unit Volume (Density)		
lb/in.^{3}	g/cm^3	2.767 990E+01
lb/in.^{3}	kg/m^3	2.767 990E+04
lb/ft^3	g/cm^3	1.601 846E−02
lb/ft^3	kg/m^3	1.601 846E+01
Pressure (Fluid)		
lbf/in.^{2} (psi)	Pa	6.894 757E+03
in. of Hg (60°F)	Pa	3.386 850E+03
atm (Standard)	Pa	1.013 250E+05

To Convert From	To	Multiply By
Stress (Force Per Unit Area)		
lbf/in.^{2} (psi)	MPa	6.894 757E−03
ksi (1000 psi)	MPa	6.894 757E+00
msi (1 000 000 psi)	MPa	6.894 757E+03
Temperature		
°F	°C	5/9 (°F−32)
K	°C	K−273.15
Thermal Conductivity		
Btu/ft−h−°F	W/m−K	1.730 735E+00
Thermal Expansion		
in./in.−°F	m/m−K	1.800 000E+00
in./in.−°C	m/m−K	1.000 000E+00
Velocity		
in./s	m/s	2.540 000E−02
ft/s	m/s	3.048 000E−01
ft/min	m/s	5.080 000E−03
ft/h	m/s	8.466 667E−05
Viscosity		
P	Pa−s	1.000 000E−01
Volume		
in.^{3}	m^3	1.638 706E−05
ft^3	m^3	2.831 685E−02

Appendix B

A Brief Review of Materials Fundamentals

Materials can be classified as either metals, ceramics, or polymers. The dominate aerospace structural materials since the mid-1920s have been metals. However, since the mid-1980s, high strength polymeric composite materials have started to replace metals in some aerospace structures. Since composites can be classified as a hybrid material with a reinforcement embedded in a matrix, they can consist of high strength fibers, or other types of reinforcements, in either polymer, metal, or ceramic matrices. The relationship among the three materials classes and composites is shown in Fig. B.1. This appendix provides a review of some of the basic aspects of these different material options.

B.1 Materials

The properties of materials are determined by both their composition and structure.[1] At a very high level, the composition of metals, ceramics, and polymers is obviously as different as their properties. However, even within these material groups, small changes in composition can have a profound effect. For example, the addition of less than one percent of carbon to iron, combined with the proper heat treatment, can increase the strength of steel in the dramatic fashion shown in Fig. B.2. The further addition of minute amounts of hydrogen, in parts per million, can then ruin the properties, resulting in sudden brittle failures (hydrogen embrittlement). Even within a specific alloy group, such as aluminum, the number of compositions or alloys to choose from is very large with over 500 registered alloys.

The concept of structure is much more subtle but every bit as complicated, or even more complicated, as composition. While composition is somewhat fixed by the starting ingredients, a material's structure is influenced through a great number of steps during its processing, or manufacturing, into a finished component. While composition determines the types and amounts of atoms present, it is processing that determines the arrangements of the atoms and the number and types of defects in the atomic structure.

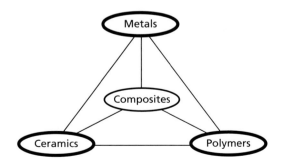

Fig. B-1. Materials Triad

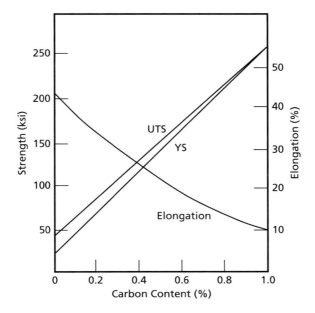

Fig. B-2. Effect of Carbon Content on Steel Strength[2]

B.2 Metallic Structure

Metals are held together by metallic bonding, as illustrated in Fig. B.3, in which the positive nuclei are surrounded by a shared electron cloud. Metals are typically close-packed and have a large number of nearest neighboring atoms (usually 8–12), which helps to explain why they have high densities and high elastic stiffness. In addition, the concept of a shared electron cloud helps to explain their high electrical and thermal conductivities. The ability to alloy, or mix several metals together in the liquid state, is one of the keys to the flexibility of metals. In the liquid state, solubility is often complete, while in the solid state, solubility is generally much more restricted. This change in solubility with temperature forms the basis for heat treatments that can vary the strength and ductility over quite wide ranges.

Metallic structures demonstrate long range order, in which the atoms are arranged in a crystalline structure. The crystalline structure contains atomic packing, which is repetitive over distances large compared to the atomic size. Almost all structural metals crystallize into one of three crystalline patterns: *face centered cubic (FCC), hexagonal close-packed (HCP)* or *body centered cubic (BCC)*, as shown in Fig. B.4. To a great extent, these crystalline patterns determine the processing characteristics of the metal. In terms of atomic packing, the FCC structure is the most efficient, with 12 nearest atom neighbors, also referred to as the coordination number (CN), i.e. the FCC structure has a

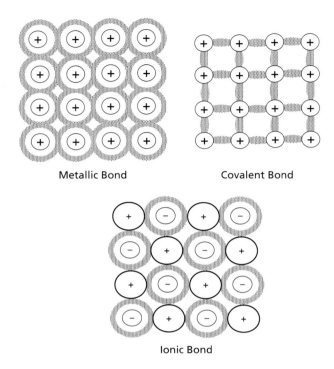

Fig. B-3. Chemical Bonding[3]

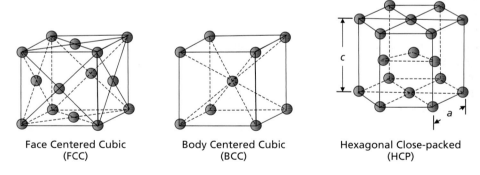

Fig. B-4. Common Metallic Crystalline Structures

CN = 12. As shown in Fig. B.5, the stacking sequence is ABCABC. The atoms in the HCP structure are also packed along close-packed planes. Atoms in the HCP planes (called the basal planes) have the same arrangement as those in the FCC close-packed planes. However, in the HCP structure, these planes repeat

Appendix-B A Brief Review of Materials Fundamentals

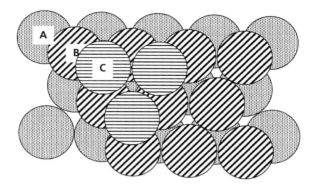

Fig. B-5. Close-packing of Planes[4]

every other layer to give a stacking sequence of ABAB. Two lattice parameters, c and a, shown in Fig. B.4, are also needed to describe the HCP unit cell. When $c/a = 1.63$, the maximum packing efficiency just described is obtained; however, few real HCP structures have this ideal ratio and maximum packing efficiency. The coordination for the BCC structure is 8, which is less than that of the FCC and HCP structures. Since the packing is less efficient in the BCC structure, the closest-packed planes are less densely packed.

The atomic structures of real metals are not perfect but contain defects. One of the most important defects is the line or edge *dislocation*. Without the presence of dislocations, plastic deformation of metals would be much more restricted. Dislocations create an atomic disruption in the lattice, which makes slippage of the planes that have dislocations much easier, as shown in Fig. B.6. The movement is much like advancing a carpet along a floor by using a wrinkle that is easily propagated down its length. The stress required to cause plastic deformation is orders of magnitude less when dislocations are present than in a dislocation-free perfect crystalline structure.

Dislocations[5] do not move with the same degree of ease on all crystallographic planes of atoms and in all crystallographic directions. Ordinarily there

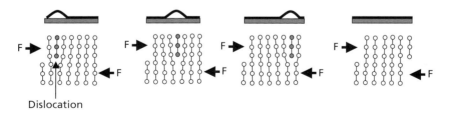

Fig. B-6. Dislocation Movement

are preferred planes, and in these planes, there are specific directions along which dislocation motion can occur. These planes are called *slip planes* and the direction of movement is known as the *slip direction*. The combination of a slip plane and a slip direction forms a *slip system*. For a particular crystal structure, the slip plane is that plane having the most dense atomic packing, i.e. it has the greatest planar density. The slip direction corresponds to the direction, in this plane, that is most closely packed with atoms, i.e. has the highest linear density.

Since *plastic deformation* takes place by slip, or sliding, on the close-packed planes, the greater the number of slip systems available, the greater the capacity for plastic deformation. FCC metals have a large number of slip systems (12) and are therefore capable of moderate to extensive plastic deformation. Although BCC systems often have up to 12 slip systems, some of them like steel exhibit a ductile-to-brittle transition as the temperature is lowered due to the strong temperature sensitivity of their yield strength, which causes them to fracture prior to undergoing significant plastic deformation. In general, the number of slip systems available for HCP metals is less than that for either the FCC or BCC metals, and their plastic deformation is much more restricted. The HCP structure normally has only 3–6 slip systems, one fourth to one half the available slip systems in FCC and BCC structures. Therefore, metals with the HCP structure have only moderate to rather poor ductility at room temperature. The HCP metals often require heating to elevated temperatures, where slip becomes much easier, to facilitate forming operations, which is usually required for alloys of magnesium, beryllium, and titanium.

While slip is required to facilitate plastic deformation, and therefore allows a metal to be formed into useful shapes, to strengthen a metal requires increasing the number of barriers to slip and reducing its ability to plastically deform. Increasing the interference to slip, and increasing the strength, can be accomplished by methods such as strain hardening, grain refinement, solid solution strengthening, precipitation hardening, phase transformation, and dispersion hardening. It should be pointed out that not all of these strengthening mechanisms are applicable to all metals. For example, grain size refinement can be used to strengthen steel but is not as nearly effective for aluminum. In addition, several mechanisms can be used for the same metal. For example, solid solution strengthening and precipitation hardening are used in combination for some nickel based superalloys.

As a metal is plastically deformed, new dislocations are created, so that the dislocation density becomes higher and higher. In addition to multiplying, the dislocations become entangled and impede each other's motion. The result is increasing resistance to plastic deformation with increases in the dislocation density. This continual increase in resistance to plastic deformation is known as *work hardening, cold working,* or *strain hardening*. Work hardening results in a simultaneous increase in strength and a decrease in ductility. Since the work hardened condition increases the stored energy in the metal and is thermodynamically unstable, the deformed metal will try to return to a state of lower energy. This generally cannot be accomplished at room temperature; elevated temperatures, in the range of 1/2–3/4 of the absolute melting point, are necessary to allow mechanisms, such as diffusion, to restore the lower energy state. The process of heating a work hardened metal to restore its original strength and ductility is called *annealing*. Metals undergoing forming operations often require intermediate anneals to restore enough ductility to continue the forming operation.

Annealing occurs in three stages: recovery, recrystallization, and grain growth. The first stage is recovery. During recovery, there is a rearrangement of dislocations to reduce the lattice strain energy, along with the annihilation of a number of dislocations. As shown in Fig. B.7, recovery results in a significant relief of internal stresses with only a moderate reduction in strength, a process called *stress relieving*. During recrystallization, the deformed grain structure is totally replaced by new unstrained grains through a nucleation and growth process, in which new stress free grains grow from nuclei in the deformed matrix. Recrystallization results in a large decrease in strength and a large increase in ductility. After recrystallization, the material can further lower its energy by reducing its total area of grain surface. With extensive annealing, the grain boundaries straighten, small grains shrink and disappear, and larger ones grow. This last stage of annealing, grain growth, is rarely desirable since fine grained structures generally have higher static and fatigue strengths than large grained structures. However, grain coarsening is important in some of the

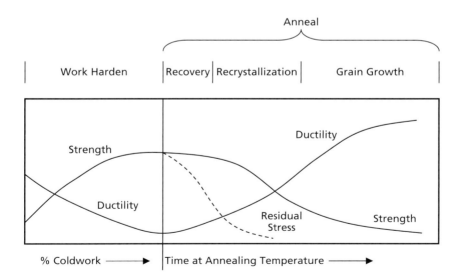

Fig. B-7. Cold working and Annealing

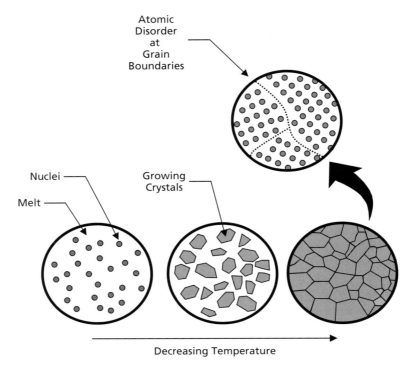

Fig. B-8. Solidification Sequence for Metal

high temperature super alloys, because large grains resist creep better than small grains due to their higher resistance to creep mechanisms.

As shown in Fig. B.8, the atoms in the grain boundaries are not aligned with the atoms in either crystal and are therefore in a higher energy state than the crystals themselves. In polycrystalline metals, the orientation of the slip planes in adjoining grains is seldom aligned, and the slip plane must change direction when traveling from one grain to another. Reducing the *grain size* produces more changes in direction of the slip path and also lengthens it, making slip more difficult; therefore, grain boundaries are effective obstacles to slip. Decreasing the grain size is effective in both increasing strength and also increasing ductility, and, as such, is one of the most effective methods of strengthening. Fracture resistance also generally improves with reductions in grain size, because the cracks formed during deformation, which are the precursors to those causing fracture, are limited in size to the grain diameter.

When a metal is alloyed with another metal, either substitutional or interstitial *solid solutions* (Fig. B.9) are usually formed. Substitutional solid solutions are those in which the solute and solvent atoms are nearly the same size, and solute

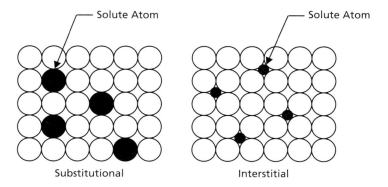

Fig. B-9. Solid Solutions

atoms simply substitute for solvent atoms on the crystalline lattice. Interstitial solid solutions are those in which the solute atoms are much smaller, and fit within the spaces between the existing solvent atoms on the crystalline structure. The insertion of both substitutional and interstitial alloying elements strains the crystalline lattice of the solvent structure. This increase in distortion, or strain, creates barriers to dislocation movement.

Precipitation hardening is used extensively to strengthen aluminum alloys, nickel based superalloys, and PH stainless steels. In precipitation hardening, an alloy is heated to a high enough temperature to take a significant amount of an alloying element into solid solution. It is then rapidly cooled (quenched) to room temperature, trapping the alloying elements in solution. On reheating to an intermediate temperature, the host metal rejects the alloying elements in the form of a fine precipitate that creates matrix strains in the lattice. The fine precipitate particles are barriers to the motion of dislocations and provide resistance to slip.

A portion of a phase diagram for an alloy system that has the characteristics required for precipitation hardening is shown in Fig. B.10. Note that the solvent metal at the left hand edge of the diagram can absorb much more of solute metal at elevated temperature than it can at room temperature. When the alloy is heated to the solution heat treating temperature and held for a sufficient length of time, the solvent metal absorbs some of solute metal. When it is rapidly cooled to room temperature, atoms of the solute metal are trapped in a supersaturated solid solution of the solvent metal. On reheating to an intermediate aging temperature, the supersaturated solution precipitates a very fine phase that acts as barriers to dislocation movement. Note the effects of different aging temperatures shown in Fig. B.10. If the metal is aged at too low a temperature (T_1), the desired strength is not obtained and the alloy is said to be underaged. On the other hand, aging at too high a temperature (T_4) also results in lower than desired strength and the

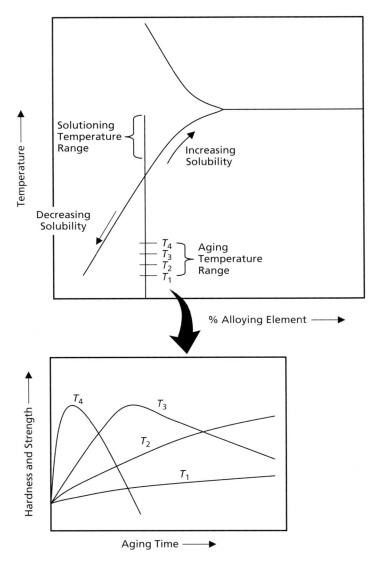

Fig. B-10. Precipitation Hardening Heat Treatment

alloy is now referred to as overaged. Commercial heat treatments are closer to T_2 and T_3, in which the optimum strength can be obtained in a reasonable aging time. The alloy used in this example is one that requires artificial or elevated temperature aging. Some alloys will age satisfactorily at room temperature, a process called natural aging.

Some metals, such as iron, have different crystalline structures at different temperatures. For example, when iron is heated above 1670° F, it transforms from a BCC structure to an FCC structure. The FCC structure can absorb up to 2% carbon, while the BCC structure can tolerate almost no carbon. If the steel, an alloy of iron and carbon, is quenched to room temperature, the large amount of carbon in the FCC structure does not have time to diffuse out and is trapped in the BCC structure. The carbon atoms severely distort the BCC structure into a body centered tetragonal (BCT) structure called *martensite*, creating significant crystalline lattice strain and tremendously increasing the strength and hardness of the steel alloy.

A typical heat treatment just described is shown on the iron–carbon phase diagram of Fig. B.11. In this heat treatment, a steel alloy with approximately 0.40% carbon is heated into the FCC austenite field (γ). As shown in Micrograph A, the structure is totally austenite (γ). On slow cooling to point B (just inside the two-phase $\alpha + \gamma$ field), ferrite (α) starts to precipitate at the austenite grain boundaries in the manner shown in Micrograph B. This ferrite is called proeutectoid ferrite since it forms at temperatures higher than the eutectoid temperature. As the alloy continues to cool through the $\alpha + \gamma$ field, additional ferrite (α) forms along the grain boundaries as shown in Micrograph C. When the alloy cools past the eutectoid temperature and enters the ferrite (α) + cementite (Fe_3C) field, the remaining austenite transforms to what is known as *pearlite*, a lamellar type structure of ferrite and cementite shown in Micrograph D.

However, if it is quenched, it was previously stated that it would transform into a BCT structure called martensite, which is not shown on the phase diagram. Martensite is not shown on the diagram because phase diagrams show equilibrium conditions, and martensite is not produced under equilibrium conditions. The diagram one now needs is called a TTT diagram, where TTT stands for time, temperature, transformation. A typical TTT diagram is shown in Fig. B.12. To produce martensite, the steel must be cooled quickly enough to miss the "nose" of the curve, as shown in the cooling path A–B. When it reaches the m_s (martensite start) temperature, it immediately starts forming the BCT martensitic structure. The martensitic structure is completed when it cools past the m_f (martensite finish) temperature. If the quench is not fast enough to miss the nose of the TTT curve, then it will form pearlite, such as shown in the cooling path A–C. The other structure shown in the TTT diagram, bainite, can be produced by the interrupted quench cycle shown in the path A–D. Molten salt baths are used to produce the intermediate temperature hold necessary for bainite formation. Like pearlite, bainite is also a mixture of ferrite and cementite, but has much better toughness than pearlite. The TTT diagram shown is for a plain carbon steel. Alloying elements, in addition to providing solid solution strengthening, move the nose of the TTT curve to the right, thus providing more time to quench a steel to the fully martensitic condition.

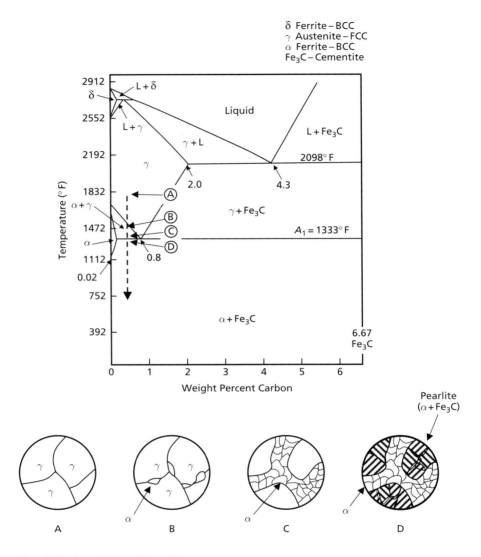

Fig. B-11. Iron–Carbon Phase Diagram

Dispersion strengthened alloys, in which small discrete hard particles block dislocation movement, are used in some nickel based superalloys. Yittria (Y_2O_3) and nickel alloy powder are ball milled together in a high energy process called mechanical alloying. When the powder is consolidated under heat and pressure to form an alloy, it displays improved high temperature strength, due to the presence of a uniform dispersion of fine refractory oxide particles in the nickel alloy matrix, which again act as barriers to the movement of dislocations.

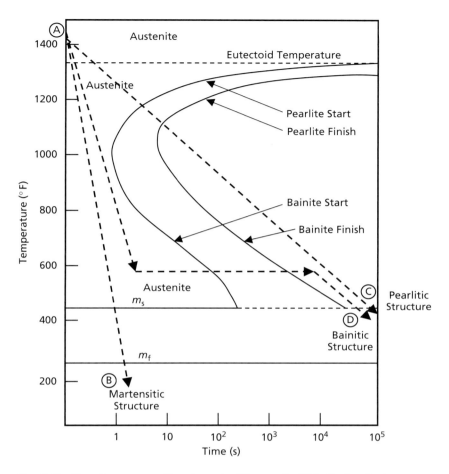

Fig. B-12. Time, Temperature, Transformation Diagram for Plain Carbon Steel

Four important reactions that can occur during processing are shown in Fig. B.13, along with representative phase diagrams. The four reactions are called eutectic, peritectic, eutectoid, and peritectoid.

- *Eutectic*: In a eutectic reaction, a liquid freezes to form two solid solutions:

    Liquid → Solid α + Solid β

- *Peritectic*: In a peritectic reaction, a solid solution and a liquid react to form a solid solution of a different composition during freezing:

    Solid α + Liquid → Solid β

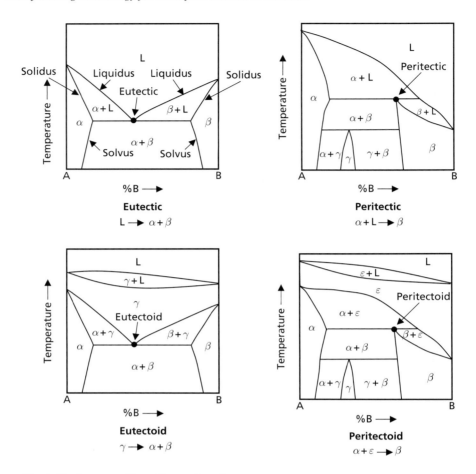

Fig. B-13. Important Phase Change Reactions

- *Eutectoid*: A solid state reaction similar to the eutectic reaction is the eutectoid reaction. In a eutectoid reaction, a solid solution of one composition transforms to two solid solutions of different compositions.

    Solid $\gamma \rightarrow$ Solid $\alpha +$ Solid β

- *Peritectoid*: Like the eutectic and eutectoid reactions, the peritectic also has a similar solid state reaction, the peritectoid. In this reaction, two solid solutions react to form a solid solution of a different composition:

    Solid $\alpha +$ Solid $\beta \rightarrow$ Solid γ

Note in Fig. B.13 that since the eutectic is the lowest melting composition in the alloy, it can often cause problems during hot working and heat treating operations. During casting of an ingot, due to non-uniform cooling rates, there is often quite a bit of segregation of the alloying elements. The low melting eutectic composition is normally the last portion of the metal to freeze, normally at the grain boundaries. On reheating the metal for hot working operations, or for heat treatment, if the lowest melting eutectic temperature is exceeded, melting can occur along the grain boundaries and the part is frequently ruined. It should also be noted that alloying systems with a number of different alloying elements will often form lower melting point eutectics than for the simple binary system shown in the figure. To help prevent some of these problems, as-cast ingots are often reheated to temperatures just below the melting point and soaked for long times (called homogenization) to create more uniform structures prior to hot working or heat treating.

A note of caution: it should be remembered that phase diagrams represent equilibrium conditions, which are often not obtained in industrial processes; nevertheless, phase diagrams are very useful tools in processing.

B.3 Ceramics

Ceramics are inorganic non-metallic materials which consist of metallic and non-metallic elements bonded together with either ionic or covalent bonds. Although ceramics can be crystalline or non-crystalline, the important engineering ceramics are all crystalline. Due to the absence of conduction electrons, ceramics are usually good electrical and thermal insulators. In addition, due to the stability of their strong bonds, they normally have high melting temperatures and high chemical stability to many hostile environments. However, ceramics are inherently hard and brittle materials that, when loaded in tension, have almost no tolerance for flaws. As a material class, few ceramics have tensile strengths above 25 ksi, while the compressive strengths may be 5–10 times higher than the tensile strengths.[6]

Under an applied tensile load at room temperature, both crystalline and non-crystalline ceramics almost always fracture before any plastic deformation can occur. Stress concentrations leading to brittle failure can be minute surface or interior cracks (microcracks), or internal pores, which are virtually impossible to eliminate or control. Plane strain fracture toughness (K_{Ic}) values for ceramic materials are much lower than for metals; typically below 9 ksi $\sqrt{\text{in}}$., while for metals they can exceed 100 ksi $\sqrt{\text{in}}$.). There is also considerable scatter in the fracture strength for ceramics, which can be explained by the dependence of fracture strength on the probability of the existence of a flaw that is capable of initiating a crack. Therefore, size or volume also influences fracture strength; the larger the size, the greater the probability for a flaw and the lower the fracture strength.

In metals, plastic flow takes place mainly by slip. In metals, due to the non-directional nature of the metallic bond, dislocations move under relatively low stresses, and because all atoms involved in the bonding have an equally distributed negative charge at their surfaces. In other words, there are no positive or negatively charged ions involved in the metallic bonding process. However, ceramics form either ionic or covalent bonds, both of which restrict dislocation motion and slip. One reason for the hardness and brittleness of ceramics is the difficulty of slip or dislocation motion. While ceramics are inherently strong, they cannot slip or plastically deform to accommodate even small cracks or imperfections, i.e. their strength is never realized in practice. They fracture in a premature brittle manner long before their inherent strength is approached.

The nature of the *ionic* and *covalent bonds* is shown in Fig. B.3. In the ionic bond, the electrons are shared by an electropositive ion (cation) and an electronegative ion (anion). The electropositive ion gives up its valence electrons, and the electronegative ion captures them to produce ions having full electron orbitals or suborbitals. As a consequence, there are no free electrons available to conduct electricity. In ionically bonded ceramics, there are very few slip systems along which dislocations may move. This is a consequence of the electrically charged nature of the ions. For slip in some directions, ions of like charge must be brought into close proximity to each other, and because of electrostatic repulsion, this mode of slip is very restricted. This is not a problem in metals, since all atoms are electrically neutral. In covalently bonded ceramics, the bonding between atoms is specific and directional, involving the exchange of electron charge between pairs of atoms. Thus, when covalent crystals are stressed to a sufficient extent, they exhibit brittle fracture due to a separation of electron pair bonds, without subsequent reformation. It should also be noted that ceramics are rarely either all ionically or covalently bonded; they usually consist of a mix of the two types of bonds. For example, silicon nitride (Si_3N_4) consists of about 70% covalent bonds and 30% ionic bonds.

Due to their extremely high melting temperatures, ceramics are often processed by powder processing methods in which powders are mixed, consolidated under pressure, and then sintered at high temperatures. During the high temperature sintering operation, the particles fuse together through diffusion processes, much in the manner as shown in Fig. B.14. Although many industrial ceramic products are sintered without external pressure, high performance ceramics, since they are extremely sensitive to all flaws, even very small voids, are usually sintered with high external pressures to minimize the amount of porosity.

B.4 Polymers

While polymers are light and some exhibit high levels of ductility, they are simply not strong enough by themselves to classify as high strength structural materials. However, when combined with high strength fibers, they become

Appendix-B A Brief Review of Materials Fundamentals

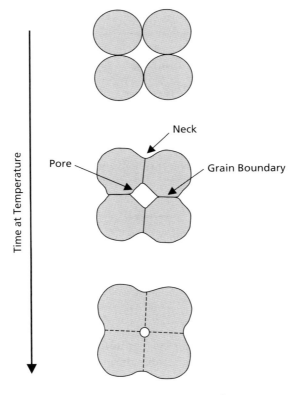

Fig. B-14. Progression of Ceramic Sintering[7]

structural materials (i.e., composites) with high strength and stiffness-to-weight ratios and are becoming increasingly important in aerospace.

Polymers can be classified as either *thermosets* or *thermoplastics*. As shown in Fig. B.15, a thermoset crosslinks during cure to form a rigid intractable solid. Prior to cure, the resin is a relatively low molecular weight semi-solid that melts and flows during the initial part of the cure process. As the molecular weight builds during cure, the viscosity increases until the resin gels, and then strong covalent bond crosslinks form during cure. Due to the high crosslink densities obtained for high performance thermoset systems, they are inherently brittle, unless steps are taken to enhance toughness. On the other hand, thermoplastics are high molecular weight resins that are fully reacted prior to processing. They melt and flow during processing but do not form crosslinking reactions. Their covalently bonded main chains are held together by relatively weak secondary bonds. However, being high molecular weight resins, the viscosities of thermoplastics during processing are orders of magnitude higher than that of thermosets. Since thermoplastics do not crosslink during processing, they can be

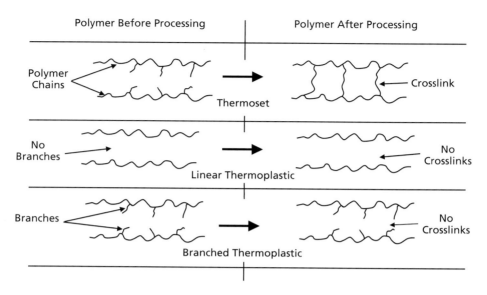

Fig. B-15. Comparison of Thermoset and Thermoplastic Polymer Structures[8]

reprocessed; for example, they can be thermoformed into structural shapes by simply reheating to the processing temperature. Thermosets, due to their highly crosslinked structures, cannot be reprocessed and, if reheated to high enough temperatures, will thermally degrade and eventually char. However, there is a limit to the number of times a thermoplastic can be reprocessed. Since the processing temperatures are often close to the polymer degradation temperatures, multiple reprocessing will eventually degrade the resin and in some cases it may crosslink.

Many thermoplastics are polymerized by what is called *addition* polymerization, as shown in Fig. B.16 for the simple thermoplastic polyethylene. Addition polymerization consists of three steps: initiation, propagation, and termination. During initiation, an active polymer capable of propagation is formed by the reaction between an initiator species and a monomer unit. In the figure, R represents the active initiator that contains an unpaired electron (·). Propagation involves the linear growth of the molecule as monomer units become attached to each other in succession, to produce a long chain molecule. Chain growth, or propagation, is very rapid, with the period required to grow a molecule consisting of 1000 repeat units on the order of 100–1000 s.[7] Propagation can terminate in one of two ways. In the first, the active ends of two propagating chains react together to form a non-reactive molecule. The second method of termination occurs when an active chain end reacts with an initiator. Polyethylene can normally have anywhere from 3500 to 25 000 of these repeat units.[9]

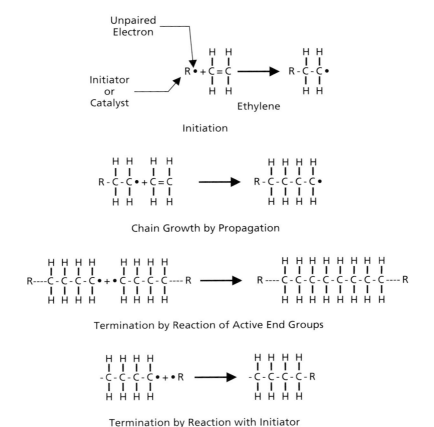

Fig. B-16. *Chain Polymerization of Polyethylene*

Another characteristic of thermoplastics is that some contain totally random chain structures, known as *amorphous*, while others contain regions of closely folded chains and are known as *semi-crystalline*. A comparison of these two structures is depicted in Fig. B.17. An amorphous thermoplastic contains a massive random array of entangled molecular chains. The chains themselves are held together by strong covalent bonds, while the bonds between the chains are much weaker secondary bonds. When the material is heated to its processing temperature, it is these weak secondary bonds that breakdown and allow the chains to move and slide past one another. Amorphous thermoplastics exhibit good elongation, toughness, and impact resistance. As the chains get longer, the molecular weight increases, resulting in higher viscosities, higher melting points, and greater chain entanglement, all leading to higher mechanical properties.

Semi-crystalline thermoplastics contain areas of tightly folded chains (crystallites) that are connected together with amorphous regions. Amorphous

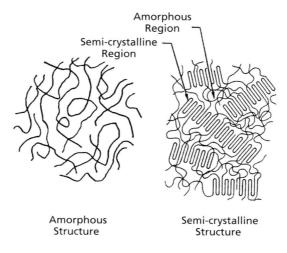

Fig. B-17. Amorphous and Semi-crystalline Structures

thermoplastics exhibit a gradual softening on heating, while semi-crystalline thermoplastics exhibit a sharp melting point when the crystalline regions start dissolving. As the polymer approaches its melting point, the crystalline lattice breaks down and the molecules are free to rotate and translate, while non-crystalline amorphous thermoplastics exhibit a more gradual transition from a solid to a liquid. Crystallinity increases strength, stiffness, creep resistance, and temperature resistance but usually decreases toughness. By decreasing and restricting chain mobility, the tightly packed crystalline structure behaves somewhat like crosslinking in thermosets. The maximum crystallinity obtainable is about 98%, whereas metallic structures are usually 100% crystalline and exhibit much more ordered structures. In general, semi-crystalline thermoplastics used for composite matrices contain about 20–35% crystallinity.

Thermoset polymers are always amorphous. They start as low molecular weight liquids that cure by either addition or condensation reactions. A comparison of these two cure mechanisms is shown in Fig. B.18. In the addition reaction shown for an epoxy reacting with an amine curing agent, the epoxy ring opens and reacts with the amine to form a crosslink. The amine shown in this example is what is known as an aliphatic amine and would produce a crosslinked structure with only moderate temperature capability. Higher temperature capabilities can be produced by curing with what are known as aromatic amines. Aromatic amines contain the large and bulky benzene ring, which helps to restrict chain movement when the network is heated. A typical curing agent, diamino diphenyl sulfone (DDS), and a typical epoxy, tetraglycidyl methylene dianiline (TGMDA), are shown in Fig. B.19. Note that the curing agent and the epoxy both contain benzene rings. Also note that the TGMDA has four

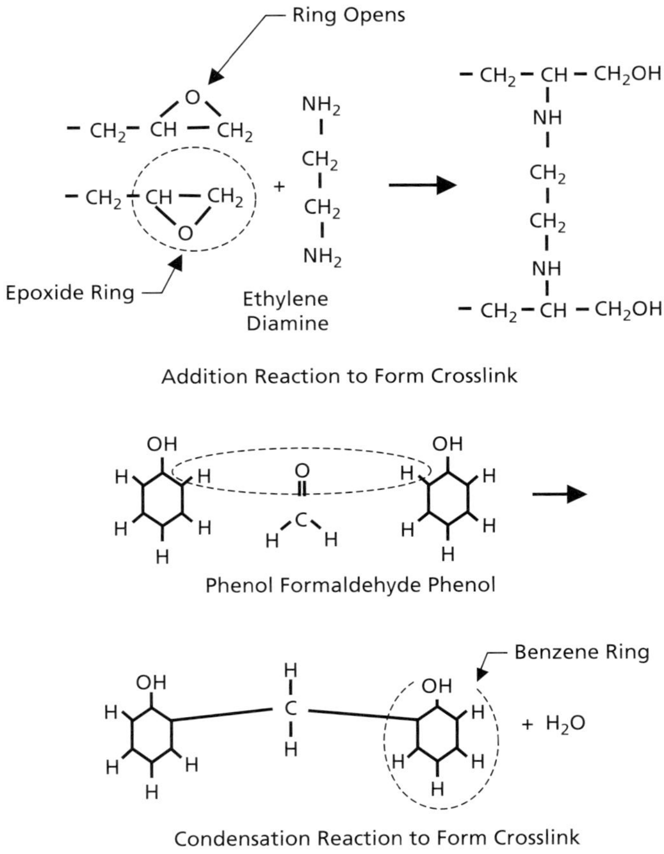

Fig. B-18. Thermoset Cure Reactions

epoxy rings, each capable of reacting. Hence, this epoxy has what is known as a functionality of four. Since it has four active epoxide rings, it is capable of forming high density crosslinked structures. This particular curing agent and epoxy are used in many of the composite matrix resin formulations.

In the lower portion of Fig. B.18, two phenol molecules are shown to react with a formaldehyde molecule to form a phenolic linkage in what is known as a *condensation* reaction. The significant feature of this reaction is the evolution of a water molecule each time the reaction occurs. When thermosets, such as phenolics and polyimides that cure by condensation reactions, are used for composite laminates, they often contain high porosity levels, due to the water or alcohol vapors created by the condensation reactions. Addition curing thermosets, such as epoxies, polyesters, vinyl esters, and bismaleimides, are generally much easier to process with low void levels.

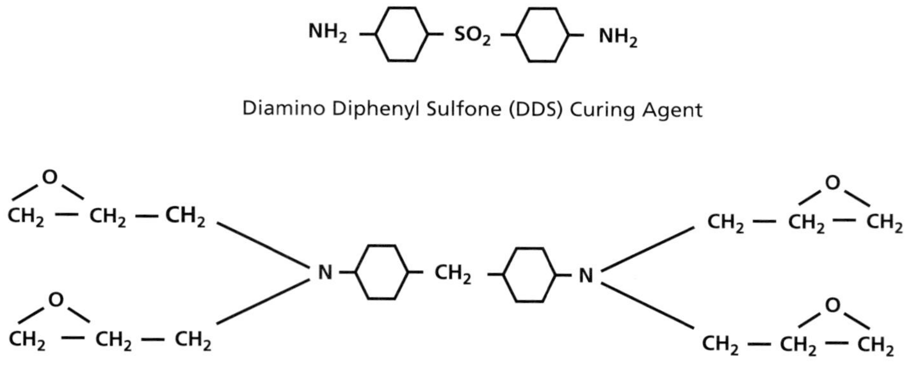

Diamino Diphenyl Sulfone (DDS) Curing Agent

Tetraglycidyl Methylene Dianiline (TGMDA) Epoxy

Fig. B-19. *Composite Matrix Resins*

Thermoset resins are currently the matrix of choice for composite materials. The matrix holds the fibers in position, protects the fibers from abrasion, transfers loads between fibers, and provides interlaminar shear strength between the layers. The advantages of thermoset resins are high thermal stability, high rigidity, high dimensional stability, resistance to creep and deformation under load, and ease of processing.

B.5 Composites

A composite material can be defined as a combination of two or more materials that results in better properties than when the individual components are used alone. As opposed to metallic alloys, each material retains its separate chemical, physical, and mechanical properties. The two constituents are normally a *fiber* and a *matrix*. Typical fibers include glass, aramid, and carbon, while matrices can be polymers, metals, or ceramics. While the fiber reinforcements can be continuous or discontinuous, aligned or random, it is the continuous aligned composite materials that provide the high strength-to-weight and stiffness-to-weight materials that are used in aerospace.

A typical composite material, shown in Fig. B.20, consists of high strength fibers embedded in a matrix. The longitudinal (0°) tension and compression loads are carried by the fibers, while the matrix distributes the loads between the fibers in tension, and stabilizes and prevents the fibers from buckling in compression. The matrix is also the primary load carrier for interlaminar shear (i.e., shear between the layers) and transverse (90°) tension. Since the matrix transfers load to the fibers at the interface through shear, the fiber-to-matrix *interface* bond is important in assuring that the load transfer is effective. For example, carbon fibers that are used in epoxy resins are chemically surface treated to maximize the bond between the fiber and matrix. For high temperature metal matrix

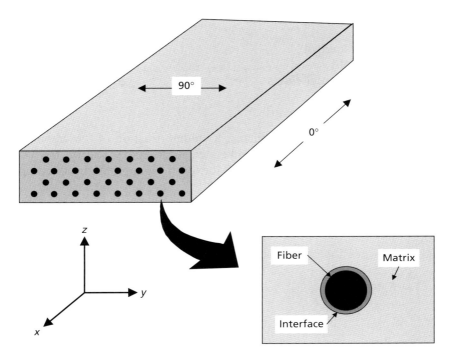

Fig. B-20. Fiber Reinforced Composite

composites, due to the high processing temperatures required to consolidate the matrix, a reaction can occur between the fiber and matrix at the interface. Therefore, protective coatings (diffusion barriers) are usually applied to the fibers to prevent this reaction which can seriously degrade the fiber strength.

Ceramic matrix composites are a little different. While the objective in polymer and metal matrix composites is to obtain a strong interfacial bond, in ceramic matrix composites, the objective is to obtain a weak bond. The strong bond in polymer and metal matrix composites is needed to strengthen the material, to allow effective load transfer between the matrix and the fibers. However, in ceramic matrix composites, where the ceramic matrix is extremely brittle, the main objective is not to strengthen the ceramic but to toughen it. The weak interfacial bond in ceramic matrix composites allows the fibers to debond, deflect cracks and pull-out of the matrix when the ceramic cracks, slowing the propagating cracks through what are called energy dissipating mechanisms.

As previously stated, the fibers provide the strength. The strength of the 0° aligned composite shown in Fig. B.20 can be calculated by the *rule of mixtures*:

$$\sigma_c = \sigma_f V_f + \sigma_m V_m$$

where

σ_c = composite strength
σ_f = fiber strength
V_f = volume fraction of fibers
σ_m = matrix strength
V_m = volume fraction of matrix

For example, if the composite contains a volume fraction of 0.6 carbon fibers with a tensile strength of 750 ksi, embedded in an epoxy matrix (0.4 volume fraction) with a tensile strength of 10 ksi, the tensile strength of this 0° composite is then about:

$$\sigma_c = (750\,\text{ksi})(0.6) + (10\,\text{ksi})(0.4) = 450\,\text{ksi} + 4\,\text{ksi} = 454\,\text{ksi}$$

A similar expression can be used for the 0° modulus of elasticity:

$$E_c = E_f V_f + E_m V_m$$

where

E_c = composite modulus
E_f = fiber modulus
V_f = volume fraction of fibers
E_m = matrix modulus
V_m = volume fraction of matrix

In our example, assume that the fiber tensile modulus is 44 msi and the matrix modulus is 0.5 msi, then:

$$E_c = (44\,\text{msi})(0.6) + (0.5\,\text{msi})(0.4) = 26.4\,\text{msi} + 0.2\,\text{msi} = 26.6\,\text{msi}$$

These rather simple calculations emphasize that the fiber properties determine the composite properties, at least when a 0° laminate is loaded in tension along the fiber axis. Referring back to Fig. B.20, it should be obvious that the 90° strength is going to be a lot less than the 0° strength, primarily because the much weaker matrix is carrying most of the load. There are also expressions for calculating the 90° strength and stiffness, but they are not as accurate as the 0° calculations, because one has to make assumptions about the strength of the interface bond between the fibers and matrix.

It should be emphasized that 0° laminates are almost never used in engineering structures, because all of the loads almost never line up with the fiber direction. This fact, combined with the low 90° properties, forces designers to use laminates with fibers aligned in multiple directions, for example +45°, −45°, 0°, and 90°.

Appendix-B A Brief Review of Materials Fundamentals

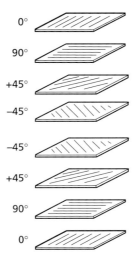

Fig. B-21. Quasi-isotropic Laminate Lay-up

If a laminate has equal numbers of layers aligned in the +45°, −45°, 0°, and 90° directions, it is called a *quasi-isotropic* laminate, as shown in Fig. B.21, because the different directions on the x–y plane have about equal strength and stiffness. By way of example, a carbon fiber/epoxy matrix composite with 50% 0° plies and 50% 90° plies will have about the same strength and modulus as a high strength aluminum alloy, but the composite laminate will weigh about 1/2 that of an equivalent thickness of the high strength aluminum alloy.

Note that if the laminate shown in Fig. B.20 is loaded in the z-direction, the strength is again largely dependent on the matrix strength, rather than the fiber strength. This is a weak link in the composite materials we use today. For the proper design of a composite structure, it is important that the strong fibers pick up the load and not the weak matrix. Therefore, composites need stiff and well-defined load paths in the x- and y-directions, with no out-of-plane loading in the z-direction. It should be noted that there is a considerable amount of work being done to put reinforcements in the z-direction, but very few of these concepts are currently used in production designs.

Recommended Reading

[1] Smith, W.F., *Principles of Materials Science and Engineering*, McGraw-Hill, 1986.
[2] Callister, W.D., *Fundamentals of Materials Science and Engineering*, 5th edition, John Wiley & Sons, Inc., 2001.
[3] Askeland, D.R., *The Science and Engineering of Materials*, 2nd edition, PWS-KENT Publishing Co., 1989.
[4] Campbell, F.C., *Manufacturing Processes for Advanced Composites*, Elsevier Ltd, 2004.

References

[1] Courtney, T.H., "Fundamental Structure-Property Relationships in Engineering Materials", in *Materials Selection and Design,* ASM Handbook Vol. 20, ASM International, 1997.
[2] Callister, W.D., "Phase Transformations", in *Fundamentals of Materials Science and Engineering*, 5th edition, John Wiley & Sons, Inc., 2001, p. 345.
[3] Thrower, P.A., "Bonding", in *Materials in Today's World*, Revised edition, McGraw-Hill, Inc., 1992, pp. 25–38.
[4] Thrower, P.A., "Crystal Structures", in *Materials in Today's World*, Revised Edition, McGraw-Hill, Inc., 1992, pp. 39–53.
[5] Callister, W.D., "Deformation and Strengthening Mechanism", in *Fundamentals of Materials Science and Engineering*, 5th edition, John Wiley & Sons, Inc., 2001, pp. 197–233.
[6] Smith, W.F., "Ceramic Materials", in *Principles of Materials Science and Engineering*, McGraw-Hill, 1986, pp. 527–595.
[7] Callister, W.D., "Synthesis, Fabrication, and Processing of Materials", in *Fundamentals of Materials Science and Engineering*, 5th edition, John Wiley & Sons, Inc., 2001.
[8] Campbell, F.C., "Thermoplastic Composites", in *Manufacturing Processes for Advanced Composites*, Elsevier Ltd, 2004, p. 358.
[9] Smith, W.F., "Polymeric Materials", in *Principles of Materials Science and Engineering*, McGraw-Hill, 1986, p. 303.

Appendix C

Mechanical and Environmental Properties

This appendix provides some definitions, or explanations, of some of the important mechanical properties and environmental degradation mechanisms that can occur in structural materials. It should be pointed out that these are very brief explanations, often simplified, and that much more extensive explanations can be found in texts dedicated to these subjects.

C.1 Static Strength Properties[1]

The tensile properties of a material are determined by applying a tension load to a specimen and measuring the elongation or extension. A typical stress–strain curve for a metal is shown in Fig. C.1. The load can be converted to engineering *stress* (σ) by dividing the load by the original cross-sectional area of the specimen.

$$\sigma = P/A_o \text{ in lb/in.}^2 \text{ or psi}$$

where

$P = $ load in lb
$A_o = $ original cross-sectional area in in.^{2}

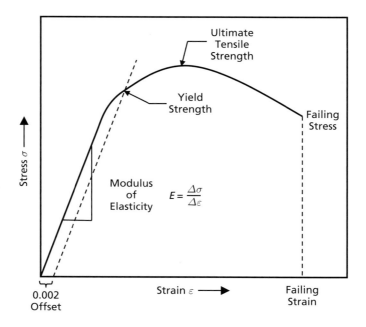

Fig. C-1. Typical Stress–Strain Curve

The engineering *strain* (ε) can be calculated dividing the change in gage length by the original gage length.

$$\varepsilon = (l - l_0)/l_0 \text{ in in./in.}$$

where

l = gage length
l_0 = original gage length

As shown in Fig. C.1, the *ultimate tensile strength* is the maximum stress that occurs during the test. For metals without a definite yield point, the *yield strength* is determined by drawing a straight line parallel to the initial straight line portion of the stress–strain curve. The line is normally offset by a strain of 0.2% (0.002). Yield strength is generally a more important design parameter than ultimate strength, since the possibility of plastic yielding is unacceptable for almost all structures.

The *modulus of elasticity* (E) is determined by taking the slope of the initial straight line portion of the stress–strain curve. The higher the modulus of elasticity, the stiffer is the material.

$$E = \Delta\sigma/\Delta\varepsilon \text{ expressed in msi (millions of pounds/in.}^2\text{)}.$$

Percent elongation and *reduction of area* are measures of ductility, with higher values indicating greater ductility. Since percent elongation is sensitive to the gage length, it is important to record the gage length used when reporting percent elongation. In general, shorter gage lengths result in higher values for percent elongation. Percent elongation can be determined by:

$$\% \text{ Elongation} = \{(l_f - l_0)/l_0\} \times 100$$

where

l_0 = original gage length
l_f = final length of the gage section

Likewise, reduction in area can be determined by the tensile test.

$$\% \text{ Reduction in Area} = \{(A_o - A_f)/A_0\} \times 100$$

where

A_0 = original area of the gage section
A_f = final area of the gage section

Static compressive properties are determined in a similar manner, except the specimen is loaded in compression rather than tension.

C.2 Failure Modes[1,2]

Failure modes can be classified into four general categories: ductile failures, brittle failures, decohesive or intergranular failures, and fatigue failures.

Ductile failures are associated with large amounts of plastic deformation. As a result of plastic deformation, localized necking or distortion is often present. Ductile failures occur by tearing of the metal with a large expenditure of energy. Microscopically, ductile failures occur by a process of microvoid nucleation and growth (Fig. C.2). Microvoids form at stress concentrations, and are most frequently initiated by constituent particles, followed by void formation and growth around the particles, or by particle cracking. In aluminum alloys, the constituent particles that most often initiate mircovoids are quite large ($> 1\,\mu$m) particles of Al_7Cu_2Fe, Mg_2Si, and $(Fe, Mn)Al_2$. Note that these particles contain iron and silicon. Many improvements in the properties of aluminum alloys have been a result of lowering iron and silicon impurity levels.

The mircovoids then coalesce, and grow, to produce larger voids until the remaining area becomes too small to support the load and failure occurs. Shear lips, due to slip mechanisms, often occur at angles approaching 45° to the applied tensile stress, to form the well-known cup and cone fracture appearance.

Brittle fractures are generally flat, with little or no evidence of localized necking. Glasses and crystalline ceramics, when fractured at room temperature, fracture in a purely brittle manner, with no evidence of plastic deformation.

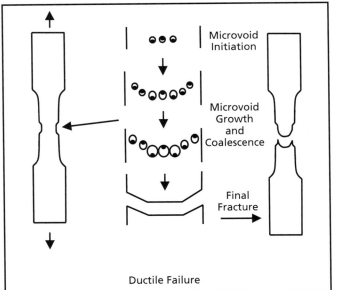

Fig. C-2. Comparison of Ductile and Brittle Failure Modes[1]

However, even the most brittle metal will exhibit some slight evidence of plastic deformation. Initiation of a crack normally occurs at a small flaw, such as a defect, notch or discontinuity, which acts as stress concentration, and rapidly propagates through the metal. Cracks resulting from machining, quenching, hydrogen embrittlement, or stress corrosion can cause brittle failures. Even normally ductile metals can fail in a brittle manner at low temperatures, in thick sections, at high strain rates such as impact loading, or when there are pre-existing flaws. Brittle failures normally initiate as a result of cleavage. Cleavage is a brittle failure mode that occurs by breaking of the atomic bonds. Brittle failures are characterized by rapid crack propagation with less energy expenditure than in ductile fractures.

Ductile fractures proceed only as long as the material is being strained, i.e. stop the deformation and crack propagation stops. At the other extreme, once a brittle crack is initiated, it propagates through the material at velocities approaching the speed of sound, with no possibility of arresting it. There is insufficient plastic deformation to blunt the crack. This makes brittle fractures extremely dangerous, i.e. there is usually no warning of impending fracture.

Some BCC and HCP metals, and steels in particular, exhibit a *ductile-to-brittle transition*, as depicted in Fig. C.3 when loaded under impact. At high temperatures, the impact energy is high and the failure modes are ductile, while at low temperatures, the impact energy absorbed is low and the failure mode changes to a brittle fracture. The *transition temperature* is sensitive to both alloy composition and microstructure. For example, reducing the grain size of

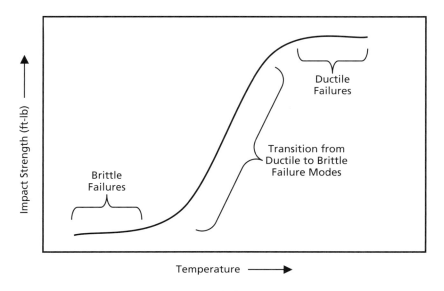

Fig. C-3. Ductile–Brittle Transition Typical of Steels

steels lowers the transition temperature. Not all metals display a ductile-to-brittle transition. Those having an FCC structure, such as aluminum, remain ductile down to even cryogenic temperatures.

Since the grain boundaries are usually stronger than the grains themselves, ductile fractures normally occur in a *transgranular* manner (i.e., through the grains) in metals having good ductility and toughness. Brittle failure modes, which exhibit little or no plastic flow, can occur along the grain boundaries, or *intergranularly*. A comparison of these two failure modes are shown schematically in Fig. C.4. The occurrence of intergranular failures at room temperature often implies some embrittling behavior, such as the formation of brittle grain boundary films, or the segregation of impurities or inclusions that cluster at the grain boundaries. However, at temperatures high enough for creep to become dominate, the reverse is true, with intergranular failures being more common than transgranular failures.

Fatigue failures generally initiate at a surface stress concentration, followed by relatively slow crack growth through the part until the remaining cross section becomes too small to support the stress, and final failure occurs. The final overload zone can be either a ductile or a brittle failure. Under the right types of light and magnification, the growth of a fatigue crack through a metal (Fig. C.5) can often be detected by observing fatigue striations, which are lines that essentially represent one fatigue cycle per striation.

C.3 Fracture Toughness[3,4]

Fracture toughness is a measure of the ability of a metal to resist fracture in the presence of a flaw. The fracture mechanics approach assumes that all real structures contain one or more sharp cracks, either as a result of manufacturing or due to material defects. The problem then becomes one of determining the level of stress that may be safely applied to a crack, before it grows to a critical size and causes failure.

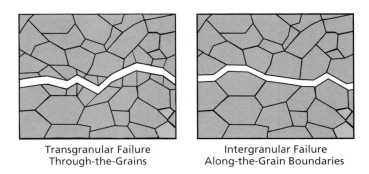

Transgranular Failure
Through-the-Grains

Intergranular Failure
Along-the-Grain Boundaries

Fig. C-4. Transgranular and Intergranular Failures

Appendix-C Mechanical and Environmental Properties

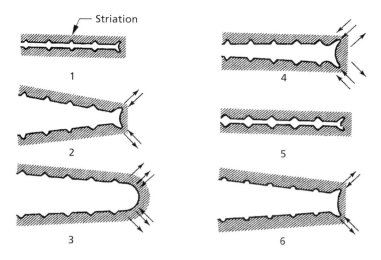

Fig. C-5. Fatigue Crack Propagation[5]

When the material behavior is brittle rather than ductile, the mechanics of fracture are quite different. Instead of the slow growth coalescence of voids associated with ductile rupture, brittle fracture proceeds by high velocity crack propagation through the loaded member. However, normally ductile materials can also fail in a brittle manner, in the presence of a crack, if the combination of crack size, part geometry, temperature, and/or loading rate lies with certain critical regions. The use of higher strength materials, the wider use of welding, and the use of thicker, highly loaded structural members have reduced the capacity of structural members to accommodate local plastic deformation without failure.

The elastic stress field in the vicinity of a crack tip can be described by a parameter called the *stress intensity factor K*, which is a function of the crack geometry and the applied stress in the immediate vicinity of the crack. As shown in Fig. C.6, a fracture toughness test is performed by applying a tensile stress to a standard specimen, with a crack of a known size and geometry.

The stress intensity factor K can be calculated using the general formula:

$$K = \alpha\sigma(a\pi)^{0.5} \text{ in ksi (in.)}^{0.5}$$

where

α = geometry factor for the specimen and flaw
σ = applied stress in ksi
a = flaw or crack size in in.

There are three crack opening modes that can be evaluated (Fig. C.7). However, Mode I, the tensile opening mode, is usually measured because it is the limiting value for K, producing the lowest values.

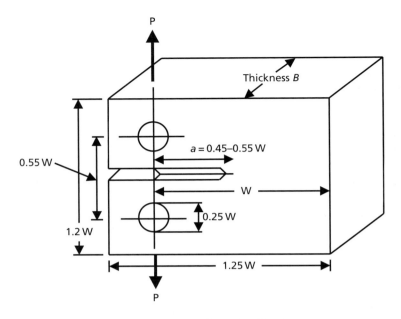

Fig. C-6. Compact Tension Fracture Toughness Test Specimen

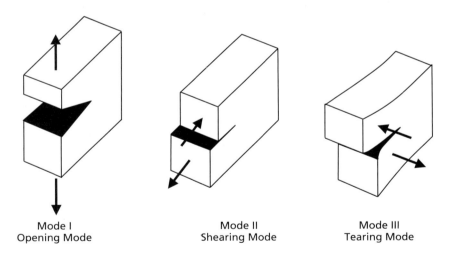

Fig. C-7. Crack Opening Displacement Modes

The value of the stress intensity factor K in which the crack propagates to failure is called the *critical stress intensity factor* K_c. If the specimen provides sufficient plastic constraint at the crack tip, a plane-strain condition is produced which gives the lowest value K_c of the material. When these conditions are met during the fracture toughness test, then K_c is equal to the K_{Ic}, the *plane-strain fracture toughness*. The significance being that any combination of stress and crack length that exceeds K_{Ic} will produce unstable crack growth leading to failure. The dependence of stress intensity factor on specimen geometry is shown in Fig. C.8. It should be noted that the plane-strain fracture toughness K_{Ic} is a material property, much like tensile and yield strength are. Similar mathematical treatments have been developed for materials that fail under ductile conditions, known as plane-stress loading.

Toughness depends on both strength and ductility. For example, the higher strength steel alloys have higher fracture toughness values than the lower strength aluminum alloys. However, within a material class, increases in strength almost always results in decreases in fracture toughness. For example, a medium carbon low alloy steel heat treated to a yield strength of 280 ksi will have a lower fracture toughness than one heat treated to only 200 ksi. On the other hand, the steel heat treated to a yield strength of 280 ksi will normally have the greatest fatigue strength.

For materials that can develop a region of plastic deformation at the crack tip, rapid crack extension is related to material ductility. Unstable crack growth occurs when the plastic zone reaches a critical size. Therefore, the larger the plastic zone ahead of the crack, the more energy that is consumed in propagating the fracture, and the tougher the material. Reducing the grain size of most alloys results in an increase in both strength and fracture toughness, and is one of the few methods available for increasing both strength and ductility.

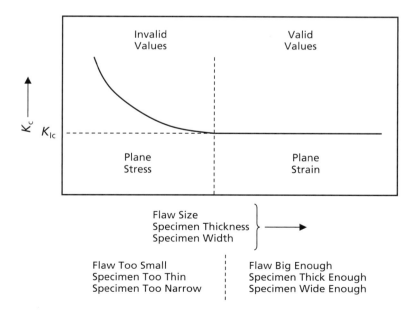

Fig. C-8. Variation of K_c with Specimen Dimensions

Fracture control consists of controlling the nominal stress and crack size so that the combination always lies below a critical value for the material. Fracture control plans for aerospace vehicles include damage tolerance analysis, tests, and subsequent inspection and repair/replacement plans. The most critical parts to the performance and safety of flight are termed *fracture critical*. They receive the most rigorous analysis and inspection frequencies. The frequency of inspections depends on the minimum size flaw that can be detected by non-destructive testing, and is always spaced so that two inspections are scheduled before any flaw could reach a critical size.

C.4 Fatigue[6,7]

Fatigue is the progressive degradation that can occur during cyclic loading of a part. Fatigue is the cause, or a contributor, to a large number of structural failures. Fatigue properties are normally reported as the total number of cycles to failure, as shown in the S–N curve in Fig. C.9, where S represents the stress and N is the number of cycles. Higher stress levels result in shorter fatigue lives, and lower stress levels allow longer lives. Fatigue tests are normally run on smooth specimens. However, if a notch is present, due to the stress concentration created by the notch, the fatigue curve will be lower (shorter lives) than for smooth specimens.

Three different loading cycles, and some standard fatigue nomenclature, are shown in Fig. C.10. In the first cycle, fully reversed cycling is shown in which the maximum tensile stress is the same as the maximum compressive stress. The second cycle has the same shape except all of the stresses are tensile in nature, because there is a tensile applied mean stress. Finally, the last cycle shows a random loading, or spectrum, that might be encountered in on a real component. Determining the correct loading cycle for a part, or large structure, is one of the main difficulties of fatigue analysis.

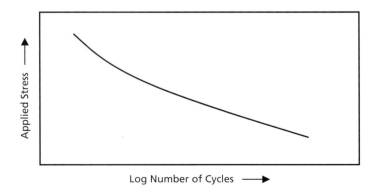

Fig. C-9. Typical S–N Fatigue Curve

Appendix-C Mechanical and Environmental Properties

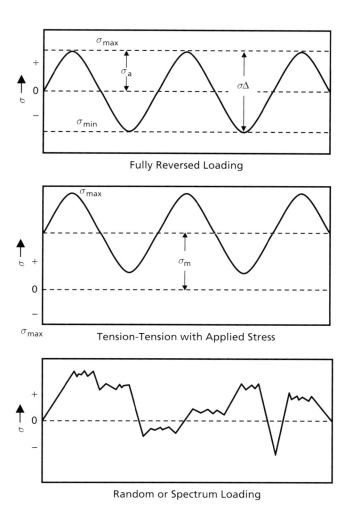

Fig. C-10. *Typical Loading Cycles*

The *fatigue strength, or fatigue life*, is the number of cycles to failure at a specified stress level. Normally, the fatigue strength increases as the static tensile strength increases. For example, high strength steels heat treated to over 200 ksi yield strengths have much higher fatigue strengths than aluminum alloys with 70 ksi yield strengths. A comparison of the S–N curves for steel and aluminum are shown in Fig. C.11. Note that steel not only has a higher fatigue strength than aluminum, it also has what is called an *endurance limit*. In other words, below a certain stress level, the steel alloy will never fail. On the other hand, aluminum does not have a true endurance limit. It will always fail if tested to a sufficient number of cycles. Therefore, the fatigue strength of aluminum is usually reported as the stress level it can survive at a large total number of cycles, usually 5×10^8 cycles.

Fatigue can be broken down into the stages of crack initiation, crack growth and final failure. The propagation of a fatigue crack, or the *fatigue crack growth rate*, is expressed as da/dn, or the change of the crack length a per cycle n. When the crack growth reaches a critical point, in which the remaining cross-section can no longer carry the load, failure occurs.

Fatigue normally starts at a surface-related stress concentration, such as a fastener hole or in radii, although internal stress concentrations can also initiate cracks. Therefore, any method which removes stress concentrations, such as smoother surfaces or blended radii, will delay, or prevent, the initiation of fatigue cracking. Surface treatments that introduce compressive stresses on the surface, such as shot peening of part surfaces or cold working of fastener holes, also reduces fatigue. The applied tensile stress must first overcome the residual compressive stress on the surface before the surface actually sees any applied tensile stress.

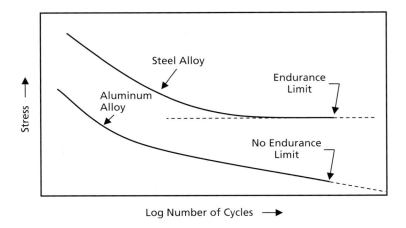

Fig. C-11. Comparison of Steel and Aluminum Fatigue Behavior

Appendix-C Mechanical and Environmental Properties

Fatigue tests can be characterized as being *high cycle fatigue* tests ($>10^5$ cycles) or *low cycle fatigue* tests. In high cycle fatigue tests, the stresses are low enough that the strains are in the elastic region. In contrast, in low cycle fatigue tests, the stresses are high enough to cause some plastic deformation. This plastic deformation results in a hysteresis loop during the unloading portion of the cycle, where there is recovery of both the elastic and plastic deformation, as shown in Fig. C.12. Due to strain hardening of the specimen, the hysteresis loop usually stabilizes after a few hundred cycles. The area of the hysteresis loop is equal to the work done or energy loss per cycle. The total strain $\Delta\varepsilon$ consists of both the elastic and the plastic components.

$$\Delta\varepsilon = \Delta\varepsilon_e + \Delta\varepsilon_p$$

where

$\Delta\varepsilon_e$ = elastic strain
$\Delta\varepsilon_p$ = plastic strain

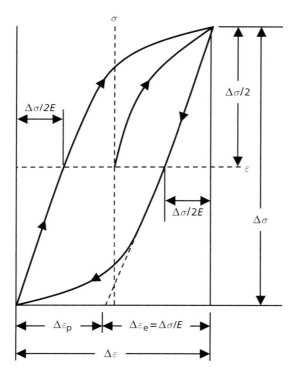

Fig. C-12. Hysteresis Loop for Cyclic Loading

Fracture mechanics has also been applied to fatigue. In the *fracture mechanics approach to fatigue crack growth*, the crack growth rate, or the amount of crack extension per loading cycle, is correlated with the stress intensity parameter K. This approach makes it possible to estimate the useful safe life and inspection intervals. An idealized *da/dn* versus ΔK curve is shown in Fig. C.13. In region I, ΔKth is the fatigue crack growth threshold, which is at the lower end of the ΔK range, where crack growth rates approach zero. In region II, the crack growth rate is stable and essentially linear, and can be modeled by the power law equations, such as the Paris equation.

$$da/dn = C(\Delta K)^m$$

where

a = flaw or crack size in inches
n = number of cycles
C and m are material parameters
$\Delta K = \Delta K_{max} - \Delta K_{min}$ is the stress intensity parameter range

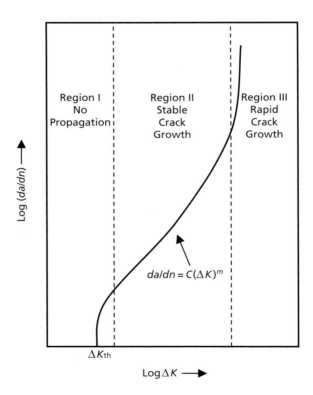

Fig. C-13. Crack Propagation Curve for Fatigue Loading

Finally, in region III, the crack growth rate accelerates since the fracture toughness of the material is approached, and there is a local tensile overload failure.

C.5 Creep and Stress Rupture[8,9]

A metal subject to a constant tensile stress at a sufficient elevated temperature will *creep*, or undergo a time-dependent increase in length. Temperatures higher than about 1/2 the melting temperature, where diffusion mechanisms become active, are usually sufficient to make creep an important consideration. A creep test measures the dimensional changes that occur at elevated temperature, while a stress rupture test measures the effects of temperature on the long-term load-bearing characteristics.

To determine the creep curve for a metal, a constant load is applied to a tensile specimen maintained at the temperature of interest, and the strain (ε), or extension, of the specimen is determined as a function of time. An idealized creep curve is shown in Fig. C.14. The slope of the curve is ($\dot{\varepsilon} = d\varepsilon/dt$) called the creep rate. The three stages of creep are shown. Following the initial elongation of the specimen (ε_o), the creep rate decreases with time during primary creep. During secondary creep, the creep rate reaches a steady state in which the creep rate changes little with time. During the third stage of creep (tertiary creep), the creep rate increases rapidly with time until fracture occurs. Creep tests are usually long time tests, with times between 2000 and 10 000 h typical, with total strains usually less than 0.5%. The purpose of the creep test is to precisely measure the creep rate and total creep strain. Creep tests are often terminated before final failure occurs.

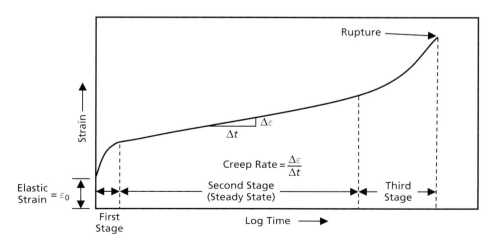

Fig. C-14. Typical Creep Curve

The *stress rupture* test is similar to the creep test except that the test is always carried out to failure. Higher stresses are used and therefore the creep rates are higher, with strains often approaching 50%. Due to the higher stress levels, stress rupture tests can often cause failures in 1000 h, or less. The higher stress levels, and greater creep levels, used in the stress rupture test cause microstructural changes (e.g., overaging) to occur in shorter times than they would in a normal creep test. A break in the stress rupture curve, as shown in Fig. C.15, is often indicative of a microstructural instability.

Creep becomes much more complicated when it is combined with cyclic loading and fatigue also becomes operative.

C.6 Corrosion[10]

Corrosion is the gradual degradation of a material due to the environment. In *chemical corrosion*, the material dissolves in a corrosive liquid until the material is consumed. In *electrochemical corrosion*, metal atoms are removed from the solid material due to an electrical circuit that is produced. In electrochemical corrosion, the anode, which is the metal (M_a) that corrodes, undergoes an oxidation reaction and gives up electrons to the circuit.

$$M_a \rightarrow M^{n+} + n^{e-}$$

A reduction reaction, which is the reverse of the anode reaction, occurs at the cathode.

$$M^{n+} + n^{e-} \rightarrow M_c$$

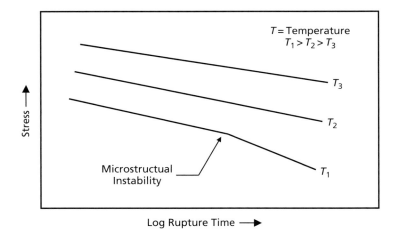

Fig. C-15. *Stress-Rupture Curves*

Appendix-C Mechanical and Environmental Properties

In oxygen free liquids, such as acid pickling baths, the most common cathodic reaction evolves hydrogen.

$$2H^+ + 2e^{e-} \rightarrow H_2 \uparrow$$

Hydrogen produced in this manner is important in both hydrogen embrittlement and stress corrosion cracking.

For electrochemical corrosion to occur, there must be a liquid electrolyte (e.g., water) to allow an electrical circuit to transport the current (i.e., electrons).

Galvanic corrosion is a quite common type of attack that occurs when two metals of different compositions are electrically coupled in the presence of an electrolyte. The less noble, or more reactive, metal becomes the anode and corrodes, while the more inert metal, or the cathode, does not corrode. The galvanic series in seawater is shown in Fig. C.16. An example of galvanic

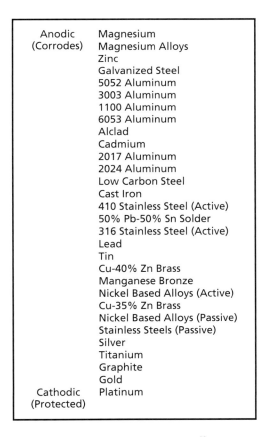

Fig. C-16. Galvanic Series in Seawater[10]

583

corrosion is aluminum in direct contact with carbon/epoxy. Since aluminum is more reactive than the carbon fibers, corrosion of the aluminum occurs. Therefore, to provide an electrical insulator, a layer of fiberglass is often bonded to the carbon/epoxy surface to prevent corrosion of the aluminum. Note the position of magnesium on the galvanic series. It is anodic to every other metal and therefore very susceptible to galvanic corrosion.

Other important types of corrosion encountered in aerospace alloys include pitting, exfoliation, and intergranular corrosion. *Pitting*, as the name implies, is the formation of small pits on the surface due to localized attack. In *exfoliation*, delamination of the surface grains, or layers, occurs under forces exerted by the corrosion products. High strength aluminum alloys are subject to both pitting and exfoliation. Iron and silicon impurities can contribute to the attack because they are cathodic with respect to the aluminum matrix.

Intergranular corrosion occurs preferentially along grain boundaries and is especially prevalent in stainless steels. In certain heat treatments, chromium carbides precipitate at the grain boundaries, depleting the areas right next to the grain boundaries of chromium, as shown in Fig. C.17. Since chromium, in amounts greater than 12%, is necessary to prevent corrosion of stainless steels, the chromium-depleted regions are susceptible to accelerated attack.

C.7 Hydrogen Embrittlement[3,11]

Hydrogen embrittlement has particularly been a problem in heat treated high strength steels. In general, the higher the strength level of the steel, the greater is the susceptibility to hydrogen embrittlement. Hydrogen embrittlement occurs primarily in BCC and HCP metals, while FCC metals are generally not susceptible.

Hydrogen embrittlement results in sudden failures at stress levels below the yield strength. It is normally a delayed failure, in which an appreciable amount

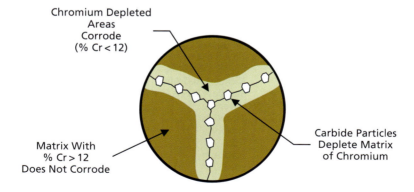

Fig. C-17. *Intergranular Corrosion of Stainless Steel*[10]

of time passes between the time hydrogen is introduced into the metal and failure occurs. Hydrogen embrittlement is a complex process, and different mechanisms may operate in different metals under different environments and operating stresses. However, hydrogen is a small molecule that can dissociate into monatomic hydrogen that readily diffuses into the crystalline structure. Very small amounts of hydrogen can cause damage, for example, as little as 0.0001% hydrogen can cause cracking in steel. Typical sources of hydrogen include melting operations, heat treatments, welding, pickling and plating. In addition, the cathodic reaction during corrosion in service can also produce hydrogen.

Characteristics of hydrogen embrittlement include a strain rate sensitivity, a temperature dependence, and delayed fracture. As opposed to many forms of brittle fracture, hydrogen embrittlement is enhanced by slow strain rates. In addition, it does not occur at low or high temperatures, but occurs at intermediate temperature ranges. For steels, the most susceptible temperature is near room temperature.

A comparison between hydrogen-free notched tensile specimens and ones charged with hydrogen in a static tensile test is shown in Fig. C.18. Note that there is a time delay before failure occurs, hence the term *static fatigue*. Also, below a certain stress level, failure does not occur. The higher the hydrogen content, the lower the stress level that can be endured before failure. There is also a large reduction in ductility associated with embrittlement. There is no single fracture mode associated with hydrogen embrittlement. Fracture can be transgranular, intergranular, and can exhibit characteristics of both brittle and ductile failure modes.

If the part is not under stress when it contains hydrogen, then the hydrogen can usually be safely removed without damage to the part by baking the part at

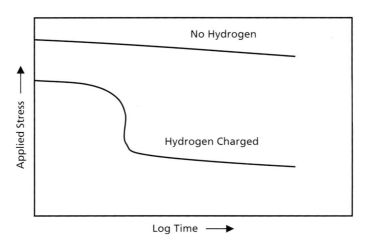

Fig. C-18. Hydrogen Effect on Static Tensile Strength

elevated temperature. The use of a vacuum during baking is even more effective. High strength steels are usually baked at 365–385° F for at least 8–24 h to remove any hydrogen after chromium or cadmium plating operations.

C.8 Stress Corrosion Cracking[3,11]

Stress corrosion cracking (SCC) is the failure of an alloy due to the combined effects of a corrosive environment and a static tensile stress below the yield strength of the alloy. The stress for SCC can be either an applied stress or residual stresses in the alloy. Only specific combinations of alloys and environments result in SCC. For example, the high strength aluminum alloy 7075-T6 will readily crack in sea water, while Ti-6Al-4V is immune to sea water. As explained in Chapter 2 on Aluminum, the use of T7 overaged tempers greatly reduces the stress corrosion cracking susceptibility to the high strength aluminum alloys. If the right combination of stress and environment are present, almost every metal can be prone to SCC. Of the main aerospace alloys, SCC is a more serious problem in high strength aluminum alloys and high strength steels. Although titanium has been made to crack in laboratory tests, SCC of titanium alloys in service has not been a serious problem.

A simplified mechanism for SCC is shown in Fig. C.19. Stress causes rupture of the oxide film at the crack tip, which exposes fresh metal that corrodes and forms another thin oxide film. The oxide ruptures again, allowing more corrosion, and the crack slowly propagates through the alloy until the crack reaches a critical length, and failure occurs. Since the cathodic reaction during corrosion can often produce hydrogen, hydrogen can contribute to SCC, often making it difficult to distinguish between SCC and hydrogen embrittlement. In general, there is a threshold stress for stress corrosion cracking, denoted by K_{ISCC}, below which crack growth is not observed. The level of K_{ISCC} with respect to K_I of a material gives a measure of its susceptibility to SCC, as shown schematically in Fig. C.20.

In SCC, failure can occur either by transgranular failure, intergranular failure or by a mixture of the two. Normally ductile metals fail in a brittle manner. Failures are often characterized by branched type failures with multiple cracking.

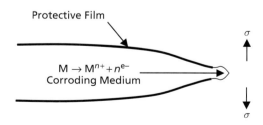

Fig. C-19. Stress Corrosion Cracking[11]

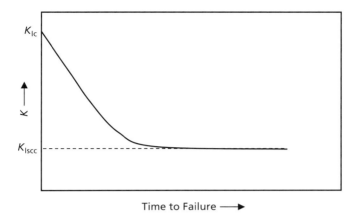

Fig. C-20. K_{Iscc} Threshold

C.9 High Temperature Oxidation and Corrosion[12]

The two major environmental effects on superalloys are oxidation and hot corrosion. At temperatures of about 1600° F and lower, *oxidation* of superalloys is not a major problem; however, at higher temperatures, oxidation can rapidly occur. Since CrO_3 forms as a protective oxide, the level of oxidation resistance at temperatures below 1800° F is a function of the chromium content. At temperatures above 1800° F, the aluminum content becomes more important as Al_2O_3 becomes the dominant oxide protector. Chromium and aluminum can contribute in an interactive manner to provide oxidation protection. For example, the higher the chromium content, the less aluminum that may be required. However, the alloy content of many superalloys is insufficient to provide long-term protection, and protective coatings are usually required to provide satisfactory life.

Hot corrosion, often referred to as sulfidation, is classified as either Type I or II depending on the temperature. Type I occurs at higher temperatures (1650–1920° F), while Type II occurs at lower temperatures (1255–1380° F). Both are triggered by the presence of sulfur in the fuel combining with salt from the environment. Hot corrosion is an accelerated, often catastrophic, surface attack of parts in the hot gas path. It is believed that the presence of alkali metal salts (i.e., Na_2SO_4) is a prerequisite for hot corrosion.

C.10 Polymeric Matrix Composite Degradation[13]

Temperature has a pronounced effect on polymeric composite mechanical properties. Typically, as the temperature increases, the matrix-dominated mechanical properties decrease. Fiber-dominated properties are somewhat affected by cold

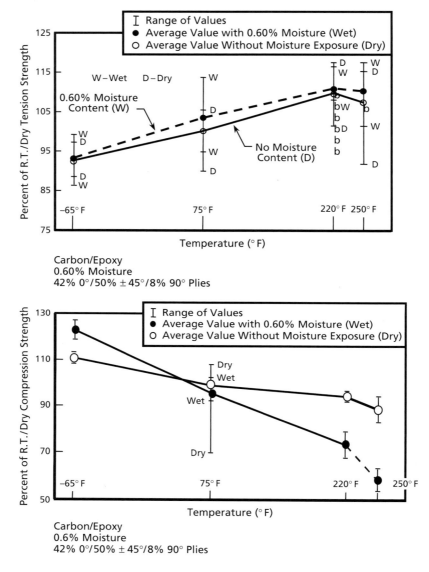

Fig. C-21. *Effects of Temperature and Moisture on Strength of Carbon/Epoxy*[13]

temperatures, but the affects are not as severe as the effects of elevated temperature on the matrix-dominated properties. As shown in Fig. C.21, the design parameters for carbon/epoxy are cold-dry tension and hot-wet compression.

The amount of *absorbed moisture* (Fig. C.22) is dependent on the matrix material and the relative humidity. Elevated temperature speeds the rate of

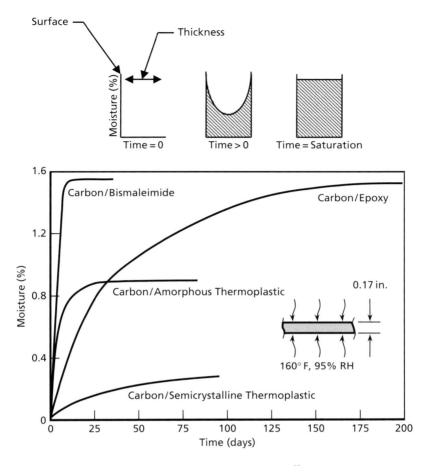

Fig. C-22. *Absorption of Moisture for Polymer Composites* [13]

moisture absorption. Absorbed moisture reduces the matrix-dominated mechanical properties. Absorbed moisture also causes the matrix to swell. This swelling relieves locked-in thermal strains from elevated temperature curing. These strains can be large and large panels, fixed at their edges, can buckle due to the swelling strains. During freeze-thaw cycles, the absorbed moisture expands during freezing and can crack the matrix. During thermal spikes, absorbed moisture can turn to steam. When the internal steam pressure exceeds the flatwise tensile strength of the composite, the laminate will delaminate.

Composites are also susceptible to *delaminations* (ply separations) during fabrication, assembly, and in-service. During fabrication, foreign materials, such as prepreg backing paper, can be inadvertently left in the lay-up. During assembly, improper part handling or incorrectly installed fasteners can cause delaminations.

When in-service, low velocity impact damage (LVID) from dropped tools, or fork lifts running into aircraft, can cause damage. The damage may appear as only a small indentation on the surface but can propagate through the laminate, forming a complex network of delaminations and matrix cracks, as depicted in Fig. C.23. Depending on the size of the delamination, it can reduce the static and fatigue strength, and the compression buckling strength. If it is large enough, it can grow under fatigue loading.

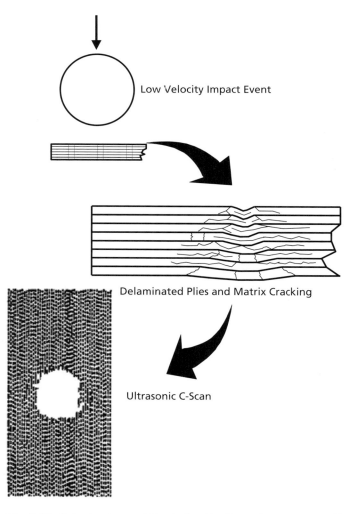

Fig. C-23. Delaminations and Matrix Cracking Due to Impact Damage [13]

Typically, damage tolerance is a resin-dominated property. The selection of a toughened resin can significantly improve the resistance to impact damage. During the design phase, it is important to recognize the potential for delaminations and use conservative enough design strains so that damaged structure can be repaired.

Recommended Reading

[1] Dieter, G.E., *Mechanical Metallurgy*, 3rd Edition, McGraw-Hill Book Co., 1986.
[2] Courtney, T.H., *Mechanical Behavior of Materials*, 2nd Edition, McGraw-Hill Book Co., 2000.
[3] Myers, A.M., Chawla, K.K., *Mechanical Metallurgy – Principles and Applications*, Prentice-Hall, Inc., 1984.
[4] Askeland, D.R., *The Science and Engineering of Materials*, 2nd Edition, PWS-KENT Publishing Co., 1989.

References

[1] Askeland, D.R., "Mechanical Testing and Properties", in *The Science and Engineering of Materials*, 2nd edition, PWS-KENT Publishing Co., 1989, pp. 145–181.
[2] Callister, W.D., "Failure", in *Fundamentals of Materials Science and Engineering*, 5th edition, John Wiley & Sons, Inc., 2001, pp. 234–280.
[3] Laird, C.M., *ASTM STP 415*, ASTM, 130, 1966.
[4] Dieter, G.E., "Brittle Fracture and Impact Testing", in *Mechanical Metallurgy*, 3rd edition, McGraw-Hill Book Co., 1986, pp. 471–500.
[5] Courtney, T.H., "Fracture Mechanics", in *Mechanical Behavior of Materials*, 2nd edition, McGraw-Hill Book Co., 2000, pp. 404–453.
[6] Dieter, G.E., "Fatigue of Metals", in *Mechanical Metallurgy*, 3rd edition, McGraw-Hill Book Co., 1986, pp. 375–431.
[7] Courtney, T.H., "Fatigue of Engineering Materials", in *Mechanical Behavior of Materials*, 2nd edition, McGraw-Hill Book Co., 2000, pp. 566–629.
[8] Dieter, G.E., "Creep and Stress Rupture", in *Mechanical Metallurgy*, 3rd edition, McGraw-Hill Book Co., 1986, pp. 432–470.
[9] Courtney, T.H., "High Temperature Deformation of Crystalline Materials", in *Mechanical Behavior of Materials*, 2nd edition, McGraw-Hill Book Co., 2000, pp. 293–353.
[10] Askeland, D.R., "Corrosion and Wear", in *The Science and Engineering of Materials*, 2nd edition, PWS-KENT Publishing Co., 1989, pp. 777–804.
[11] Courtney, T.H., "Embrittlement", in *Mechanical Behavior of Materials*, 2nd edition, McGraw-Hill Book Co., 2000, pp. 630–685.
[12] Smith, W.F., "Nickel and Cobalt Alloys", in *Structure and Properties of Engineering Alloys*, 2nd edition, McGraw-Hill, Inc., 1993, pp. 487–536.
[13] Campbell, F.C., *Manufacturing Processes for Advanced Composites*, Elsevier Ltd, 2004, p. 30–33.

Index

Adhesive bonding, 10–11, 370, 415–16
  advantages, 370–1
  bonding procedures, 385
    adhesive application, 387–8
    bond line thickness control, 388
    bonding, 388–90
    prefit evaluation, 386–7
    prekitting of adherends, 385–6
  disadvantages, 371–2
  epoxy adhesives, 383–4
    film, 385
    two-part room temperature curing liquid/paste, 384
  joint design, 372–7
  sandwich structures, 390–3
    balsa wood, 403–404
    foam cores, 404–406
    honeycomb core, 393–9
    honeycomb processing, 399–403
    inspection, 407–408
    syntactic core, 406–407
  surface preparation, 378–9
    aluminum and titanium, 380–3
    principles, 380
    protection during handling, 383
    techniques, 379–80
  testing, 377–8
  theory of adhesion, 372
Air travel, 2
Airbus A320, A330, A340, 10
Airframe durability, 2
Alloys, 6
Aluminum, 4–6, 89–90
  advantages, 16–17
  alloy designation, 23–5
  alloys, 25–31
  casting:
    alloys, 57–8
    chemical compositions, 57
    contamination, 59–60
    die casting, 64
    evaporative pattern casting, 64–5
    furnaces, 58–9
    grain size control, 58
    heat treatment, 65
    investment casting, 64
    permanent mold casting, 63–4
    plaster/shell molding, 62–3
    premium quality, 58
    properties, 65–6
    sand casting, 60–2
    sludge formation/settling, 60
    temperature control, 59
    uses, 57
  chemical finishing, 88–9
  disadvantages, 17
  forging, 43–6
    blocker, 45–6
    conventional, 46
    high definition, 46
    precision, 46
  forming, 46–7
    blanking/piercing, 47–8
    brake forming, 48–9
    deep drawing, 49–50
    rubber pad forming, 51
    stretch forming, 50–1
    superplastic forming, 51–7
  heat treating, 37
    annealing, 42–3
    solution heat treating/aging, 37–42

Aluminum, (*Continued*)
  joining, 76
  machining, 66–8
    chemical milling, 76
    high speed, 68–76
  major attributes of wrought alloys, 18
  melting/primary fabrication, 31–3
    extrusion, 37
    rolling plate/sheet, 33–6
  metallurgical considerations, 17–23
  strengthening solution, 96
  welding, 77–8
    friction stir, 83–8
    gas metal/gas tungsten arc, 78–80
    laser, 81–2
    plasma arc, 80–1
    resistance, 82–3
Aluminum MMCs, 435–40
Aluminum–beryllium alloys, 116
Aluminum–copper alloy (2XXX series), 6
Aluminum–zinc alloy (7XXX series), 6
Aramid fiber, 278
Assembly *see* Structural assembly
Automated tape laying (ATL), 295–8
Automated variable polarity plasma arc (VPPA), 80–1
AV-8B Harrier, 2, 10

B-2 bomber, 10
Balsa wood, 403–404
Beryllium, 6–7, 94–5, 109, 116–18
  alloys, 110
  aluminum–beryllium alloys, 116
  fabrication:
    forming, 114–15
    joining, 116
    machining, 115–16
  metallurgical considerations, 109
    corrosion resistance, 109
    toxicity, 110
  powder metallurgy, 111–14
Boeing aircraft, 7, 10
Boron, 2
Boron fiber, 279–80

Carbon fiber, 2, 278–9
Carbon–carbon (C–C) composites, 12, 463
Ceramic matrix composites, 11–12, 460–4, 490–2, 563
  chemical vapor infiltration (CVI), 482–7
  directed metal oxidation (DMO), 487–8
  fabrication methods, 472
  fiber architectures, 471
  interfacial coatings, 470
  liquid silicon infiltration (LSI), 488–90
  materials, 467–70
  polymer infiltration/pyrolysis (PIP), 476–82
  powder processing, 472–4
  reinforcements, 464–7
  slurry infiltration/consolidation, 474–6
Ceramics, 555–6
  ionic/covalent bonds, 556
Chemical vapor infiltration (CVI), 482–7
Cobalt, 8
Cold hearth melting, 7
Commercial aircraft, 4
Compocasting (slurry casting), 427
Composites, 2, 4, 9–10, 562–5
  fiber, 562
  interface, 562–3
  matrix, 562
  rule of mixtures, 563–5
  see also Ceramic matrix composites; Metal matrix composites (MMCs); Polymer matrix composites
Contour tape laying machines (CTLM), 295
Corrosion, 582
  chemical, 582
  electrochemical, 582–3
  exfoliation, 584
  galvanic, 583–4
  intergranular, 584
  pitting, 584
Crack growth rate, 125–6
Creep, 581

Didymium, 96–7
Direct current electrode positive (DCEP) arrangement, 78, 79
Directed metal oxidation (DMO), 487–8
Directionally solidified (DS) casting, 9

E-glass fiber, 277
Electron beam (EB) welding, 166, 168–9
Electroslag remelting (ESR), 226–7
Environmental properties
  see Mechanical/environmental properties

Fatigue, 576–7
  crack growth rate, 578–81
  crack initiation/growth, 6
  endurance limit, 578
  fracture mechanics approach, 580
  high cycle tests, 579
  strength, 125–6
  strength/life, 578
Fiber metal laminates, 452–4
Fighter aircraft, 2, 7
Flat tape laying machines (FTLM), 295
Foam cores, 404–406
Fracture toughness, 6
Friction stir welding (FSW), 6, 83–8

Gas metal arc welding (GMAW), 78–80, 166, 168, 259
Gas tungsten arc welding (GTAW), 78–80, 166, 167–8, 258–9
Glass fiber, 277
Glass fiber reinforced aluminum laminates (GLARE), 11, 453–4
Graphite fiber, 278–9

High fracture toughness steels, 198–200
High strength steels, 8, 176
  high fracture toughness steels, 198–200
  maraging steels, 200–202
  medium carbon low alloy steels, 182–6
    fabrication, 186–91
    heat treatment, 191–7
    metallurgical considerations, 176–82
  precipitation hardening stainless steels, 202–207
  quality levels, 185
  stress corrosion cracking, 190–1
High temperature oxidation/corrosion, 587
  hot corrosion, 587
  oxidation, 587
Honeycomb core, 393–4
  advantages, 396–7
  cell configurations, 394–5
  comparative properties, 396
  corrosion protection, 397
  expansion process, 395–6
  freeze-thaw cycles, 397–8
  liquid damage, 397–9
  processing, 399
    bonding, 401
    cleaning/drying, 401
    cocuring, 402–403
    forming, 399
    migration/crushing, 403
    potting, 399–401
    pressure selection, 401–402
    splicing, 399
    trimming, 399
Hot isostatic pressing (HIP), 8
Hydrogen embrittlement, 584–6

Impurities, 6
Integrally cocured structure, 10–11, 408–10
  advantages, 410
  cobonding, 413, 415
  disadvantages, 410
  hat, 410
  spring-in, 410
  terminations, 411, 413
Investment casting, 8
Iron–nickel, 8

Laser beam welding (LBW), 81–2, 169
Liquid metal infiltration (squeeze casting), 427–30
Liquid molding, 327–8

Index

Liquid silicon infiltration (LI), 488–90
Low velocity impact damage (LVID), 590

Magnesium, 6–7, 94, 95
  corrosion protection, 108–109
  fabrication, 103
    forming, 103–104
    heat treating, 106–107
    joining, 107–108
    machining, 107
    sand casting, 104–105
  metallurgical considerations:
    HCP crystalline structure, 95–6
    melting point, 95–6
    strengthening solution, 96–7
Magnesium alloys, 97
  casting alloys, 99
    Mg–Ag–Rare Earth, 102–103
    Mg–Al/Mg–Zn, 99–101
    Mg–Zn–Zr/Mg–Rare Earth–Zr, 101–102
  wrought alloys, 97–9
Manganese, 96
Maraging steels, 200–202
Material density, 2
Materials, 542
Mechanical alloying (MA), 230–2
Mechanical/environmental properties, 568
  failure modes, 570
    brittle fractures, 570–1
    ductile, 570
    ductile-to-brittle transition, 571
    fatigue, 572
    intergranularly, 572
    transgranular, 572
    transition temperature, 571–2
  fracture control, 576
  fracture critical, 576
  fracture toughness, 572–4
    critical stress intensity factor, 574
    plane-strain, 574–5
  static strength, 568–9

Medium carbon low alloy steels, 182
  43XX class, 183–4
  classification, 183
  elements, 182–3
  fabrication:
    annealed condition, 188
    forging, 186–8
    grinding, 189
    machinability ratings, 188–9
    welded/brazed, 189–90
  hardening, 183, 192
    austenitizing, 195–6
    quenching, 196
    tempering, 197
  heat treatment, 191–4
  one-step temper embrittlement, 197
  stress relieving, 192
  susceptible to decarburization, 192
  two-step temper embrittlement, 197
Metal matrix composites (MMCs), 11–12, 420–4, 455–6
  continuous fiber aluminum MMCs, 435–40
  continuous fiber reinforced titanium matrix composites, 440–7
  discontinuously reinforced, 424
  liquid metal infiltration (squeeze casting), 427–30
  powder metallurgy methods, 432–4
  pressure infiltration casting, 430–1
  secondary fabrication of titanium matrix composites, 447–51
    fiber metal laminates, 452–4
  secondary processing of discontinuous MMCs, 434–5
  slurry casting (compocasting), 427
  spray deposition, 431–2
  stir casting, 424–7
Metallic structure, 543–55
  annealing, 547
  body centered cubic (BCC), 543
  dislocation, 545–6
  dispersion strengthened alloys, 552
  eutectic reaction, 553
  eutectoid reaction, 554
  face centered cubic (FCC), 543
  grain size, 548

hexagonal close-packed (HCP), 543
martensite, 551
pearlite, 551
peritectic reaction, 553
peritectoid reaction, 554
plastic deformation, 546
precipitation hardening, 549
slip direction, 546
slip planes, 546
slip system, 546
stress relieving, 547–8
substitutional/interstitial solid
    solutions, 548–9
work hardening, 546–7
Metric conversions, 540
Mischmetal, 96
Multiaxial warp knits (MWKs), 331

Nickel, 8

Plasma arc welding (PAW), 166, 168
Polymer infiltration and pyrolysis (PIP), 476–7
    conventional processes, 479–80
    sol-gel infiltration, 480–2
    space shuttle C–C, 477–9
Polymer matrix composites, 364–6
    advantages, 274
    automated tape laying, 295–8
    cost drivers, 275–6
    cure tooling, 286
        considerations, 286–91
        expansion/contraction, 289–90
        inside/outside moldline
            (IML/OML), 287
        material selection, 287–9
        spring-in, 290–1
    curing, 311–13
        condensation curing system, 323–4
        epoxy composite, 313–14
        hydrostatic resin pressure, 318–22
        residual curing stresses, 324–7
        resin/prepreg variables, 322–3
        theory of void formation, 314–18
    disadvantages, 274–5
    fabrication processes, 286

fiber placement, 304–307
filament winding, 298–300
    autoclave curing, 304
    choice of mandrel material, 303
    equipment, 300
    fiber orientation, 300
    helical, 300
    hoop, 301–302
    polar, 301
    prepregs, 303
    viscosity/pot life, 302
    wet, 302–303
liquid molding, 327–8
materials, 276–7
    fibers, 277–80
    hybrids, 285
    matrices, 280–2
    preform, 285–6
    prepregs, 282–3
    product forms, 282–6
    rovings, tows, yarns, 282
    stitched fabric, 284–5
    woven fabric, 283–4
ply collation, 291
    flat ply collation/vacuum
        forming, 294–5
    manual lay-up, 291–4
preform technology, 328–9
    braiding, 333–4
    fibers, 329–30
    multiaxial warp knits, 331
    preform handling, 334–5
    stitching, 331–3
    woven fabrics, 330–1
pultrusion, 341–3
resin injection, 336–8
    RTM curing, 338
    RTM tooling, 338–9
thermoplastic composites, 343–5
    consolidation, 345–51
    joining, 355–61
    thermoforming, 351–5
trimming/machining operations, 361–4
vacuum assisted resin transfer
    molding, 339–41
vacuum bagging, 307–11

Polymeric matrix composite degradation:
  absorbed moisture, 588–9
  delaminations, 589–91
  temperature, 587–8
Polymers, 556–7
  thermosets/thermoplastics, 557–62
Polymethylmethacrylimides (PMIs), 406
Polystyrene cores, 405
Polyurethane foams, 405
Polyvinyl chloride (PVC) foam, 406
Powder metallurgy (PM), 228, 432–4
  forged alloys, 228–30
  mechanical alloying, 230–2
Pressure infiltration casting (PIC), 430–1

Quartz fiber, 277

Rare earths (RE), 96
Resin transfer molding (RTM), 327–8
  curing, 338
  tooling, 338–9

Self-forming technique (SFT), 454
Silver, 96
Single crystal (SC) casting, 9
Slurry casting (compocasting), 427
Space shuttle, 477–9
Spray deposition, 431
Squeeze casting (liquid metal infiltration), 427–30
Stir casting, 424–7
Stress corrosion cracking (SCC), 6, 586
Stress rupture, 582
Structural assembly, 12, 496, 535–6
  fastener selection/installation, 515–18
    blind fasteners, 527–8
    bolts/nuts, 525–7
    fatigue improvement/interference fit fasteners, 528–32
    pin/collar fasteners, 523–5
    solid rivets, 520–3
    special considerations for composite joints, 518–20
  framing, 496–8

hole drilling, 499–500
  automated, 505–508
  automated riveting equipment, 508–509
  countersinking, 514–15
  drill bit geometries, 509–14
  manual, 500–504
  power feed, 504–505
  reaming, 514
painting, 534–5
sealing, 533–4
shimming, 498–9
Superalloys, 8–9, 212–13, 266–70
  coating technology, 264
    diffusion, 264–5
    overlay, 265–6
    thermal barrier, 266
  commercial, 219–21
    cobalt based, 225
    iron–nickel based, 224–5
    nickel based, 221–4
  forging, 232–3
    die lubrication, 234
    furnace heated, 233
    isothermal/hot die, 233, 235–6
    open die, 233
    plastic deformation, 234
    quality, 235
    recrystallization, 234
    ring rolling, 233
    roll, 233
    slow strain rates, 234
    upset/extrusion, 233
  forming, 236
    annealed condition, 237–8
    cold operations, 236–7
    hot, 237
  heat treatment, 243
    cast superalloy heat treatment, 247–8
    precipitation strengthened iron–nickel base, 246–7
    precipitation strengthened nickel base, 244–6
    solution strengthened, 243–4

investment casting, 238–9
  directional solidification (DS) casting, 240–2
  polycrystalline, 239–40
  single crystal (SC) casting, 242–3
joining, 256
  brazing, 260–3
  transient liquid phase (TLP) bonding, 263–4
  welding, 256–60
machining, 248–50
  grinding, 254
  milling, 252–4
  turning, 251–2
melting/primary fabrication, 225–6
  electroslag remelting, 226–7
  vacuum arc melting, 226–7
  vacuum induction melting, 226
metallurgical considerations, 213
  compositions, 215–16
  creep failures, 218
  forms/usage, 217–18
  powder metallurgy (PM), 218
  processes, 218
  strengthening, 213–15
  topologically closed-packed (TCP) phases, 216–17
powder metallurgy, 228–32
Superplastic forming, 51–2
  advantages, 52–3
  Ashby and Verral model, 53–4
  cavitation, 55–6
  gas pressure, 56–7
  requirements, 53
  single sheet process, 54–5
Superplastic forming/diffusion bonding (SPF/DB), 8

Thermal barrier coatings (TBC), 9
Thermomechanically affect zone (TMAZ), 85
Thermoplastic composites, 343, 557–62
  addition polymerization, 558
  advantages, 344–5
  amorphous, 559, 560–1
  condensation reaction, 561
  consolidation, 345–6

autoclave, 349
autoconsolidation/in-situ placement, 349–51
Autohesion process, 347–8
  continuous, 348–9
  film stacking, 346
  processing temperature, 346–7
  two press process, 348
joining, 355–61
  adhesive bonding, 356
  dual resin bonding, 356
  induction welding, 358–9
  mechanical fastening, 356
  melt fusion, 356
  resistance welding, 357
  ultrasonic welding, 358
semi-crystalline, 559–60
thermoforming, 351–2
  diaphragm forming, 354
  matched metal dies, 352
  preheating methods, 352
  pultrusion, 354–5
  resin transfer molding, 355
  transfer time, 352–4
thermoset/thermoplastic difference, 343–4
Titanium, 7–8, 120, 171–2
  alloys, 126
  brazing, 170
  directed metal deposition (laser powder, laser direct manufacturing, electron beam free form fabrication), 140–3
  forging, 137–8
    alpha–beta defects, 138–9
    beta, 139–40
    hot die/isothermal, 140
  forming:
    hot formed, 143–5
    springback, 143
    vacuum/creep forming, 145
  heat treating, 150–1
    annealing, 152
    solution treating and aging, 152–4
    stress relief, 151–2
  investment casting, 154–8

Titanium, (*Continued*)
  joining, 165
  machining:
    chemical milling, 164
    cutting fluids, 160
    cutting tools, 160
    damage to surface, 163–4
    difficulties, 158–9
    flood coolant, 164
    improper, 159
    milling and drilling, 160–3
    rigid machine tools, 159–60
    successful, 159
  melting/primary fabrication:
    as-cast ingot conditioning, 136
    cold hearth melting, 133
    consumable vacuum arc melting, 132
    defects, 134–5
    equiaxed structure, 136
    hot rolling, 136–7
    Hunter process, 132
    Kroll process, 132
    primary, 135–6
  metallurgical considerations, 120
    affinity for interstitial elements, 123
    alpha/beta phases, 120–1
    classification of alloys, 121–3
    melting point, 126
    microstructure/mechanical property development, 124–6
    strength, 123–4
  superplastic forming:
    advantages, 145–6
    four-sheet process, 149–50
    single-sheet process, 146–7
    three-sheet process, 147–9
    two-sheet process, 147
  welding, 165–6
    cleanliness, 166–7
    diffusion bonding, 169–70
    electron beam welding, 166, 168–9
    gas metal arc welding, 166, 168
    gas tungsten arc welding, 166, 167–8
    laser beam welding, 169
    plasma arc welding, 166, 168
    spot/seam welding, 169
    types, 166
Titanium alloys, 126
  alpha–beta, 128–31
    beta anneal (BA), 129
    mill annealed (MA), 129
    recrystallization anneal (RA), 129
    solution treated and aged (STA), 129
  alpha/near-alpha, 127–8
  beta, 131–2
  commercially pure, 126–7
Titanium matrix composites (TMCs):
  continuous fiber reinforced, 440–7
  secondary fabrication, 447–51
Transient liquid phase (TLP) bonding, 263–4
Turbine blades, 8

Vacuum arc remelting (VAR), 132–3, 226–7
Vacuum assisted resin transfer molding (VARTM), 338, 339–41
Vacuum induction melting (VIM), 226
Vacuum melting, 7

Zinc, 96
Zirconium, 96